Red Sea Pilot

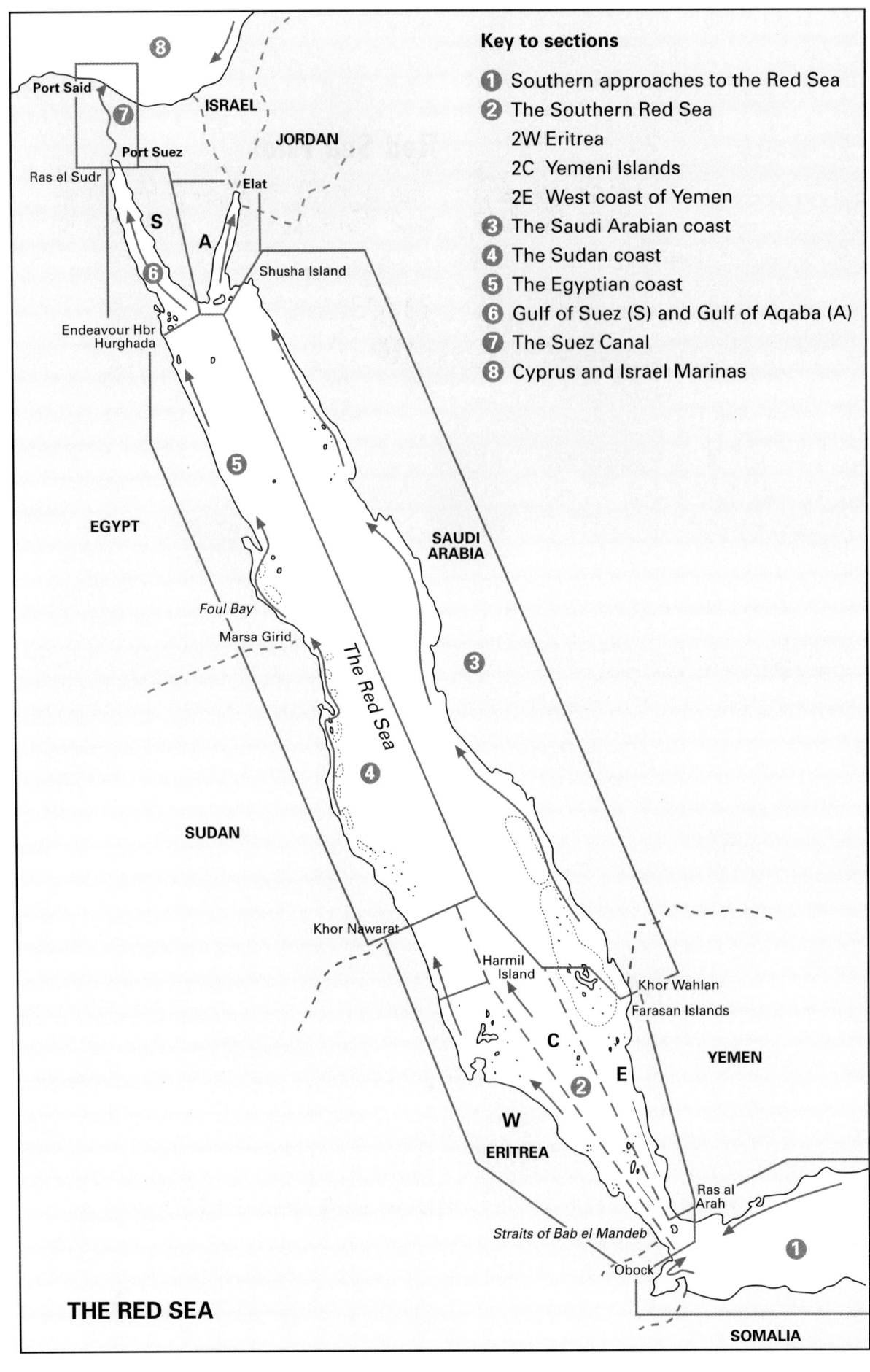

Red Sea Pilot

STEPHEN DAVIES AND
ELAINE MORGAN

Imray Laurie Norie and Wilson Ltd

Published by
Imray Laurie Norie and Wilson Ltd
Wych House St Ives Huntingdon
Cambridgeshire PE27 5BT England
☎ +44(0)1480 462114 *Fax* +44(0)1480 496109
E-mail ilnw@imray.com
www.imray.com
2002

All rights reserved. No part of this publication may be reproduced, transmitted or used in any form by any means – graphic, electronic or mechanical, including photocopying, recording, taping or information storage and retrieval systems or otherwise – without the written permission of the publishers.

1st edition 1995
2nd edition 2002

© S. Davies and E. Morgan 2002
British Library Cataloguing in Publication Data

A catalogue record for this book is available from the British Library.

ISBN 0 85288 554 7

CAUTION
Every effort has been made to ensure the accuracy of this book. It contains selected information, and thus is not definitive and does not include all known information on the subject in hand; this is particularly relevant to the plans, which should not be used for navigation. The authors and publishers believe that it is a useful aid to prudent navigation, but the safety of a vessel depends ultimately on the judgement of the navigator, who should assess all information, published or unpublished, available to him.

CORRECTIONS
The editors would be glad to receive any corrections, information or suggestions which readers may consider would improve the book, as new impressions will be required from time to time. Letters should be addressed to the Editor, *Red Sea Pilot*, care of the publishers. The more precise the information the better, but even partial or doubtful information is helpful, if it is made clear what the doubts are. Supplementary information is published on our web site www.imray.com and is regularly updated.

This work is based on research completed in 2001.

Printed in Great Britain by
Imray, Laurie, Norie and Wilson Ltd

For those of you using a pirate photocopy of this book please remember a simple truth. You're a petty thief. So when officials or businessmen in Red Sea countries demand *baksheesh* or exorbitant fees they are only trying to rip you off in exactly the same way as you have ripped us off. So don't complain. You deserve it.

Contents

Foreword to the 2nd edition, vii
Acknowledgements, vii
Key to symbols, viii

Part I. Introduction
Islam, 1
Country information
　Oman, 2
　Yemen, 3
　Somalia, 5
　Djibouti, 5
　Eritrea, 6
　Saudi Arabia, 7
　Sudan, 8
　Egypt, 9
　Israel, 11
　Jordan, 13

Part II. Planning
Routes, 15
　Timing, 15
　Approach and exit routes, 18
　Routes in the Red Sea, 22-3
Weather, 24
　Climate, 24
　Rules of thumb, 30
　Forecasting services, 31
Sea, tide and currents, 34
　Tides and tidal streams, 40
Navigation, 41
　Charts and position fixing, 41
　Reefs and other hazards, 43
　Anchorages, 46
　Aids to navigation, 46
Yacht and equipment, 48
　Hull, rig and sails, 48
　Ground tackle, 49
　Engine, 50
　Electrics, 50
　Spares, 50
　Ancillariesm, 50
　Tenders, 50
　Radios and services, 51
Environment, 55
　Reefs, 55
　Marine life, 55
　Diving and dive sites, 56
　Bird life, 57
　The desert and offshore islands, 57
　Pollution, 58
Problems, 58
　Terrorism and war, 58
　Piracy: rumours and risks, 59
　Other cultures, other ways, 62
　Medical and health, 63

Part III. Pilotage
Introduction to pilotage notes and plans, 66
　Orientation diagram for Red Sea and S approaches, 67
1 Southern approaches to the Red Sea, 66
　Key map: Southern approaches to the Red Sea, 68
Oman, 66
　South Yemen coast to Bab el Mandeb, 78
　Key map: Straits of Bab el Mandeb, 78
Yemeni islands E of the Horn of Africa & N Somalia, 79
　Key map: Anchorages on the North Somalian coast, 80
Djibouti and the Gulf of Tadjoura, 83
　Key map: Anchorages in the Gulf of Tadjoura, 82
2. Southern Red Sea, 88
Eritrea, 88
　Key maps:
　　Eritrea, 89
　　Howakil Bay, 94
　　Dahlak Bank, 99
Yemeni islands including the Hanish and Zubayr Groups, 102
　Key maps:
　　Southern Red Sea Central Part, 102
　　Jabal Zuqar & Hanish Islands, 102
　　Zubayr Group, 107
Yemen Red Sea coast to the Farasan Islands, 111
　Key maps:
　　Yemen coast to Farasan Is, 111
　　Farasan Is, 117
3. The Saudi Arabian coast, 119
　Key maps: Ports and anchorages on the Saudi Arabian coast
　　A, 119
　　B, 120
　　C, 124
4. The Sudan coast, 126
South of Port Sudan, 126
　Key maps:
　　Anchorages & routes in the Suakin Group & Southern Sudan, 127
　　Shubuk Channel, 132
North of Port Sudan, 145
　Fold out charts . . .
　Key maps:
　　Anchorages in North Sudan, 146
　　Inner Channel: Anchorages & dive sites, 157
5. The Egyptian coast
Foul Bay to Hurghada, 181
　Key maps:
　　Foul Bay anchorages Marsa Girid to Ras Baniyas, 182
　　Anchorages and dive sites in southern Egypt, 192
El Gouna to Suez, 213
　Key maps:
　　Anchorages, dive sites and routes in the Strait of Gubal, 214
　　Anchorages in the Gulf of Suez, 226
6. The Gulf of Aqaba and approaches, 236
　Key maps:
　　Dive sites and anchorages in the Gulf of Aqaba, 240
　　Marinas, anchorages and dive sites in the North Gulf of Aqaba, 242
7. The Suez Canal and approaches, 246
　Key map:
　　Suez Canal, 247
8. Northern approaches to the Red Sea, 258
　Key map:
　　Ports in the Northern Approaches to Port Said, 258
Part IV. Appendix
　I. Charts, 267
　II. Books & websites, 273
　III. Food, 274
　IV. Glossaries, 276
Index, 278

Arabic ship's letter (for photocopying) and translation, End papers

In memoriam
John Henry Morgan
1923–1999

Foreword

The number of yachts in the Red Sea each year continues to increase. Given the intractable Palestine/Israel stand-off, the regular flare-ups of disputes left over from the past, the basket case of Somalia and, more generally, political Islam (a better term for the main problem than Islamic Fundamentalism), this is both surprising and gratifying. It shows that Red Sea sailors have their heads on straight. Most of us know the picture portrayed by the international, and especially the yachting media, and fostered by too many ministries for foreign affairs is seldom what one finds 'on the ground'. For every al-Qaeda fanatic, there are several million kind, friendly, welcoming people in the Red Sea and Gulf of Aden. Just as westerners would be saddened to find themselves and their countries condemned by the antics of ETA, the IRA and Una bombers so are the people of this area. It is good to see that Red Sea sailors judge others as they would be judged themselves.

The Red Sea itself is suffering more from developed world self-indulgence than local upsets. Global warming means the reefs are suffering. Expanding tourism is adding its damage. The coast of Egypt is less and less remote. Reefs and islands are increasingly polluted by dive boat garbage.

The positive side is that slowly the distance you have to sail without help is decreasing. In the first edition there was but one marina, at Elat. Now there are seven – or maybe that's six and a half. With the completion of the first phase of Port Ghalib, there is now a marina only a day's sail north of Foul Bay.

This edition has benefited hugely from the time we spent in the Red Sea and the Gulf of Aden in 2000/2001 using the book, as best we could, as if we were new to the area. We re-surveyed 72 anchorages and gathered information on several new reef anchorages. The result we hope is better illustrated, more concise and easier to use. We have added sketch maps where we have new information and replaced or improved the old ones. We would love to have included aerial photos of *marsa* entrances but economics ruled them out even supposing security paranoia would have allowed them to be taken.

We have much enlarged our coverage of diving and dive sites. For detailed descriptions of exactly where to dive and what to look for, a dive guide will still be useful, but you can now find between these covers the location of almost all the known 'sites' from Bab el Mandeb to Elat.

We have also added more information, here and there, on the history of pilotage and charting of the Red Sea. It is a curious tale and goes a long way to explaining why large-scale coverage is so very unsatisfactory. The first full survey, by the Indian Naval officers Commander Thomas Elwon and Lt Robert Moresby between 1830 and 1834 was a staggering achievement. Much of their survey was to a scale of one inch to the mile and some to ten inches to the mile. The results were the first widely available, proper charts of the Red Sea. They were to a scale of 1:600,000 or so, but the survey data would have laid the basis for more detailed work over time. Sadly the Indian Navy was the Cinderella of the Raj. Its second Superintendent, Capt Robert Oliver RN, a martinet, seemed to have thought surveying a waste of time. The result was that the Survey Office was banished to a corner of the main sail loft in Bombay and eventually all of the original survey material was lost or eaten by white ants and cockroaches! All that was left were the charts. An equivalent re-survey would have commanded the same devotion of time, money, ships and men. That was not forthcoming so it never happened. Instead, in 1870, as steam ships took over the seas, the French and British hydrographers agreed that detailed charting of the coasts was unnecessary. That has remained the policy ever since.

In 2001, though still to relatively small scales, the British Admiralty published the first series of WGS84 adjusted charts for the entire Red Sea. At the same time many of the large-scale old fathoms plans for places in the Gulf of Aden and Red Sea were withdrawn. And to compound problems, metrication has removed or blurred detail, especially with respect to reefs. This makes official charts less and less helpful to coast-hopping small craft. This book is now the sole source of detailed pilotage information to these remote and beautiful coasts.

As with the first edition, if you find errors or discover somewhere new, do please let us know. That way we shall be able to keep improving for future Red Sea cohorts what we gratefully understand has become The Pilot.

Acknowledgements

We are particularly grateful to our companions in the Red Sea in 2000/2001, *Mintaka* and *Freedom*, who were able to solve our occasional need to be in several places at once. A special thank you to Manfred for good suggestions on how to make the book more user-friendly. Philip Jones of Abu Tig Marina has been a huge help and a constant and indefatigable informant. Without Renato

Red Sea Pilot

Marchesan of *l'Ernesto Leoni* our knowledge of the many reef anchorages and official charting errors would be much the poorer. Captain Farid Roushdy showed us Ismailia and told us of his plans for the yacht club there. Many of the officers who lead a lonely and probably unrewarding life in posts at remote anchorages in Egypt, Sudan and the Yemen were unfailingly polite and did their best to answer our barrage of queries. Bashir el Sheikh al Mustafa el Amin looked after us royally in Port Sudan. Mohammed was a tower of strength in Suakin as was Mike in Massawa. In Aden Captains Ali and Hussein were extremely kind and helpful, especially over the problem of piracy. And at all times countless kindly fellow yachties, whose boats' names have appeared in the first edition and the supplements over the years, and are truly too many to list here, have kept us regularly in touch with changes. Mr Ferhat Benyahia in Chester has done a beautiful translation of the ship's letter which will greatly assist communication between the book's readers and the people they meet in the Red Sea. To all of you thank you.

Finally, many thanks to Willie Wilson and the staff of Imray's.

Elaine Morgan and Stephen Davies
Corneilla-de-la-Rivière, France

KEY TO SYMBOLS

Plans show depths and heights in metres. Bearings are true unless indicated otherwise.

- (on large scale sketches) areas, not impassable, but within which sailing only in daylight is feasible, and with a vigilant lookout aloft.
- direction of flood tide or regular current
- established GPS position (WGS 84) with co-ordinates
- shallow or hazardous waters
- mangroves
- sand beach
- watchtower/lookout
- rocks
- coral reef
- beacon, with topmark where relevant
- stake
- above water rock or boulder
- 'bommie' or isolated reef/coral head
- offshore rig or production platform
- rock below surface but with less than 2m over
- buoy (lit) (various symbols following BA conventions)
- wreck exposed at all states of tide
- quay; mole/breakwater; jetty
- large radio aerial
- cliff (usually low)
- recommended anchorage
- untried, daytime or emergency anchorage
- mosque
- isolated building
- ruined building
- conspicuous palm tree
- customs
- authority (nature specified, e.g Port Authority)
- water
- fuel
- post office
- telephone

Other abbreviations

- s   sand
- m   mud
- r   rock
- c   coral
- w   weed
- f   fine
- M   miles (nautical), magnetic (bearings)
- light and characteristic, e.g Fl.G.3s
- F fixed Fl flashing Fl(2) group flashing
- Oc occulting R red G green Y yellow
- dive site

Part I
Introduction

> 'The water of this red see is not redde of his owne kynde, the colour of it is by reson of the costes and the botom of it which be redde grounde, and with contynually betyng and dasshing of the water upon the costes and incresying of it and decresying, it maketh the water redde. Notwithstonding ther be other opynyons...'
> Roger Barlow *A Brief Summe of Geographie*, c.1540

2002

That was a pragmatic, 16th century English view. But why on earth should it have been called the Red Sea in the first place? That's a question that has provoked speculation since 113BCE when Agatharkides of Cnidos wrote his *On the Erythraean Sea*. Agatharkides, a grammarian and philosopher of the Peripatetic school, offers five possible explanations and plumps for the one which concludes, triumphantly, that it isn't in fact called the Red Sea anyway!

The Greek word for red was *erythros*. It still is if you look at Cypriot red wine bottles. So, all commentators have assumed that *Thalattan Erythran* can be translated as Mare Rubrum, la Mer Rouge, Rotes Meer, or the Red Sea. But what sort of redness? Pigmentation? Atmospheric? Geological? Alginous? Or why? Did it come from the name of the Nabataeans who lived near rose-coloured Petra and whose name means 'Red People' in ancient Hebrew? Or was it linked to the supposedly 'red Homeritae' of the ancient Yemen. Agatharkides dismisses any such simple explanation. It was nothing to do with the reddish mountains lit by a glowing sun, colouring the sea in shape-shifting reds. Nor the lurid sunsets, streaming through an atmosphere of desert dust. He doesn't even consider the favourite Roman candidate, avowed by Pliny, of alginous blooms of red tide. Leave alone GWB Huntingford's eccentric idea that it's a reference to what navigators' skin looked like after suffering a common Red Sea summer gift, prickly heat.

Agatharkides' tale is a slice of local history which he says he got from a Persian called Boxos. Another Persian called Muozaios had a courageous and wealthy son called Eruthras – you can see where this one's going. Eruthras had a springtime home on the Red Sea where he kept prize mares. One day lions attacked them and promptly killed a couple. Terrified by the lions and simultaneously plagued by gadflies, the rest galloped into the sea. A typically strong Red Sea wind was blowing and before they knew it, taken by wind driven currents, the mares beached on an island offshore from Eruthras' stud farm. Their loyal herder followed them. When Eruthras showed up, he built the first ever raft in the area – rafts were still in wide use in the southern Red Sea in the first century CE according to the *Periplus* – waited for a weather window, and popped across to rescue horses and herder. Once at the island he was much taken with it. He built a port there, peopling it from the mainland. Word of his deeds got round and everyone began talking of Eruthras' Sea. And since in those times the entire sea area east of Suez to the Gulf of Aden, the Arabian Sea and the Persian Gulf was called the Red Sea, perhaps a Persian origin to the name isn't wholly fanciful.

Agatharkides the Grammarian was thus able to point to a simple copying error as the origin of the name. Eruthras' Sea got mistaken for Erythraean Sea and that's why we call it as we do.

Then came Islam. After its triumphant inclusion of the Red Sea behind the 'Muslim Curtain' (as medieval writers labelled the eastern marches of Christendom), Islam ignored the old tradition altogether. The faithful called it the Sea of the Hijaz or the sea of the pilgrimage route. It's also been called the Arab Sea. The Ottoman Empire wavered between the South Sea, south of Istanbul of course, and the Reef-encumbered Sea. But these days even the Arabic name, Al Bahr al-Ahmar, means the Red Sea. At least we agree on that.

Islam

Islam is today the official religion of about 45 nations, with 800,000,000 believers worldwide. The following notes are intended merely to brief anyone who plans to cruise in the area on aspects of Islam likely to impinge upon him or her during a short visit.

Dress

Loose-fitting clothes with maximum covering consonant with comfort are essential for women but head-covering is not obligatory. The treatment of

Islamic fundamentalism

Don't worry overmuch about Islamic fundamentalism. If you observe the rules of common politeness, you will give no offence. Ninety per cent and more of the people you meet are not fanatics. In countries where there have been troubles, these have seldom affected yachts, except where the troubles have been of a political or military rather than religious nature. Keep a low profile and be sensitive to Islamic beliefs, practices and rules.

women in Islamic society is very different from that in developed western countries. Sexual harassment of 'infidel' women, particularly those who do not dress conservatively and act demurely, is not unheard of. Men should wear long trousers and a shirt with sleeves when going ashore, especially when dealing with officials or businessmen when a shirt with a collar is properly polite.

Mosques

Yachts at anchor near any village or town will become familiar with the call of the muezzin from the mosque, largely because it is usually electronically amplified but also because the faithful are called to prayer five times a day. You can visit some mosques, when it is not time for prayers, but you should ask first, dress modestly, and take off your shoes before entering. Visitors are not allowed in any mosques in Oman, nor many in Yemen. Be very discreet about photography in the vicinity of a mosque.

Ramadan

Muslims fast from sunrise to sunset during the month of Ramadan, which is the ninth month of the lunar year. Fasting means abstaining from food, drink, smoking and sex during daylight hours. This means many cafés and restaurants close but food is still available. Markets and shops are usually open, though busier in the late afternoon. The daily breaking of the fast, the *fitr*, is just after sundown and morning breakfast has to be before daybreak. Eid al Fitr is the holiday which immediately follows Ramadan. Celebrations usually last 4 or 5 days.

Pilgrimage

All Muslims try to make the pilgrimage (*hajj*) to Mecca (Makkah) at least once. The time specified is the last month of the lunar calendar. Ferries taking pilgrims to Jeddah, the closest port to Mecca, may be seen in the Gulf of Aden and the Red Sea itself at that time. About half a million Muslims make the hajj each year.

Food and drink

Muslims are forbidden to consume alcohol, pork, blood or the meat of any animal not slaughtered in the prescribed manner. Be careful what you offer visitors and remember that you won't be able to buy alcohol except in Eritrea or in western style hotels, so stock up on your favourite drinks in advance.

Country information

Oman

Ports of entry: Salalah, Muscat

History and politics

Arab settlers arrived in Oman around the first century AD. At this time the trade in frankincense was booming and Pliny reported the area to be the richest in the world. The centre of trade was at Sumhuram, known as Moscha by the Greeks and now called Khor Ruri (Khawr Rawri), 20M E of Salalah. This is one of the few areas in the world where the trees of the *Boswellia* genera, which produce frankincense, grow.

In the 7th century AD local tribal society converted to Islam and elected its first Imam. Oman's history thereafter was tied up with three things. First, struggles with external aggressors, mainly the Portuguese and the Persians. Second, struggles between the coastal areas and the tribal interior, not finally settled until British military action was followed by a brokered peace in 1950. But above all a long, successful and far ranging history of engagement with the sea. It's worth remembering that an Omani ruler controlled Zanzibar, now part of Tanzania, until 1964! Omani ships ranged far and wide, dominating early Arabian Sea trade and reaching China before the 15th century.

In 1938 the rule of Said bin Taimur began, during which the country, once amongst the Arab world's most powerful, became isolated and its economy stagnated. A bloodless palace coup in 1970 was led by the Sultan's son, Qaboos bin Said, who introduced the present era of modernisation and development made possible by oil revenues. Oman became fully independent in 1971 and in the same year was admitted to the Arab League. In recent years the Sultan has increased popular participation in his government. The main consultative council is an elected body. Women, who both have the vote and stand for election, work in most professions and hold the Sultan in high esteem.

Land area and climate

Oman has an area of 310,000km^2 and extends inland from a fertile S coastal plain, where the population is concentrated, over rugged mountain ranges, dropping to the borders of the Rub al Khali ('the Empty Quarter'), the huge desert of the S central Arabian peninsula. Although the climate is hot and humid from April to October, there is pleasant spring weather in Salalah, and rainfall in summer brings relief from the heat.

Population and language

There are about 2·5 million people whose official language is Arabic but English is widely spoken. People are very friendly, helpful and courteous. In comparison with other countries we cover, Oman is affluent.

Visas and embassies

Yacht crew may enter Salalah without visas and are given day shore passes allowing them to stay locally for about one week. For travel further afield you will need a visa as will most crew arriving by air or overland. Visas issued overseas cost about US$40. Allow a couple of weeks for processing. If you apply for one in Salalah officials will suggest you use an agent and pay a fee of approx US$75–100. An alternative is to find a sponsor who will sign your visa application. Sponsors include local business companies or hotels. If you choose this method get a visa application form, usually available from Immigration in the port, type out the details and take it with 2 passport photos to the Immigration Dept near the airport. This will greatly reduce fees but will take much longer to organise. Visas are usually valid for 3 weeks. If your passport shows evidence of visiting Israel you will not get one.

Foreign embassies and consulates in Muscat include Germany, India, New Zealand, UK[1], USA and Yemen.

1. The UK embassy will handle emergencies for Commonwealth citizens. There are reciprocal agreements between the diplomatic missions of member nations of the European Union regarding the handling of diplomatic affairs for European nationals.

Money

The currency is the Omani rial, divided into 1000 baiza. You can exchange money at banks and moneychangers. ATMs take international credit and debit cards such as *Visa, MasterCard, Plus, Cirrus* etc. Stock up on US$ here because, other than in Djibouti, Salalah is the last convenient place linked into the international banking network until Egypt. As a guide, you should allow at least US$1000 in cash. Get new, unworn notes if possible. The most useful are $20 bills, then $10, then $5.

Communications

There is a GSM mobile phone network. Cardphones are easy to find and phonecard outlets are ubiquitous. Oman's IDD code is 968. There are several Internet cafés in Salalah. Mail can be sent from the PO in the town centre where there is also a poste restante service but mail will be returned to sender if not collected within 15 days. Various shops and hotels in the town will send and receive faxes. There are several daily bus and air services to Muscat. Local transport consists of car hire, taxis and hitching.

LMT, opening hours and holidays

Oman is 4 hours fast on UTC.

Government offices and embassies: 0730–1430 Saturday to Wednesday.

Shops: 0800/0900–1300 and 1600–1900
PO: 0730–1330 Saturday to Wednesday, 0900–1100 on Thursdays.
Banks: 0800–1200 Saturday to Wednesday, 0800–1100 on Thursdays.
Moneychangers usually also 1600–1900.

Public holidays include 1 January, 18 November (National Day), 19 November (Sultan's birthday), all Muslim religious festivals and all Fridays.

National flag

Red, with a white panel in the upper fly and a green one in the lower fly and the national emblem, consisting of two *jambiyas* (Arab daggers) crossed over a vertical sword, in white on the upper hoist.

Yemen

Ports of entry: Aden, Al Hudaydah, Al Mukalla

History and politics

The first Yemeni kingdoms emerged and consolidated between 1500BCE and the 7th century CE, thanks to trade in frankincense and myrrh. The Romans powerfully influenced the trade but never succeeded in bringing Yemen within the Roman Empire for all that it was known as 'Arabia Felix', Happy or Fortunate Araby. The great Ethiopic Kingdom of Aksum rose as the power of Rome declined. This led to the christianisation of the Yemen until a rebellion in 543CE. Aksum's power collapsed and in the 7th century CE the Yemen converted to Islam. It remained firmly within the orbit of Islamic hegemony until the first European incursions in the 16th century CE. That's when the history of the modern country begins.

The British East India Company captured Aden in 1839. Over the ensuing years a series of agreements with local sultans created a loose federation of British protectorates in S Yemen. By 1965 there was strong opposition to British colonial rule in the affluent Aden protectorate. Meanwhile in the N, an area never subjected to foreign rule, a vicious civil war was being played out between the supporters of the nascent Yemen Arab Republic (YAR), backed by Egypt and the old Soviet bloc, and that of the traditionally royalist Imamate backed by Saudi Arabia, the Shah's Iran, Britain and the USA. By 1970 the war in the YAR was over and a compromise peace concluded. The resulting state, despite a military coup and presidential assassinations, inclined towards a market economy.

Meanwhile anti-British insurgency in the S ended colonial rule in 1967. But the creation of the People's Republic of South Yemen (renamed the People's Democratic Republic of Yemen in 1970) under an extreme leftist leadership, orientated to the old Soviet bloc, merely added to the larger Yemeni problem. The obvious path to a single Yemen was now not only bedevilled by internal factionalism in each component country, but by the deep ideological divide between them.

During the 1980s internal differences in the S dominated. A fierce civil war broke out in early 1986. The sinking of a yacht in Aden harbour and the rescue of the crew by a foreign ship were widely reported in the international press. Thousands were killed and the country virtually closed to foreigners for about six months.

In 1990 the two Yemens unified to form the Republic of Yemen. The union was a makeshift affair. The surface appearance of success held until the spring of 1994 when deep-rooted differences between the N and S led to open conflict. 5,000 people died. Since then Yemen has begun to settle down.

The inevitable legacy of the imposition of a modern state system on historically independent local fiefdoms remains a problem in the Yemen. There are endemic outbreaks of occasional, localised lawlessness such as the kidnapping of westerners, the attacking of the USS *Cole* by suicide bombers in 2000 and incidents of armed robbery* in the Yemeni waters of the Gulf of Aden. The authorities seem unable to deal with the problems and disinclined to make the effort. From a visiting yacht's point of view, the main thing to remember is that these problems are generated by a recalcitrant few.

*For more detail, see 'Problems' below and our piracy warning on Imray's website www.imray.com

Land area and climate

Yemen has a total area of about 550,000km^2, including the Hanish group, Kamaran I, Mayyun (Perim I) and Suqutra (Socotra). In 1995 Eritrean forces invaded the Hanish Is. This was a legacy of the collapse of the Ottoman Empire and the absence of any clear title of sovereignty. The dispute was settled in Yemeni favour by international arbitration in the Hague in 1998. Military surveillance has been relaxed a little since then but if you stop be sure to fly the Yemeni courtesy flag, and expect to be visited and asked to show your papers.

Suqutra is now considered safe for yachts though it lacks secure anchorages. It is home to many rare species and is to become a national park, you should arrange a visit via Mukalla or Nishtun.

On the Gulf of Aden coast there are dramatic cliffs especially near Aden and Nishtun.

The S has a tropical climate whereas the capital, San'a up at about 2,400m, has low humidity and comfortable temperatures. The plains on the Red Sea coast rise to mountains of over 3,000m in the centre.

Population and language

The population is estimated at 17 million and rising fast. Arabic is the official language, but English is also spoken. Yemen is a very poor country but its people offer visitors an extraordinarily warm and generous welcome.

Visas and embassies

You only need a visa if you want to travel inland. They cost US$60 for European nationals. Otherwise crew are treated as seamen. Shore passes are issued in exchange for passports on arrival. Passports can be reclaimed for visits to embassies and so forth. Take crew lists and the boat's stamp when you check in. Signing crew off or on requires a new crew list and a letter of confirmation to immigration. Signed off crew in Aden without visas are given a police guard from the port to the airport. Any crew member who intends to join you in Aden should obtain a visa in advance, and should ask the embassy that issues it to check whether it will be necessary to prove that they are joining a yacht. Otherwise they may be asked to produce a return air ticket on arrival. You may call at Yemeni ports along the coast without a visa but again you will have to exchange passports for shore passes.

There are Yemeni diplomatic missions in many major European and American cities at which to apply for a visa. There are also Yemeni embassies in almost all of the capital cities of the Middle East and N and E Africa.

Foreign embassies and consulates in San'a include Oman, Djibouti, France, Germany, India, Italy, Japan, Netherlands, UK, USA. Former embassies in Aden have generally become consulates or trade commissions. The UK and Egypt are still represented but the former Indian embassy has closed. If you get an Egyptian visa here make sure that it is valid for three months. Check the date in Arabic figures. The visa's validity will begin with the date of issue. Costs vary according to nationality.

Money

The currency is the Yemeni rial. Official moneychangers generally offer more favourable rates than banks. Black-market rates, widely available, fluctuate, but are often better than the legal rate. The main problem is surreptitiously counting out all the notes. There are banks and moneychangers in Tawahi, Ma'ala and Crater. Note that large purchases may be checked against bank receipts by customs at the gate to the jetty as you go back to your boat. Discretion is advised.

Communications

The easiest way to call overseas is from one of the many international call centres. Calls are metered and rates are official. The IDD country code is 967. There is a GSM mobile phone network, but it is not reliable. Internet access is available and cheap but slow. Mail can be sent to poste restante at the Tawahi PO, Port Aden, Yemen. Letters take about a week from the UK but we don't suggest you have anything valuable sent. The PO is a short walk from the Prince of Wales Pier towards the harbourmaster's office. For sending mail from Aden we have used both the stall on the Pier and the Tawahi PO and frankly, recommend neither. Important mail has gone missing from both, though poste restante at Tawahi seemed reliable, if slow. For faxes, try the shipping agents and shopkeepers near the Pier. If you are expecting a reply you may be expected to pay for it.

There is an international airport in Aden with direct flights to London but most overseas flights are routed via San'a. For local and inland travel, buses and shared taxis are cheap. Individual taxi fares should be negotiated in advance and if you take a taxi to San'a beware that it may be in poor condition, as may its driver!

LMT, opening hours and holidays

Yemeni time is 3 hours fast on UTC.

Immigration and Customs in the ports open 24 hours a day.

Other offices: 0800–1300 except Thursday pm and Friday.

Banks: 0730–1230.

PO: 0800–1300 and 1600–2000, closed Friday and Saturday.

Shops: 0800–1200 and evenings.

Public holidays include 1 January, 1 May, 13 June, 26 September (Revolution Day – N Yemen), 14 October (Revolution Day – S Yemen), 30 November (Independence Day – S Yemen), all Muslim religious festivals, and every Friday.

National flag

Three horizontal stripes of red, white and black.

Somalia

We include Somalia because it has a coastline on the Gulf of Aden. We are not suggesting you call there except in an emergency. Cautions still apply within 50M of the coast. Since UN troops withdrew in 1995, two new Somali states have emerged in the N, but the situation is fluid. Neither has gained international recognition and there are regular outbreaks of violence. Most of the country is divided by rival clan factions but we optimistically include some details about the country and its anchorages.

The Republic of Somaliland, once run by the British, on the NW coast of the Gulf of Aden declared independence in 1991. In 1998 part of NE Somalia, the old Italian Somaliland, became the Somali State of Puntland, with its main port at Boosaaso. The president, Mr Abulahi Yusuf, has been employing a Fishery Protection force and Coast Guard on contract. Very expensive armed escorts for yachts passing through the region can be arranged by a company called Hart Nimrod. See the Pilotage section, page 81 for details.

Ports of entry: Berbera, Boosaaso

Language

Somali and Arabic. English and Italian are also spoken.

(Old) National flag

Light blue, with a white star in the centre.

Djibouti

Port of entry: Djibouti

History and politics

The territory was annexed by the French in 1862 and Djibouti became its capital in 1892. At the time it was known as French Somaliland. French cruisers will know Djibouti as the base from which Henri de Montfreid cruised the Red Sea. His accounts of his voyages in small trading boats paint an almost incredible picture of these waters between the wars. In 1958 the country was renamed the French Territory of Afars and Issas, after the names of the two dominant tribal groups. Djibouti became independent in 1977 but there is still a French garrison, albeit scaled down from its former size. Ethnic tensions and conflicts in neighbouring countries affect the country's peace from time to time though a superficial calm normally reigns and the port has benefited by increased throughput. There was a brief outbreak of violence in early December 2000 when a sacked police chief attempted a coup d'état. The incident was subject to a news blackout although it was all over in a day. Djibouti is normally viewed as a haven of calm in the tormented Horn of Africa but it has to bear the economic burden of refugees fleeing from Somalia, Ethiopia and Eritrea.

Land area and climate

The republic forms the shoreline of the Gulf of Tadjoura at the W end of the Gulf of Aden. The land area, including offshore islands, is 23,200km^2. There is a protected underwater marine park E of the Iles Moucha. The weather is hot all year round with very low rainfall. June and August are the hottest and most humid months but from October to April temperatures average 25°C.

Population and language

Djibouti has about 750,000 inhabitants but at least a quarter are refugees. Official languages are French and Arabic though half the people are Afars and half are of Somali origin. Many Djiboutians are multilingual and Somali is the most widely spoken language.

Visas and embassies

Visas, costing approx US$20–30 and valid for 10 days are issued on arrival. If you want to make enquiries in advance, there are embassies in San'a, Cairo, France and the US. Some other Red Sea countries have embassies in Djibouti and you can usually get Egyptian visas here for about US$20. Yemen and Sudan also have embassies in Djibouti if you want visas, as does Eritrea when diplomatic relations are not strained. European countries and the US have a diplomatic presence.

Money

The currency is the Djibouti franc, divided into 100 centimes and it's pegged to the US$. Cash from ATM's is available with *Visa* cards at one or two banks and you can use some credit cards to obtain cash over the counter on production of your passport. Major currencies can be exchanged easily but there is a large commission on travellers' cheques. Beware of pickpockets.

Communications

The PO offers poste restante service and handles air mail at a satisfactory speed. International calls can be made from there. You must make a deposit. There are no cheap rate periods. Phonecards are available for public phones. There is a GSM mobile phone network. The IDD country code is 253. Email and fax are available but the connections to the Internet are slow. There is a railway line to Addis Ababa, an international airport 3M from the city centre and both ferry and air services to Obock. Long-distance buses are the cheapest way to travel inland. Minibuses and taxis ply for hire in the city.

LMT, opening hours and public holidays

Local time is 3 hours fast on UTC.

Government offices in the port: 24 hours daily except Thursday pm and Friday till 1400.

Offices: 0730–1330 Sunday–Thursday am.

Banks: 0700–1130 Sunday–Thursday.

Shop: 0700–1230, 1600–1830 Sunday–Thursday.

There are also moneychangers and shops that stay open in the afternoon.

Holidays include 1 January, 1 May, 27 June (Independence Day), 15 August, 1 November, 25 December, 31 December, every Friday, and Muslim religious holidays.

National flag

Horizontally blue over green, with a white triangle based on the hoist charged with a red star.

Eritrea

Ports of entry: Massawa, Assab

History and politics

Eritrea is an old land with a new name. In ancient times it was part of the great Kingdom of Aksum, the dominant S Red Sea power in the 4th century CE. Since then it has periodically both dominated and been dominated by the Ethiopic peoples of the interior and occupied by Egyptians, Turks, Italians and the British. It was the Italians of the 19th century who called their new Red Sea colony Eritrea. A flight of nostalgia for the classical Mediterranean world of Roman glory. *Erythros* means 'red' in classical Greek.

When the Italians were defeated in 1941, the British took over until 1952. In that year, without the Eritreans being asked, their country was handed over to Emperor Haile Selassie's Ethiopia, in accordance with a UN resolution. The idea was for a federal state within which Eritrea would be autonomous. It failed. In 1962 the Ethiopians declared a unitary state. The Eritrean parliament formally voted itself out of existence. And Eritrea became merely a province. Almost immediately the Eritrean People's Liberation Front (EPLF) formed. It was the beginning of nearly three decades of war, drought and famine for Ethiopia and Eritrea.

In 1974 Haile Selassie was deposed and Ethiopia became a radical, leftist country, governed by a military organisation, known as the Dergue, closely allied with the Soviet Union. Colonel Mengitsu, its leader, continued the war against the Eritrean rebels. In 1989 Col. Mengitsu stepped down and preliminary peace talks began. The cease-fire was temporary and the Eritrean secessionists were joined by a group of new combatants, the Ethiopian People's Revolutionary Democratic Front (EPRDF). By the early 90's it was clear that the central Ethiopian government was losing ground and further peace talks began in London. Finally in 1993 a UN-supervised referendum on self-determination led to an independent Eritrea.

Since then there have been disputes with the Yemen over the Hanish Is and with Djibouti over their common border. Two more years of fighting against Ethiopia began in 1998, ostensibly over a patch of barren desert. In July 2000 the UN brokered a peace agreement in Algiers and in November 2000 peacekeeping troops (UNMEE) arrived to patrol a buffer zone between the two countries. However, the loss of direct Ethiopian access to the Red Sea, a consequence of Eritrean independence and the real cause of the recent conflict, remains unresolved.

Land area and climate

Eritrea has a land area of 125,000km^2 with highlands W of the coastal plain. The mountains are lower and more broken towards the Sudanese border. The archipelago of the Dahlak Bank contains about 100 islands that have been designated a marine park.

Assab is hot and dry most of the year, whereas Massawa is cooler and more humid and it even rains in winter. In Asmara, at 2,400m, the average annual temperature is 18°C.

Population and language

There are about 3·5 million people who speak about nine different languages, but Tigrinya is predominant in Massawa. Arabic, Afar, English and Italian are also used. The Eritreans' commitment to their young country is everywhere apparent. Yachtsmen and women continue to be charmed by the courtesy and warmth with which they are received. However, there are now signs of increasing government authoritarianism and intolerance of dissent. Sad but probably inevitable.

Visas and embassies

Visas are not required in advance. Shore passes are issued on arrival. For stays longer than a few days and travel inland, one-month visas are easily obtainable in Massawa for US$40. Egyptian visas are available from the embassy in Via Derg,

Afawark, Asmara. Other diplomatic missions in the capital include France, Germany, Italy, Sudan, Yemen, the UK and the USA.

If you intend to cruise on the Dahlak Bank for an extended period you should apply for a permit from the Ministry of Tourism once you arrive in Massawa. Fees are US$20 per person, unlimited stay.

Money

The unit of currency is the nakfa, sometimes spelt nagfa and abbreviated nfa. In Massawa you will need foreign exchange or travellers' cheques as the banks have no facilities for issuing cash on credit cards. In Asmara you can arrange for a transfer of funds with Western Union or with Himbol. The latter accept cheques drawn on personal accounts but clearance takes at least a week.

Communications

A GSM mobile phone network was begun in 2001 and should fairly soon cover Asmara and the main ports.

International mail is efficient, reliable and extremely good value. International phone calls can be made from the PO but a deposit of nfa300 is required before the call is put through and rates are quite high. The IDD country code is 291. Calls within Eritrea can be made using a phone card which you can buy from the PO. Fax services are also available at the PO. There is an Internet facility in Massawa but it is usually out of service. Internet access at cybercafés was available in Asmara in 2001 but it was very slow.

Buses to the capital take 3–4 hours and are cheap. Taxis are also available. Asmara airport is 9km out of the town and has direct flights to Middle Eastern and European countries. The railway from Massawa to Asmara is completely defunct.

LMT, opening hours and holidays

Eritrea is 3 hours fast on UTC.

Port offices: 0700–1400 daily.

Banks: 0800–1100, 1400–1600 Monday–Friday.

Other business hours: 0800–1200, 1600–1800, Monday–Friday, 0800–1300 Saturday.

Public holidays include 1 January, Timket (variable, in January), 8 March (International Women's Day), Easter, 1 May (Worker's Day), 24 May (Liberation Day), 20 June (Martyrs' Day), 1 September (Start of the Armed Struggle), Christmas, and Muslim religious festivals.

National flag

Green over blue, with a red triangle based on the hoist bearing the symbol of the Liberation Front, a yellow wreath of leaves around a 6 branched tree.

Country information

Saudi Arabia

History and politics

The modern history of Saudi Arabia dates from the collapse of the Ottoman Empire at the end of World War I. A murky chapter in modern history follows, especially with respect to British influence and activities, as readers of T E Lawrence (Lawrence of Arabia) will know. Little regarded at the time, but now of international importance, was the fact that the dominant status of the Sa'ud family began with their adoption of the fundamentalist Wahhabite sect of Islam in 1744.

In 1927 the Treaty of Jeddah was signed with the British and the Sa'ud family recognised as the rulers of the new kingdom. International recognition followed in 1932 and the economic basis of the new state, the exploitation of oil resources, began in 1938. After World War II when ARAMCO (the Arabian American Oil Company) was formed, Saudi Arabia was recognised as having the largest oil reserves in the world. There was phenomenal internal growth in the 1960s and 1970s, matched by waste and corruption within the royal family. The creation of OPEC (Organisation of Petroleum-Exporting Countries) represented another milestone. With the embargo on oil exports to the USA and the Netherlands in 1973–74 and the subsequent fourfold price increases, Saudi Arabia achieved a very high international profile. The entire Saudi economy, and hence social and political stability in the long term, are critically dependent on continuously rising demand for oil and, therefore, steady or rising prices.

The Saudi government walks a swaying tightrope. At one end is the fundamentalist Islam upon which the ruling dynasty's fortunes are founded and which it uses to sustain its legitimacy. At the other is uncritical western economic, military and diplomatic support for its oil milch-cow, without which the anachronistic, quasi-feudal regime would probably implode.

It follows that Saudi Arabia is no model of an open society and that visiting yachts are not yet generally welcome. That said, there are many refuges along its Red Sea coast and Jeddah offers repair facilities that, while very expensive, outclass anything else in these waters. If you go into Jeddah for repairs you will be courteously received but probably confined to the port area. Photography, fishing, swimming and the consumption of alcohol are prohibited anywhere in harbour and you will be subject to strictly enforced regulations and standards concerning behaviour and demeanour, particularly irksome if you are a woman.

Major ports include Gizan, Jeddah, Sherm Rabegh, Mina al Malik Fahd, Yanbu al Bahr, Mina Duba and Port Sharma.

Land area and climate

Saudi Arabia occupies 80% of the Arabian Peninsula, with an area of about 2,200,000km^2.

The coastal strip consists largely of steppe, with occasional oases. The inland areas are mostly desert or semi-desert, with scrub. Mountains in the W run parallel to the Red Sea dropping sharply to the coast in some areas.

The coastal climate is hot in summer but winters are cooler with some rain.

Population and language

There is an estimated population of 20·2 million. About 45% of this number live in urban areas.

The official language is Arabic, but English is widely spoken.

Visas and embassies

In an emergency, you may be issued with a transit visa. There are no tourist visas at present. Foreign embassies and consulates in Jeddah include Australia, Canada, Egypt, France, Germany, New Zealand, UK, USA.

Money

The currency is the rial, divided into 100 halalas. 5 halalas equal 1 qurush.

Banks include the Saudi British Bank and the National Commercial Bank. Moneychangers are common. *American Express* is represented by Ace Travel in Jeddah, where card holders can cash personal cheques and collect mail.

Communications

There is an excellent telecommunications system including a GSM mobile phone network, and there are public call boxes for making international calls. The IDD country code is 966. There are no poste restante facilities, but outward mail is efficient.

There are regular international and domestic flights from the major airport at Jeddah and daily buses to Cairo via Aqaba. Women cannot travel alone on these buses, and they are also banned from driving in Saudi Arabia. It is possible to travel by ferry from Jeddah to Port Sudan, Suakin and Suez.

LMT, opening hours and holidays

Local time is 3 hours fast on UTC.

Banks: 0830–1200 Saturday–Thursday

PO: 0700–2100 Saturday–Thursday

Shops: 0830–1300, 1600–1900 Saturday–Wednesday and Thursday am

At prayer time (5 times a day) everything closes. The religious police enforce Islamic orthodoxy. Only Eid al Fitr and Eid al Adha are holidays, but many businesses and government offices may be closed for 2–3 weeks at the time of the hajj each year, especially in Jeddah.

National flag

Green, with the text (in white Arabic script) 'There is no God but Allah and Mohammed is his prophet', and beneath this a white sabre. The hilt of the sword is towards the hoist. There are fines and prison sentences for those who do not fly the flag while in Saudi waters.

Sudan

Port of entry: Port Sudan. Clearance is also possible at Suakin.

History and politics

Sudan's history has been tied to that of Egypt since ancient times. Modern history began with the establishment of the Turko-Egyptian regime in 1820. By the 1870s the Egyptian Khedive Ismail was encouraging British involvement in the Sudan.

A successful, fanatical Islamic revolt led by Muhammed Ahmed ibn Abdullah, known as the Mahdi, began in 1881. In effect this was an anti-Egyptian rebellion as much as a religious war, or jihad. The Mahdi's forces had early success against the Egyptians and the British advised Egyptian evacuation. General 'Chinese' Gordon came out of retirement to supervise events and try to mediate with the Mahdi. He failed. The Mahdi besieged and took Khartoum in 1885, and Gordon was killed.

The Mahdi also died later that year. He was succeeded by his son, who ruled, increasingly weakly, until 1896. In that year an Anglo-Egyptian force, under General Kitchener, set out to re-conquer the Sudan. The Mahdist forces were defeated at the Battle of Omdurman in 1898.

The following year the Anglo-Egyptian Condominium was created. This gave de facto power to the British, who administered in colonial fashion for the next half-century. By pushing S to the borders of Uganda and Zaire (present day Congo) and E to Chad, they created the territory of present-day Sudan, and by extension a potential disaster. Sudan became a sovereign independent state in 1956.

N Sudan is primarily Arab and Muslim while the S is predominantly Christian or animist, divisions which caused civil war between N and S from the outset. There have also always been unresolved border disputes between Sudan and Egypt, a matter of some interest to Red Sea voyagers.

There was a military coup in 1958, but the revolt in the S and general discontent led to the reinstatement of civilian government in 1966 and the rise to power of Sadiq al-Mahdi. However, economic and social conditions continued to decline until power was seized in another, but bloodless, military coup by Major-General Numeiri in 1969.

Numeiri ruled for the next 16 years, his major achievement was a peace agreement with the S in 1972. Nevertheless, political instability and economic crises continued to plague Sudan. An army coup in 1985, led by Lieutenant General Siwar al-Dhahab, deposed Numeiri. The following year the first elections for nearly 20 years were held and Sadiq al-Mahdi came to power again. In 1989 there was another military coup. President Omar Bashir took charge of a nation fraught with political turmoil, economic chaos, civil war, widespread famine, and a chronic refugee crisis caused by refugees from war and famine in Ethiopia and Eritrea.

Not much changed until 1999. A new constitutional law introduced multiparty politics, though the change has at best been cosmetic. In the same year peace agreements were signed with Egypt and Eritrea, though border tensions continue. Meanwhile the regime had deftly managed to become an international pariah through support for Islamic terrorist groups like al-Qaeda. The one bright spot, though given Sudanese politics even that is doubtful, was the discovery of oil in W Sudan. The problem is simple. Many of the oil deposits are in rebel-held territory and separatists also threaten the 2000-mile long pipeline to the coast between Port Sudan and Suakin. The regime has begun massaging its international reputation. It has attached itself to the coat-tails of the international anti-terrorist alliance, aware that western oil interests, not renowned for the slightest environmental or political probity, will bat for it in return for a share of the goodies. Maybe good will come. Looking at the streets of Port Sudan, it is desperately needed.

Land area and climate

Sudan, with a land area of 2,500,000km² is the largest country in Africa and can be divided into several physical regions. The Libyan and Nubian deserts in the N are split by the valley of the Nile. Red Sea Province is on the E side of this region. Central Sudan consists of the steppes from Khartoum to El-Obeid and the savannah region S of the steppes. Further S again is the equatorial rainforest, in the centre of which lies the Sudd, a vast marshland of over 12,000km². Sudan has a continental tropical climate, though the influence of the sea on the coast results in some rain in winter and cool night-time temperatures. Moderate to strong N winds on the coast are the norm.

Population and language

There are about 35 million people in over 300 distinct tribes, with more than 100 languages and dialects. The official language is Arabic, which is spoken by over half the population. Dinka is prevalent in S Sudan. English is widely spoken.

Visas and embassies

Visas are not essential for yacht crew but you will need Immigration clearance to go ashore in Suakin or Port Sudan. Shore passes are issued in exchange for passports. This is normally arranged by agents in Port Sudan and Suakin. They will meet you when you arrive and ask for crew lists and a passport photo of each crew. The agent also organises a health official to visit you. When you leave they will arrange Customs clearance and sailing permission after you have paid port dues. See the Pilotage section for further information.

If you need to leave your boat in Sudan to fly home, first buy your air ticket and then apply for a visa in Port Sudan. Consult an agent to short-cut red tape. If you plan any independent inland travel, apply at the Tourism and Hospitality Department for the obligatory permit. You will need passport photographs as well as a visa, and there will be a fee. Allow for several days' processing delay. If your agent makes the arrangements life is much simpler. You should also apply here for a (free) permit to take photographs from which any objects of conceivable military interest will be excluded.

If your passport shows evidence of a visit to Israel you may be denied a visa, and South African nationals may also have problems.

Money

The currency is the Sudanese dinar. Exchange for foreign currency and travellers' cheques is possible at all the large banks, though there have been problems changing US travellers' cheques. Inflation means you get a lot of dinars for your dollars so take a large bag with you. Most fees for formalities are payable only in US$. There are no credit card facilities or ATM's in Sudan.

Communications

There is a good GSM mobile phone network in the Port Sudan area but coverage doesn't reach Suakin. There are telecoms 'centres' in Suakin but international lines are usually busy. There are public card phones and the Internet is accessible at cafés in Port Sudan but not yet in Suakin. The IDD Code is 249.

Outward mail from Port Sudan seems reliable, if slow and expensive. Inward mail is even slower and the PO is a genuine antique. Although poste restante is available we can't recommend it. If you have anything sent in by courier, use an individual or business address instead.

Sudan Airways in principle has daily flights between Port Sudan and Khartoum, where there is an international airport. Flights may be cancelled for any number of reasons and often are. If you buy an airline ticket in Sudan you must produce a bank receipt as proof of legal exchange. You may be able to pay with a credit card. Consult your agent.

Buses run from Port Sudan to Khartoum. Taxis in the towns are cheap, but negotiate rates before you set off. There are good roads immediately around Port Sudan. You can take a bus between Port Sudan and Suakin if you make arrangements with an agent for security clearance.

LMT, opening hours and holidays

Sudan is 2 hours fast on UTC.

Government offices: 0800–1400 Saturday–Thursday

Banks: 0930–1300 Saturday–Thursday

PO: 0730–1300

Shops and markets: 0700–1300, 1700–2000 Saturday–Thursday, some also Friday am

Public holidays: 1 January (Independence Day), 3

March (Unity Day), 25 May (Revolution Day), 25 December (Christmas Day), every Friday, and Muslim religious festivals.

National flag
Three horizontal stripes (red, white and black), with a green triangle based on the hoist.

Egypt

Ports of entry: El Quseir, Safaga, Hurghada, Sharm el Sheikh, Port Tewfiq (Suez), Port Said, El Arish, Alexandria.

History and politics
The oldest records of humanity are Egyptian. They include a piece of pottery with the earliest known picture of a sailing boat. It follows that the world's first known sailors, about 8,000 years ago, were Egyptians. We give only the barest outline of pre-modern times, followed by a brief account of the country since the Ottoman Empire.

3200BCE onwards The 30 dynasties of native Egyptian rule.

525BCE Collapse of the last Pharaonic dynasty.

332BCE Conquered by Alexander the Great. Thereafter Greek rule under the Ptolemies, the last of whom was Queen Cleopatra.

30BC–638CE Roman rule until the collapse of the Western Roman Empire. Thereafter Byzantine rule from Constantinople.

640CE Conquered by the Arabs under 'Amr ibn al-'As. The beginning of 11 centuries of Islamisation and Arabisation.

640–1250CE Rule from Medina and later Baghdad by governors of ruling caliphate.

1250–1517CE The rule of the Mamluk dynasty during which Egypt was the centre of the Arab Islamic world.

1517–1798CE The period of the Ottoman Empire when Egypt was ruled from Istanbul, though by increasingly autonomous Mamluk viceroys or pashas.

1798CE French Napoleonic occupation and the beginning of western influence.

By 1803 a tenuous Ottoman control had been restored. However the ruling Pashas, beginning with the Albanian Muhammed Ali, had almost complete independence from the sultanate. Muhammed Ali reorganised the administration, agriculture, commerce, industry, education and the army, with the aid of French advisers, and made Egypt into the leading power not only in the E Mediterranean, but in the Red Sea as far S as the Yemen. Muhammed Ali, who ceded power in 1848 and died in 1849, can reasonably be called the father of modern Egypt.

Muhammed Ali's successors turned away from the French and more towards the British who built Egypt's first railways. Nonetheless in 1854 Muhammed Ali's old friend Ferdinand de Lesseps was given the concession to build a canal from the Mediterranean to Suez. (For more on the history of the Suez Canal, see the relevant entry in the Pilotage section.) During the next 20 years, dominated by the dynamic Ismail Pasha, Egyptian state finances became increasingly shaky. Looming bankruptcy provoked French and British intervention. Then, in 1882 the British occupied Egypt in order to snuff out the nationalist revolt led by Arabi Pasha. They continued to hold de facto control for the next 40 years.

In 1914 Egypt officially became a British protectorate. Nominal independence under a constitutional monarchy was granted in 1922. The Anglo-Egyptian treaty of alliance of 1936 permitted a British garrison to be maintained in Egypt. In the post-World War II period this became a constant bone of contention.

After King Farouk led the country to defeat in the 1948 Palestine War with Israel, the monarchy was overthrown by a military coup in 1952. A brief power struggle ensued and Colonel Gamal Abd-ul-Nasser became Egypt's leader in 1953. A Revolutionary Command Council was established and Egypt declared a republic. The coup was partly a consequence of the 1948 defeat, but as much a function of the continued British military presence, which was seen as quasi-colonialism. A particular cause of grievance was the British insistence on garrisoning the Canal Zone, which implied that the Egyptian government could not be trusted. In 1954 the British agreed to withdraw all forces within 20 months.

From 1948 onwards, arguably in contravention of the 1888 Constantinople Convention, the Egyptians refused passage through the Suez Canal to Israeli ships. By 1956 the closure, combined with increasing Arab and Egyptian bellicosity, led to the Sinai War with Israel. Meanwhile, Colonel Nasser nationalised the Suez Canal in retaliation for western interference. His action provoked an Anglo-French invasion, later revealed to have been in direct if covert support of Israeli forces. During the ensuing crisis the Egyptians deliberately blocked the canal by sinking ships. Egypt's absolute military defeat was averted largely by US and Soviet intervention. The arrival of UN peacekeeping troops guaranteed safe passage through the canal, if not for Israel!

Egypt, under Colonel Nasser, dominated the Arab world for the next ten years, accepting Soviet military and economic aid while launching a programme of domestic modernisation. Arab-Israeli tension escalated till 1967 when Egypt was crushed by Israel in the Six Day War. Colonel Nasser died in 1970 and was succeeded by Anwar Sadat, who maintained similar policies towards Israel until war was again unavoidable. In 1973 Egyptian and Syrian forces were once more defeated.

After another UN-brokered cease-fire, it was clear that the costs of a militant anti-Israel stance were too high. Sadat sought a settlement and alliance with the West, and in 1979 signed a peace treaty with Israel on the basis of the Camp David Agreement. Sadat was assassinated by Islamic fundamentalists in 1981. He was succeeded by President Hosni Mubarak, who has continued a policy of moderation and reconciliation. By 1982 most of Sinai had been returned to Egypt, relations with Israel had been normalised, and alignment with western nations was well under way. However, peace with Israel led to the ostracism of Egypt by the rest of the Arab world and by most socialist and developing countries.

Mubarak, now into his fourth term as leader, has managed a gradual rapprochement with other Arab countries while also acting as a mediator in Arab-Israeli peace talks. This success has been overshadowed by domestic problems including sluggish economic growth and a sharp increase in population. Militant Muslim fundamentalists continue to attack foreign residents, public figures and tourists. There has been an inevitable downturn in tourism, though you would hardly guess that from the mushrooming of dive resorts along the Red Sea coast.

Land area and climate

Egypt proper has an area of over 1,000,000km^2. Gubal, Shaker and the Gifatin Is and Geziret Zabargad are also Egyptian territory.

The inhabited and cultivated areas of the Nile valley, its delta and a few oases, make up only 3·5% of the total. The rest is desert intersected by mountain ranges, the highest of which are close to the Red Sea coast. The Nile vies with the Amazon for the title of longest river in the world. It is an estimated 4,160 miles long, and has no tributary for its entire 960 mile course through Egypt. The Sinai Peninsula is largely mountainous desert. The climate is dry, with comfortable daytime winter temperatures and very hot summers. It can get quite cold at night in winter, especially at sea or in the mountains.

Population and language

The population is estimated at 66 million. The official language is Arabic, although English and French are both widely spoken, especially in the N.

Visas and embassies

Although it is advisable to obtain Egyptian visas in advance, immigration restrictions have been somewhat relaxed and you can also quite easily get one at your port of entry. Otherwise, visas are available from Egyptian diplomatic missions in Aden, Asmara, Djibouti and Port Sudan. You will need a passport photograph. The problem with getting your visa well ahead if you are headed N is that unless you make swift progress you might find it is out of date on arrival. Visas are normally valid for three months from date of issue and one month from date of arrival. They cannot be post-dated.

Foreign embassies and consulates in Cairo include Canada, France, Germany, Israel, Sudan, UK and the USA. Several EU countries as well as India and the USA also have consulates in Port Said.

Money

The currency is the Egyptian pound (E£), divided into 100 piastres. Commercial banks and hotels will change currency and usually travellers' cheques. Cash advances on credit cards may only be obtained at banks in the larger tourist centres. There are ATM's in Port Suez and Port Said.

Communications

Card phones have been installed in the N Egyptian ports and phonecards are widely available. The problem is buying one with a high enough stored value for overseas calls once you are S of Hurghada. An alternative is to use a metered phone in a shop or make an expensive call from a hotel. Off-peak rates apply 2000–0800. Agree on the tariff beforehand. There is a good GSM mobile phone network along the Red Sea coast of N Egypt and the Gulf of Suez and on the Sinai Peninsula. It gets patchy S of Safaga and non-existent S of Marsa Alam. The country code is 20. Internet access is easy enough in the towns. It gets cheaper the further N you go. Be wary about the Egyptian postal service. We have had mail go astray.

There are international airports at Hurghada and Sharm el Sheikh. Domestic flights can be taken from Port Suez and Port Said. A coast road runs along the entire Red Sea coast and buses run between all the major towns. There are taxis and minibuses in the towns. Establish the fare before you get in.

Note: Theft is not uncommon in Egypt. It is advisable to lock up before leaving your boat. In some ports, such as Suez and Ismailia, it is worth taking all easily removable items below, especially at night. Ashore, take care of your belongings.

LMT, opening hours and public holidays

2 hours fast on UTC with a +1 hr daylight saving adjustment in the summer.

Government offices: 0800–1400 Saturday–Thursday
(You will avoid any suggestion of overtime charges, either directly or via your agent's fees, if you check in and out early in the morning of a working day)
Banks: 0830–1400 Saturday–Thursday. The bigger banks, usually those with credit card facilities, also open for a couple of hours on Friday morning.
PO: 0830–1700 daily, except Fridays
Shops: summer 0900–1230 and 1600–2000; winter 0900–1800
Public holidays: 1 January, 7 January (Coptic Christmas), 19 January (Epiphany), 22 February (Unity Day), 23 March (Annunciation), Easter, 1 May, 18 June (Evacuation Day), 23 July (Revolution Day), 6 October (National Day), 23 October (Suez National Day), 23 December (Victory Day), Muslim religious festivals, and every Friday.

National flag

Three horizontal stripes (red, white and black), with the national emblem of an eagle with shield in gold in the centre.

Israel

Ports of entry: Tel Aviv, Haifa, Ashkelon, Elat.

History and politics

The modern state of Israel developed from the Zionist campaign for a Jewish state in Palestine. The Balfour Declaration, made by Britain in 1917, was in favour of a Jewish national home in Palestine, without prejudice to the civil and religious rights of the non-Jewish peoples of the region. It was so ambiguous that both Jews and Palestinians have used it to back their claims.

After World War II, increasing opposition to the British presence in Palestine led to the referral of the problem to the UN, which reiterated the recommendation made by the Peel Commission in 1937 that Palestine should be divided into separate Jewish and Arab states. Britain withdrew in 1948 and an independent Jewish state of Israel was established. The result was that Palestine as a political entity ceased to exist and the seeds of the present intractable mess were irrevocably sown.

The first of many Arab-Israeli wars ensued between the new country and the Palestinian Arabs, supported by Egypt, Syria, Lebanon and Jordan. Israel won this War of Independence and, while agreeing to a UN armistice, considerably enlarged its territory by what we would now call a deliberate policy of 'ethnic cleansing'. A substantial Palestinian refugee problem was created as many were forced to leave their homes and property in Israeli-controlled territory.

The next major conflict was the 1956 Suez (Sinai) War. This involved British, French and Israeli troops in an offensive against Egypt in retaliation for her belligerence towards Israel and, probably more important to Britain and France, her nationalisation of the Suez Canal. Israel attacked Sinai, and only withdrew the following year when a UN emergency force was stationed along the Gaza Strip.

The next ten years were relatively calm, but the Six-Day War (June War) broke out in 1967 when the UN force in Sinai withdrew at Egypt's request. Egypt blockaded the Strait of Tiran for the second time. Israel responded by taking the Gaza Strip, the Sinai Desert, W Jordan and the Golan Heights, and reunifying Jerusalem. Jewish settlers moved in, expelling more Palestinian inhabitants.

War broke out again in 1973 (the Yom Kippur or Ramadan War) when Egyptian forces took the offensive by crossing the Suez Canal. Syrian forces simultaneously invaded the Golan Heights. Israel retaliated but later agreed, under pressure from the USA and the Soviet Union, to a phased withdrawal from the Sinai Peninsula. UN troops were reinstated in buffer zones until 1979, when the Israeli-Egyptian accord was signed by President Sadat of Egypt and Prime Minister Begin of Israel. By 1982 Israel had withdrawn from Sinai.

The Lebanon War of 1982 was the next major outbreak of violence, in which Israeli troops attacked Palestine Liberation Organisation (PLO) bases on the border and occupied part of S Lebanon. PLO-sponsored terrorism throughout the Arab-Israeli conflict has earned worldwide condemnation, although the Palestinian cause has also gained widespread recognition. In the late 1980s and again in the early years of the 21st century, a popular uprising, the Intifada, in the Occupied Territories has focused international attention on the problem. The PLO gained further sympathy by announcing in 1988 that it recognised the state of Israel, and began peace talks with the USA as intermediary.

Five years later, in September 1993, an Arab-Israeli peace accord was signed in Washington, and negotiations for the withdrawal of Israel from the Occupied Territories began. In 1995 Prime Minister Yitzhak Rabin was shot dead in Tel Aviv by a Jewish fundamentalist because of his conciliatory posture towards the Palestinians. Since then Israeli-Palestinian relations have continued steadily downhill. The Israelis have become more thuggishly hawkish; the Palestinians more frantically desperate. There is no light at the end of the tunnel.

Land area and climate

Israel's land area is about 21,000km^2. The fertile coastal strip is paralleled by a series of low mountain ranges and by the Rift Valley, which runs the length of the country, with the Sea of Galilee at one end and the Dead Sea at the other. The two are joined by the River Jordan. On the Mediterranean coast, summers are hot, but the winters can be quite cold and wet. Elat is much drier, with winter temperatures similar to those in the Gulf of Suez. Summer temperatures there are much higher.

Population and language

Its population is 6 million and the official languages are Hebrew and Arabic. English is widely spoken.

Visas and embassies

Unless you plan to stay longer than three months you will not need a visa. There are a few exceptions, for example if you hold an Arab passport. If you expect shortly to be entering an Arab country you should tell the Israeli authorities, who will not stamp your passport.

Foreign embassies and consulates (in Tel Aviv) include Australia, Canada, Egypt, Japan, South Africa, UK, USA and many European countries. There are UK and Egyptian consulates in Elat and French and US consulates in Haifa.

Money

The currency is the shekel, divided into 100 agorot. There are ATMs at many banks which also accept most credit cards for cash advances. Use the PO or moneychangers for exchange.

Communications

The telecommunications system is good. There is a GSM mobile phone network. The Internet is everywhere. The IDD country code is 972. Use a phone office or buy a phonecard at a PO and use a card phone. Faxes can be sent from POs or hotels. Poste restante is available in the towns.

There are international air terminals at Tel Aviv and Elat, and domestic flights to and from Haifa. There is a good road system and a much smaller railway network.

LMT, opening hours and public holidays

2 hours fast on UTC.

Banks: 0830–1230 and 1600–1730 Sunday, Tuesday and Thursday but mornings only Monday, Wednesday and Friday

PO: 0830–1230 and 1530–1800 Sunday–Tuesday and Thursday; Wednesday 0800–1400; Friday 0800–1300

Shops: 0800–1300 and 1600–1900 Monday to Thursday, Fridays 0800–1400

Public holidays: Independence Day (a moveable feast in April or May), 1 May, Jewish New Year, Yom Kippur and all Jewish religious festivals plus all Saturdays. Muslim and Christian religious festivals are also celebrated by the minority groups.

National flag

White, with two horizontal blue stripes, and the blue Star of David in the centre.

Jordan

Port of entry: Aqaba

History and politics

Jordan's modern history begins in 1918, when the country briefly came under the control of King Faisal in Damascus at the fall of the Ottoman Empire. In 1920 it was made part of the British mandate of Palestine. A year later the British recognised Abdullah ibn Hussein, son of the Sharif of Mecca, as the first Hashemite ruler of Jordan. Full independence followed the Second World War in 1946. And in 1948–49, after the first Arab-Israeli War, the country was considerably enlarged, through a secret deal with the Israelis at the expense of the Palestinians, by occupation of the West Bank and East Jerusalem. It was an action which was approved only by Britain and Pakistan.

Jordan trod a delicate line in the Arab-Israeli stand-off over the next 30 years. It couldn't last. In order to avoid being toppled by Palestinian forces, King Hussein took the Arab side in the 1967, Six-Day War. The result was catastrophic. Israel occupied the West Bank. And the Palestinians became an even greater destabilising factor. They established a commando force in Jordan, the fedayyin, to mount raids into Israel. And Al Fatah, the organisation which assumed leadership of the Palestine Liberation Organisation (PLO) in 1969, was based in Jordan. In 1970 hostility to the government's policy of moderation towards Israel led to civil war between the PLO, supported by Syrian tanks, and troops loyal to the King. With US, British and Israeli assistance, the regime only just survived. By 1971 the PLO had moved its base of operation to Syria and the Lebanon, and Jordan was ostracised by and isolated from much of the Arab world.

During the middle and later 1970s relations between Jordan and the other Arab states improved at the cost of severed ties with Egypt. At the heart of these diplomatic ebbs and flows lay two things: on the one hand the continuing Palestinian problem, and on the other King Hussein's intention to maintain relations with the western powers at a time when Syria, Egypt and Iraq were more in the Soviet camp. The tensions continued, and Jordan's newly minted rapprochement with Syria collapsed in 1979. However, by the late 1970s relations with the PLO were improving and King Hussein took a key role in subsequent Middle Eastern peace negotiations. He maintained Jordan's unhappy balancing act with great skill, even managing to side with Iraq in the Gulf War without wholly losing ties with the west.

King Hussein died in 1999 and his son King Abdullah succeeded him. Under King Abdullah's rule the country maintains remarkable political stability despite a serious unemployment problem. The politics of Jordan remain both internally and externally precariously balanced.

Land area and climate

Jordan, correctly known as the Hashemite Kingdom of Jordan, has about 13 miles of coastline in the Gulf of Aqaba. Its land area is about 89,200km^2. The climate inland is predominantly Mediterranean. Aqaba has a comfortable winter climate with cool nights while summer is dry and hot.

Population and language

The total estimated population is 5 million. The official language is Arabic. English is the second language.

Visas and embassies

2-week visas are issued on arrival and can be extended at any police station for up to 3 months at no charge. There is an Egyptian consulate in Aqaba and foreign embassies in Amman include Australia, Canada, Egypt, Israel, UK and USA. There is no problem travelling between Israel and Jordan but be sure that the Israeli officials do not stamp your passport if you are heading S to other Arab countries.

Money

The currency is the dinar, divided into 1,000 fils. Aqaba has many banks in the centre of town, with good facilities, including cash advances on credit cards though not always via ATMs. *Visa* and *MasterCard* are the most common. There are moneychangers with good rates outside banking hours.

Communications

Telecoms are good but international lines can get overloaded. There is now a GSM mobile phone network operating in Aqaba. Otherwise there is a choice of phonecards and phone offices in town for overseas calls. Rates are lower after 8pm and all day Friday. The IDD country code is 962 and the area code for Aqaba is 03. Internet access is ubiquitous. Shops are often cheaper than cybercafés for sending email. Postal services are usually reliable if rather slow. Quicker courier services are available, amongst which EMS is the most reasonable. The Royal Jordanian Yacht Club will hold mail for you. See the Pilotage section for details. There is an international airport at Aqaba. Buses and service taxis run between the main towns and there is a railway to the Syrian border. Transport into town from the port area is by taxi only.

LMT, opening hours and holidays

2 hours fast on UTC.

Offices: 0800–1300, 1700–1900 Saturday–Thursday.

Banks: 0830–1500 Saturday–Thursday.

PO: 0700–1700 Saturday–Thursday, 0800–1300 Friday.

Shops: 0900–1300, 1600–1900.

Public holidays: 15 January, 22 March, 1 May, 25 May (Independence Day), 10 June, 11 August, 14 November, and all Fridays and Muslim religious festivals.

National flag

Three horizontal stripes (black, white and green), with a red triangle based on the hoist bearing a white seven-pointed star.

Part II
Planning guide

Routes
Timing
The sailing season in the Red Sea is September–May. The best time for you depends on how direct a route you choose, whether you are headed S or N and where you are going next. That said, the high summer months N of Foul Bay are no worse than a hot summer in S Turkey but S of that latitude can be infernal until S of Massawa. For a through run, summer temperatures over the open sea are less extreme than on the coasts, though you risk a higher percentage of calms and occasional sandstorms. For divers summer is the time of greatest water clarity.

We shall assume you are coasting and are using the Red Sea as a through route because 80% of you are likely to be cruising from or to the Indian Ocean. Unless you intend to sail direct allow at least 2 months.

Northbound
Arrival: Get to Aden or Djibouti by early March. Leave Port Said after late March. The NE monsoon in the Arabian Sea doesn't kick in reliably until December. November is the second highest risk cyclone month in the N Indian Ocean (see Weather below). By March adverse winds and currents can start appearing near Suqutra. So, enter the Red Sea December–mid/end February.

Moving N: In the S Red Sea SE-quadrant winds start in October. The convergence zone, S of which the such winds blow, moves N during

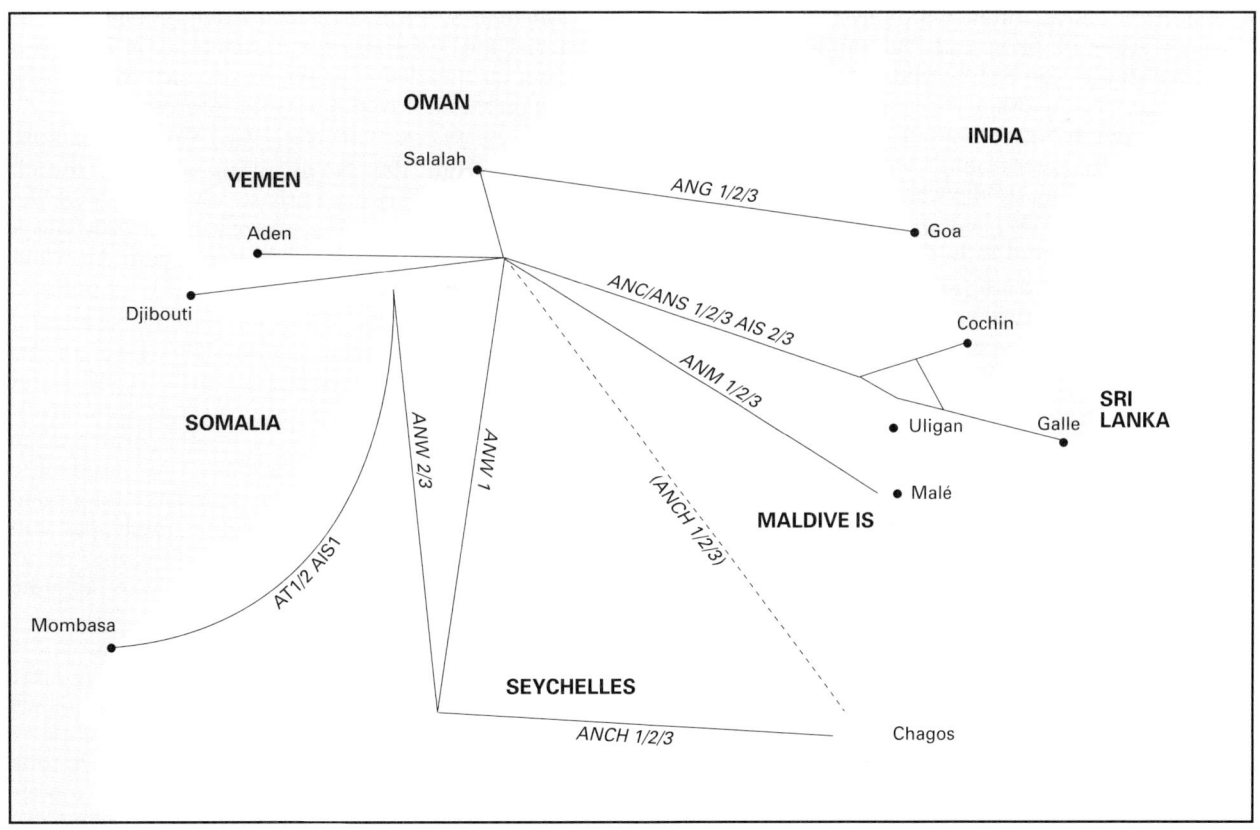

INDIAN OCEAN APPROACH ROUTES

Red Sea Pilot

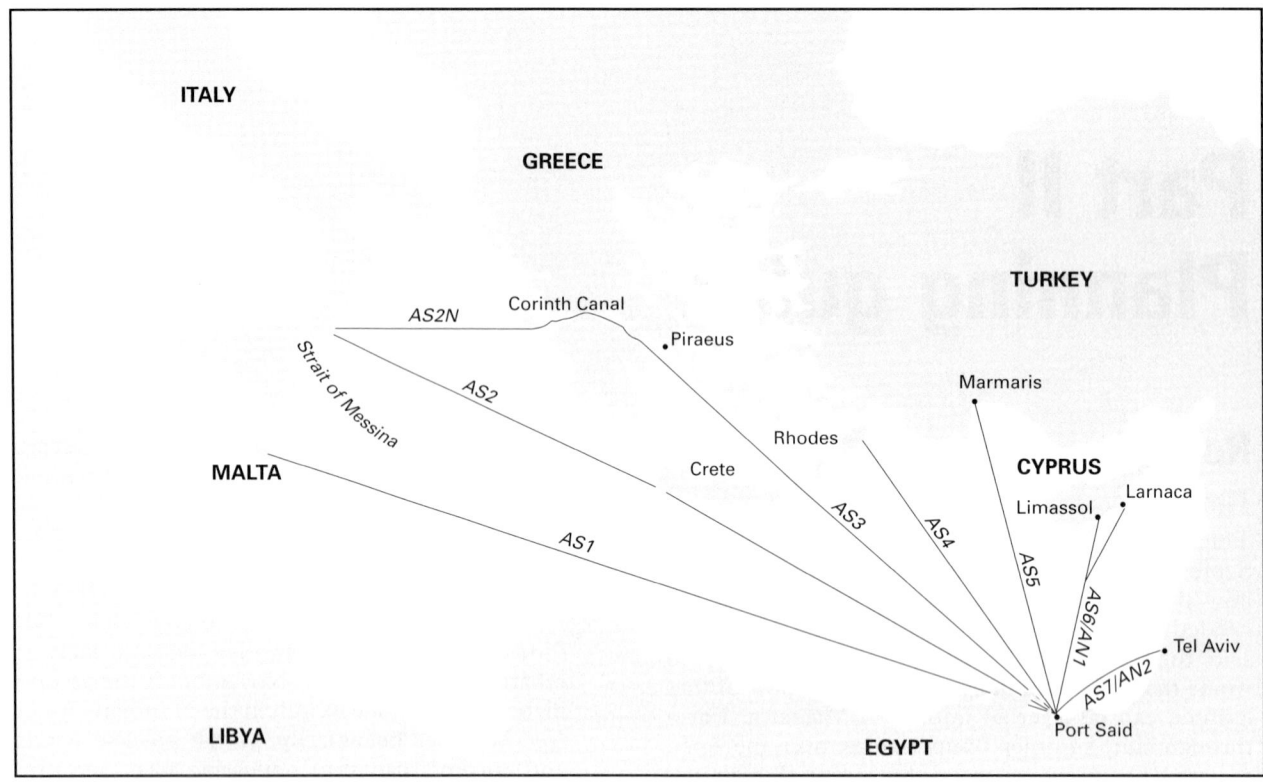

EAST MEDITERRANEAN APPROACH ROUTES

November and December. It reaches Port Sudan, occasionally further N, in January and February. N-quadrant winds are consistently strongest in the N half of the Red Sea in January and February. N-quadrant winds in the N Red Sea get weaker in April and May and extend S to cover all the Red Sea by mid-June. Note that the strongest N-quadrant winds in the Gulf of Suez blow from June to September, continuing with reduced frequency until December. Therefore reach Foul Bay at the earliest in early March; Suez by early May, later if you are happy to wait for a weather window in the Strait of Gubal area.

Leaving: The NW European sailing season is short, the more so the further N you are headed. In the Mediterranean the comfortable season (when the Mediterranean front, with its fast-building depressions, is least active) is May to September. So, enter the Med at the earliest in early April. Note that the later one reaches the E Mediterranean, the shorter the time left for onward movement, and the stronger and more persistent the summer N–NW winds.

Southbound

If headed as fast as possible for India and the Far East, aim to transit the Red Sea towards the end of August to ride the failing SW Monsoon E in September.

Arrival: Arrive in Port Said any time in the summer and in any event by late October. E Mediterranean late-autumn and winter weather changes quickly and can be vicious. Save for a direct passage to Bab el Mandeb, the earlier your arrival in Port Said, the slower should be your advance S. Time arrival in Aden or Djibouti for latest late February for destinations in E Africa, later if headed further E. Avoid May, July, August and November in the Arabian Sea.

Moving S: The N half of the Red Sea has 60–80% winds from the N quadrant all year round. October–January has fairly steady SE'lies over the S half of the Red Sea, decreasing in N'ly reach and frequency February–March. From April on, N winds are at their strongest in the Gulf of Suez.

Leaving: In the full NE monsoon the adverse current in the Gulf of Aden can run 2kts close to the Yemen coast, decreasing in the deeper water offshore. Calms are common so fuel state needs monitoring. If headed E, cross the Arabian Sea before July and August. In those two months the SW monsoon blows at up to gale force for a good percentage of the time, bringing rough seas and putting many Indian coast ports off limits.

Avoid May and November; these are the cyclone months (though see Weather below).

In the Arabian Sea in February and March the NE monsoon is not strong, except through the Palk Straits. Force 3–4 is normal, and a port-tack passage E'bound is not taxing. In late February 2001, a year of very fitful NE Monsoon, we took 11 days from Salalah to the Maldives, mostly under sail.

Planning Guide

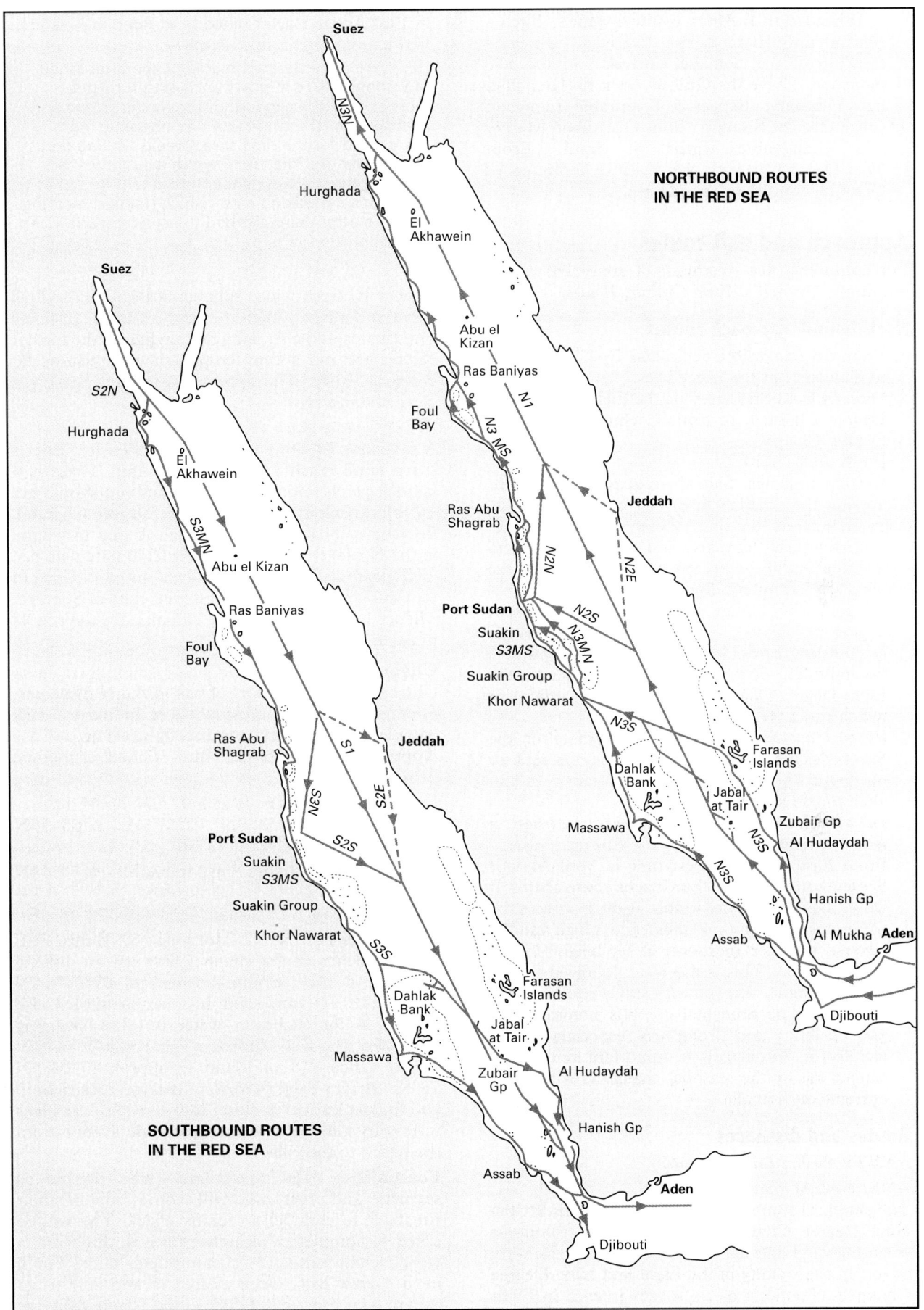

Red Sea Pilot

If headed to E Africa you will want to hitch a ride on the NE monsoon, and avoid any early showing of the Somalia Current and the SW monsoon. Leave the Gulf of Aden no later than late February. Expect a favourable monsoon (once clear of Suqutra) from December–March, and changeable winds in April. From May–December, with the possible exception of October, conditions are adverse.

Approach and exit routes

For comprehensive treatment of approach routes, see Jimmy Cornell's *World Cruising Routes*.

Northbound approach routes

1. From Sri Lanka to Cochin and Goa, and thence to Salalah N of the Laccadives.
2. Directly from Sri Lanka via the Eight or the Nine Degree Channel, or from Cochin via the Nine Degree Channel.
3. From the Maldives

   Approaching via Salalah reduces time spent crossing the Arabian Sea. The counterclockwise circulation in the N Arabian Sea sometimes begins in early February, and can push down to 10°N. The NE monsoon seems lighter on the more NW'ly route to Salalah.

   Along the Oman coast an adverse current begins in February and spreads further W as the season progresses. Winds are usually favourable but light. The direct route from Sri Lanka via the Eight Degree Channel to Aden usually offers fair winds and a favourable current.
4. From Chagos, head for the longitude of the Seychelles before turning NE to cross the equator on about 53°E. When the NE monsoon is felt, alter course for the Gulf of Aden.

   In all circumstances at present, keep a minimum of 60 miles off the Somali coast.
5. From East Africa the best time is April–May or September, following the Chagos route above. It would be very uncomfortable at the height of the SW monsoon when gale incidence is high and the adverse summer monsoon at its height in the Gulf of Aden. This is the route followed by the old *dhow* trade, and you may still see some *dhows* en route. The Somalian coast is notorious for poor visibility and strong sets, especially in the vicinity of Suqutra. It is important not to get caught out by the seasonal change in winds and currents off Somalia.

Routes and distances

ANS From Sri Lanka (Galle)

ANS1 Galle to Salalah (1,680M)
The standard route, to avoid shipping, aims for the Nine Degree Channel. You may need to motor when the wind falls light in the lee of India. Be wary of sets in the vicinity of the Eight and Nine Degree Channels. Traffic is by no means intense in Eight Degree Channel towards which you'll be pushed

> In 1951 Adrian Hayter sailed from Aden in June and had a very rough ride to Bombay:
> '... I had ... entered the area of the established Monsoon, where it blows unceasing for three months over the expanse of the Indian Ocean and Arabian Sea. The great lazy swell became more vicious and for the next three weeks *Sheila's* decks were never dry. The wind, which averaged near gale force was always enough to tease the top of a roller into a breaking crest, which flooded seething over the after-deck, climbed the coaming and filled the cockpit.'

anyway by fresh winds funnelling through the Palk Strait and a S set down the coast of India. In many years winds in the Arabian Sea are lighter the further N you are and a counter-current can push as far down as 10°N. The approach to Mina Salalah is clear of dangers.

ANS2 Galle to Aden (2,130M)
A dull first 50–60M under Sri Lanka's lee, then a sharp close reach past the Palk Strait. There is a lightish patch under India's lee, but you should get the wind back in or around Eight Degree Channel. Be wary of N sets in the channel. A stop in Uligan in the N Maldives is now wholly legitimate with full CIQ clearance. It is also a marvellous spot. Once in the Arabian Sea, shape to pass well clear of Suqutra, whence keep in amongst the shipping for the run W to Aden.

Uligan
Uligan is the easiest port of call in the N Maldives. Pilotage details are included here because it falls outside the Red Sea but is directly relevant.

Approach Clear passes into Thiladhunmathee Atoll are:

From N, the deep pass is at 7°07'·2N 72°54'·6E.
From SW via 6°53'·0N 72°57'·4E, 6°55'·55N 72°59'·25E, 6°57'·0N 72°58'·3E.
From SE, the pass close S of Mulhadhoo, 6°59'·44N 72°59'·35E.
From E the pass S of Uligan, 7°04'·92N 72°56'·4E.

Anchorage For the NE Monsoon (NNE thru ESE winds) anchor in the channel between an offlying patch and the lagoon border at 07°04'·75N 72°55'·27E, 10–14m, sand. It's possible to tuck a bit further N for shallower water but less swinging room. For the SW Monsoon, (generally W to NW winds), anchor S of island at approx 07°04'·5N 72°55'·9E. It's deep (<25m) unless you tuck right in and find a clear patch closer to the beach. The place is easy to spot because there's a wide avenue from the beach to the village.

Formalities The coastguard (who double as security and customs) will come out in their runabout to conduct a security check. You will be asked to complete a clearance form in duplicate, a vessel details form and a customs declaration. You'll need 3 crew lists. Once cleared in and able to go ashore, you must pay US$5 to the Island Office as an anchorage fee. The rules are:

Planning Guide

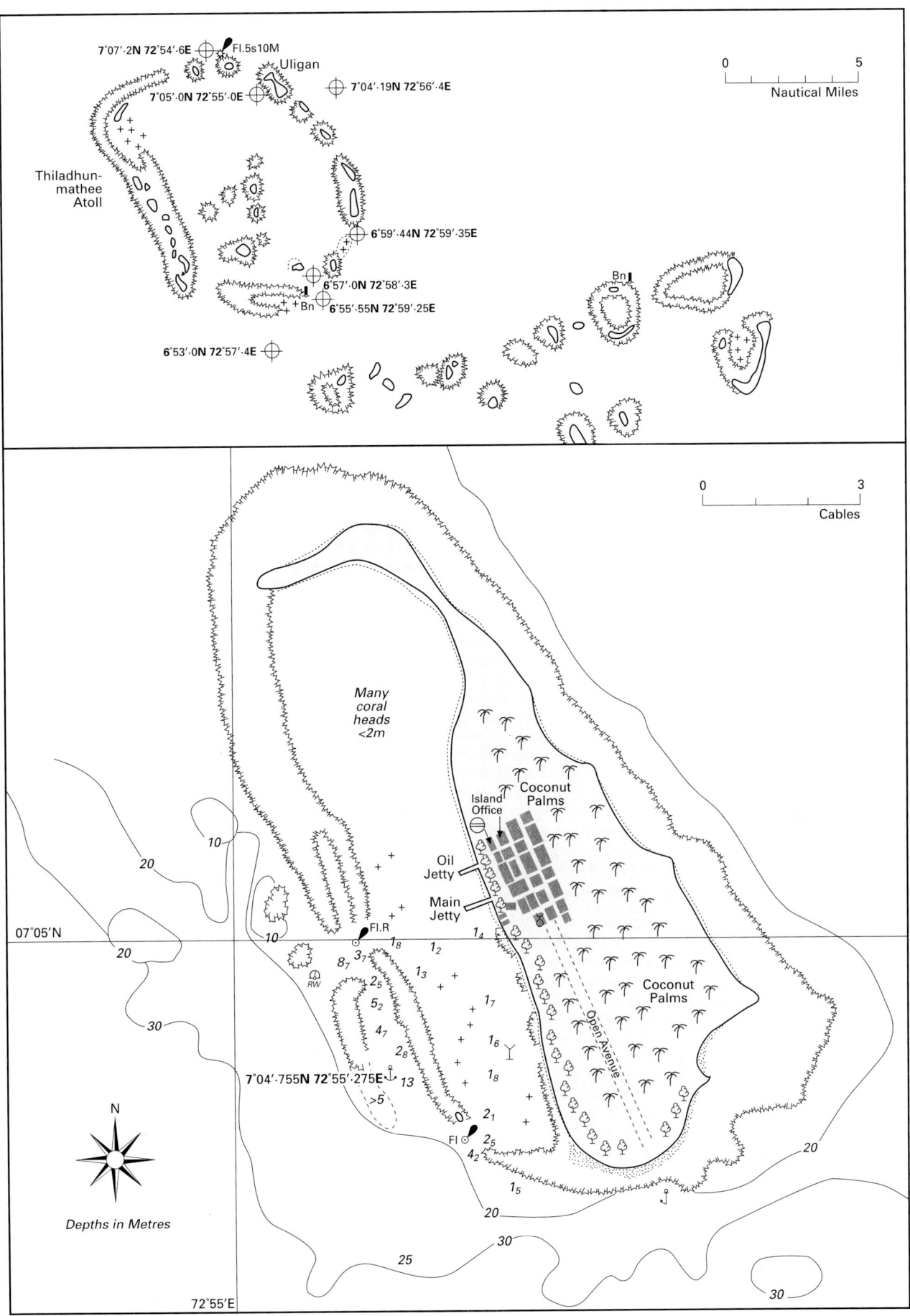

ULIGAN I. AND APPROACHES

- An anchor light is compulsory.
- No member of the local population is allowed aboard.
- No one to be ashore from 2200–0600.
- No (major) repair work to be done.
- Any gifts for people ashore must be shown to and approved by the customs officials.
- No alcohol to be taken ashore.
- Dress ashore to be modest (shorts OK for ladies if not super short and skin tight).
- The day before leaving inform the Coastguard. Produce your receipt for the US$5 anchorage fee and DON'T LEAVE until a security check has been completed.

This sounds bureaucratic and complicated. It isn't; the officials are delightful and anxious to make things easy.

Facilities Hassan of Sailor's Choice store will probably meet you when you first go ashore and he may call you on VHF beforehand. The store (owned by Ahmed Naseer, the local magistrate) has a small range of basics and souvenirs and can arrange limited quantities of diesel @ US$50c/litre by jerry jug. If you want it delivered customs clearance is necessary. There's a phone near the jetty and phonecards (US$3 and US$10) are on sale. Collect calls are expensive. Mail can be sent from Sailor's Choice. They also offer showers and laundry. A supply boat brings replenishment of stores every 7–10 days. Fresh water in limited quantities is available by jerry-jug for a small fee. AMSCO, the main agent in Male, also has an office here and you can arrange to buy diesel through them. There is one other small store, Hafolhu, which has much the same range of provisions. Maldivian style meals can be prepared with advance booking. It is possible to take a boat to Hanimaadhoo, (6°45'N 73°10'E) where there is an airfield with flights to Male.

General Uligan is an idyllic place not yet affected by the regular stream of cruising visitors. There are plans for a new marina at Haa Dhaalu, Nolhivaranfaru run by Aussies. Uligan lost the vote. It'll be near Hanimaadhoo because the airfield was instrumental in swinging the decision. 3·5M SW of Hanimaadhoo there's a likely looking lagoon.

Whatever you hear on the gossip net PLEASE DO NOT call at Filadhu I (approx 6°55'·5N 73°13'E). With the all too usual developed world tourist's bad manners, yachts have called here in the past because they preferred the more sheltered anchorage. Filadhu is NOT a permitted anchorage, although the Maldivians are far too polite to tell invading yachts that they are an intrusive embarrassment.

ANS3 Galle to Djibouti (2,240M)
As for previous entry. Be prepared for freshening and backing winds as you get deeper into the Gulf of Aden. At night it is safer to leave the Iles Moucha to port before shaping for Djibouti.

ANC from Cochin

ANC1 Cochin to Salalah (1,380M)
It is best to head slightly S to pass through Nine Degree Channel rather than N of the Laccadives. Thereafter, as for routes from Galle.
ANC2 Cochin to Aden (1,850M)
ANC3 Cochin to Djibouti (2,000M)

ANG from Goa

ANG1 Goa to Salalah (1,140M)
A clear and direct course with favourable, though probably light winds and some calms. Possibility of some adverse current.
ANG2 Goa to Aden (1,805M)
ANG3 Goa to Djibouti (1,940M)

ANM From Male (Maldives)

ANM1 Male to Salalah (1,450M)
ANM2 Male to Aden (1,780M)
ANM3 Male to Djibouti (1,860M)

ANW From Mahe (Seychelles)

ANW1 Mahe to Salalah (1,380M)
ANW2 Mahe to Aden (1,460M)
ANW3 Mahe to Djibouti (1,570M)

ANCH From Chagos

ANCH1 Chagos to Salalah (1,750M; via Seychelles 2,400M)
ANCH2 Chagos to Aden via Seychelles (2,480M)
ANCH3 Chagos to Djibouti via Seychelles (2,590M)

ANT Transitional season

ANT1 Mombasa to Aden (1,640M)
ANT2 Mombasa to Djibouti (1,750M)

Northbound exit routes

AN1 Port Said to Larnaca/Limassol (230/210M)
AN2 Port Said to Tel Aviv (140M)

Heading N from Port Said is easy at any time, though winter to late March calls for careful timing. If headed for Israel, remember to let the Israeli navy know who you are and where you are going when 50M from the Israeli coast.

Southbound approach routes

Bound for Port Said you can approach from anywhere in the E Mediterranean at any time of year. Keep well off the Libyan and Egyptian shores unless closing a port of refuge, in effect there's only Alexandria and then not always. The routes can be sailed in any season but summer is much easier.

AS1 Malta to Port Said (940M)
AS2 Straits of Messina to Port Said (935M)

A simple route, best done towards the end of summer. Keep N of the rhumb line until E of Crete, especially in winter. This keeps one N of shipping traffic and within reach of harbours of refuge. Consider using the (expensive) Corinth Canal. With a strong summer *meltemi*, this gives a fair slant once clear of the Saronic Gulf. Keep a check on your course made good as you close the Egyptian coast. The E-setting current can run at up to 1 knot. Keep outside the 20m line until you sight the shipping concentrated in the waiting anchorage W of the buoyed channel into Port Said.

Planning Guide

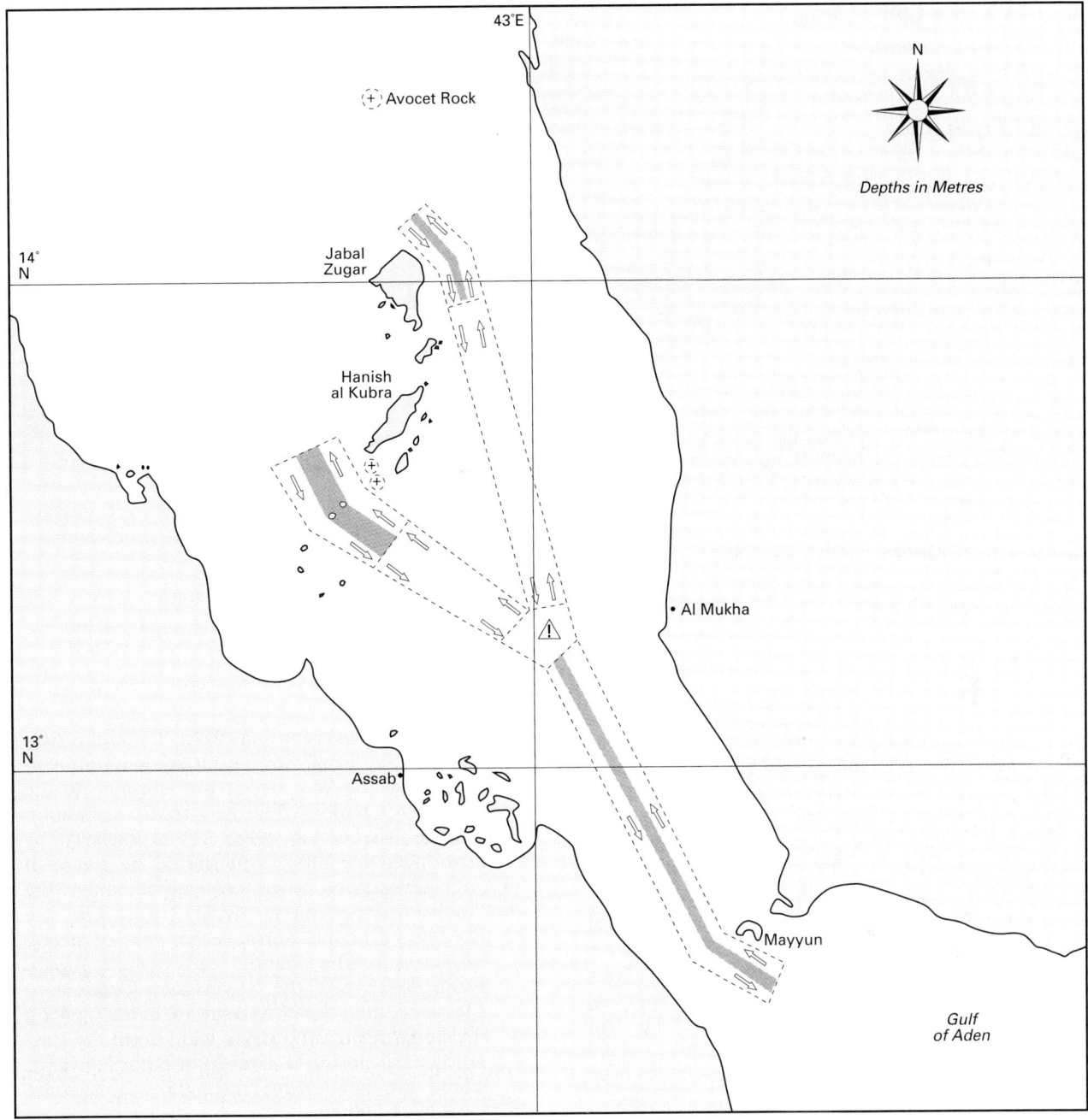

NEW TRAFFIC SEPARATION SCHEME: SOUTHERN RED SEA

If you prefer company you might contact the East Mediterranean Yacht Rally, c/o Kemer Marina, Antalya, POB 627, Antalya 07000, Turkey. They offer support for cruising in N Egypt and Israel.

AS3 Piraeus to Port Said (670M)
AS4 Rhodes to Port Said (350M)
AS5 Marmaris to Port Said (365M)
AS6 Larnaca/Limassol to Port Said (230/210M)
AS7 Tel Aviv to Port Said (140M)

Southbound exit routes

AIS1 Aden to Mombasa (1,640M)
AIS2 Aden to Galle (2,130M)
AIS3 Aden to Salalah to Cochin (600M and 1,380M)

Exit requires careful timing unless you like boisterous downwind sailing with heavy squalls and rain. The transitional seasons are best. If headed for E Africa, aim for no later than the end of February. A NE-monsoon crossing to India is life at an angle, but better than the strong winds, grey skies, squalls and heavy rain of the transition season.

Routes in the Gulf of Aden

ANAS from Salalah
ANAS1 Salalah to Aden (600M)
ANAS2 Salalah to Djibouti (720M)

ANAA From Aden
ANAA1 Aden to Bab el Mandeb (98M)
ANAA2 Aden to Djibouti (140M)
ANAA3 Djibouti to Bab el Mandeb (71M)

Once in the Gulf of Aden Djibouti is an alternative to Aden and has a good reputation amongst cruisers with deep pockets who feel a need for French wine and other such provisions. The disadvantage is the added distance. The Gulf of Tadjoura is an interesting cruising ground.

We have included in Part III ('Pilotage') information on anchorages in the islands between Suqutra and the African mainland, and along the N coast of Somalia. At present no cruising in these waters can be recommended, and we include no routeing information.

Routes northbound in the Red Sea

N1 The direct route: Bab el Mandeb to Suez (approximately 1,200M)

Note: The Authorities in the Yemen have told us that in principle the Small Strait is a restricted military training area. They said that if the International Convention on the Territorial Sea had allowed it (it doesn't), they would have extended the prohibited area around Mayyun (Perim I) to include all of the Small Strait and the Yemeni shore. They reluctantly accept direct passage through the Small Strait. They wish to discourage any anchoring except, presumably, in a genuine emergency.

Pass through the Small Strait (E of Mayyun Perim I), but see note above and leave the Hanish group well to port to avoid the many unlit hazards. Note the new shipping separation scheme (see sketch, page 21) that will be in force by 2003 and keep clear of it. Leave the Zubayr group to starboard to avoid East Rocks. This is the main shipping route, so keep a good lookout. Jazirat at Tair, Geziret Zabargad (St John's Island), Abu El Kizan (Daedalus Reef) and El Akhawein (The Brothers) are waypoints en route for the Strait of Gubal. Note that the traditional N'bound sailing-ship route always held the E (Saudi Arabian) coast until opposite Mina (Port) Safaga.

The rough seas, shipping traffic, and strong tides of the Strait of Gubal can be avoided by using the Shadwan, the Tawila, or the Zeit Channel. The first is navigationally the easiest. In the Gulf of Suez S of Ras Gharib at night or in poor visibility it is safest to stay on the edge of the shipping channels. Wind and sea are easier on the E side of the Gulf of Suez, especially N of El Tur (Tor Harbour).

N2 The direct route, stocking up at Port Sudan

N2S Bab el Mandeb to Port Sudan (approximately 700M)
Unless approaching from Djibouti, follow the direct route to Jazirat at Tair, noting the caution about the new traffic separation scheme above between Bab el Mandeb and the N of the Hanish Islands, whence shape to pass well E of Harmil I, then NE of Masamirit I, and subsequently N of Hindi Gidir and N Towartit Reef to Port Sudan (see Pilotage for Port Sudan approach waypoint details).

N2N Port Sudan to Suez (approximately 700M)
From Port Sudan, leave Salayet (Wingate Reef) and Sanganeb Reef to port. Thence, giving Shab Rumi a good berth, pass to seaward of Abington Reef light, shaping course for Abu el Kizan (Daedalus Reef), and continue as for the direct route.

N2E Via Jeddah Stocking up here is possible but the authorities are only sympathetic if one is calling to effect repairs or refuel. You must use an agent which is very expensive and you will be confined to the port. The only recognised tourists in Saudi Arabia are Muslim pilgrims, though changes are in the air. The main advantage is that the windward passage may be easier on the Saudi Arabian shore and the distances are much the same.

N3 Central and W coast routes

These are all variations on a theme, breaking the offshore passage by anchoring.

N3S Bab el Mandeb to Khor Nawarat (approx 500M)

I. *W coast* Unless coming from Djibouti, pass through the Small Strait (but see note above), then close the W coast at the entrance to the Rubetino Channel. The new traffic separation scheme (see sketch, page 21) is likely to be introduced by 2003. You should be aware of your obligations when crossing it under the Collision Regulations, Rule 10(c). Off the Rubetino Channel entrance be wary of strong SE currents. The Channel leads towards Assab. From Assab take the S Massawa Channel to Massawa, then the N Massawa Channel, giving the Sudan/Eritrea border a wide berth (at least 10M) until closing Gazirat Iri or Khor Nawarat. There are anchorages throughout this route until N of Difnein I.

II. *Intermediate* From Small Strait head for the Hanish group then via the Zubayr group and/or Harmil I to Gazirat Iri/Khor Nawarat. Alternatively, after the Hanish Is use the S Massawa Channel and then the W coast as above. For further details see III ii below.

III. *E then W coast* From Small Strait (but see note above), hold to the E shore as far as Al Mukha. Head for the Hanish Is, being careful to observe regulations crossing the new traffic separation scheme then,
  i. follow II above,
  ii. head for the S Massawa Channel S of the Dahlak Bank. The clearest passage goes between Seven Fathom Banks and Shab Shakhs. Strong SE-setting currents are common in the Channel. From Massawa, follow the N section of I above. Alternatively, keep on the E shore to Al Hudaydah. Thence

iii. as for III.
iv. cross via the Zubayr group to Harmil I/NE of the latter to Gazirat Iri/Khor Nawarat
v. keep on the E coast via Uqban I and the outer Farasan Is before crossing to Gazirat Iri/Khor Nawarat.

N3MS Khor Nawarat to Port Sudan (approximately 200M)
Three routes:
I. Follow the Shubuk Channel to the inshore channel leading N from Marsa Esh Sheikh.
II. Follow the Inner Channel via Harorayeet (Two Islets) N of Shab el Shubuk to Suakin.
III. (For keen divers) Pick your way N through the reefs and islets of the Suakin group as weather and inclination suggests. Move into the inshore channel via a number of clear, deep breaks in the main offshore reef.

N3MN Port Sudan to Safaga (approximately 500M)
Two routes which may be combined in various ways:
I. *The Inner Channel* This is clear, deep and beaconed. Anchorages are 10–30M apart. The main passage ends at Marsa Halaib. If you have plenty of time and are used to navigation in coral, you can coast-hop tortuously to Port Berenice. If you're confident and like diving, Foul Bay is worth exploring. Proceed via the Egyptian coast to Safaga.
II. *The offshore reefs* From Port Sudan head for Sanganeb Reef, then Shab Rumi, Jazirat Bayer, Shab Shinab, and Shab Qumeira to the N Sudan *marsas*. In settled weather Gaziret Zabargad makes an interesting stop. Proceed via the Egyptian coast to Safaga. In settled weather stop at either Abu el Kizan (Daedalus Reef) or Al Akhawein (The Brothers).

N3N Safaga to Suez (approximately 200M)
Make N to the Gubal Strait through the reefs and islands S and N of Hurghada. Once N of Shab Ali, in strong weather seas and winds are less on the E shore of the Gulf of Suez. On either shore winds and seas decrease N of El Tur. Tidal currents can be used to advantage (see 'Sea, tide and currents' below). There are three inshore routes from Hurghada to the Strait of Gubal, via the Zeit, Tawila, and Shadwan Channels. The last is navigationally the easiest and has excellent anchorages. Other more intricate channels may repay effort. There are three onward routes:
I. *W shore* Take an inshore channel (see Pilotage section) to Marsa Zeitiya. From Marsa Zeitiya, in strong N-quadrant winds and if time presses, keep some 50m from the fringing reef in depths <15m. This keeps you in relatively calm waters, but is only feasible for boats >10m LOA with powerful auxiliaries. This route, with off-liers in various places N of Ras Shukeir, runs to Suez. Stop at any of the W coast anchorages.
II. *E shore* As W shore above until Bluff Point on Gubal Saghira. Carry on to Shab Ali and El Tur, and thence N to Suez, playing the E shore anchorages.
III. *Playing both shores* The best option for small, low-powered cruisers on a tight schedule and unable to wait for a calm or S'ly.

N4 E coast route

As the N'bound single-stop E coast route above. This used to be the chosen inshore route (not without its excitements, as readers of W. A. Robinson's *Deep water and shoal* will know). Yachts are not yet welcome in Saudi Arabian waters unless obliged to seek refuge by damage, stress of weather or similar emergency. Given unobtrusive conduct and the greatest restraint and good manners, a yacht which took the inshore route, anchoring in uninhabited places, might be unobjectionable. At present the only written permission acceptable to the authorities is a letter of introduction from a Saudi Arabian diplomat or member of the Saudi royal family.

N5 The Gulf of Aqaba

Strait of Tiran to Elat/Aqaba (approximately 110M)
Any passage up the Gulf of Aqaba will be hard, windward work, with relatively few places of shelter. Unless one gets one's yacht trucked over to a Mediterranean port (a feasible though very expensive operation), or bases oneself in the Red Sea, a stay in the Gulf of Aqaba subsequently entails a bash up the Gulf of Suez.

Routes southbound in the Red Sea

S1 The direct route: Suez to Bab el Mandeb (approximately 1,200M)
The Gulf of Suez has many hazards. Keeping a good lookout, stay in the S'bound shipping lane towards the W shore. Shipping in the Strait of Gubal can be avoided by using the Shadwan Channel to pass W of Shaker (Shadwan) I. Thence El Akhawein (The Brothers), Abu El Kizan (Daedalus Reef), Geziret Zabargad (St. John's I) and Jazirat at Tair form natural waypoints. Pass the Zubayr group to port (to avoid East Rocks) and the Hanish group to starboard. There is a new traffic separation scheme (see plan, page 21), which extends around the Hanish Islands as far as S of Mayyun (Perim Island). It is best to keep out of the lanes used by ships. Exit through the Small Strait E of Mayyun (Perim I) unless bound for Djibouti.

S2 The direct route, stocking up at Port Sudan

S2N Suez to Port Sudan (approximately 700M)
As above until abeam of Ras Abu Shagrab (Shagara), when Abington Reef light should be left to starboard. Ensure a good clearance of Angarosh, Qita el Banna, Merlo Reef and Shab Rumi to pass Sanganeb Reef to starboard. Thence S of Salayet (Wingate Reef) to Port Sudan.

S2S Port Sudan to Bab el Mandeb (approximately 700M)
From Port Sudan leave Hindi Gidir and Masamirit to starboard. Proceed as for the direct route above.
S2E An alternative is to stage at Jeddah, by prior official arrangement.

S3 Central and W coast routes
The object is to break the offshore passage by anchoring.
S3N Suez to El Gouna/Safaga (approximately 200M)
As for the direct route until closing the Ashrafi Is. The inshore route to El Gouna uses the S Qeisum and Zeit Channels. If pressing straight on to Safaga, use the Shadwan Channel.
S3MN El Gouna/Safaga to Port Sudan (approximately 500M)
There are plenty of anchorages on the Egyptian coast S to Foul Bay. If you have time and are used to navigating in coral, Foul Bay is worth exploring. Enter the Sudan Inner Channel between Ras Hadarba and Elba Reef. It's well beaconed to Port Sudan, with anchorages 10–20M apart.
S3MS Port Sudan to Khor Nawarat (approximately 200M)
Three routes:
I. Inner Channel then Shubuk Channel.
II. Inner Channel to Suakin, continuing N of Shab el Shubuk via Harorayeet (Two Islets).
III. (For keen divers) Pick your way S through the Suakin group as weather and inclination suggest.

S3S Khor Nawarat to Bab el Mandeb (approximately 500M)
I. *W coast* Keeping clear of the Sudan/Eritrea border (at least 10M offshore), take the N Massawa Channel to Massawa. From Massawa take the S Massawa Channel towards Assab. From Assab take the Rubetino Channel until clear of Ras Mukwar, and carry on to the Small Strait unless headed for Djibouti. In the Massawa Channels there are anchorages in the Dahlak Is and on the Eritrean shore.
II. *Intermediate* From Khor Nawarat, head so as to pass NE of Harmil I unless you plan to stop there. Continue to the Zubayr and Hanish groups before passing through Small Strait (but see note on page 22).
III. *E coast* As II, but from the Zubayr group cross to the E shore to make landfall at Ras Katib. Enter the Yemen at Al Hudaydah (Hodeidah) and follow the E coast via Al Mukha to Small Strait.

S4 E-coast routes
Full routeing is given in NP64, *Red Sea and Gulf of Aden Pilot*. All the reservations expressed in Route N4 above apply. From the Farasan Is S'wards the route lies in Yemeni waters. The Saudi Arabia/Yemen border area is best avoided. Complete Yemen formalities at Al Hudaydah and proceed via Al Mukha to the Small Strait.

Weather
The Red Sea is notorious for its weather. Something for everyone to hate. We'll explain why the reputation is so bad and then show you how to diminish the fearful weather bogey. The worst of the reputation comes from yachts headed N as fast as possible against a high percentage of headwinds in short, steep seas. There can be rapid and unpredictable changes. Rough seas, poor visibility, sand and dust storms can alternate with mirror calms. Some boats are ill-prepared. Others are all but unable to sail well to windward. After months of trade-wind passage-making this makes for hard sailing. For yachts heading S life is easier, though in winter they too face headwinds over the last few hundred miles.

Anyone used to sailing to windward will not find the Red Sea markedly worse than anywhere else. In some respects, such as the absence of severe squalls in winter, it is better. If you choose to sail straight up the middle your boat must be up to the strains this will impose (see page 48, 'Yacht and equipment'). Once out of the favourable SE quadrant winds, which with good timing can get you to Port Sudan, on average you will be fetching, sometimes close reaching, into about 20kts of wind during daylight hours, usually easing during the night. Every few days a considerably stronger blow will come through.

If you can't or don't want to go to weather, the sole answer is patience. For yachts which take their time, winter Red Sea weather has much to recommend it. Clear blue skies, warm days, cool nights, and sparkling sailing wearisome only if you sail too far in a day, or try to. Summer weather can be close to unbearable, but if that is when you sail, patience still pays.

This section is in three parts. The broad climatological picture, covering knowledge helpful in planning your cruise, then rules of thumb and finally forecasting services.

Climate
Global weather patterns are changing. Our climatological statistics have been collected over the last 150 years or so, a period of comparatively stable weather. Whether because of global warming or a reversion to climatologically more 'normal' unstable weather, traditional projections of average weather, routeing charts, etc, must be treated with caution. Today's weather patterns seem more biased toward extremes.

The Red Sea is hot and dry; its average annual temperature is one of the highest in the world. Winds blow mainly from NW to SE along the axis of the sea, except during winter S of 18–20°N, when they often blow SE to NW. Gradient winds are affected by sea breezes. On the W coast when the gradient wind is slight the sea breeze starts WNW or NW, veering steadily NNE–NE, not usually more than Force 4–5, by mid-afternoon. On the E coast

Weather

the pattern is E or NE backing NNW–NW or WNW. Gradient winds above Force 4 are enhanced, though little affected in direction, by sea breezes during the day. Apart from in the convergence zone during the autumn and early winter, rain is rare. There are seasonal differences. In the N, winter temperatures can be markedly cool. Winter winds tend to be stronger, except in the Gulf of Suez.

The Red Sea's climate is governed by the monsoons of the Arabian Sea, thus:

November–March NE monsoon winter weather
April–early May Transition period
June–August SW Monsoon summer weather
September–early October Transition period

Barometric pressure and its effect on wind

The diagrams summarise the difference between winter and summer.

Winter

The winter NE monsoon results from the winter Siberian high (1035mb), and low pressure near the equator. For the Red Sea this means relatively high but fluctuating pressure to the N from the interaction of two components: an E-pushing ridge of the winter-weakened (1020mb) Azores-Mid Atlantic high, and a W and SW-pushing ridge of the Siberian high. The latter divides, one lobe extending over the Balkans, the other over Saudi Arabia. The waxing and waning of these high-pressure zones dictate the ebb and flow of strong NW–N winds. The strongest winds occur when the leading edge of an E-moving cell of relatively high pressure passes over Egypt and the Sinai Peninsula. This seems connected to a temporary weakening of the Siberian high.

Another way of looking at it is as a breaking and reconnection of the 1020mb isobar. The E-moving cells are the first sign of the 1020mb isobar re-establishing a continuous, relatively high-pressure field over the N Red Sea.

Winter depressions passing through the E Mediterranean, if deep, can also affect winds in the N Red Sea (see 'Fronts', below).

In the S, a trough of low pressure pushes up from central Africa along the axis of the Red Sea. How far S–SE winds push up the Red Sea from the Straits of Bab el Mandeb is a function of the N'ward extent of the trough and the E/W location of its axis. The strongest and furthest-N-pushing S–SE winds in the S Red Sea are associated with a N and E extension of the trough; weaker winds follow its retreat S and W.

Winter winds in the Red Sea are thus a function of the relative weakening and strengthening of the main pressure belts to N and S.

Summer

The summer pattern is complex. Over the Mediterranean, the Azores anticyclone extends a

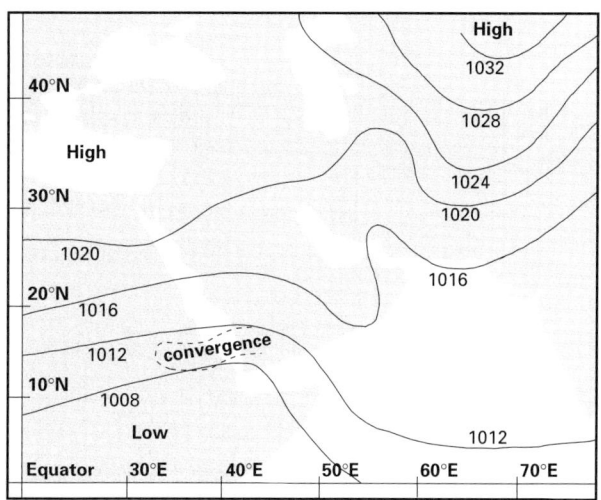

Winter episode with southeasterlies retreated S, just after a northerly surge has passed through. Winter convergence zone at, or S, of Khor Nawarat.

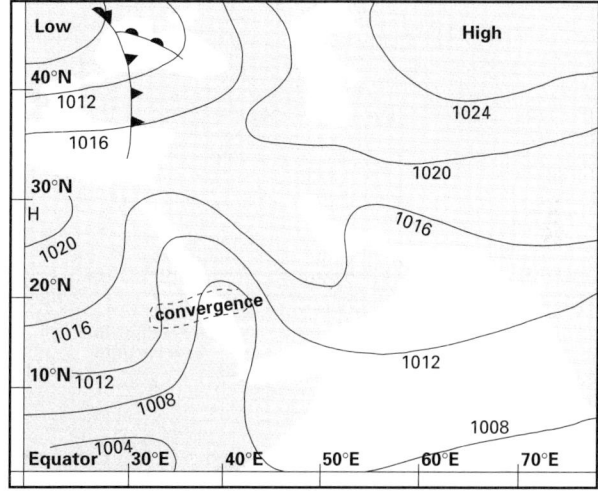

Winter episode with southeasterlies at furthest N extent, a freshening of northerlies and retreat of southeasterlies imminent. Winter convergence zone at or N of Khor Nawarat.

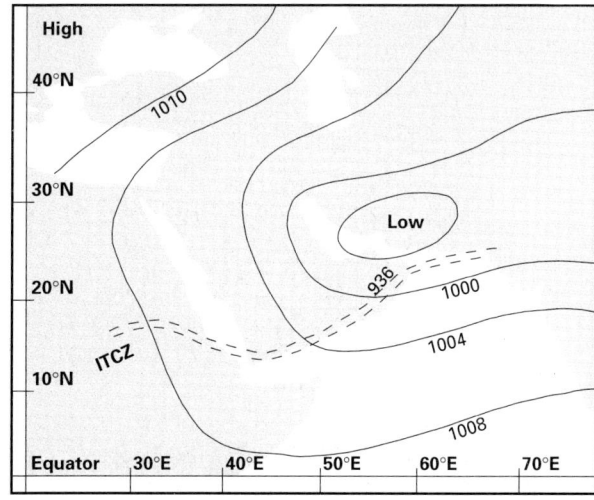

TYPICAL MID-SUMMER PRESSURE DISTRIBUTION

Red Sea Pilot

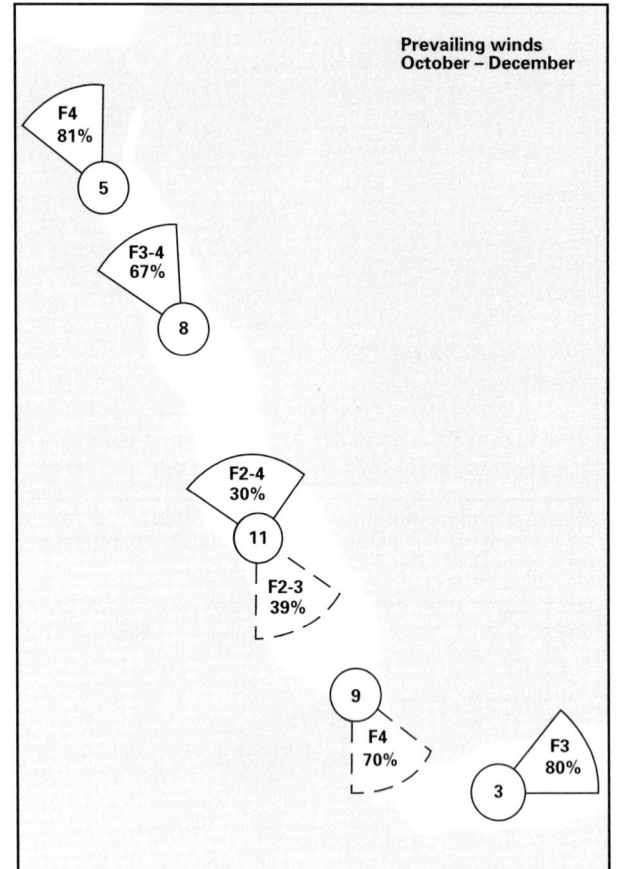
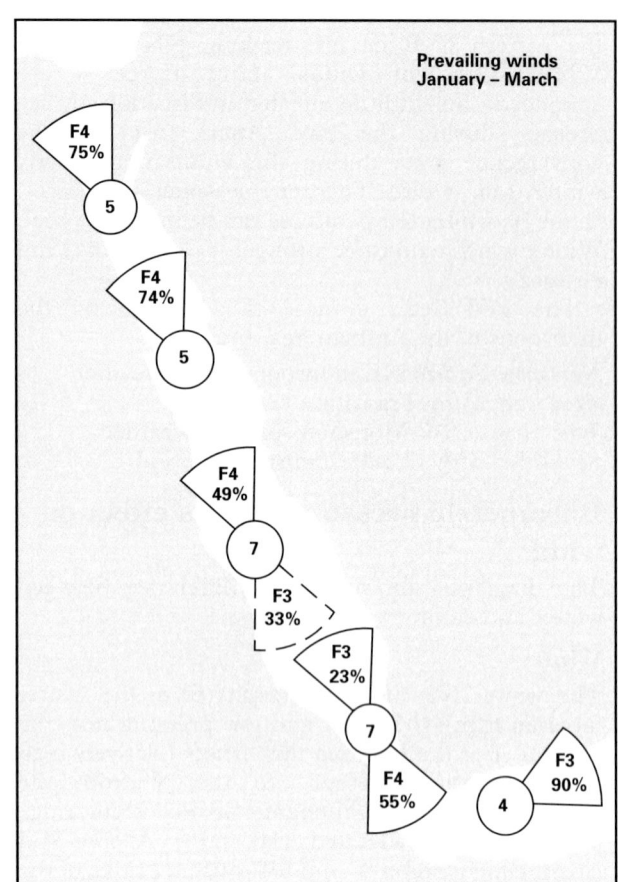
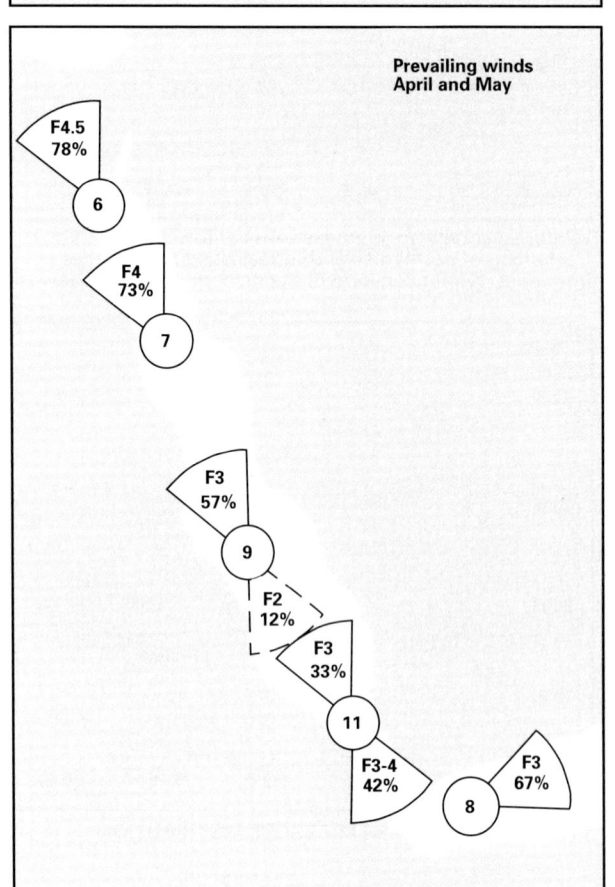
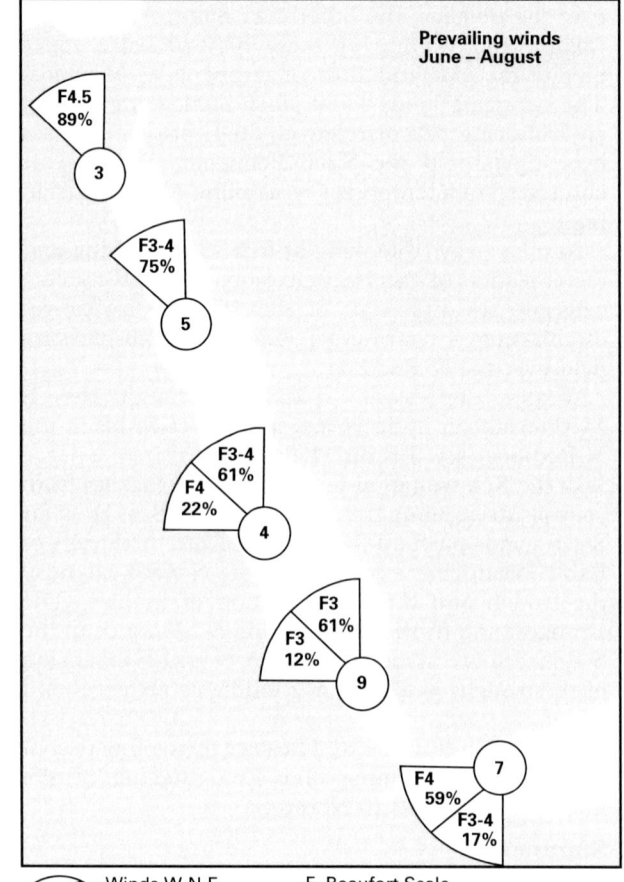

PREVAILING, SEASONAL WINDS (FIGURES IN CIRCLE % CALM)

Weather

Prevailing winds September

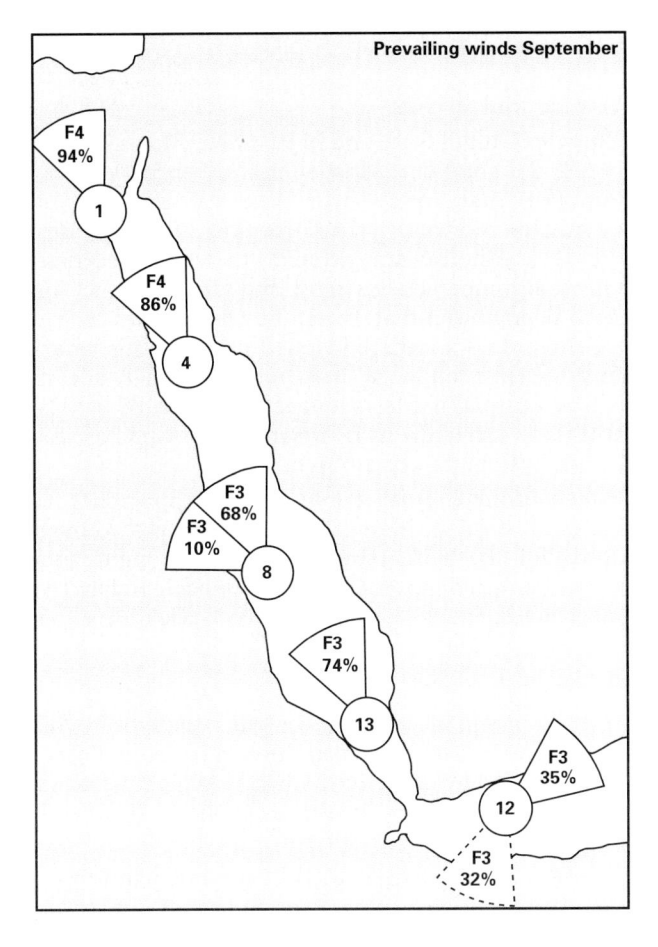

ridge of high pressure. To the E of the Red Sea a low-pressure area begins over the Persian Gulf and extends E over the Himalayas. This low is the engine of the SW monsoon. To the S is the Indian Ocean anticyclone. The SW monsoon in the S approaches to the Red Sea is caused by the pressure gradient between the Indian Ocean high and the Himalayan low. Over the Red Sea the prevailing N to S flow is a product of the interaction of the Mediterranean high and that part of the Himalayan low extending over the Persian Gulf. The Red Sea pressure gradient is comparatively slight, which is why summer winds are more fitful. They strengthen and weaken in the N in relation to high cells intensifying over the Balkans, then weakening as they move E across the Black Sea. In the S, winds strengthen and weaken with the retreat and advance over the Persian Gulf of the W end of the Himalayan low.

Fronts

There is no frontal system in the Red Sea, but the N Red Sea can be affected by the winter Mediterranean front. Deep depressions passing across the E Mediterranean can trail a cold front or associated shear-line across the N half of the Red Sea. These may be preceded by *khamsins* (see below) and succeeded by enhanced N winds. The effects of shallower depressions are usually felt only in the Gulf of Suez.

More significant are those creatures of meteorological metaphysics called convergence zones or, where what comes together are similar air masses, confluences. There are two such zones of interest in the Red Sea.

Winter

The winter convergence zone moves up and down the S Red Sea. It separates the prevailing NW wind of the N Red Sea from the S or SE wind of the S. It is marked by a belt of cloud, sometimes with heavy thunderstorms, rain and reduced visibility. It can move as far N as Port Sudan (occasionally Ras Abu Shagrab) or as far S as Assab. Its movement is a function of the relative strengths of the two airstreams. S of it, winds are S or SE. N of it, N or NW winds prevail. At the convergence, winds are light.

Summer

In the summer the Intertropical Convergence Zone (ITCZ), which divides the monsoon airstreams of the N and S hemispheres, moves N as it follows the sun. In July it lies just N of the S coast of the Arabian Peninsula, bending NW over the Hanish Is to lie along the coast of Eritrea, crossing inland at about 15°N. It is associated with unstable weather (see also 'Haboob', page 28).

Prevailing winds

The broad pattern is shown by the diagrams. Note two major influences. The first is topography. The steep-sided Rift Valley, in the bottom of which the Red Sea lies, channels wind along its axis. This is most marked where the sides of the valley close in – the Gulfs of Suez and Aqaba, and the Straits of Bab el Mandeb and approaches. The second influence is daytime heating, which slants the wind, making one tack more favourable than the other; change occurs between day and night virtually all over the N two-thirds of the Red Sea.

Winter

In the Red Sea and the E Mediterranean, winds in winter are stronger than those in summer. In the N Red Sea the prevailing NW wind blows at Force 4–5; it is stronger in the Gulfs of Suez and Aqaba, especially in the Straits of Gubal and Tiran. In the Gulf of Aden winter winds are NE through E to SE Force 2–4, occasionally stronger, especially along the coast, and strengthening and veering E to SE closer to Bab el Mandeb.

From Bab el Mandeb to S of the Zubayr group S–SE winds are prevalent and can blow very strongly (Force 6–7, occasionally 8) from Bab el Mandeb to the Hanish group. This is mainly a product of funnelling, though a sea breeze sometimes adds its bit. Further N, S–SE winds will blow 30–40% of the time between the Zubayr group and Port Sudan, the proportion diminishing as the season advances. The balance divides between light to moderate N–NW and calms. S–SE winds are at their most constant October–January, blowing 50% of that time in the Hanish group and 75% in

27

Red Sea Pilot

January.

N of Port Sudan N or NW winds prevail, becoming more constant and fresher (Force 4–5, occasionally 6–8, especially when enhanced by sea breezes) as one moves N. Diurnal effects are marked even in the middle of the sea. When the gradient N winds are light, NNE–NE sea breezes dominate in coastal waters. Between Port Sudan and Ras Baniyas S winds can occur, more frequently in the S, but not more than 10–15% of the time. They are rare N of Ras Baniyas, though not unknown. You should not count on any S winds from Ras Baniyas N'wards, although you may get lucky. A late E–NE-moving depression passing S of Cyprus is most likely to bring S winds from El Quseir to Suez, though the effects are usually only felt in the Gulf of Suez. Calms throughout are short-lived, and 24 hours is the maximum you should hope for.

Do not forget that percentages are only statistics. '75% of winds from the N quadrant' can mean three weeks of continuous N winds rather than five days followed by two days of something else! '5% calms' may mean three brief 12-hour calm spells between blasts of N wind, or a three-day calm after a month of unrelenting N'lies.

Summer

N–NW winds prevail. They are not as strong as in winter, the Gulf of Suez excepted. In the S prolonged periods of calm can be expected August–September. In the Gulf of Aden the prevailing wind in summer is W through SW, occasionally S. It is usually Force 3–4, blowing more freshly both in the Straits of Bab el Mandeb and as one moves E along the S coast of the Arabian Peninsula. The *haboob*, the *kharif*, sand and dust storms, whirlwinds and waterspouts are most common in summer.

On the S Saudi Arabian and Red Sea Yemeni coasts strong E or NE winds associated with electrical storms occur. They don't last long, but arrive with scant warning.

Transition seasons

April–early May and September–early October, during the change of monsoon, you may get weaker and more variable winds and a higher percentage of calms, especially in the S in August and September when calms can be prolonged.

If you intend to sail with as few stops as possible from Bab el Mandeb to Suez, April is the best month. The N winds in the N half of the Red Sea are not as strong as in January or February, yet one may still pick up a S in the Gulf of Suez during the last episodes of frontal activity in the E Mediterranean. The risk is that you may meet stronger N winds in the Gulf of Suez in late April–early May.

Local winds

Winter and spring

Khamsin Gulf of Suez and N Red Sea. Blows from S or SW. Hot and dry. Most common February–May, but not that many each year. Caused by a depression moving E or NE from the N Sahara into the E Mediterranean, which the wind precedes. Often accompanied by sand and dust storms. An Arab saying has it that 'when the *khamsin* has blown for three days a man is justified in killing his wife'. In non-sexist language, it is horrid weather, and it will be unusual if it lasts longer than a day or two. It is usually ushered out by strong NW winds as the frontal system passes through.

Belat S of the Arabian Peninsula and approaches to the Gulf of Aden. N or NW katabatic wind coming down off the mountains. Strongest at night and around dawn, when it can reach Force 7. Relatively short-lived (not more than two days). Kicks up a big sea very quickly.

Summer

Haboob S Sudan coast. A violent squall from SE through W. Gusts can reach Force 8 or more. Often accompanied by dense dust storms. At anchor this is a wind to be wary of. One hit us in Marsa Fijab and gusted 45 knots for 45 minutes.

Kharif Gulf of Aden. S or SW katabatic wind from the African shore, the mirror image of the *belat*. Blows up during the night, reaching maximum strength (up to Force 7) around dawn and during the morning. Strength depends on the force of the SW monsoon. Associated with dust storms.

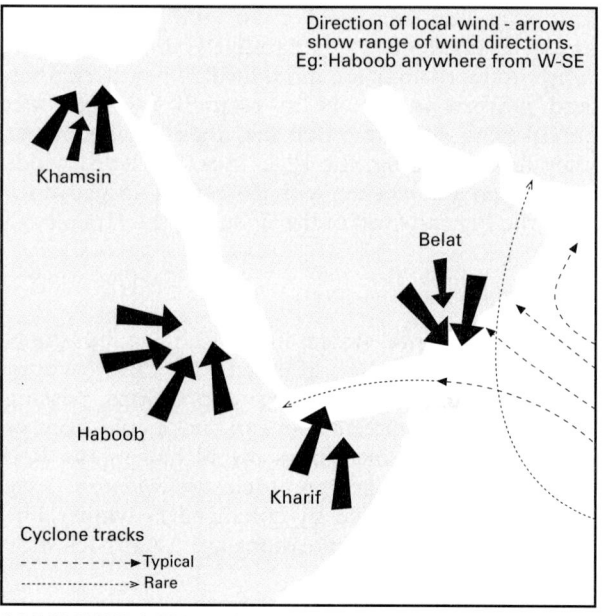

LOCAL WINDS IN RED SEA AND GULF OF ADEN

Tropical revolving storms

This is the generic name for typhoons, hurricanes and cyclones. They are rare in the Red Sea and the Gulf of Aden, but not unknown. Most likely to occur in the transition periods, April–June and September–November. By the time any errant cyclone gets to the Red Sea it has diminished in force, but what is left is bad enough. Kamaran I was hit by a bad one in 1936. In late spring or early autumn keep a more careful eye on your barometer. If the corrected reading (see 'Rules of thumb', page 30) is 3mb or more below the mean pressure for the time of year where you are (mean seasonal pressures for selected stations can be found in the Admiralty Pilot), something nasty is brewing.

Visibility

Visibility is usually good or excellent though in summer, and to a lesser extent in winter, refraction and mirage effects are common. Dust haze often obscures features more than a few miles inland of the coast.

Fog is rare, but not unknown. We were once making towards Hurghada when night radiation fog reduced visibility to some 50–100m. The fog lasted about 2 hours, burning off rapidly after sunrise.

Air temperature and humidity

The Red Sea is generally hot. The average annual temperature range is 22–30°C (72–86°F).

Winter

Red Sea nights can be cool. Mean minimum temperatures can get down to 19°C (66°F) at the S end and 13°C (55°F) around Port Sudan. In the N it can get sufficiently cold to require warm clothing. Daytime minimum temperatures in the Gulf of Suez (Hurghada N'wards) can get down to 9–10°C (48–50°F) in February. Wind chill must also be considered. Early morning starts during a late April transit of the Suez Canal in brilliantly clear weather required windproofs and a couple of layers.

Humidity close to the coast or in *marsas* can be very, very low.

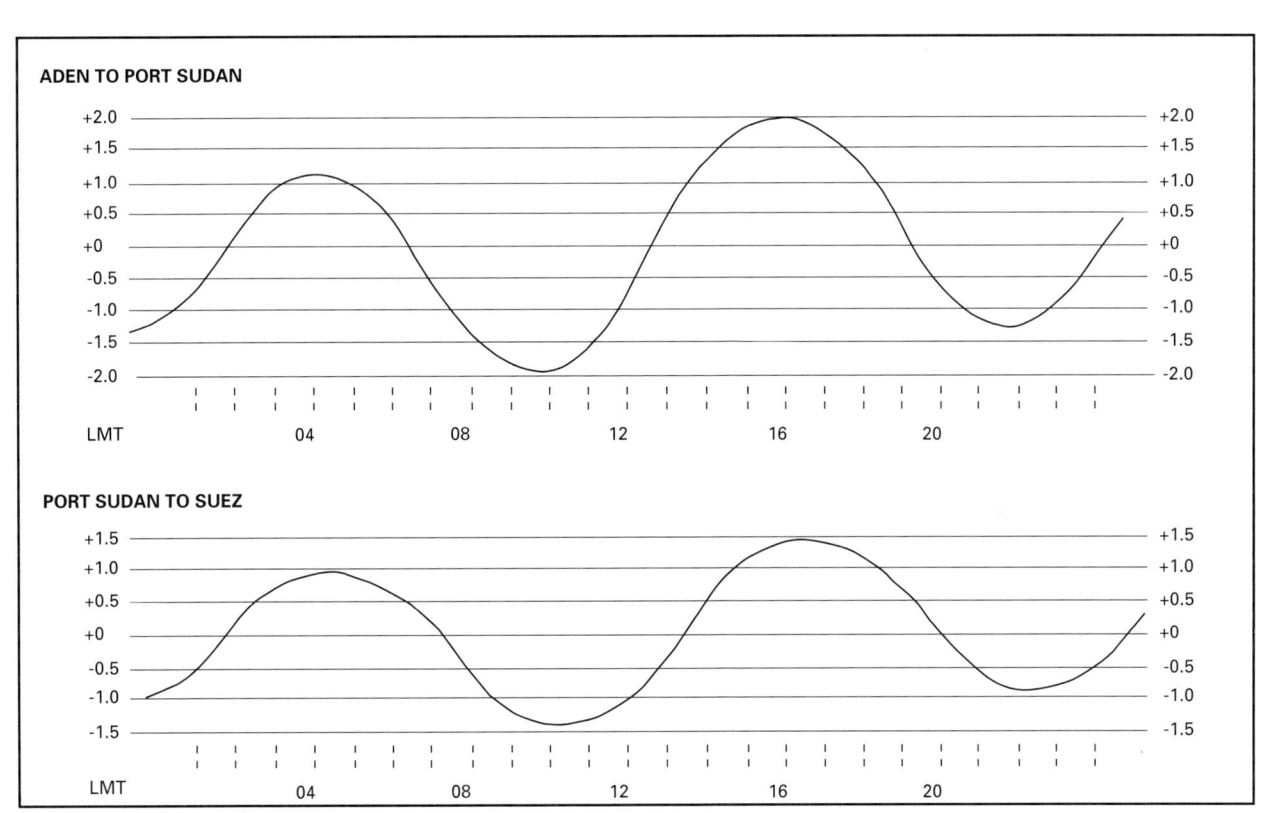

TABLES FOR DIURNAL VARIATION IN PRESSURE
To correct your barometer reading: look along the bottom of the relevant chart for the local *time* of the reading; move vertically up to meet the *curve*; look left or right, level with where vertical from *time* met *curve*, to scales at sides; read off *correction* (+ or -) and apply to *barometer reading*.

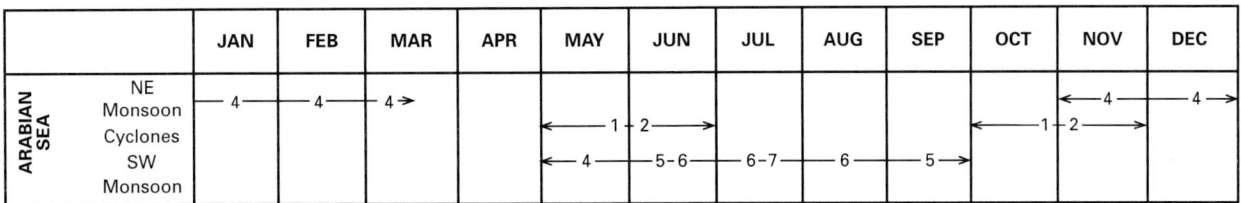

WIND PLANNING CHART - ARABIAN SEA
Monsoon figures: Beaufort Scale. Cyclone figures: Average number per year

Summer

The Red Sea is furnace-like. Mean daytime shade temperatures of 37–38°C (98–100°F) are quite usual. In the S Red Sea that rises to 42–43°C (108–110°F). Daily maxima of 44°C (111°F) have been recorded in the Gulf of Suez, and around Khor Nawarat that gets up to 55°C (131°F). By July it can be like living in a sauna.

Humidity is lower in summer and higher in winter in the S Red Sea (the opposite is the case in the N) and in general higher in the S than in the N. It is higher first thing in the morning than later in the day, and higher at sea than on the coast. The following table gives some average values.

	NE Monsoon	Transition	SW Monsoon	Transition
Gulf of Suez	59%	61%	68%	60%
El Quseir	49%	45%	48%	50%
Port Sudan	68%	57%	42%	66%
Mitsiwa	74%	74%	57%	67%
At sea	66%	73%	71%	75%

Rules of thumb

The majority of cruisers coast-hop N'wards along the Sudan and Egyptian shores. The following are primarily for them.

Any decision to sail introduces the first two rough rules of thumb:

- Daytime winds 30M offshore are usually 5–15kt less than coastal winds. Though there is less diurnal variation, it still exists.
- Diurnal wind slants can be used to good effect. Off the W coast, stbd tack is generally favoured during the day, port at night. Off the E coast the opposite is the case.

In winter S or SE winds can be hoped for at least as far as the Zubayr Is, and with luck to the Sudan border or further. If luck is not favouring you, this rule of thumb may help:

- S winds may hold in further N (as far as the Farasan Is) on the E coast.

If you coast-hop, you have to contend with a sea-breeze enhancement of the prevailing wind. Further rules of thumb are:

- If there is no wind or a light W land breeze overnight, the sea breeze will begin to build later (after 1000). The wind will gradually veer from W through NW and N to NNE or NE by early afternoon, and is unlikely to blow at more than Force 4, lower 5.

  NB: This rule of thumb applies to the W coast. On the Saudi Arabian coast a light E'ly early in the morning means a day of light sea breezes, the sea breeze being likely to back NNW or NW.

- If you have dew on the deck in the morning, expect a day of S winds or light to moderate sea breezes.
- If there is a N or NW wind overnight, the sea breeze will set in by 0800–0900, will be fresh by 1000 and by 1200–1300 will be blowing at 25–35kt NNW to NNE.
- Therefore a further simple rule: If day-sailing, start as soon as it is light enough to clear your anchorage and finish by midday. This is most important on the W coast, since *marsa* entrances are hard to spot once the sun is past the meridian, and next to impossible after 1500.

What about forecasting what is coming? The most fundamental rule of all follows from the nature of the Red Sea's climate:

- Keep in touch with what's happening ahead of you by listening in to radio skeds.

Every year 100 or more yachts sail N between January and May. There is usually at least one radio net (on marine single sideband/or amateur (ham) band) with a daily or more frequent schedule (sked). Whatever is happening 100M N of you will happen to you in about 24 hours. Join a sked if you have the equipment (see 'Radios'). If you don't, buddy via VHF with someone who has. If you can, relay your weather conditions to help those behind you.

Watch the barometer. Small fluctuations, as small as 2mb, have big consequences. The isobaric interval on ordinary barograph barograms is too coarse. The best plan is to keep a graph of barometric pressure with an exaggerated vertical scale, correcting for diurnal variation (some 3mb in the N Red Sea and 4mb in the S) by using the charts opposite.

A crude pattern of the relation between barometric pressure and what the wind does is summarised as follows:

- In winter just after the barometer starts to rise expect a N'ly. It will come later on the rise as one moves N. Just after pressure peaks and begins to fall, expect a S'ly in the S Red Sea and decreased winds or a calm in the N. The barometer movement in question is seldom more than 2mb, but the bigger the movement, the bigger the wind. (This is NOT infallible!)

If you're getting desperate, watch the moon. There does seem to be a connection between the phases of the moon and changes in the weather:

- The strongest N–NW winds come with the full moon; lulls come with the new moon.

  Moon-spotting is less reliable than most other indicators, but it's worked for us quite well.

Other forecasting rules of thumb tend to depend on where in the Red Sea you are. Cloud formations are a classic source. In the N Red Sea there is that old standby, the evening sky:

- High cirrus in W through N quadrants forecasts stronger N winds within 12–48 hours.

In the S Red Sea there is a related recourse:

- Increasing lower-atmosphere cumulus from N and W means a N'ly is imminent. Cloud is the first indication that the convergence zone is moving S towards you.

In the S Red Sea even the nose can help out!

- If at sea you smell dust during a calm or light airs, be prepared for a fresh S–SE'ly.

There are other rules of thumb relevant to particular areas:
- If it's blowing 25kt at Aden, it's blowing 35–40kt from Bab el Mandeb to the Hanish Is.
- In the Strait of Gubal especially, and the Gulf of Suez more generally, an ebb tide causes fresher winds and a flood tide steeper seas.
- In the Gulf of Suez winds and seas are easier N of El Tur (Tor Harbour) and easier on the E coast than on the W.

For the hard ones bashing up or down the middle: you will seldom have sustained winds of over 30kt, though it may gust at 40+. 15–25kt is normal in daytime, less at night. Expect seas steeper and shorter than those in open ocean. If you've sailed in the Mediterranean, or gone to weather on the Queensland coast in Australia during the SE trades, you'll know what to expect. Calms are rare. One day in ten would be about it. With good timing (a.k.a. good luck) it is possible to get up the Red Sea in under three weeks without dealing with any really hard headwinds. Good luck!

Local knowledge

Fishermen, soldiers who spend months and years in lonely outposts, lighthouse keepers and charter boat skippers know their local weather. The chatter of the many rig-service vessels on VHF channels can also give useful tips.

Forecasting services

Radio and weatherfax services are few. For what they are worth here are the marine-band radio weather forecasts, weather facsimile forecasts, and amateur radio (ham) and other informal sources.

Marine band MF/HF radio weather forecasts

Scheduled voice forecasts are rapidly disappearing because of the spread of satcoms, Navtex and GMDSS. Voice forecasts have an accepted international format: gale or storm warnings, then a summary of the general synoptic situation, then area forecasts for the next 12 or 24 hours. After the area forecasts (sometimes after each area forecast) comes the outlook for the next 12 hours+ beyond the forecast period.

Not all of the forecast services can be relied upon to broadcast on time, or at all. For the forecast areas mentioned below, see the accompanying diagrams.

From Bab el Mandeb onwards. VHF reception in the Red Sea can be startling. Once you are in the Trinkitat to Marsa Umbeila area, try monitoring Ch 25.

In the Red Sea, the Jeddah forecast is the most reliable. There are awkward silences when the N wind is about to blow hard! Gossip suggests that the operators are obliged to broadcast hourly if they have forecast winds of over 28kt. So they don't forecast them. Whatever the truth, if there's no forecast, be wary. The operators' English is good, but what they have to say is always much the same. And what feels a strong wind and rough sea to us is often no worse than a gradient force 5–6 with moderate seas (1·25–2m seas).

Haifa is useful from Suez on, and helps one to time one's exit from the Suez Canal, especially early in the season when the Mediterranean front is still active.

Note: Morse services are in almost total eclipse for the same reason. We didn't find any of the services which still worked and could be de-coded by a demodulator working to software in our laptop of the slightest use.

Weatherfax

The number of publicly accessible fax broadcasts listed in the Admiralty *List of Radio Signals, Vol 3(1)* has much diminished. Whether one can pick anything up of what's left is another question. Reception is not always good, and is sometimes hopeless. Stations don't always come up as scheduled, and when they do they are not always punctual.

New Delhi's 1:20,000,000 scale coverage of the Indian Ocean includes the Red Sea. Frequencies are 7403kHz (1430–0230 transmissions) and 14840kHz (0230–1430 transmissions). Useful times/maps are: Surface Analysis 0011 0634 1211 1820; 500hPa analysis 0734 1946; Surface prognosis 0812 2022; 500hPa prognosis 0936 2223;

Cairo's 1:20,000,000 scale includes a vast area within which the Red Sea is a small patch. Frequencies are 4526·5kHz and 10123kHz. Useful times/maps are: Surface prognosis 0000 0600 0640 1200 1800 1840; Surface analysis 0020 1220; 500hPa prognosis 0340 0940 1540 2140; 500hPa analysis 0740 1940; satellite imagery with wind analysis 1240.

Ankara's 1:10,000,000 chart is useful N of Port Sudan if you can get it. It is also helpful for the E Med. Frequencies are 3377kHz and 6790kHz. Useful times/maps are: 24hr and 36hr 500hPa analysis 0330; Surface analysis 0430 0610 0940 1240 1610 1840 2152; 500hPa analysis 0640 1910; 24hr significant weather 0740 1710; 48 and 72hr 500hPa analysis 0910; Significant weather 1010; Maximum wind analysis 1040.

For the complete weather-fax enthusiast, every conceivable transmission one can find (military, civilian and even amateur) is listed in the *Guide to Facsimile Stations* available from Klingenfuss Publications, Hagenloher Strasse 14, D-72070, Tübingen, Germany. This is arguably a better guide than the Admiralty *List of Radio Signals, Vol 3*.

Websites

There are various Internet weather websites and one can waste a lot of time chasing around them. We have found Wetteronline (www.wetteronline.de) consistently good. It is in German (for some reason the English version is dumbed down), but not difficult to find one's way around. Choose keyword

Red Sea Pilot

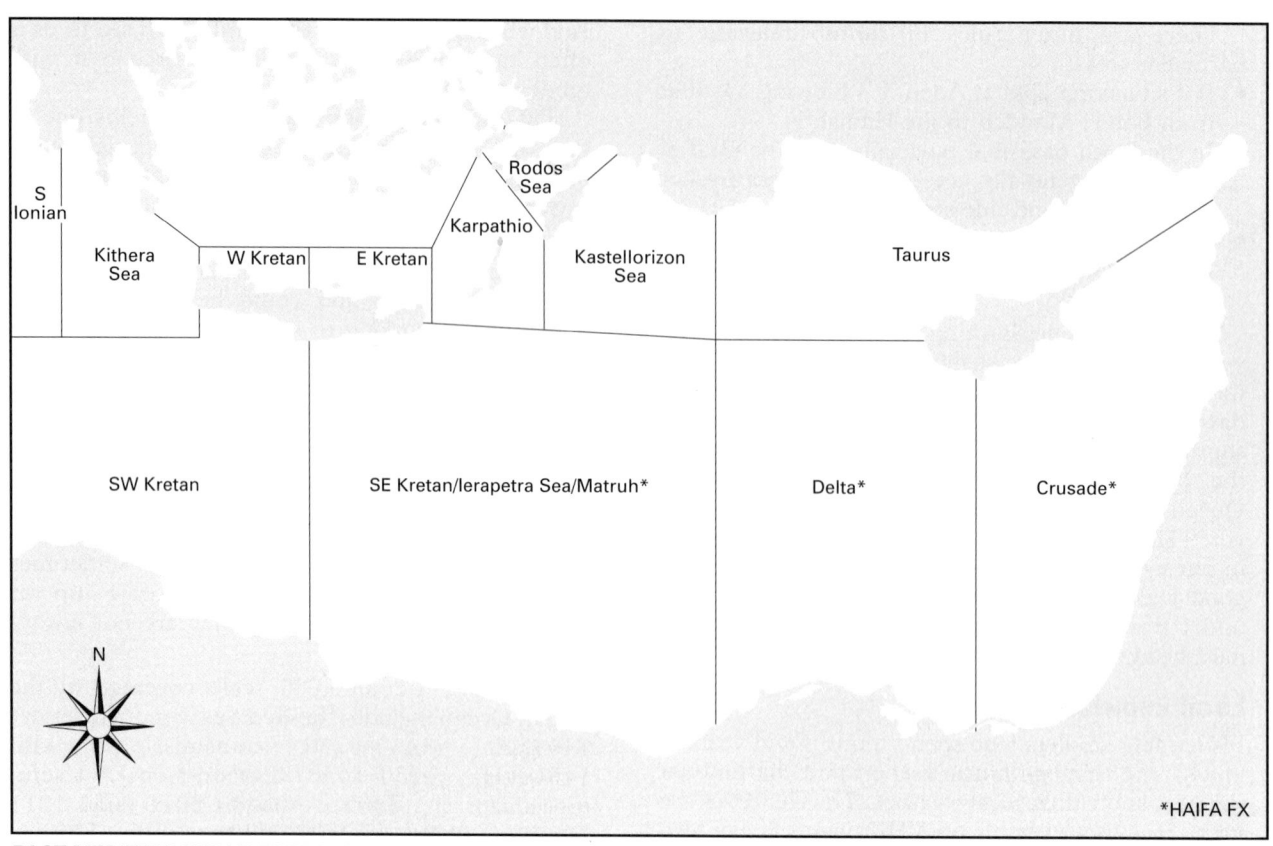

EAST MEDITERRANEAN FORECAST AREAS

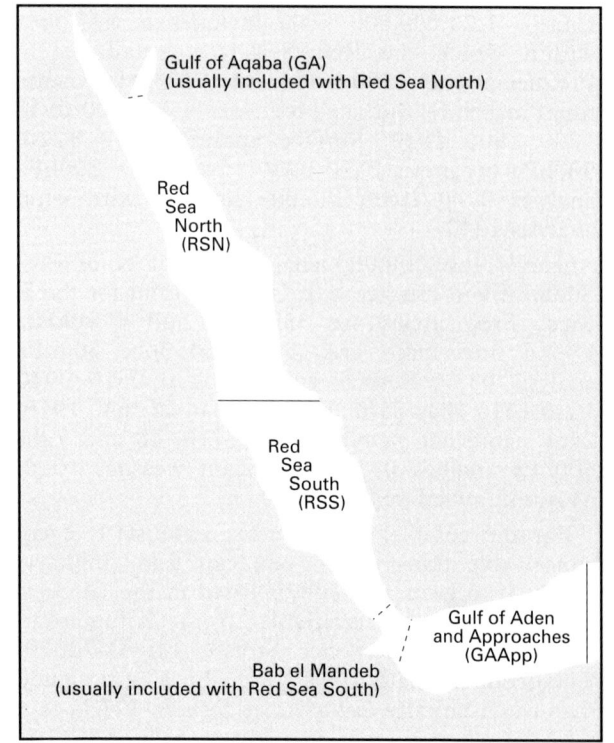

JEDDAH RADIO FORECAST AREAS

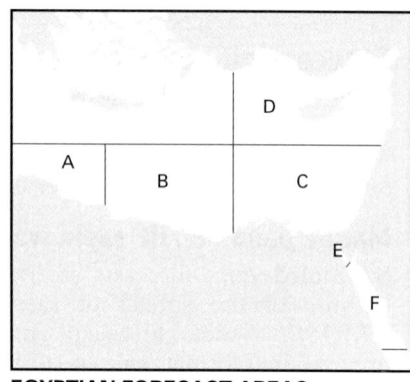

EGYPTIAN FORECAST AREAS (ALEXANDRIA AND QUSEIR)

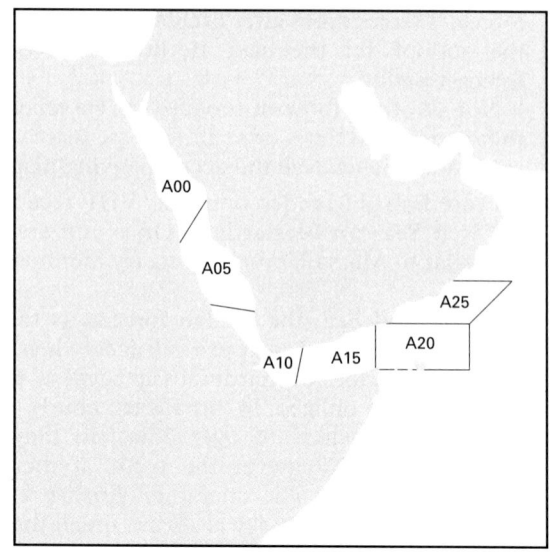

DJIBOUTI FORECAST AREAS

Weather

In S approaches and Red Sea

Station	Callsign	Freq kHz	Morse/voice	Mode	Time UTC	Areas
Djibouti	J2A	464	M	A1A	0430/0900/1700	A00–A25
		8682	M	A1A	0430/1700	
		12728	M	A1A	0900	
Jeddah	HZ	1726	V	J3E	0503/1703	RSN, RSS
		Ch 25	V	VHF	0503/1703	GA, GA App
		436	M	A1A	0520/1720	
Pt Sudan	STP	500	M	A1A	0810	Red Sea 18°N–23°N
Quseir	SUK	435·5	M	A1A	1230	E, F
Serapeum	SUZ	515·5	M	A1A	0648/1048/ 1448/1848	Med, Egyptian Red Sea Waters
		4255	M	A1A	0648/1048	
		8450	M	A1A	1048/1448	
		12670·5	M	A1A	1048/1448	
Aqaba	JYO	Ch 22	V	VHF	0815/1815	Gulf of Aqaba

In E Mediterranean

Station	Callsign	Freq kHz	Morse/voice	Mode	Time UTC	Areas
Haifa	4XO	2649	V	H3E/J3E	0303/1903	Matruh/Delta/Crusade
		Ch 25	V	VHF	0303/1903	
Alexandria	SUH	444	M	A1A	0100/1300	A/B/C/D

NAVTEX

Navtex is now quite useful in the N Red Sea and Mediterranean

Saudi Arabia

Jeddah	H	518kHz	0100/0500/0900/ 1300/1700/2100	24hr FX
Dammam	G	518kHz	0110/0510/0919/ 1310/1710/2110	

Egypt

Quseir	V	518kHz	0330/1530	Bulletin for N Red Sea
Alexandria	N	518kHz	0210/1410	12hr FX for E Med

Israel

Haifa	P	518kHz	0020/0420/0820/ 1220/1620/2020	12hr FX and further 12hr outlook for E Med

'Segel', then choose Rotes Meer. The forecast divides the Red Sea into three regions, south, centre and north. It offers a one week forecast with prognostications every six hours each day. To access these, you click on the relevant hour on the time and date calendar at the bottom right of the screen. We were truly impressed. The service forecast a shift from SE quadrant to N quadrant winds in the S Red Sea with an accuracy of 2 hours! It got the subsequent duration of the N wind exactly right.

The problem is how to access this if you don't have Inmarsat A, B or mini-M and very deep pockets. The Inmarsat C equipped boat from whom we benefited had a simple system. Their son back in Germany digested a forecast and sent them an email. Obviously a similar scheme is open to any Inmarsat C equipped boat with a helpful, wired friend or relative. It is also a route for hams with the excellent ham-based email system, subscribers to one of the marine band SSB email systems, your laptop and modem when in harbour if you can find a phone line, or to anyone with access to a cybercafé. There are cybercafés in Galle, Salalah, Mukallah, Aden, Djibouti, Asmara (and maybe Massawa), Port Sudan, Safaga, Hurghada, El Gouna, Port Suez, Ismailia, Port Said, Sharm el Sheikh, Elat and Aqaba. With some thought you can weather plan at least some of your stages.

With the exception of Wetteronline, Red Sea website weather suffers from the same problem as does weatherfax. The standard coverage areas nearly always have the Red Sea up in a corner and, at the scales used, there is scant useful detail.

www.weather.yahoo.com – a good satellite picture of the Red Sea and Gulf of Aden, excellent for convergence spotting.

www.oceanweather.com – the public access site has an hourly update with a synoptic chart and ship reports.

www.nlmoc.navy.mil – the US Navy site, the automatic observation reporting area which covers the N Indian Ocean, Gulf of Aden and Red Sea is vulnerable to events in the Middle East and was not accessible in late 2001. When it is available, it's amongst the best sources.

www.nottingham.ac.uk – has a direct click access to the latest satellite pictures from Meteosat, areas D3 and D6 are the ones to look at.

www.rsmas.miami.edu – not much good for weather, but the programme being run has excellent infra-red satellite imagery of oceanic currents and is good for spotting an early change of direction to the Somali Current.

www.fnmoc.navy.mil – a sexy new animated 36 hour forecast. A bit short on detail (usually only two wind arrows for the Red Sea) but not bad for spotting the convergence.

Red Sea Pilot

Amateur (HAM) radio

DIY services can help a lot. In the Red Sea proper your best bet is the yachts ahead of you. Failing such very local nets, forecasts over larger areas can sometimes be got from the very organised amateur-radio networks worked every day by enthusiastic hams around the world.

In the S approaches, try:
- The Perth, Australia Indian Ocean Net operated by VK7BO 1115–1130 UTC on 14332kHz, USB.
- For the E Mediterranean there is the E Med net on 7096kHz LSB on Mondays, Wednesdays and Fridays at 0515UTC. There is often good copy on boats in the N Red Sea. There is another yacht sked on SSB at 0500 UTC on 8101kHz USB, Mondays, Wednesdays and Fridays. Depending on propagation, both nets are usable from N Sudan N'wards. This, if you can hear it, is helpful for timing your transit of the Suez Canal. Note too the weather forecast system by email mentioned in the previous section.

Sea, tide and currents

Topography and bathymetry

The Red Sea's unique topography forms part of the Great Rift Valley, which snakes from the Israeli/Syrian border to Malawi in southern Africa. Over 260M E–W in the latitude of Port Sudan the valley's profile runs from 1,700m coastal mountains through the sea's central trough (over 2,000m deep) to the 2,600m mountains backing the Saudi Arabian shore. It does so in a series of steps each about 50–60M wide, with a 15M-wide trench at the bottom.

At its widest, towards the S, the Red Sea proper is 190M across, but towards the N, at Ras Baniyas, for example, it is only 90M wide. The navigable width of open sea is reduced by the large coralline banks at their biggest on the Dahlak and Farasan Banks in the S and by narrower banks in the N. S of Ras Baniyas the banks drop in two descents, first to a 500m shelf then to a narrow trench 1,500m–3,030m deep. N of Ras Baniyas the central trough is wider and seldom deeper than 2,000m. In the latitude of the Hanish Is there is a natural 'sill', nowhere deeper than 50m, separating the main Red Sea basin from the narrow, <300m deep trench through the Straits of Bab el Mandeb.

The Red Sea is open only at the 17M Strait of Babel Mandeb making it a long, narrow, mostly enclosed sea, edged by steep mountains. Most of the W shore is formed by a raised coral platform about 2–3m above mean sea level. This varies from a few yards to several miles wide. Where the coastal range does not rise directly from the shore, a low plain intersected by shallow wadis (occasional watercourses) and low coastal hills (at most a few hundred metres high) fronts the foothills. The coastal mountains are 1,000–1,500m high in the N, rising to over 3,000m around Massawa, declining in height towards Bab el Mandeb. The pattern is similar on the E shore except there are higher, rocky cliffs in place of the low coral cliffs of the W shore.

The Gulfs of Suez and Aqaba are divided by the Sinai Peninsula, a dramatic, mountainous desert region with, in the S, the >2,500m high massif of Gebel Katherina. The Gulf of Suez is some 150M long, averages about 15M wide and is 25M at its widest, near Ras Abu Zenima. It is shallow, averaging 40m and at its deepest 64m. The 95M long Gulf of Aqaba is only 10M wide but, because it is the rift valley extension, at its deepest over 1,800m with steep-to shores. Depths of over 100m within half a mile of the shore are the norm.

Salinity

The Red Sea is the most saline of the world's open seas averaging 38ppt (parts per thousand) compared to the oceanic average of 35ppt. The S (36ppt) is less saline, salinity increasing as one goes N to the Gulfs of Suez and Aqaba (41ppt). This is because no permanent rivers flow into the Red Sea, there is little rain and there is unceasing solar evaporation hoovering up >900 billion m^3 of sea water annually. In addition the summer monsoon reverses surface currents reducing the inflow of less saline water through Bab el Mandeb. In summer sea level can be <0·7m lower than in winter. Recent research has also found pools of hot, dense brine at the bottom of two deep, isolated basins in the central Red Sea. This is the product of leakage of intense heat through the volcanic walls of the central trough.

A result is that surface and deep-water flows oppose each other during the main sailing season. To replace water lost by evaporation, water enters via Bab el Mandeb. Since it is of average salinity it is relatively buoyant. Pushed by the SE winds of winter, it flows N. Below the surface, denser water flows outwards into the Gulf of Aden. This, along with tides and the effects of wind, explains the circulation of currents in the Red Sea, for details of which see below.

Bioluminescence

The Red Sea has occasional startling displays of bioluminescence which, the British Admiralty Pilot observes, gives a vivid visual impression of approaching and passing through very rough water despite knowing it's flat calm. It is fortunately rare. A related phenomenon, known as 'milky sea' creates an even, diffuse glow at or close below the sea's surface.

Temperature

The Red Sea is the hottest sea in the world though there's quite a difference between summer and winter temperatures. The charts on page 36 give a rough guide (°C):

Sea, tide and currents

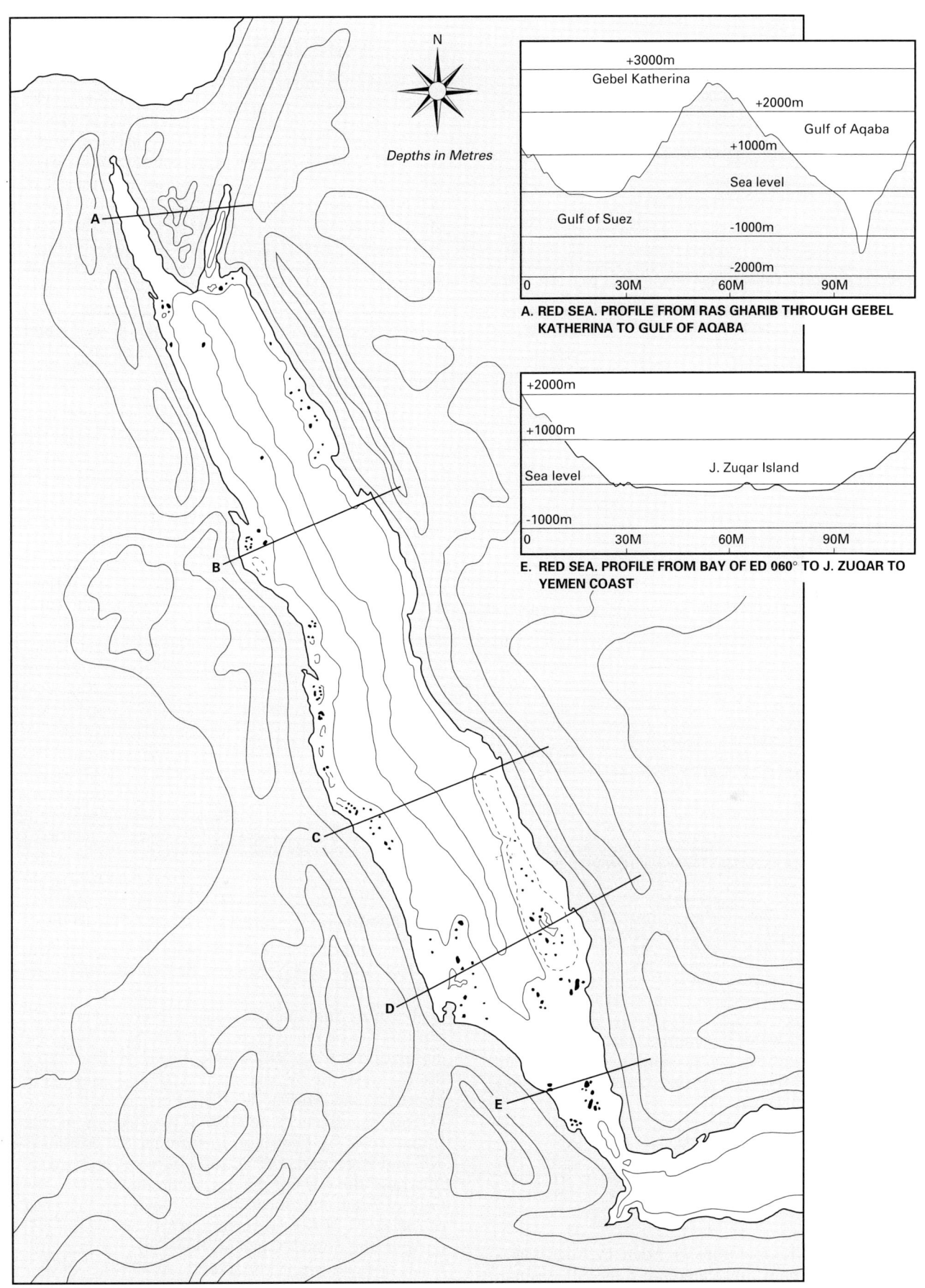

A. RED SEA. PROFILE FROM RAS GHARIB THROUGH GEBEL KATHERINA TO GULF OF AQABA

E. RED SEA. PROFILE FROM BAY OF ED 060° TO J. ZUQAR TO YEMEN COAST

Depths in Metres

RED SEA – A SIMPLIFIED TOPOGRAPHY, SHOWING LINES OF PROFILES A–E

35

Red Sea Pilot

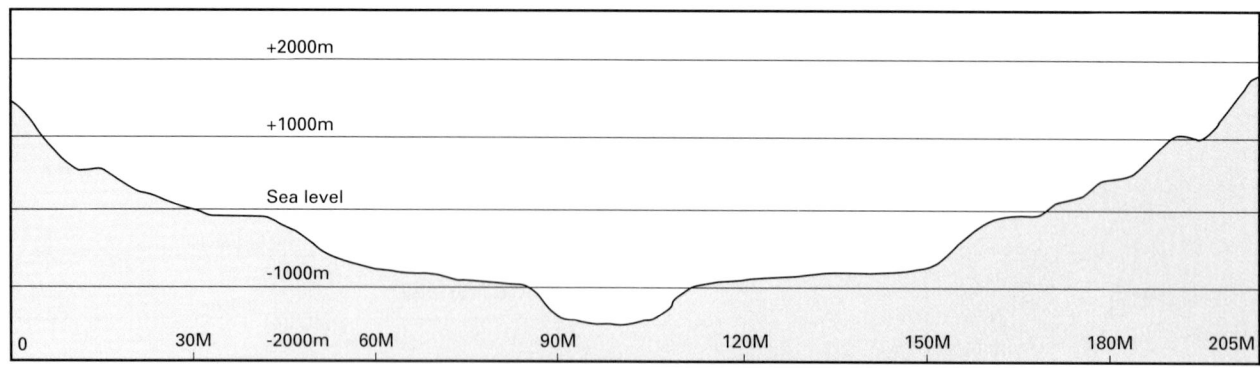

B. RED SEA. PROFILE FROM MARSA SHAAB (FOUL BAY) 060° TOWARDS YANBU AREA

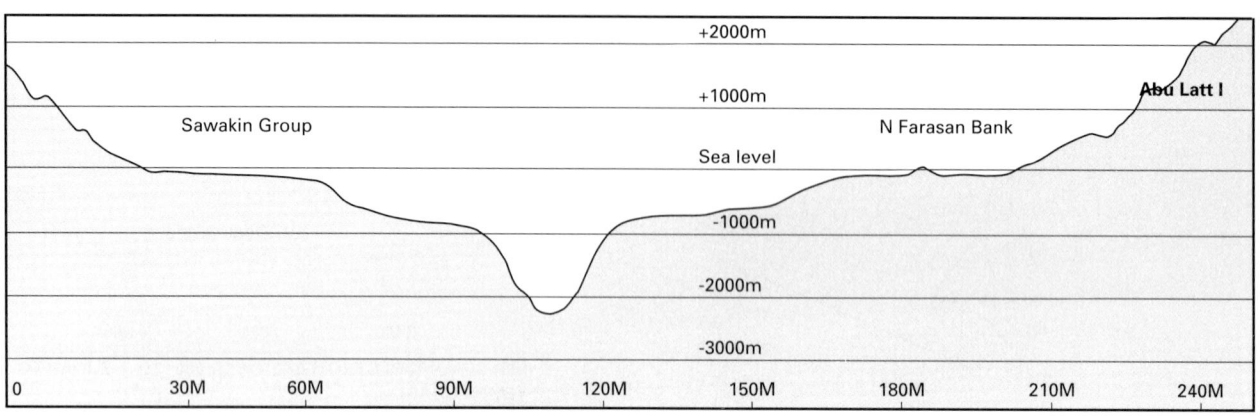

C. RED SEA. PROFILE FROM TRINKITAT HARBOUR 060° TOWARDS ABU LATT I

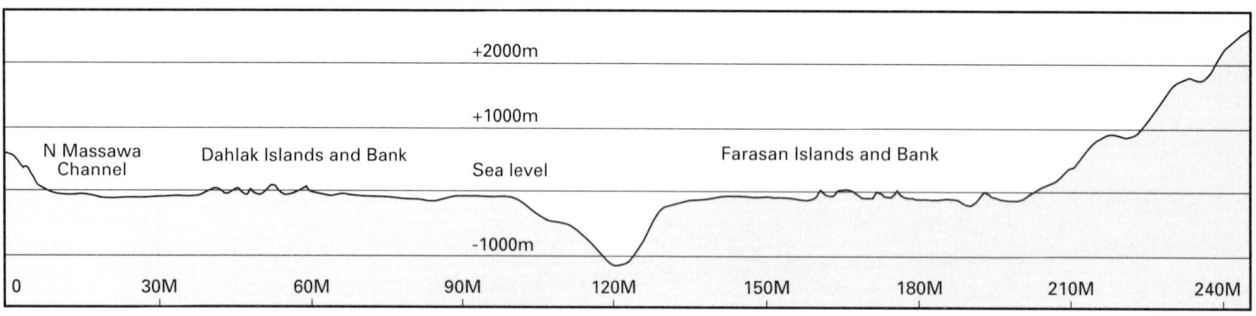

D. RED SEA. PROFILE FROM MASSAWA 070° TOWARDS JIZAN

	Feb	May	Aug	Nov
30°N	17	21	27	23
25°N	22	25	28·5	25·5
20°N	25	28	30	29·5
15°N	26	>29	>31	28

The warmest water in summer is in the S third of the Red Sea, with the temperature decreasing again slightly as one moves S to Bab el Mandeb and out into the Gulf of Aden:

	Feb	May	Aug	Nov
Aden	25	>29	28	26·5
Bab el Mandeb	25·5	>29	30	27
Hanish Is	26	>29	31	27

The cooler temperature in Aden in August is a function of a cold upwelling along the SE Arabian coast. At this time sea water temperatures off Oman are down to 21°C.

Currents

The seasonal patterns of currents in the Red Sea, Gulf of Aden and N Arabian Sea are shown in the diagrams.

N Arabian Sea The prevailing winter current is W, but this can be relied on only S of 14°N. A clockwise circulation setting NE or ENE begins to flow along the E part of the S coast of the Arabian Peninsula in February and March. S of the Persian Gulf it can extend its effects S to 10°N, penetrating to the Indian coast by February. In summer the current is NE and is general. Off Suqutra currents are strong. In winter the current divides, some flowing into the Gulf of Aden, but most flowing down the SE coast of the Horn of Africa. In summer a very strong NE current, the Somali Current, flows up past Suqutra. It has been known to run at 6kts. At all times of year the currents around Suqutra and in the channels between Suqutra and the African coast are strong

36

Sea, tide and currents

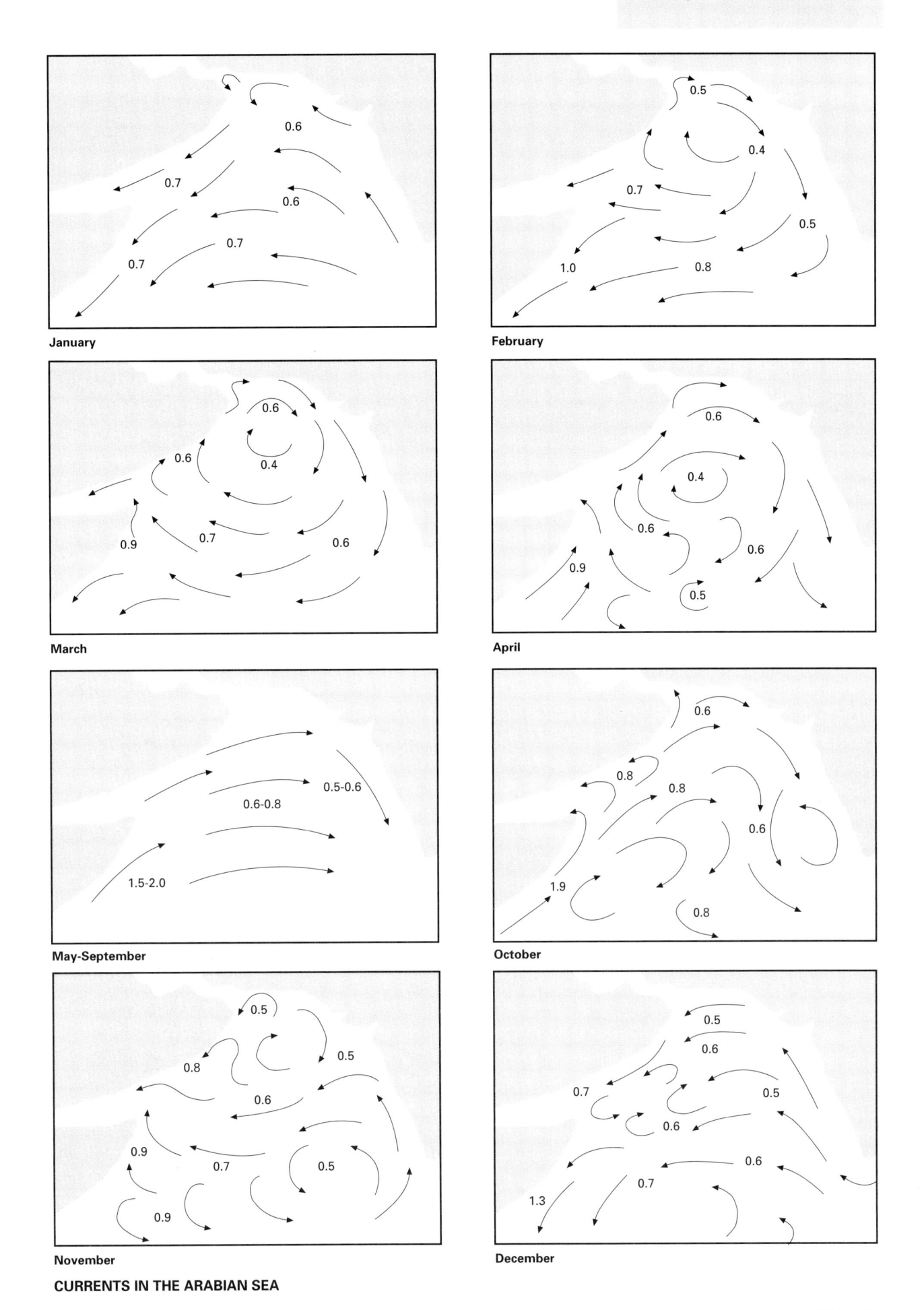

January

February

March

April

May-September

October

November

December

CURRENTS IN THE ARABIAN SEA

37

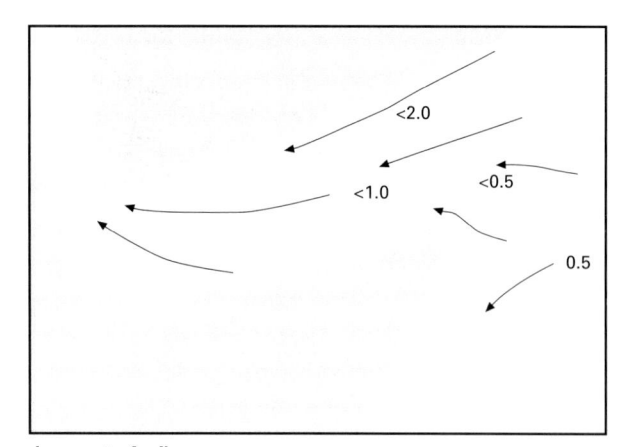

Red Sea Pilot

January to April

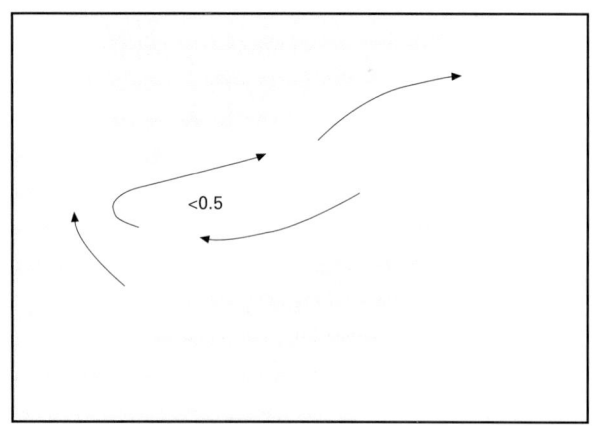

October to December

May

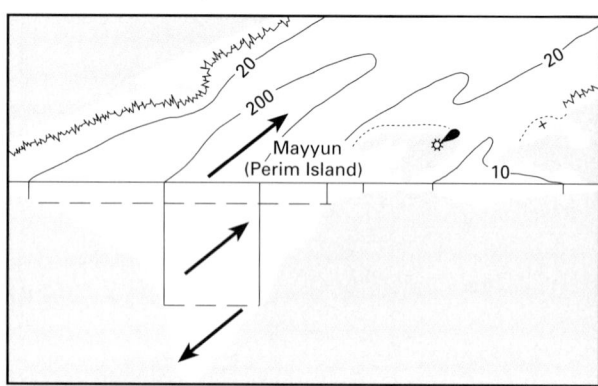

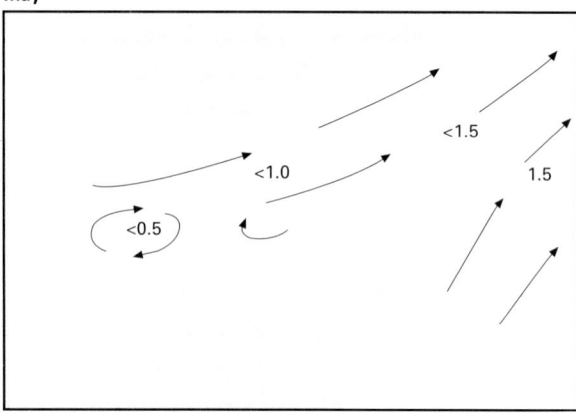

October to April
N flowing surface and mid-level currents (average salinity)
S flowing bottom current (high salinity)

June to August

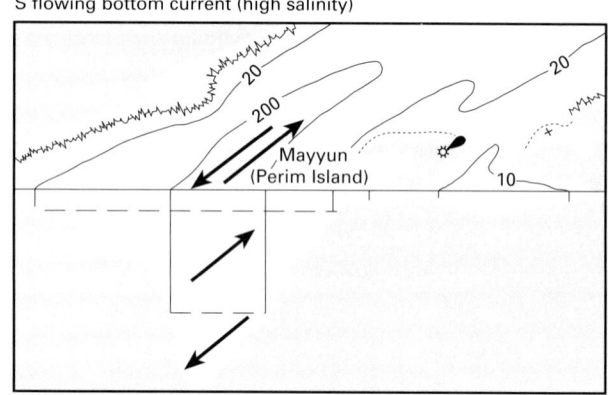

Transitional
Variable (often tidal) surface currents (average salinity)
N flowing mid level current (average salinity)
S flowing bottom current (high salinity)

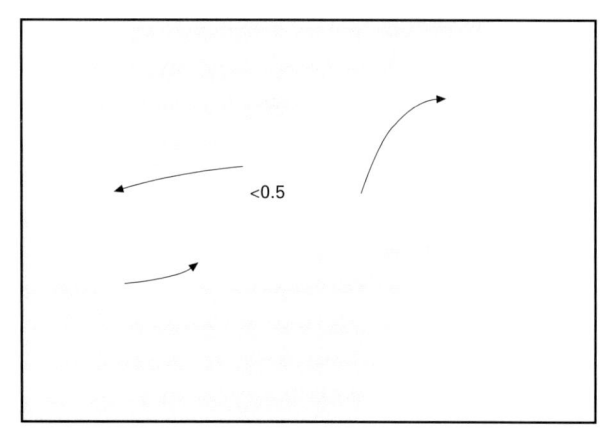

September

CURRENTS IN THE GULF OF ADEN

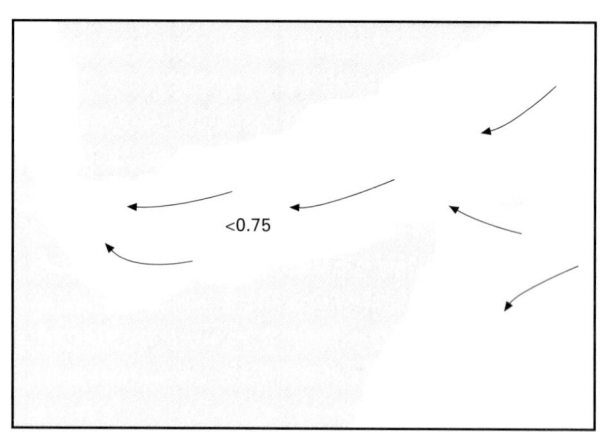

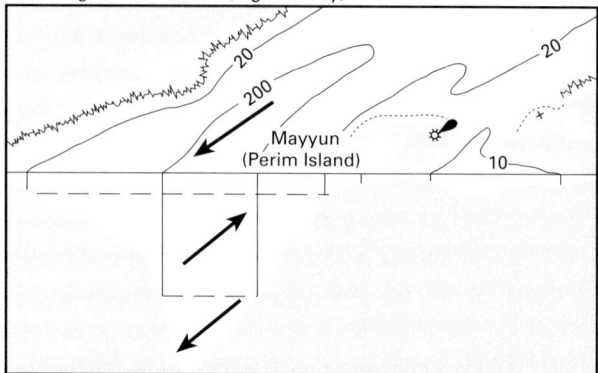

June to August
S flowing surface and bottom currents (average and high salinity)
N flowing mid-level current (average salinity)

Sea, tide and currents

January to April

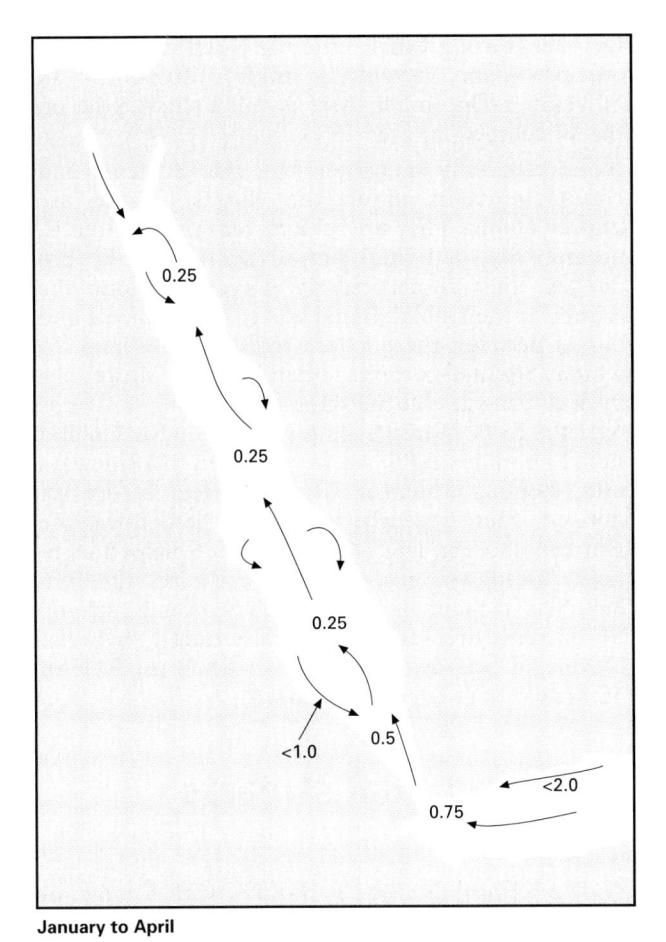

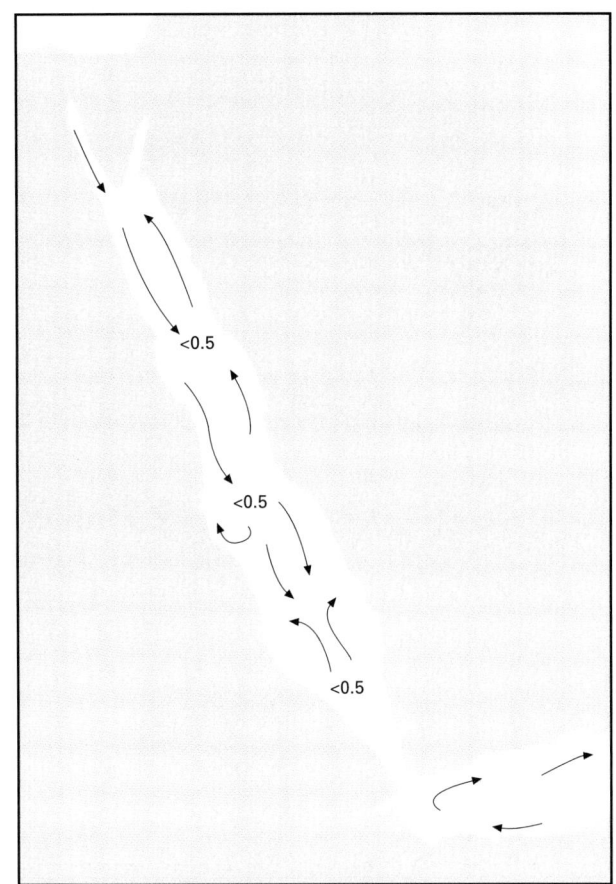

May

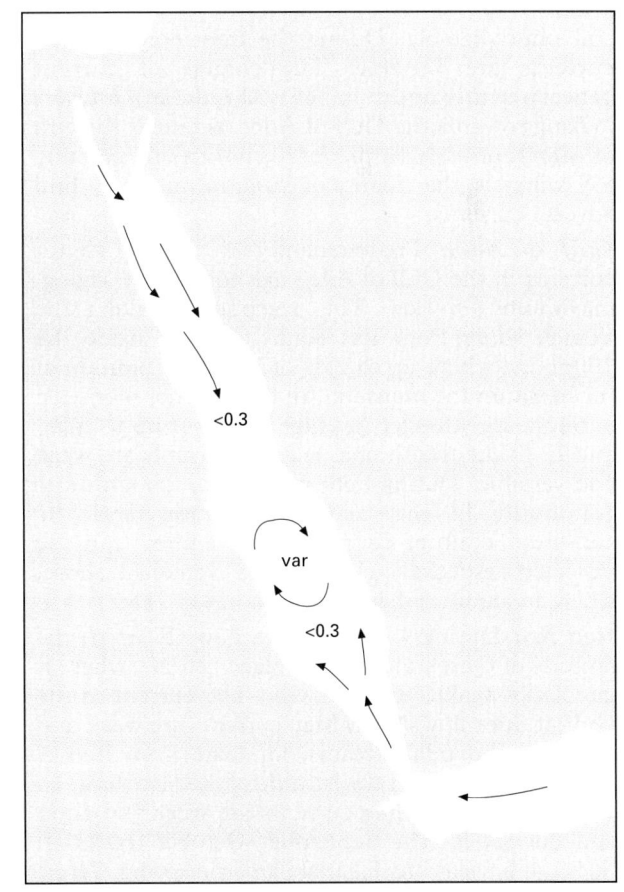

June to September

October

SEASONAL CURRENTS IN THE RED SEA

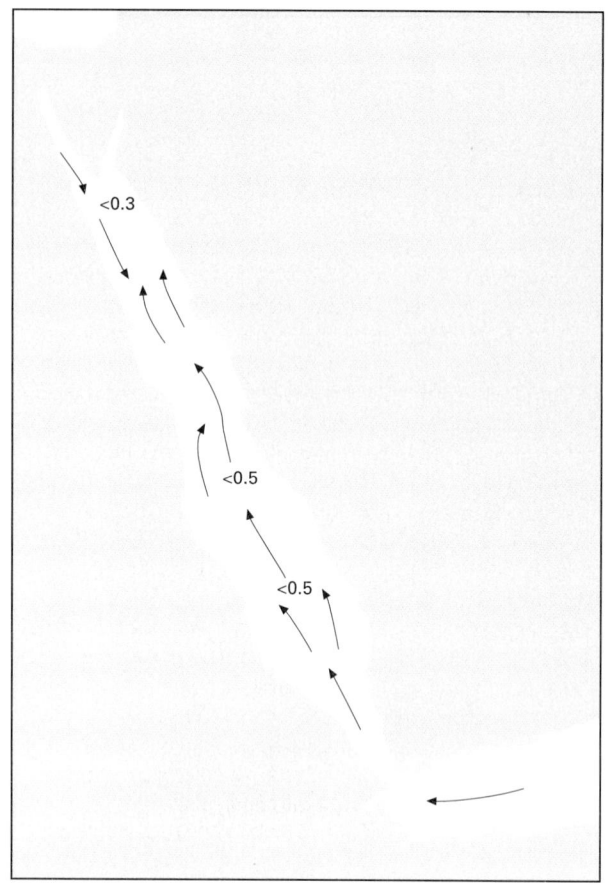

November to December

and unpredictable. During the transitional season currents are variable. The change of current between winter and summer is abrupt, and cruisers making towards the Gulf of Aden as late as the end of March or early April, apart from risking meeting SW winds in the vicinity of Suqutra, may also find adverse currents.

Gulf of Aden Predominantly W setting winter currents in the Gulf of Aden are not usually strong, maximum 0·5–1kts. The exception is along the Yemen shore from Ras Kalb to Aden inside the 200m line where up to 2kts of W set is common at the height of the monsoon. In the summer monsoon currents are stronger (average >1kt but up to 3kts) and E. In the transitional seasons currents are weak and variable. During both monsoons, though more commonly in the transition months and the summer, counter-currents and eddies can be expected, especially close to the Somalian shore, where in summer they can run at up to 2kts.

Red Sea During winter the sea flows N up to the latitude of Quseir and Mina Safaga (26°N), where it circulates weakly anticlockwise. The current in the Gulf of Suez flows S. In May currents are weak and variable, and tidal streams dominate from Bab el Mandeb up to about the latitude of the Hanish Is. In summer the prevailing currents are weak (<0·6kts) and flow N–S. The September–October transition does not have as much variability as in spring. There is a clockwise circulation in the S central Red Sea in October, during which time the S-tending summer currents are reversing their direction. In November–December there is still a slight S set on the W coast as far S as 25°N.

Note, especially in winter, the side currents and counter-currents along the shores. These are unpredictable. They are seldom very strong (that is, no more than 1–1·5kts), but can set one rapidly into danger. This is particularly the case among the islands of the Dahlak Bank, and in the channels and passes between the offshore reefs and the land. In winter counter-currents often flow S down the inshore passages, at the same time setting outwards over the reefs. Whilst sailing N between Suakin and Port Sudan pre-GPS, we were carried 2M sideways into shoaling water on Towartit Reef before we noticed. More recently, when beating S down the Eritrean shore in late December, GPS helped us to find a useful winter counter-current inside the 30m line. Sets in channels in the S Red Sea can be strong. SE currents of 2–3kts around Shumma I (between Dehalak Deset and the Penisola Di Buri, ESE of Massawa) have been experienced.

Tides and tidal streams

Tides

Gulf of Aden The tide is diurnal, with a range of about 2·7m (Aden) to 3·0m (Djibouti), becoming more semidiurnal near Salalah. Sadly, details of High Water Full and Change are no longer readily available, so carry tide tables.

Red Sea The tide is semidiurnal, and entirely contained within the Red Sea proper. When the tide is high at Hurghada it is low at Massawa, and vice versa. Since the wave oscillates around the central Red Sea, tidal range is greatest at the latitudes of Shadwan (Shaker I) with 0·7m at springs and Massawa with 0·9m springs. In the middle of the Red Sea in the latitudes of Suakin and Jeddah there is hardly any tidal movement at all. There's just a small diurnal tide with an oscillation different from that of the main semidiurnal wave.

Gulf of Suez Here there is a separate and peculiar tidal regime which mirrors the Red Sea's. When it is HW at Suez it is LW in the Strait of Gubal, and vice versa. The range at springs in Suez is between 1·4 and 2·0m. This has an effect on currents in the Suez Canal (for details see Pilotage). Note too that the wind can significantly affect sea levels. A strong, prolonged S'ly in the Gulf of Suez can raise the water level in Suez Bay by over 2m.

Gulf of Aqaba HW and LW occur more or less simultaneously all along the gulf. HW is about 1–1·5hrs after HW at Shadwan (Shaker I). Spring range is 0·6–1·2m.

Tidal streams

Gulf of Aden These are weak, subordinate to currents and set obliquely onshore and off. The flood sets SE W of 54°42'E, NE E of that longitude.

Bab el Mandeb In principle the stream sets NW during the rising tide and SE during the falling tide in a 12hr/12hr diurnal pattern, with a rate of 1·25kt each way. But as the BA Pilot points out, it is in fact irregular in rate and duration, sometimes being negligible, sometimes hitting 4kts! The main agent of this is the prevailing wind. In strong NW winds the SE (ebb) stream is extended, sometimes up to 16 hours. In strong SE winds the reverse is true. The stream is influenced by Mayyun (Perim I) and by overfalls. Counter-sets can be expected close to. Felix Normen, in his *Guide pratique voile et plongée de la Mer Rouge et du Golfe d'Aden*, suggests that during the summer months, when the level of the Red Sea is lower, tidal streams predominate over currents. Our sole advice is that whatever the tide tables predict, don't bet on it. A new tidal atlas is in preparation for Bab el Mandeb which, when it is ready, will be a useful planning supplement, especially if you are southbound.

Red Sea There are tidal streams, but they seem irregular and unpredictable. They fall into the same category as the currents, namely gremlins against the appearance of which one must be on one's guard. One generalisation is that any tidal stream not cancelled by surface currents floods into reef passes and across reefs towards the shore or lagoon. During the ebb the reverse.

Gulf of Suez The tidal stream can be appreciable, especially in the Gubal Strait and Suez Bay. The rule is: the stream sets towards Suez during the flood at Suez and sets S during the ebb. Get this right because rates can be 1·5kts, and up to 3kts off Bluff Point in the narrows of the Gubal Strait, even in brisk N winds. The direction and rate of the stream in certain parts of the Gulf is uncertain, especially at Ras Abu el-Darag (29°23'N 32°34'E) and the Sheratib Shoals (28°35'N 33°05'E).

Gulf of Aqaba The timing of the tidal streams is uncertain. Whatever the time, they seem stronger at the head of the Gulf (1·7kts) than at the entrance (0·5kts). More significant is the irregularity of the stream in the Strait of Tiran. The direction is unpredictable even though rates can be up to 3·25kts. The wind sometimes seems to have an effect and sometimes does not!

Navigation

Charts and position fixing

A full list of charts appears in the Appendix. The scale and quality of charting over major stretches of the Yemeni, Eritrean, Sudanese, Saudi Arabian and Egyptian coasts make reliable position-fixing sometimes close to impossible even with GPS. This is particularly so on the NW Yemeni coast, the Eritrean coast and Dahlak Bank, N of Port Sudan, within Foul Bay, and between Ras Baniyas and Hurghada. A scale of 1:382,000 (except over selected small areas) may be the best available for the first two zones, and, with some exceptions around Port Sudan, 1:750,000 or not much larger for the rest. Admiralty coverage is the most comprehensive and was finally reconciled to WGS84 in 2001. However, on such small scale charts the average plotted position covers up to a square mile, so don't expect too much precision! Because the only charts available have such small scales, and because the only full coastal survey was done in 1830–34, eyeball navigation is the order of the day and you must keep a sharp lookout.

Some of the older large-scale Admiralty charts, now withdrawn, are worth copying if you can find someone who still has them. The metrication of charts has over-simplified and older fathoms charts often give detail now omitted which is useful to small boats. Worse, the Admiralty has confirmed with us that in default of survey data to modern exacting standards, their policy, if satellite pictures suggest shoal ground of any sort, is to show solid reef. In the Shubuk Channel, for example, this has led to hydrographic nonsense.

Electronic charts

If you have an electronic-chart package aboard, especially if it is interfaced with your GPS and autopilot, proceed with great care when in pilotage waters, like paper charts, they lack detail. Check above all what the compilation source is and when the chart was digitised. Your new electronic package may not be based on the latest paper charts. In any case, if your electronic chart enlarges scale progressively, in theory adding more information from its database as it moves upscale, be wary. Your system may offer a larger scale, but it is unlikely to include more data. All you get are larger blank spaces! We have used C-Map in the Red Sea and it is good. That said, it also replicates faithfully every shortcoming of the paper charts. With direct raster copies this is obvious. With vectored packages using the same data, it is not. Remember, there have been virtually no surveys of up to 70% of the Red Sea's shores since 1834. An electronic chart cannot and does not compensate for this. Be wary.

Visual fixing

Traditional navigation with a hand-bearing compass is worth the effort. Charts have mostly been reconciled to GPS but no WGS84-based surveys have been carried out, reconciliation being a

Red Sea Pilot

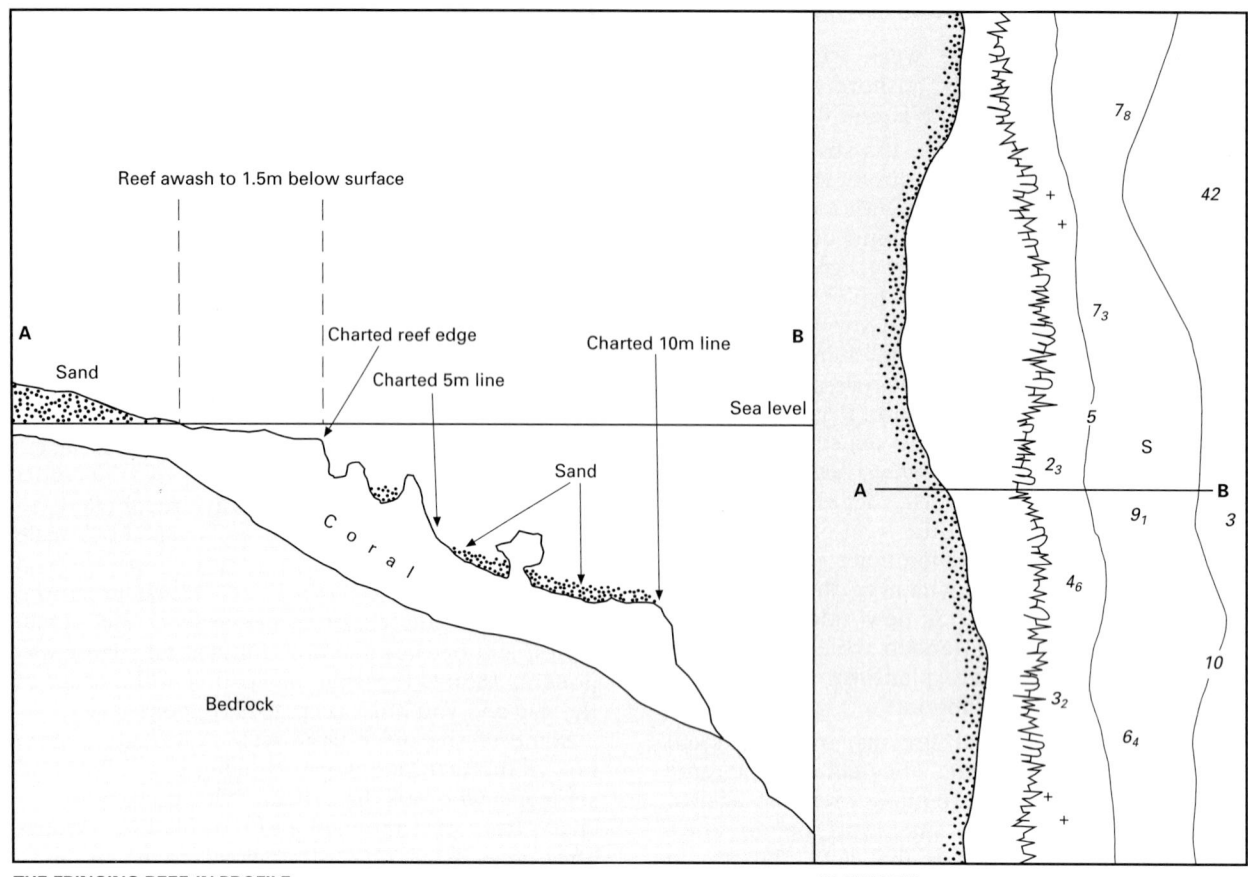

THE FRINGING REEF. IN PROFILE

- Reef awash to 1.5m below surface
- Charted reef edge
- Charted 5m line
- Charted 10m line
- Sand
- Coral
- Bedrock

AS CHARTED

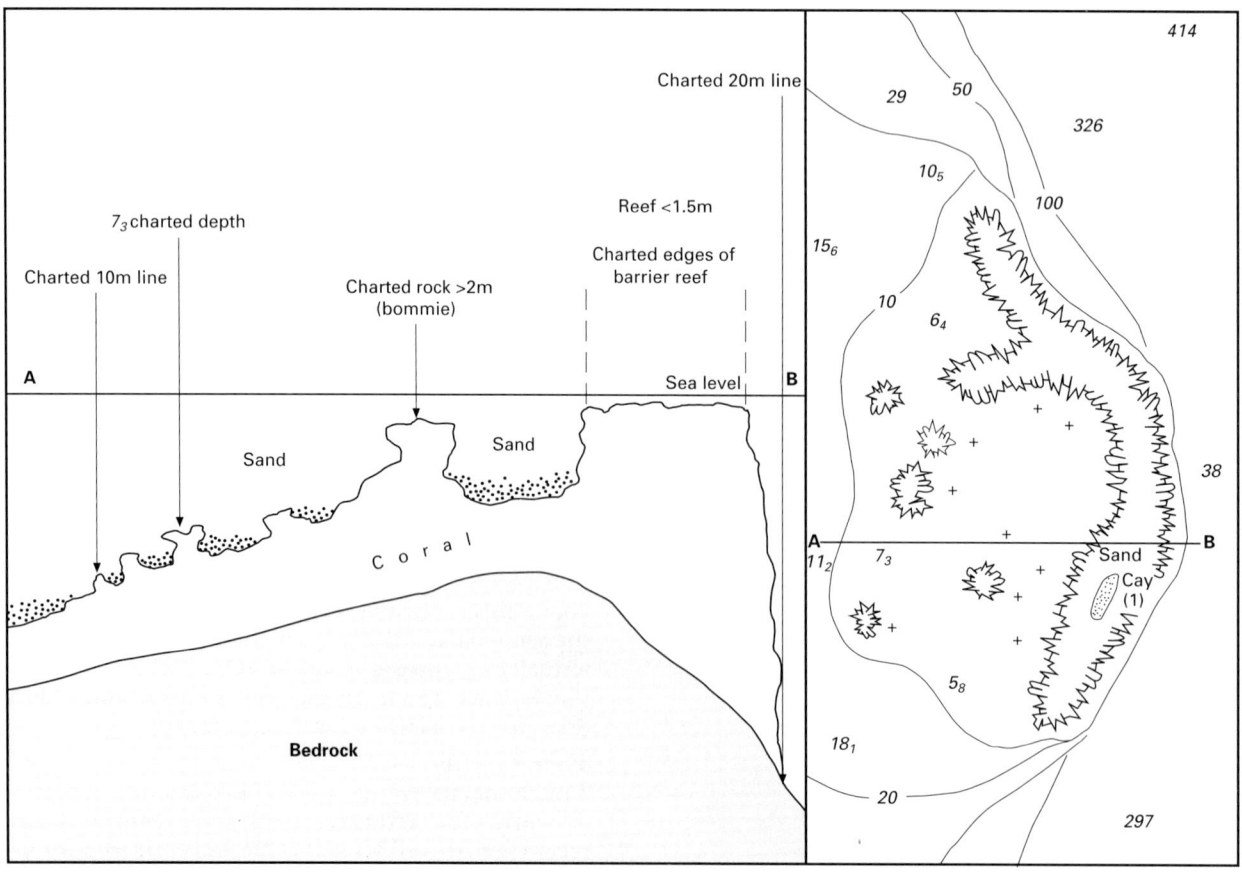

THE BARRIER REEF. IN PROFILE

- Charted 10m line
- 7_3 charted depth
- Charted rock >2m (bommie)
- Reef <1.5m
- Charted edges of barrier reef
- Charted 20m line
- Sand
- Coral
- Bedrock

AS CHARTED

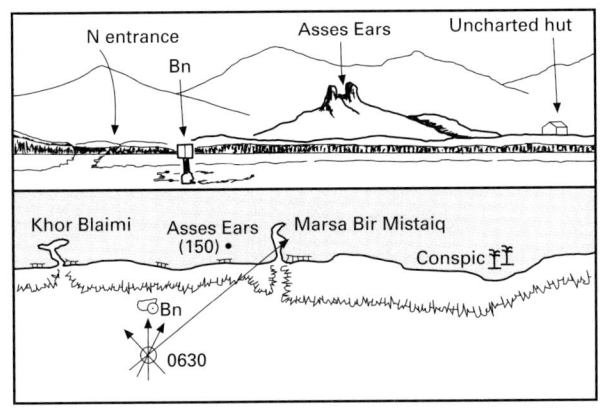

Navigation

1. Take bearing of 3 charted objects and the uncharted object. Plot fix from 3 uncharted objects. From fix draw line of bearing of uncharted object.

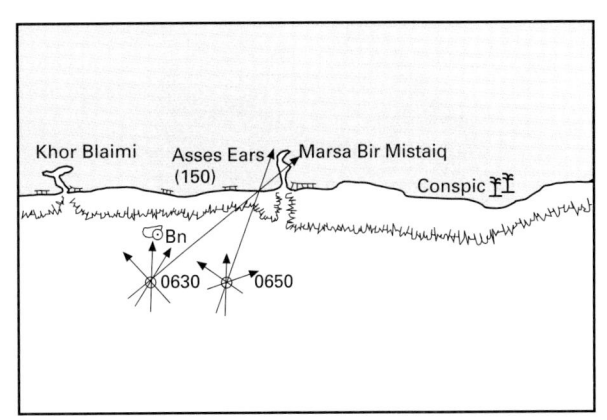

2. Repeat when bearing of uncharted object has changed at least 30°

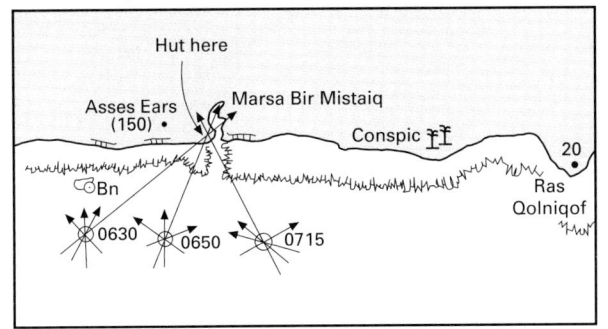

3. Repeat step 2. Position of uncharted object established.

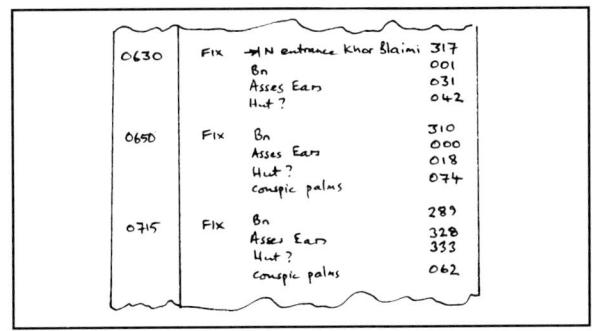

SHOOTING UP AN UNKNOWN OBJECT

Notebook entries

product of satellite photogrammetry, individual ship or harbour reports and block correction. That means most lulus from original surveys remain. Visual fixing keeps you familiar with which hill or reef is which. The coast is low, uniform and lacks conspicuous features. If visibility allows, use the hills near the coast or the coastal mountain ranges some 20M inland. Remember that the atmosphere makes judging heights of hills difficult, which can lead to misidentification. Errors usually involve overestimating the heights of uncharted features 1–2M from the coast and hence mistaking them for higher, charted features further inland obscured by haze. Because of the small scales of the charts you'll be using many of the low hills close to the coast are not charted.

Inventiveness over what to use for fixing is well repaid. In good visibility and with caution you can use reef edges. A fisherman's lean-to can be visible for miles. Use four-point fixing to 'shoot up' conspicuous but uncharted objects, or to verify the positions of objects charted in the wrong place (see diagram). Once you have plotted its position you can use it for further fixing.

Reefs and other hazards

The central passage of the Red Sea is free of dangers N of the Zubayr group. However, the coastal areas are littered with reefs, some of them uncharted. Elsewhere some charted reefs don't exist, and few charted reefs are of the shape and extent shown. The only answer is to keep a sharp lookout. Avoid closing the coastal areas, the Dahlak and Farasan Banks or the Suakin group at night or in poor visibility. If you do, watch your echo sounder like a hawk, and proceed very cautiously once it registers 70m. Radar can help, but not with any kind of sea running. In summer life is easier, since the sea level is then about 0·5–0·7m lower than in winter. That makes reefs close to the surface more visible both to the eye and to radar. You must have some means of getting someone aloft – at least 5m above sea level. See 'Yacht Equipment' for how to rig instant ratlines.

In the Strait of Gubal and in the Gulf of Suez in general there are oil industry hazards and unlit structures. Especially dangerous are those which have been cut off at or close to water level, leaving unmarked stumps – sharp, invisible, and deadly. It is NOT prudent to navigate at night in the Gulf of Suez S of Ras Gharib unless you stick firmly to the charted shipping channel.

Reef sailing

The majority of yachts sailing in the Red Sea have come from coral waters. For those who haven't, the following notes may help.

Coral lives in waters up to 55m deep or possibly deeper in the Red Sea because of the even temperature gradient. So you are unlikely to find coral in deep water, except on off-lying or isolated rock outcrops and pinnacles. Popular belief is that reefs grow both upwards and outwards quickly. It is true that reefs spread and coral waters get shallower, but the process is slower than gossip supposes (see

Red Sea Pilot

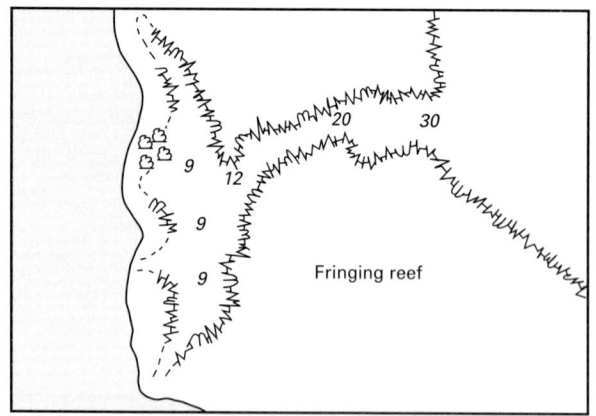

A SHARM – A NARROW DEEP INLET IN FRINGING REEF OR LAND

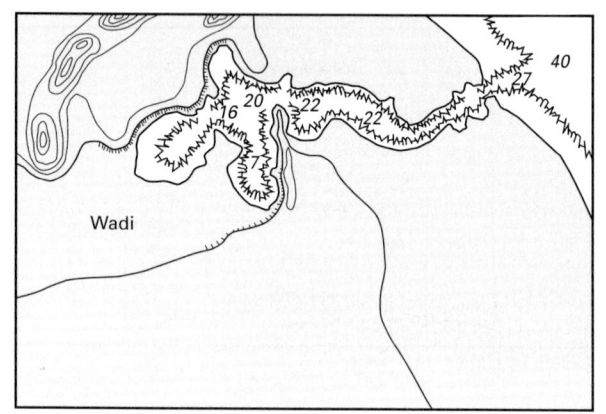

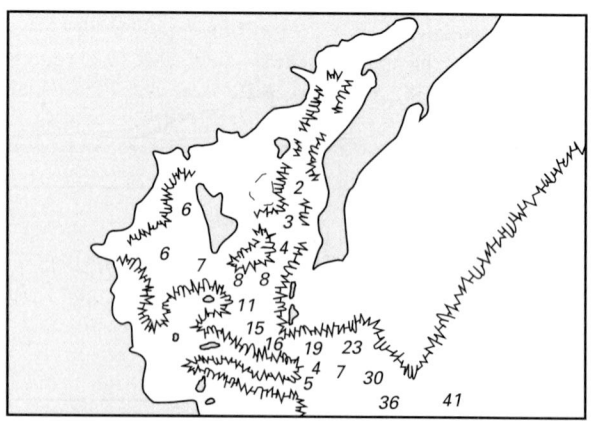

A MARSA – A NATURAL BAY

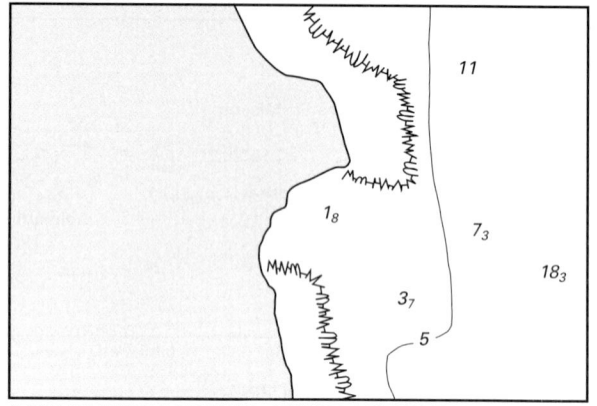

A KHOR (SHARM AT THE END OF A WADI)

A RAS (ALMOST A MARSA!) – BEHIND A HEADLAND

'Environment' for more detail). In our years in tropical waters, we have not yet found any reef to have spread conspicuously beyond what's charted, even in areas where the updating of charts is at best slow and often non-existent. Passages with less than 5m charted depths are an exception. Coral grows slowly in deep passes through reefs. In shallow water, slight growth plus some erosion detritus can make a lot of difference!

The slow growth of coral and its preference for shallower water mean that, unless the bottom is very uneven, if your echo sounder reads >50m you are comparatively safe. Most reefs dangerous to draughts <3m are clearly visible in fair weather. Stirred-up coral sand in their lee gives a whitish, glittery appearance to the water and the seaward side breaks, even if only patchily, but see below for comments on reefs in Eritrea and around Jeddah. The reef patches themselves are a clear greenish brown, often with black isolated coral boulders thrown up onto the main reef towards the seaward edge. In general Red Sea reefs have between 0·5m and 1·5m of water over them in winter. Since the sea level drops by up to 0·7m in summer, parts of the reefs dry in the summer months.

The structure of the offshore reefs is uniform. The seaward side tends to be steep-to, and the drop off to deep water vertiginous. On the land or leeward side, the main platform gives way to a shallow area with isolated coral heads (*bommies*), sometimes numerous, getting fewer and deeper towards deeper water. Sometimes these coral heads coalesce, connect with the main reef, and form a lagoon, which may or may not be accessible, Shab Qumeira, for example. More often they lie within a rough crescent or question mark formed by the main reef (Shab Suadi, Dangerous Reef, Dolphin Reef).

Note that the lee or landward side of a reef is seldom marked by disturbed water. It is easy to stray into this area, and the first warning may be your echo sounder as you go over one of the deeper outlying bommies. It is more alarming when a crew member suddenly says, 'I can see the bottom'. Don't panic. It can be easy to see the bottom when the water is still 20m or deeper. If using the inshore channels, be aware that the ebb tide or the current sets to seaward across the reefs. The closer you are to the inside edge of the reef or to a channel through the reef, the more strongly the effect is felt. Watch the 'cross-track error' read-out on your GPS or your visual plot.

Some of the reefs, either all year round or only in summer, support small sand *cays* (pronounced 'key' as in Key Largo), which shine a brilliant white in the sun. With average Red Sea visibility, from >5m above deck, a large reef with no wreck, island, beacon or *cay* is visible a mile or so away, though the sharp-eyed may pick it up earlier. If the sand *cays* are permanent, vegetation may grow on them (usually low scrub, occasionally trees) and you can hope to see them from about 4M off. Wrecks also give early warning. Never expect to see any sign of a reef at much over 4M. Isolated outcrops, on the other hand, may not show until you're a few hundred metres off.

Land is fringed by reef which may stretch anything from a few metres to a mile or more to seaward. The seaward edge is marked by a surf line, except in unusually calm weather. The edge is steep-to, but off-liers are common, and approaching closer than 100m requires a sharp lookout. Try to keep the echo sounder reading more than 25m. The exception is the W shore of the Gulf of Suez, where it is all right to stay on the 10m line, about 25m from the surf line, most of the way from Marsa Zeitiya to Suez Bay, and certainly as far as Ras Shukeir.

The N two thirds of the Red Sea are free of silt and other impediments to underwater visibility. The result is clear water and readily visible dangers. Experience will teach you what colour of water signals what depth, but roughly:

Deep aquamarine blue – safe, deep water >20m
Deep blue green with darker mottled patches – water >15m over coral
Turquoise green – water >3m over sand
Mottled green – water 3–5m over coral/rock
Greeny-brown to brown – water <3m over mud/weed/coral/rock (though note that in very clear water this colour can indicate coral at depths of 10m or more)

There are five caveats:
- On the coast of S Eritrea and the Dahlak Bank LITTLE OF THE ABOVE APPLIES. The greenish, algae-laden seawater and the peculiar island topography make this area navigationally tricky. Reefs >6m down don't show unless there's a sea running and not always even then. The pale greeny-white that looks like shallow water over sand is often water <16m deep over the sand-covered rock shelf on which many of the islands 'sit' (e.g. Sheik el Abu). The conventional chart symbol at a quick glance looks like the symbol for coral. The subtle difference between the spikiness of the latter and the blunt crenellations of the former is easy to miss. On some electronic charting packages this distinction is not only hard to spot, it doesn't exist.
- On the Saudi Arabian side the reefs are brown/green rather than the paler white/green of the W shore.
- On both shores, patches of weed and algae, especially in summer, can look like reefs. This is a threat to safety if you assume you've identified a reef when in fact you haven't.
- Judging depths and locations of shoal water in reef areas depends upon the quality of light. Overcast weather reduces colour contrast, except close to the boat. Glare in mirror-calm weather makes reef-spotting difficult and reasonable judgements can only be made very close to the boat. In both cases go very slowly. With broken cloud on a sunny day, dark cloud shadows on the water mimic reef patches, making life very trying. Visibility is best on a clear, sunny day, with a slight to moderate breeze ruffling the water. These conditions are quite common during the first part of the average Red Sea winter day.
- The visibility of reefs is affected by the position of the sun. This is important coast-hopping on the W coast. A low sun ahead or on the bow makes distinguishing clear water almost impossible until the last moment. The best conditions are with the sun behind you, >20° and <70° above the horizon.

All this leads to a few rules for getting into and out of anchorages in the Red Sea in fair safety:

1. Get going early if day-sailing.
2. Do not plan to go further than you can beat, motor-sail, or motor in 6–7 hours in standard conditions (30M is about the limit).
3. Aim to arrive at the entrance to your anchorage around midday, and no later than 1400 local time, unless the entry is open and completely free of dangers.
4. On your way in, keep a record of courses, distances and depths (easy with the waypoint-memory facilities of most GPS receivers). You'll need them in order to get away shortly after first light as recommended. (Note: If your GPS has a Tracback facility, be wary. Check it against the memorised track. You'll find that in tortuous channels it often cuts corners.)
5. Have a spotter as high in the rigging as you can manage when in amongst reefs or closing a *marsa* entrance.
6. Use polarised sunglasses.

Shipping

Shipping in the Red Sea is an exaggerated hazard. For anyone used to the English Channel, the Singapore Straits, Hong Kong, or the approaches to New York, even the Gulf of Suez could not be said to have really heavy traffic. In one two-day passage tacking across the Red Sea we saw only five ships, and we have a watch on deck at all times. Closing Bab el Mandeb at night we saw only 6 ships heading N and 4 heading S. That said, the Red Sea is a major commercial artery and a good lookout is vital. There is a cautionary and probably apocryphal tale occasionally related by Suez Canal pilots. A large commercial ship was approaching Suez. The pilot launch came alongside and the pilot made his way to the bridge. When he saw the captain, the pilot said, 'When did you last use your starboard anchor, Captain?' The skipper, thinking this a standard check, reflected a moment and said, 'Bunkering in Aden, I think.' 'We should go and take a look,' replied the pilot. They made their way to the forecastle. The pilot leaned out over the rail above the anchor and motioned the captain to his side. 'See that?' he asked. The captain saw. Tangled in his anchor was the complete mast, rigging and sails of a yacht. Like all fables, as any anthropologist knows, this is a recognised truth dressed up to make a point. Big ships these days are too reliant on automation, under-manned and the crews often under-trained. If the *RMS Queen Mary* didn't notice cutting the cruiser *HMS Curacao* in two, a ship won't notice hitting you. Be warned.

On the Yemeni, Eritrean, Sudanese and Egyptian coasts there is little shipping traffic. The exceptions are the approaches to Aden, Al Hudaydah, Assab, Massawa and Port Sudan. Even here traffic is not heavy. There are often badly-lit small coasters and fishing vessels even in the most remote areas, so don't assume that there is no traffic. Small-craft traffic is fishermen or military. Neither will be aggressive, though the latter will expect you to stop when approached, possibly accept a boarding party, and provide evidence (passports) of your bona fides (see also 'Problems' page 58). It is good manners to offer a courtesy like a drink, cigarettes, etc. . .when in Rome.

On the Saudi Arabian coast traffic in the approaches to Jeddah is quite heavy, especially during the pilgrimage season. Considerable development of the kingdom's ports has been going on over the last few years, notably at Yanbu, Mina al Malik Fahd (King Fahd Port), Sharm Rabegh and Gizan.

Anchorages

Island anchorages These are like coral-island anchorages elsewhere. Bottoms are sand or sand and coral; holding is reasonable to good, with depths of <10m, sometimes more. Shelter is good from winds from one quadrant, but you must move if the wind changes and changes in the Red Sea can be abrupt. Some island anchorages, for example in the Suakin group, require dropping one's hook on the fringing reef and falling back into deep water. Settled weather and good nerves are required! In anchorages in the lee of steep-sided islands such as those in the Hanish group, there are very strong gusts in strong winds, so select position and scope accordingly.

Reef anchorages Also like those elsewhere, bottoms are of sand or sand and coral. Clear-water spots are usually deep – 20m or so. If shallow (<5m), holding can be excellent in sand or fair in sand and coral. THIS IS NOT TRUE OF THE DAHLAK BANK where the sand is often a thin layer over rock and you need masses of chain. In reasonable depths (<10m) swinging room is often restricted. Shelter is usually all round and good, but only from the sea. One feels the full force of the wind, and a reef anchorage in a really strong blow calls for strong nerves!

Marsas/Khors/Sharms In his comprehensive French *Guide pratique* Felix Normen sketches the differences between these. Sharms/sherms are narrow, deep, often winding and push several kilometres inland or into the fringing reef. Normen likens them to *rias* in Spain, Portugal, SW Britain, etc. *Marsas/mersas* are natural bays, sometimes with creeks leading off them, formed by a headland (*ras* in Arabic) or projection of the fringing reef. Normen adds that the term *khor* can be substituted for either *sharm* or *marsa* if these constitute the 'estuary' of a wadi (the valley of an occasional watercourse). It follows that these terms are not always properly used on western charts. For example, most of the Sudanese *marsas* are actually *sharms* and *khor*-like *sharms* at that! (See diagram, page 44.)

Whatever their name, these starkly beautiful anchorages are amongst the best the authors have found anywhere, sheltered from sea and swell, but because the land is usually low, not protected from the wind. They are deep (15–20m), shoaling towards their heads (8–15m), with good holding in mud or sand and mud. Shallower anchorage (<6m) can sometimes be found in sand or sand and coral in the occasional small bay. If the wind is steady, you can close the shore and 'put the bow on the sand', drop the anchor and then fall back into deeper water. Use a stern anchor if necessary to avoid bumping a nearby reef. In strong winds allow for fetch in the larger *marsas* and always be prepared to shift berth.

Harbours Most of the harbours are in *marsas* which have been developed. You may find yourself lying alongside, anchored off, or Mediterranean-moored with an anchor down and stern lines ashore. Holding is usually excellent, and shelter good. The exceptions are Al Mukha and, to a lesser extent, Quseir.

Ras anchorages Most common in the Gulf of Suez, though there are examples in the S Red Sea. These are formally vestigial *marsas*. They offer little protection from the wind, some protection from the sea, but not much from swell. Holding fair to good in sand, mud, or sand and coral.

Aids to navigation

Lights Do not rely on lights in Eritrea or in Sudan except in the approaches to Port Sudan. The authorities do their best with the funds they have, but face an uphill job. In Yemen and Djibouti the lights in the ports usually work. Major lights on the Red Sea through route, in the Gulf of Suez and in the N approaches to the Suez Canal are reliable, though not always as powerful as listed. Minor lights are less trustworthy. Now and then lights appear where none is charted or listed. Be prepared for different characteristics to those charted. Saudi Arabian, Israeli and Cypriot lights are reliable and as listed, bar the obvious chance of recent changes.

Beacons and other conspicuous objects Some charted beacons may be missing, may not have the topmark charted, have a different one, or have no topmark at all. Sometimes most of a beacon is missing, but the stump remains. For the most part charted beacons or visible remains of them exist and are in good order. They can be less conspicuous than you might expect and often have no remaining coloured paint. On the older charts 'cairns' and 'astronomical pillars' are sometimes marked on hilltops and headlands. These seem mostly to have disappeared or are inconspicuous. Be very wary of cairns, trees or piles of boulders marked on charts as indicating entrances to *marsas*, even when they are qualified as 'conspic'. They are frequently non-existent, or visible only with the aid of a magnifying glass at close range.

GPS This is a wonderful tool which has made navigational life in the Red Sea much easier. You should nonetheless remember to allow for the scale of Red Sea charting and don't work to levels of accuracy the chart cannot match. Assuming you have a very sharp pencil and can manage perfect rectilinearity in plotting, on a 1:500,000 scale chart your 'fix', where two 0·3mm lines cross at right angles, represents a square the size of a modest marina. That means on most of the charts you'll be using you cannot accurately plot positions to better than one decimal place. The math as to why is simple. One minute of latitude is one nautical mile or 1852m. It follows that 0'·1 i.e. one decimal place, is a tenth of that distance, 185·2m. Two decimal places are not much more than the length of your boat, three the length of a settee berth and four about the length of your hand. Even where you are dealing in larger scales remember, according to the British Admiralty, no chart no matter how modern, is made with accuracy better than ±13m. Call it two decimal places.

If your GPS allows a choice of datums, make sure yours is set to the same as that of the chart you are using. With the most recent charts or electronic charts this will be WGS84. IF YOU ARE USING OLD CHARTS NOTE TWO THINGS:

- if they are British, the correction factors for WGS84 on the Yemeni coast and Red Sea west coast are unknown and differences of up to 5M between GPS and charted positions are possible.
- if they are US/French/German charts, the block corrections for GPS position plotting incorporate all the old survey errors, some up to 2M.

In the body of this pilot, unless otherwise indicated, we have given WGS84 GPS positions where these are known.

Hyperbolic radio-navigation systems

Loran C There is now virtually complete coverage, using the Saudi Arabian N chain 8830 and S chain 7030. There is also ground wave coverage in the Gulf of Aden from roughly Salalah W to Ras al Arah. Coverage is sky wave only from about Ras al Arah to the latitude of Al Mukha including all the Gulf of Tadjoura and Bab el Mandeb.

Echo sounders and electronic logs The Red Sea's higher salinity and very even temperature distribution throughout its depth affect the speed of sound in sea water, which is what electronic echo sounders and Doppler logs use as the basis for their measurements. Such instruments are usually calibrated for normal salinity values. Depths recorded on echo sounders calibrated to the normal value for the speed of sound in water (4,800ft/sec or 1,463m/sec) should be increased by 5% to agree with charted depths. That's a quarter of a metre in every five metres! This applies mostly N of Ras Baniyas, but will be noticeable from about the Suakin group on. Don't worry though, since water appears shallower and your log thinks you're going faster, the errors are on the safe side. Any small discrepancies in charted depths or speeds through the water you note are probably this Red Sea Effect.

Don't rely on your echo-sounder to warn you of dangers when closing reefs. You are generally safe in depths of 70m or more, fairly safe between 50m and 25m, and dicing with danger in under 25m.

Radar, radar detectors and reflectors Many Red Sea islands, for example the Hanish group, are excellent radar targets. Don't rely on radar among reefs. Reefs which break may be picked up at short range in calm or light weather, or they may not. In brisker weather surface clutter makes radar unreliable. Beacons may give a reasonable echo at close range if there is not too much of a sea running. Some wrecks are excellent radar targets.

The raised coral platform constituting most of the Sudanese and Egyptian coast usually gives a good echo, though coral rock is not always a reliable reflector. The cliffs of the Saudi Arabian coast give good echoes. Low, sandy islets cannot be relied upon. Be doubly wary if proceeding by radar in the Suakin group. The entrances to *marsas* on the Sudan coast are apparently easy to spot on radar no matter what your direction of approach.

Radar reflectors and detectors are useful offshore in poor visibility near major traffic concentrations, but neither is to be relied on. Whilst most traffic will probably have radar on in such conditions, not all will. In clearer conditions 75% or more of commercial shipping may operate without radar, except at choke points like Bab el Mandeb, the Strait of Gubal, and the Gulf of Suez, or when closing the coast.

There are useful racons if you have radar aboard. The Gulf of Aden is worst endowed with only Mina Salalah. The approaches to the main Saudi Arabian ports are well served. Egypt's Red Sea coast has nothing until the approaches to Mina Safaga except racons on Abu el Kizan and El Akhawein. However, from there N there are Racons on many hazards and headlands.

Astronavigation
You can navigate up the Red Sea using only traditional means. The doyens of the cruising world – people like W A Robinson and the Hiscocks – did it. If you try it, what follows may help.

When astronavigating be cautious. The Red Sea's atmosphere plays tricks. A good horizon is difficult to get. Worse, when you think you have one it may be a trick of the light. Errors of 20' longitude and 10' latitude are possible. Three tips:

- Use star sights at dawn and dusk (when refraction is least) rather than sun sights.

- Treat results with caution and use generous circles of probable error when plotting them.
- When the state of the sea allows, take sights from as low down as possible to reduce atmospheric oddities. However, because the horizon will be only about a mile away, beware of other errors, especially if there is any swell.

RDF This is an obsolete navigational tool and yacht RDF sets almost collectors' pieces. Most previously listed stations have been deleted from lists of radio signals, though many, which are aerobeacons, are still usable if you can discover the details.

Yacht and equipment

The same applies in the Red Sea as in blue-water sailing anywhere. Self-reliance is the name of the game. Where there are differences, they have to do with four things: the need to work to weather, the Red Sea's salinity and dustiness, the absence of repair facilities and difficulty in finding spares, and, especially when N'bound, the need to pass through the Suez Canal.

Hull, rig and sails

Above all evaluate the windward performance of your boat. Even sailing N to S you may face SE winds over the last few hundred miles. If your boat does not make better than 55° to the true wind at 4kt or more, or a VMG of at least 2kts, with a force 4–5 headwind in a short, steep sea, you are going to have to:

1. Coast-hop.
2. Do a lot of motoring (see 'Engine' below, and 'Routes' above).

If your boat is both close-winded and quick we recommend slowing down a bit and not sailing as close to the wind as you can. That way you'll get stopped less, suffer less damage, get less salt everywhere, and probably make better VMG as well.

Hull

You've got the hull you've got! Even if your boat isn't an upwind greyhound, the following may help make life easier:

1. Keep the hull clean.
2. Overhaul your steering gear and check it regularly.

Running gear

This includes stoppers/jammers, sheaves, snap shackles and winches, as well as running rigging.

Red Sea spray and sea air are very saline and corrosive. Salt build-up is considerable. Regularly wipe-down and huck-out small corners with bad drainage. Pay special attention to gear you seldom use, which, unserviced, will jam solid. You'll inevitably spend some time hard on the wind in moderate to rough seas and your winches will get a larger than usual dose of sea water. Give them a service before you get to the Red Sea, and again about halfway up.

Salt on deck stays moist and airborne dust and grit rapidly build it to a hard-to-shift, crusty, cake-like deposit. The windward side of all running rigging outside the mast or on deck, turns reddish brown and demands hard scrubbing and washing. The windward side of standing rigging accumulates dust which transfers to sailcloth or cordage against it. Dust gets into stoppers/jammers, sheaves, shackles and other deck gear. Salt and dust build up in the threads of rigging screws and other threaded deck gear. Regularly clean winches and running gear, wipe down rigging, stanchions, guard rails, etc. to mitigate the worst effects.

An old military tip may also help. Desert warriors have found that much or even any lubricant on running gear is a major mistake. Once oil and grease pick up dust you've got a good grinding paste, so in desert warfare a weapon's working parts are kept dry, not oiled, to minimise unwelcome stoppages in battle. Most running gear, like a soldier's rifle, mostly sits idle. You'll do little damage leaving sheaves, stoppers and winches un-lubricated during your time in the Red Sea. Just concentrate on keeping running surfaces clean to minimise the build-up of corrosive deposits.

A second tip, if salt and dust build-up is severe, wash down with sea water then dry with a cloth. Don't resort to this method on running rigging, or anything you can't dry with a cloth.

Rig

Most yachts entering the Red Sea do so after one or two years of mostly downhill, trade-wind sailing. Not much thought is usually given to the stresses and strains of windward work. Look over your rig carefully with an eye to its ability to handle 15–30kts of wind, hard on the wind, occasionally being stopped dead by a sea. Before entering the Red Sea, and at regular intervals on passage:

1. Inspect all shrouds, stays, swages and rigging screws. Check spreader angle. If you have a double-spreader rig, have a good look at your upper-spreader angle. A poorly angled upper spreader can easily lead to a broken mast.
2. Overhaul and inspect halyards, splices, and (snap) shackles if you use them. You'll be using more halyard tension than when sailing downwind.
3. If fitted, give your roller-furler gear, especially the upper swivel, a thorough check. If you want to be super-prudent and haven't had a look for a while, check the condition of your forestay and swages.
4. Make sure your reef lines are rove and running freely.
5. If you have a separate trysail track and halyard and haven't used them recently, check they are in good order, especially the track rivets/screws and, if you have a gate, the swivel/hinge.
6. If you don't have a rod-type kicker (boom vang) or a fixed-boom gallows (and anyway!), check

Yacht and equipment

your main boom topping lift where it passes over the masthead sheave.

7. Check your chainplates and rigging-screw attachments – the former for fatigue and through-deck leaks, the latter for fatigue and wear.
8. Take up your rigging tension a turn or so. When properly canvassed, hard on the wind, your lee rigging should not be in full tension, but should never be slack.

If you plan coast-hopping in the Red Sea, you must be able to get a spotter into the rigging when navigating in coral. If you have no mast steps try our easy-to-rig-and-unrig ratlines, based on the climber's prusik knot, using your lowers. You need to wear shoes to climb them!

Sails

Because so many yachts have spent one or two years downwind sailing before arriving at the Red Sea, sails are often not at their best. 10,000–20,000 miles in tropical sunlight take their toll. Before you set off into the Red Sea, for example in Phuket where there's access to a good sail loft:

1. Look over all of the stitching of your main working sails, especially:
   a. mainsail seams just above reefing points, and
   b. headsail seams along the leach
   c. the stitching of any u/v strips
2. Check cringles at tack, head, and clew for wear.
3. Check all the reef tacks and clews, and all reefing cringles/eyelets and pendants.
4. Check your leach line, especially near upper-reef-clew jamming-points.

Finally, check over your sailmaker's kit (see also 'Spares' below) so you can make running repairs.

Ground tackle

Anchors

The dominant bottoms in the Red Sea are mud, sand, sand and coral, and coral. In general the anchorages are on the deep side, often over 10m.

In the *marsas* and *sharms* the holding is usually excellent in sand, or sand and mud, and a good burying hook (CQR/Plough, Delta, Bruce, FOB/Brittany or Danforth) works well. There is little fetch, and because the surrounding land is flat the wind tends to be fairly constant. It follows that snatch-loading is not a problem, so that the deepness of the anchorages is less of a worry when riding to a shorter than ideal scope.

In a few of the *marsas/sharms*, in reef anchorages, behind some of the headlands (e.g. Ras Baniyas) and in some of the island anchorages the holding is in coral, rock and coral or sand and coral, and the standard burying hooks can have problems setting. A fisherman anchor is useful in such circumstances though carrying one of bower weight isn't really worth the hassle. If you feel the need, try using a lighter fisherman in tandem with your standard bower (see diagram), between them they should hold.

The anchorages are frequently confined and there is little room to leeward if you start to drag – often as little as a boat length or two. In the stronger blows, lying to two anchors gives peace of mind. If you anchor in shallow water on the shores of a *marsa* or *sharm*, 'putting the bow on the sand' as it's known, you may need to lay a light kedge from astern to hold you off.

In reef anchorages, buoying your anchors makes sense since both rodes and anchors readily get fouled on coral heads. Since the anchorages are often deep, diving to free things requires full equipment. However, do remember that reefs are fragile and you should avoid anchoring on the coral. See 'Diving', on page 56, for reef moorings in Egyptian waters.

There is little need for storm anchors in the Red Sea.

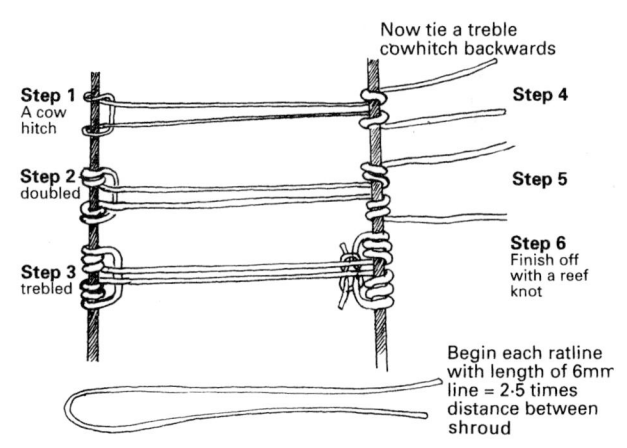

Prusik knot ratlines – 1 ratline only shown

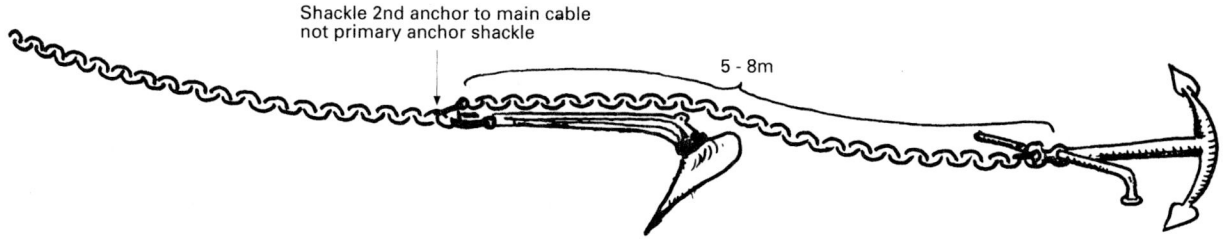

Step 1 – A cow hitch
Step 2 – doubled
Step 3 – trebled
Now tie a treble cowhitch backwards
Step 4
Step 5
Step 6 – Finish off with a reef knot

Begin each ratline with length of 6mm line = 2·5 times distance between shroud

Tandem anchoring - very secure on doubtful bottoms in strong winds

Shackle 2nd anchor to main cable not primary anchor shackle

5 - 8m

Rodes and scope

The general preference is for all-chain rodes. However, the anchorages are often deep and winds may gust 45kts so you must be able to veer adequate scope. In some of the *marsas/sharms* with depths of >15m, if you have not found a shallow patch with good holding, this means 100–150m of chain, or chain and rope. If you don't carry that much chain, one ruse, since the holding in the deep *marsas* is excellent in mud, is to use a good burying hook, 10–15m of chain and a long rope rode.

Engine

You need your engine to keep you going for the Red Sea's 1200M and then, for those N'bound, do two days' non-stop work in the canal. Sailing purist or motor sailing coast-hopper, your engine's health is critical for the Suez Canal. A tow through costs up to US$2000. Service and spares for small (<100HP) marine auxiliaries are as common as hens' teeth. If you break down you have to get to Massawa, Jeddah, Port Sudan, or the Gubal Straits area. Then wait for spares to arrive. So...

1. Have aboard:
    a. gaskets to give you two full sets, and/or gasket paper and a sheet of heavy gasket material plus gasket cement
    b. two more of each drive belt than you usually carry
    c. at least 4–5 impellers for your seawater-cooling pump
    d. if your engine is freshwater cooled with a heat exchanger, especially if it is long in the tooth, a spare freshwater-circulating pump and heat-exchanger element
    e. spare clutch plates and damper plate if yours are long in the tooth
    f. if you have a saildrive, spare membrane and possibly a spare drive leg
    g. enough lube oil and gearbox oil (especially if your gearbox requires automatic-transmission fluid) for two more changes than you think you'll need
    h. plenty of oil and fuel-filter elements
    i. a spare starter motor, and, if you have a pre-engaged starter, a spare solenoid (unless your engine is small enough to hand-crank)
    j. extra jerry cans for transporting fuel and above all for stowing spare fuel on deck
    k. a spare alternator or alternative means of generating enough power to keep your batteries charged up in normal use
2. Check your engine alignment (most gearbox and engine-mounting problems result from poor alignment). Check it again at about the halfway point, and in Suez or Port Said before the canal transit.
3. If you've ever had doubts about your propeller pitch, sort it out.

If you have problems you'll find your fellow cruisers (a.k.a. the Red Sea Mutual Aid Society) wonderful. Experts in most things are on the nets and usually bend over backwards to help.

Electrics

Unaccustomed windward work is going to find deck leaks you didn't know about, or you've forgotten you had. These may drip water onto wiring and junctions and cause problems. The movement of the mast as you thump into a head sea may find out weaknesses in wiring runs up the mast. That goes for wiring vulnerable to chafe from your hull, liner (if you have one), and joinery, which will work more than usual. Carry a big range of electrical spares, especially fuses, crimp connectors, heatshrink tubing and marine grade wire in relevant gauges.

Spares

What applies to engines applies to other spares. Most blue-water cruisers are well equipped, but look over your:

1. Spares for repairing rigging failure or damage.
2. Sail-repair materials, including lots of self-adhesive cloth and tape.

A good idea is to check your spares inventory against the excellent list in Appendix III of Nigel Calder's indispensable *Boatowner's Mechanical and Electrical Manual*.

Ancillaries

If your yacht has refrigeration, the same applies to it as to engines. Carry lots of refrigerant in case of leaks. It is not that easily come by; most boat systems use R-12, and are damaged if you substitute R-22, the common car air-conditioner refrigerant which is about all you'd be likely to find somewhere like Port Sudan.

Water makers are a boon. Look after yours and if the water looks doubtful, as it will in every port, on the Dahlak Bank, and in some of the shallower *marsas*, don't use it. But at the same time, practise water-economy. If the water maker goes toes-up, you'll often be several days from the nearest tap or well.

Generators need the same sort of care as engines (see above).

Tenders

Make sure you have the same range of spares and maintenance supplies for your outboard as for your main engine. If your tender is a RIB/inflatable, check your repair kit, especially the glue which, if it's been untouched for years, may be useless. You can't buy the stuff once you're in the Red Sea. Carry a spare pump!

Yacht and equipment

Radios and services

You can sail the Red Sea without radio. People have done it and do it still. But life is easier and maybe safer if you are radio equipped. Mechanics, electricians, spares, sailmakers and riggers are few and far between. A radio keeps you in touch with the Red Sea's mutual aid society.

Being part of a radio schedule (sked) can be fun, useful, and if things go wrong, a godsend. But beware crowd psychology. The Red Sea is best tackled by taking your time. The daily sked can hook you on the insidious Red Sea Rush syndrome and have you moving faster than is prudent.

The sked is also good for learning of places to seek out, places to avoid and pilotage information. But don't treat what you hear as something on Navtex. It isn't. What's the news source? Someone you know, trust and whose views you share?

Marine band SSB (MF/HF)

Most of the authorities in and around the Red Sea like, and sometimes require, yachts to contact them by radio to give notice of their arrival. Marine band SSB transmitting on 2182kHz will nearly always get through when VHF won't, e.g. when approaching Aden from E'wards. Second, not all yachts have a qualified radio amateur (ham) aboard, so 'skeds' (usually in the 4MHz and 6MHz bands) tend to be on marine bands.

If you haven't joined a net by the time you arrive in Salalah, Djibouti or Aden ask other boats what nets are running. There will usually be one main sked and any number of little private ones keeping four or five boats in touch with each other. The main sked will have a 'controller', some kindly soul who volunteers to call the roll each day. Come up at the end of the day's roll call and ask to be added to the list.

MF/HF propagation in the Red Sea is normally good. Range on 6MHz is 600–800M. If you're having trouble, give a thought to your aerial orientation, particularly if you have a long wire (backstay) antenna. Lying to a N wind, you radiate wonderfully to people S of you; straight through the mast and rigging towards the sea to people N. Obviously the quality of your ground is important, and if you haven't checked connections recently, give them a going over.

Propagation and reception can be locally appalling. There are two main culprits. Cloud, which tends to have a high dust content. Or electrical storms. There's nothing you can do.

Inmarsat

Those with Inmarsat will need no advice from us. All we will say is that coming down the Red Sea in company with friends who had Inmarsat C was a revelation. Two things stood out. One good, one bad. The good one was the accessibility of weather information by concise emails from wired friends and relations. The bad one was the GMDSS panic button. With little experience of the developing world, our friends were nervous of other cultures and their ways. Every local boat was a potential pirate. When group-fed panic set in and the button was pressed naturally no help was forthcoming (see GMDSS below), but Elaine's octogenarian mother was roused out by Falmouth MRCC and asked whether she knew with what boats we were sailing in company since *Fiddler's Green II's* name had been mentioned! Fortunately we'd just rung her from Mukalla, so she wasn't unnecessarily alarmed.

Amateur (ham) radio

Many yachts choose amateur (ham) radios instead of marine single sideband. Amateur nets, which exist worldwide, are splendid means for acquiring information, keeping in touch with far flung friends, and (see section on weather) for getting weather forecasts. Radio hams are enthusiasts. That means that someone, somewhere, is usually on air. This cannot be said for marine SSB services, whatever the Admiralty *List of Radio Signals* may say, not least because marine band radio is dying thanks to GMDSS, satcoms and mobile phones. Many yachtsmen think that if you're in trouble, you're more likely to get the word out on ham radio than marine SSB.

In any year a good proportion of yachts in the Red Sea are licensed hams, and there is usually at least one ham sked in operation. If you are yourself a qualified amateur you will have easy access to details of all official networks. You will also be able to join any net whilst you are in its area. Radio 'hams' are very friendly people who make newcomers welcome.

A cautionary tale

To keep this pilot up to date we've gratefully received information from fellow cruisers. But we always double or treble check the data. Too much of what we read or hear turns out to be badly misleading and sometimes plain wrong. Much more than it should be in an age of GPS. For example several boats have reported uncharted hazards. But few of our correspondents have re-worked the plot from their logs to find out exactly where they were when they had their hearts in their mouths. In almost all cases we've analysed the reported hazard has been charted. Careless navigation has been followed by face-saving. The worst case is Marsa Thelemet. We received several reports over a number of years telling us the diagram in the first edition was badly misleading because there had been major changes. We made corrections in supplements. As a result the British Admiralty altered the relevant BA chart. In 2000 we re-surveyed. Nothing significant had in fact changed. Our informants were either very poor observers or weren't where they thought they were.

So remember, few yachties will ever tell you they hit or nearly hit the bricks because they've been careless or inadequately prudent. The fault always lies in the charting, the weather or the lap of the gods. Treat sked generated information cautiously.

Red Sea Pilot

VHF

The main reason for having a VHF for a Red Sea passage is that either an MF/HF set, or a VHF is, in theory anyway, obligatory for the passage of the Suez Canal. VHF is the best means of calling port authorities to inform them of your arrival or departure (often a requirement, as in Aden). It is very good for calling up passing ships if, for example, you're worried that they haven't seen you. (Though be prepared, merchant ship radio operators are often bored and can keep you chatting for ages.) Finally, VHF is the best way of keeping in touch with boats with which you are sailing in company.

VHF propagation in the Red Sea is usually excellent, occasionally astounding. Effective ranges of over 50 miles are quite common and, now and then, contact can be made out to 100 miles. However, with the advent of GMDSS, mobile phones and Satcoms, there is now much less traffic than there used to be.

Other receivers

If you have no MF/HF set, marine band or ham, give some thought to acquiring an all-band receiver which can listen on upper and lower sideband. For good reception it will need a long wire antenna. With this you can listen in on most MF/HF skeds, and monitor weather forecasts.

Mobile phones

All the countries in this guide bar Somalia have GSM mobile phone systems. That said, they are not always usable. Whether they are will depend on agreements between your service and the local operator. Where your phone will work depends also on the reach of the local service and on local political conditions. Eritrea's system is only just starting and is unlikely to extend far beyond the capital and main ports for a while. The Port Sudan GSM network is accessible up to 5 or 6M N and S, but not in Suakin, Marsa Ata, or Marsa Fijab. Patchy GSM coverage in Egypt extends from Marsa Alam to Safaga. Coverage is good from Safaga to Port Said and all along the Mediterranean coast. There is also good coverage on the Sinai Peninsula. In late 2001 the Yemeni system was switched off by the government because armed anti-government activists were using it to co-ordinate their attacks!

Distress, GMDSS and EPIRBs

DO NOT RELY ON INMARSAT, EPIRB, VHF DSC, MF/HF DSC, STANDARD VHF OR SSB DISTRESS ALERTING.

Thanks to GMDSS ships are no longer required to keep watch on 2182MHz. They are still required to keep watch on VHF Ch 16, but that may cease to be the case on 1st Feb 2005. Because of GMDSS most ships no longer see themselves as primary respondents to a distress alert. That's an MRCC and MRSC job. The result is that most of them ignore steam radio distress calls.

This is particularly true if you are victim of armed robbery at sea. Ships may ignore you even if they are only half a mile away. An acquaintance, one of the rare and unlucky victims of an attack by an armed robber in the Gulf of Aden, had been talking to a ship on VHF five minutes before he was set upon. Although still only half a mile away the ship refused to acknowledge our acquaintance's distress calls. Another friend nearly sank in the Gubal Strait a year or so back. His flares and maydays on VHF and SSB were ignored by shipping and helicopters in sight. Luckily, and eventually, he was rescued thanks to Philip Jones of Abu Tug Marina.

Whatever the sea's code of honour, the code of practice of the bean counters in head office rules. Ships don't stop.

Satellite alerting is not much better. Whether on an EPIRB or via Inmarsat, it wakes the chaps up in Europe or wherever. In cases of armed robbery they ring your nearest and dearest and may alert the consul in the nearest consulate. In our one experience of such a GMDSS alert nothing else happens. We don't know what would be the case if you were sinking, but don't hope for much.

The Red Sea S of Foul Bay is a GMDSS black hole. The relevant diagram in the Admiralty *List of Radio Signals* shows that you are on your own. Sea Area A1 services (VHF DSC out to 35M) are operational only for Saudi Arabia and Jordan, with planned services trialling in Quseir (remotely controlling Safaga, Hurghada, Ras Gharib and Dahab), Alexandria and Port Said (remotely controlling stations on Egypt's Mediterranean coast). For Sea Area A2 services (MF DSC) there is virtually nothing. Egypt's Alexandria and Quseir stations are only on trial and otherwise that's it. Area A3 (HF DSC) is a bit better, with an operating service in Alexandria and one planned for Muscat in Oman.

EPIRBs using 121·5MHz, 121·5/243MHz and 406MHz are all detectable by COSPAS-SARSAT satellites, though remember the SAR service on 121·5MHz is to be phased out altogether on 1st Feb 2009. Remember too that 243MHz is a NATO aircraft distress frequency and not properly part of GMDSS.

In principle Jeddah is a GMDSS MRCC (Maritime Rescue Co-ordination Centre) and Port Sudan an MRSC (Maritime Rescue Co-ordination Sub-centre). Port Authorities throughout are tasked with SAR duties, but, with the possible exception of Saudi Arabia, Jordan and Israel, they have pathetic communications facilities and no SAR resources (boats, helicopters, etc.) to deploy. So in practice they can't help unless there's a nearby ship which is contactable, can be directed to help and will do what it's told. So, although an MRCC may know of your squawk for help, only in the N Red Sea is official SAR help possible, though even then not probable. In the S Red Sea and Gulf of Aden there are no SAR services. In principle the gap is filled by shipping. However, the false alarm rate with GMDSS is still so great that alerts are frequently

Yacht and equipment

disregarded and in some ships the audible alarm silenced.

GMDSS is a developed world, desk-bound bureaucrats' conceit that left out of consideration the 99% of mariners who go to sea in non-SOLAS small craft in developing world waters. Don't rely on it.

Radio services

In the list of useful radio frequencies which follow, the first will be port authorities, the second coast radio stations, the last marinas. In each case the order of entries follows that of the main body of the text.

Port Authority frequencies

Many yachts ignore port authorities, and any requirement to give notice of arrival. Nothing much usually seems to follow from this although, in Massawa for example, you may be inconvenienced. Letting the authorities know you're on your way is good manners and will ease your stay. Be patient, many operators are doing their best with indifferent equipment and a foreign language.

Southern approaches to Red Sea

Oman
Mina Salalah Call *Port Salalah Control* 2182, 4073·7, 8258·4kHz, Ch 16, 6, 10, 12. Call when 3–6hrs out with ETA, LOA, draught, reg. tonnage. Call again when close to entering.

Yemen
Nishtun Call *Nishtun Port Control* Ch 16. Operates only 0700–1400LMT.
Al Mukalla Call *Al Mukalla Port Control* 2182kHz, Ch 16, 8, 13.
Call as soon as within range.
Aden Call *Aden Port Control.* Ch 16, 06, 08, 13. Call when 3–6hrs out. Again when approaching harbour. Give vessel name, name of skipper, port of registry, reg. tonnage, last port of call, next port of call.

Djibouti
Djibouti Call *Comport Djibouti* Ch 12, 16. Listen on approach.

Red Sea

West coast
Eritrea
Assab Call *Port Assab Port* Ch 16, 12, 14. Call when 2–3hrs out. Give ETA and any other information requested.
Massawa Call *Massawa Port* Ch 16, 12, 14. Call when 2–3hrs out. Give ETA and any other information requested. The radio is manned only 0700–1800LMT.

East coast
Yemen
Al Mukha Call *Al Mukha Harbourmaster* Ch 16, 14. Call on approach. (Does not maintain a listening watch.)
Al Hudaydah. Call *Al Hudaydah Port* 4WD. Ch 16, 12, 14. Call when approaching Ras Katib.
As Salif Call *As Salif Port* Ch 14, 16.

Saudi Arabian coast
Saudi Arabia
Jizan Call *Jizan Port Control* Ch 16, 09, 11, 71, 72, 73
Jeddah Call *Jeddah Port Control* Ch 16, 9, 12, 14. Call when in VHF range, again when entering. On first call give ETA, name of vessel, flag, tonnage, name of skipper, and ask for berthing instructions.
Yanbu Call *Yanbu Port Control* 2182, 2025kHz, Ch 16, 9, 12, 14. Call when 2hrs out.

Sudan coast
Sudan
Suakin Call *Suakin Port* Ch 16.
Port Sudan Call *Port Sudan Control* Ch 16, 14, 19. Call on approach, give ETA and any other information requested. Listen on Ch 14 from 2M out.

Egyptian coast
Egypt
Mina Safaga Call *Port Safaga.* Ch 16. Call on approach. Operates 0500–0600LMT.
Sharm el Sheikh Call *Harbourmaster Sharm el Sheikh.* 8207kHz, Ch 16.

Gulf of Suez and Gulf of Aqaba
Strait of Tiran Call *VTS Gulf of Aqaba.* Ch 16, 08, 09, 10 and/or call *Salam* (the radar station at Nabq) 2182kHz Ch 13, 16. Both like to be called by vessels approaching the strait. Will give any recent navigational information. May not answer!

Israel
NOTE: In theory, when 50M out you are supposed to call/telex the Israeli Ministry of Transport (IMOT) via Haifa Radio (call sign *4XO*) with vessel name, call sign, flag, port of registry, IMO number, MMSI, year of build, dead-weight, ship type, number of crew, agent's name, last and previous ports of call, destination, position, course, speed and ETA. This does not seem to be enforced for yachts, but might be diplomatic to try. In any case, when 50M out, you MUST call the navy. Call *Israeli Navy.* Ch 16. Give name of vessel, flag, port of registry, no. of crew, colour of hull, date/time/position of report, course and speed, last port of call, port of destination in Israel, ETA. Maintain a listening watch on Ch 16 and call again when 25M from Israeli coast.
Elat Call *Yamit Elat.* Ch 16, 14, 12, 13. This is the port authority, call if no reply from Elat Marina to let the authority know you are around.

Jordan
Aqaba Call *Aqaba Port Control* Ch 16, 12. Call on 12 when 6 or so hours out with vessel name, flag, last port of call, next port of call, LOA, reg tonnage, skipper's name, ETA, and any other information requested.

Suez Canal
Theoretically you should contact the port authorities. No one expects you to do this or requires it. Keeping out of everyone's way, head for the Suez Yacht Club, or the Port Fouad Yacht Centre and take things from there via an agent. Note that the Prince of the Red Sea likes you to call him on your approach to Suez on Ch 16. See Pilotage section.

On approach to Port Tewfiq or Port Said keep a listening watch on Ch 16. If you are contacted, request permission to proceed direct to the Suez Yacht Club or the Port Fouad Yacht Centre.

Red Sea Pilot

N approaches to Red Sea
Israel
See NOTE on Israel above.
Haifa Call *Tatzpit Haifa* Ch 16, 12, 14. Call to let them know who you are and that you're headed to the Carmel YC in Qishon.
Contact marinas on the frequencies given in the pilotage section.

Coast radio stations
Radiotelephony and VHF only, working frequency underlined. Your need for these will probably be slight to non-existent. With the advent of GMDSS, Inmarsat and mobile phones, coast radio stations are an endangered species. Most developed world systems have now closed down and the rest of the world isn't likely to be far behind. Entries for RT(MF) and RT(HF) show TX/RX/times of operation (where relevant), single entries indicate TX only.

Oman
Muscat (A4M)
RT (MF) 2182/2182/24hrs, 2604, 2607, 3742, 3745kHz.
RT (HF) 4366/4074/1500–0300, 4417/4125/1500–0300, 8779/8255/0300–1500, 8788/8264/0300–1500.
Traffic lists on 2182kHz at 0403, 0703, 1103, 1703.
VHF Salalah Ch 16, 26, 27 24hrs.
Muscat Ch 01, 03, 04, 16, 26, 27 24hrs.

Yemen
Aden (70A)
RT (MF) 2182/2182/24hrs, 2595, 3628kHz.
RT (HF) 4390/4098, 8773/8249 kHz.

Djibouti
Djibouti (J2A)
RT (MF) 1813, 2182/2182/24hrs, 2586kHz.
RT (HF) 4408/4116, 8797/8273, 13104/12257 kHz.
VHF Ch 16, 24, 26.

Red Sea
Assab (ETC)
RT (MF) 2182/2182kHz/0500–2100.
RT (HF) 4363·6/4069·2, 6518·8/6212·4, 8731·3/8207·4kHz.
Traffic lists 4363·6, 6518·8, 8731·3kHz every even H+15.

Saudi Arabia
Jeddah (HZH)
RT (MF) 1726/1965·6, 1856/2037, 2182/2182kHz 24hrs.
Traffic lists on 1726kHz every odd H+03.
VHF Ch 16, 25, 27, 24hrs.
VHF stations remotely controlled from Jeddah (with channels) are: Al Birk (16, 25, 28); Al Lith (16, 23, 26); Al Qunfidah (16, 24, 27); Al Shaqiq (16, 23, 26); Al Shoabih (16, 24, 28); Al Wejh (16, 24, 27); Duba (16, 23, 26), Gizan (16, 25, 28), Obhur (16, 23, 26); Umm Lajj (16, 25, 28); Rabigh (16, 24, 27); Yanbu (16, 23, 26).

Sudan
Port Sudan (STP)
RT (MF) (0400–1800) 1750·5, 2182/2182/24hrs, 2740kHz.
VHF (0400–1800) Transmits and receives on Ch 16 (24hrs), 01, 03, 04, 05, 19, 20.

Egypt
Quseir (SUK)
RT (MF) 1680, 1849, 2182/2182kHz/0400–2200.
Traffic lists on 1849kHz at every even H+50.

Israel
Elat (4XA)
RT (MF) 2182/2182, 2649, 2656, 3656, 3765kHz. On request.
Traffic lists on 2656kHz at 0605, 0905, 1305.
VHF Transmits and receives on Ch 16, 24, 25. On request.

Jordan
Aqaba (JYO)
VHF DSC Public correspondence Ch 70/24hrs.
MF DSC Public Correspondence 2177/2189·5kHz/24hrs.
RT (MF) 2182/2182 (24hrs), 2590/2051, 2750/2054, 2813/2057kHz.
RT (HF) 4414/4122/24hrs, 8728/8204/24hrs, 13134/12287/24hrs, 17275/16393kHz/24hrs.
Traffic lists on 2182, 2590kHz every even H+15, 8728, 13134kHz every even H+15.
VHF Transmits and receives on Ch 16/24hrs, 22, 24, 27, 61.
Traffic lists on Ch 16, 27 every even H+15.

N approaches to Red Sea
Alexandria (SUH)
RT (MF) 2182/2182/24hrs, 2576, 2817kHz.
RT (HF) 4408/4116/0500–0800, 1600–2200, 6513/6212, 8767/8243/24hrs, 13122/12275, 17269/16387/ 0800–1200, 22771/22075.
Traffic lists on 2817kHz at every even H+35
VHF Ch 16/24hrs, 02, 19, 23, 25, 27, 60, 64, 66, 79, 87.

Port Said (SUP)
RT (MF) 2182/2182/0400–2200, 2840, 2860kHz.
VHF Ch 16/24hrs, 02, 04, 25, 28, 60, 66.

Israel
Haifa (4XO)
VHF DSC Public correspondence Ch 70/24hrs.
MF DSC Public Correspondence 2177/2189·5kHz/24hrs.
RT (MF) 2182/2182/24hrs (distress and safety), 2649/2225kHz.
RT (HF) 4384/4092, 8719/8195, 8746/8222/1800–0400, 13119/12272, 13185/12338, 17290/16408/0400–1800, 17323/16441, 19776/18801, 22705/22009kHz.
Traffic lists on request on 2649kHz at 0303, 0703, 1103, 1503, 1903, 2303
VHF Ch 16/24hrs, 24, 25, 26.
Traffic lists Ch 25 0303, 0703, 1103, 1503, 1903, 2303 and on request.

Cyprus
RT (MF) 2182/2182/24hrs, 2670, 2700, 3690kHz.
Traffic lists on 2700kHz every odd H+33.
RT (HF) 4372/4080, 4396/4104, 4432/4140, 6507/6206, 8737/8213, 8770/8246, 8776/8252, 8803/8279, 13077/12330, 13098/12251, 13164/12317, 14564/16366, 17248/16366, 17335/16453, 17386/16504/0530–1730, 22729/22033, 22729/16354, 22747/22051kHz.
Traffic lists on 8737kHz every odd H+33.
VHF (using relays along the coast) Ch 16, 24, 25, 26, 27.
Traffic lists on request.

Marinas

Most marinas have a VHF service. It is worth calling them up to let them know you're coming. This way, if they're full, they can let you know and, if you're lucky, suggest alternatives. If they've room, they'll tell you what to do, where to go when you arrive and organise CIQ.

Egypt
Port Ghalib Marina try Ch 16 on approach.
Soma Bay Marina Ch 12, call on approach.
Abu Tig Marina Ch 16 or 73, call on approach.
Taba Heights Marina try Ch 16 on approach.
SCA Ismailiya Yacht Club, Ch 08 for a berth.

Jordan
Royal Jordanian Yacht Club Marina Ch 67 on approach.

Israel
Elat Marina Ch 16, 11. Intermittent watch only.
Akko (Acre) Marina Ch 16, 11. Intermittent watch only.
Marina Herzlia Ch 16, 11. Intermittent watch only.
Tel Aviv Marina Ch 16, 11. Office hours.
Yafo (Jaffa) Marina Ch 16, 11. Intermittent. Can be contacted through Tel Aviv Radio.
Ashdod Marina Ch 16 or 11.
Ashkelon Marina Ch 16 or 11, on approach.

Cyprus
Larnaca Marina VHF Ch 16, 8. Call on approach. Tell them you wish to complete entry formalities.
St Raphael Marina, Limassol VHF Ch 16, 09. Call on approach. Tell them you wish to complete entry formalities.

Environment

Reefs

The Red Sea has some of the most extensive coral reefs in the world. They used to be its underwater glory. No more. Part of the damage has been caused by the explosion of dive tourism, which is poorly, if at all regulated. But that pales compared to the massive damage caused by global warming. Seven years ago reefs were 90% flourishing. In 2000 diver friends reported >50% dead coral from Port Sudan to Foul Bay and worse further N. The dead coral is being covered with algae. The N Red Sea corals are used to a winter cooling. They are getting less of it. Further S the glories of the Zubayr Is are no more. The same is true of much of the Hanish Is. The impact is not the same everywhere. Some areas, the Shubuk Channel and Dahlak Bank for example, are much as they always were. The only positive note is that with the efflorescence of algae, fish life has exploded.

Diving visibility in the Red Sea is usually good, thanks to slight tidal streams and surface currents, modest or negligible vertical tidal movement, no river outflow and desert shores. Visibility is best in summer but can be poor around reefs deep in the *marsas/sharms*, in the lee of a reef barrier in boisterous weather, or deep within the reefs and islands of the Dahlak and Farasan banks or the island complexes on the shores of Eritrea. In the Gulf of Aden and the Gulf of Tadjoura visibility is not as good and, because of greater depths and colder water from upwellings, the coral is also less extensive.

The Red Sea has coastal fringing reefs, offshore barrier reefs and isolated reefs associated with offshore islands and rocks. Three conditions affect the rate of coral growth. The water must be no colder than 20°C and between 25°C and 29°C for optimum growth. Although coral can live in water up to 55m deep, the ideal is between 3m and 30m. The depth limit is greater in the Red Sea because of the slight temperature gradient. Coral dies if exposed to the air for too long, so normally only grows to the level of mean neap tides. In the Red Sea mean sea level in the summer months is the datum, which is why most reefs have >1m over them in winter.

Coral grows very slowly close to the surface and fastest about 5m down. Branch coral grows fastest, at about 0·1m each year; large reefs at about half that rate. Erosion by wind, waves and predators just about balances growth. The eroded material tends slowly to fill passes in reefs, lagoons, and the leeward sides of reefs unless strong currents carry it away. The fastest recorded decrease in depth from any cause has been recorded as 0·3m a year.

Conditions in the Red Sea are ideal for coral. The shallow (50m) shelves along the shores have fostered the growth of the barrier reef and of the shoreline. In the section on reef navigation you will find diagrams showing the main forms of reef.

Coral reefs are composed of billions of minute animals of the phylum *Coelenterata*. The corals, with over 6,000 species, are part of a major subphylum of which sea anemones form the other part. Corals are, if you like, sea anemones with an external skeleton. The living part of a coral reef is only a few millimetres thick. The skeletons of trillions of dead coral polyps form the 'rock' which is so damaging to your boat. It seems paradoxical that something so hard should be so fragile, but it is. Each polyp of the living 'skin' of the reef feeds through a tiny mouth. Block the mouth and it starves. Reefs can therefore be damaged by stirred-up sand, silt or sediment, jettisoned rubbish, the touch of hands, fins and tanks and, of course, hugely damaging anchors and chains. Therefore:

1. Always anchor in a sandy clearing.
2. Make sure your chain does not lead over or around living coral.
3. Keep your tender away from living coral.
4. Do not jettison rubbish (which is illegal anyway).

The coral will, of course, gradually repair any damage, but that will take decades at least.

Marine life

The Red Sea is rich in fish and other marine animals, especially in the S half, which with the Gulf of Aden and the N Arabian Sea forms one of the world's richest fishing grounds. It is unusual in that it has not yet been overfished.

Sharks The Red Sea has many sharks, fewest in the N, increasing as one goes S, and most numerous in the Gulf of Aden. Common pelagic species include the mako, great white, tiger and scalloped hammerhead sharks. Around the reefs one is more likely to see white tip, black tip and grey reef sharks. The hammerhead has a reputation for awkwardness if you are spear fishing. (See 'Medical and health' for more on sharks.)

Rays Spotted eagle rays and manta rays are not uncommon. The latter can be seen breeding in the Zubayr group in spring.

Other fish Barracuda are seen frequently, as are sailfish and marlin, tuna and bonito. Flying fish are as ubiquitous as one might expect. There is the usual range of tropical and reef fishes, including coral trout, some quite enormous groupers, jacks and Spanish mackerel. The large tropical dorado or dolphin fish is less common than the smaller Mediterranean dorado (*daurade*).

Turtles All five of the major species of turtle can be found in the Red Sea: hawksbill, loggerhead, Ridley, leatherback and green. The hawksbill is still hunted for its shell. Tortoiseshell jewellery, usually made from hawksbill shell, is on sale in the tourist shops in Hurghada and elsewhere. There are many turtle nesting sites and we can only urge, should you find one, to keep it to yourself and leave it undisturbed.

Marine mammals Because the Red Sea is closed at its N end, it is not on any major migratory route for the larger whales. Even so sperm whales have been spotted, as have pygmy and dwarf sperm whales, killer whales, and false and pygmy killer whales. The major *Cetacea* in the Red Sea are pilot whales (the shortfin) and dolphins. The bottlenose, common, spinner, bridled and striped dolphins, amongst the commoner beaked whales, can all be seen. The authors have spotted several Risso's dolphins and Indo-Pacific humpback dolphins are common in *marsas*. Amongst the rarer species are the Fraser's and rough-toothed dolphins, and the densebeak and Cuvier's whales.

Amongst the rarest of the denizens of the Red Sea are dugongs or sea cows. These gentle creatures are thought, somewhat improbably for they are no beauties, to have given rise to the mermaid myth. They suckle their young at the breast while floating upright, which to sex-starved sailors in more benighted times was probably more than enough for the imagination to work on. They are protected, but protection seldom works where there is little policing and lots of poverty. However, one source suggests that dugongs can be spotted in the shallows of Khor el Marob, and Marsa Arus derives its name, Arus el-Bahr, from the Arabic for dugong.

Crustaceans and shellfish Crayfish (langouste) can be caught on the fringing reef in most places. Night fishing with a torch is the only way to go, but you've got to be quick, and either wear heavy gloves or have some sort of net. Where there are extensive mangroves, some quite large green crabs can be found, though you need skill and hunger. In the Sudanese *marsas/sharms* you can dig clams in the shallows. It helps to have a good shovel, and the advice of someone skilled in the art. Clams can dig downwards at an incredible speed. There are mussels in the Gulf of Tadjoura, and oysters on the rocks in the Gulf of Aden, mainly E of Aden and on the Somali shore.

For dangerous species see 'Medical and health'.

Diving and dive sites

The Gulf of Aqaba and the Sinai Peninsula are already seething with dive-boat operators. Charter operators out of Hurghada and Safaga have now spread their activities as far as Foul Bay. It is a rare reef which does not have a few attendant dive boats in its lee. There are dive hotels or camps of varying sophistication almost every 20 or 30km from Ras Baniyas to El Gouna. There are a few commercial charter operations out of Port Sudan, almost all Italian owned and run. The dive charter business in Massawa is just beginning and the peace with both Ethiopia and the Yemen will help. As a preparatory move the Dahlak Bank has been designated a marine park. The Yemen is as yet largely unexploited but Djibouti has a couple of dive outfits.

There are plenty of untouched, unspoiled dive sites to be explored in addition to the known ones. Where an anchorage offers particularly good diving, we usually say so in the main body of the text. All the areas below are shown in greater detail on the key maps. An asterisk indicates that political reasons may make diving inadvisable.

Sites

Abd el Kuri and the Brothers (between Ras Felug and Suqutra)*

N Somalia (towards Ras Felug)*

Gulf of Tadjoura

Dumeira I, Ras Siyyan and the Seven Brothers

Zubayr group and Hanish group

Farasan Bank*

Dahlak Bank

Shab el-Shubuk and Shubuk Channel

Reefs and islands in the Sawakin group

Sanganeb Reef and the Umbria wreck (Silayet Reef)

Barrier and offshore reefs N of Sanganeb, especially Shab Rumi (Jacques Cousteau's habitats), Shab Suadi, Gurna Reef, Jazirat Bayer, Qita el Banna, Angarosh, Abington Reef and the reefs of the area around Ras Abu Shagrab

N Sudan *marsas* (entrances) and barrier reef, especially Shab Qumeira, Shab Shinab and Elba Reef

Foul Bay, especially Shab Abu Fendera, St John's Reef, White Rock, Jez. Mikauwa, Horseshoe Reef and Ras Baniyas

Geziret Zabargad (St John's I) and Rocky I

Erg Abu Diab, Eroug Abu Diab and Petrol Tanker

Fury Shoal: Dolphin Reef, Shab Claude, Shab Mansour and Abu Galawa

Abu el Kizan (Daedalus Reef)

El Akhawein (The Brothers)

Coastal reefs from Wadi Lahami N to Safaga

Reefs and islands near Hurghada and S of Bluff Point

Sinai Peninsula from Ras Muhammad to Strait of Tiran

Strait of Tiran and Tiran I

Gulf of Aqaba

Dive moorings
From Gezirat Zabargad N to the Hurghada area the Hurghada Environmental Protection and Conservation Association (HEPCA), helped by the Egyptian Environmental Affairs Agency and with USAID assistance, has laid 536 reef moorings using the tried and tested Manta Ray system. The object has been to encourage dive boats to use them instead of their own dreadfully damaging ground tackle. The programme has been a success, but there is a continuing and chronic shortage of funds for maintenance. Not all reefs or dives have moorings, nor do we have accurate locations for all there are. We mention those we know of in the Pilotage section. There are only 137 for the long coastline from Gez Zabargad to Safaga, 72 in the Safaga area, a massive concentration of 277 around Hurghada and 50 in the waters N of Hurghada. There are equivalents in the Ras Mohammed National Park. The same is planned in the Elat/Aqaba area.

The moorings are either marked by an orange buoy about ½m in diameter or by floating pick-up ropes. Each is designed to take several large boats. They are not regularly serviced, so have a very good look before you trust your boat to one. Note too that these are dive moorings. They have been laid tight up in the lee of reefs on the supposition that prevailing winds will hold boats clear. It would be wildly imprudent to leave your yacht untended. The average yachtie will feel vulnerable.

Bird life

For birders, and even for people to whom birds are of only passing interest, the Red Sea is a small paradise. Elat is a major centre for observing raptor migrations and the birds of the Red Sea littoral are quite unused to humans and so remarkably tolerant of close approach. The Sudanese *marsas/sharms* are wonderful observation sites. The small islets within the *marsas/sharms* are nesting sites for marabou storks, African skimmers, egrets, ibis, spoonbills and ospreys, all living hugger-mugger. In the shallows troops of flamingos shuffle along, heads down for food. Around mangroves herons fish the shallows. In the autumn islands like Gez. Wadi Gimal are literally carpeted with nesting Eleonora's falcons aggressively defending their territory. Offshore over the reefs the bird life is not as rich as that in colder waters, but there are still a number of species of tern and seagull.

Shore birds are prolific and numerous species can be seen in *marsas* which have shallow inshore ends, and inshore lagoons (e.g. S of Trinkitat Harbour). We've never seen so many nesting ospreys, with several pairs in any one *marsa/sharm*, sometimes several on one islet. Once you get within about 25m of a nest the ospreys will take to the air and may dive-bomb you quite aggressively. However, if you press on they'll probably retire to a nearby spot and mutter loudly, allowing you to go right up to the nest. We watched one photographer set up his camera, tripod and hood, pick up a fledgling to pose it properly and then take his photos! He'd been doing this off and on since the eggs were laid and when he'd finished the parents returned to the nest as if nothing had happened.

Take a short walk into the desert or up into the hills in the Hanish Is and you'll see hawks and eagles quartering the hills. Underfoot small chats will zip out of the scrub. Hoopoes, looking as startled as always, are common.

Other fauna

Antelopes are said still to exist in the Zubayr and Hanish groups. No doubt they are also present, along with desert foxes, jerboas and much else besides, in the country around the *marsas*, and a patient wait could prove very rewarding. The Farasan Is are reputed to have so much wildlife they may be designated a World Heritage Site.

Insect and reptilian life is rich but, as with the mammals, it is not obvious.

The desert and offshore islands

The offshore islands are either barren or covered with a light scrub, a few scattered mangroves and pandanus. Stunted mangroves are common in the coastal shallows and quite extensive in some places such as the wonderful area around Marsa Ata. The Zubayr and Hanish groups are remarkable for their spectacular lava flows and volcanic cones. The walking is rough but marvellous, a good way of spending the day if weather-bound. Check first that there are no military around to give you a hard time.

The desert of the Sudan coast repays exploration. This is the outer edge of the great Nubian Desert, which separates the Red Sea from the Nile Valley. The desert by the shore is what used to be called madrepore, or a raised coral platform which is now 2–3m above sea level. Underfoot are shells and broken coral. It's as if, not long ago, sea-level suddenly dropped. As morning gives way to midday, or afternoon to evening, the desert goes through its wonderful palette of colours: the pinks and greys of early morning, the dazzling whites, pale silver-golds and deep black shadows of noon, and the glorious old golds, ochres, browns and reds of evening. To seaward, and slashing deep into the desert along the *marsas*, are the startling turquoises, browns and greens of the reefs, and the brilliant sparkling blues of the deeps.

Near many anchorages such as Khor Shinab, Marsa Abu Imama and Khor el Marob there are low coastal hills, seldom more than a few hundred metres in height. Getting to the top is a scrambly, slithery exercise, the broken, friable stone and ever-present sand offering little by way of traction or holds. But the views are phenomenal early in the day before the breeze creates a dust haze.

There are four dangers which you should bear in mind if you intend wandering in the desert:

1. *Flash floods* In the rainy season, such as it is, there are heavy rainstorms in the mountains which can cause flash floods. These roar down wadis in a wave a metre or more high, carrying all before them. It is unlikely that anything other than a severe storm would send a flood as far as the shore; usually it peters out in the foothills. Still, the wadis exist. Flash floods do happen.
2. *Getting lost* One stretch of desert looks pretty much like another. The wadis are deep enough to restrict your horizon. There is often a dust haze. If you are not a practised back-country walker always carry a pocket compass or handheld GPS and keep track of your location.
3. *Dust storms* Keep an eye out for dust storms, particularly during the summer when you shouldn't walk far anyway. If you see one coming and can't get back to your boat, hunker down, set your back to the wind, organise your shirt, a cloth or something to help you breathe freely, and wait for the storm to abate. Don't head for your boat. You'll get dehydrated and lost.
4. *Dehydration* The desert is dry and you lose far more water than you think. Carry at least 1 litre per person for a 2-hour walk.

Pollution

The Red Sea has been declared a Special Area for the purposes of the International Convention for the Prevention of Pollution from Ships (1973). Annex V contains special provisions relevant to the Red Sea and the Gulf of Aden on discharging garbage.

This is a vulnerable environment. Just as smokers don't actually intend to poison anyone, tourists don't intend to damage what they have come to enjoy. Cruising people are tourists. We bring to the remote places we visit the accoutrements of a comfortable urban life. Including waste.

The shores of the more open *marsas* and of most offshore islands are littered with plastic and other junk thrown from ships passing up and down the Red Sea. They also feature plastic sacks of garbage left behind by dive boats and even cruising yachts.

Please, take all your garbage with you. If you don't have space aboard and can't lead a less waste-producing existence, make a proper fire pit and burn everything that is combustible. Be sure you put the fire completely out. Take away everything large and unsightly that's left. Fill in the pit and efface all the scars of your activity. The Red Sea is too narrow for you to throw any garbage overboard which is not rapidly biodegradable and guaranteed to sink immediately. Don't forget the stringent international regulations governing the dumping of rubbish at sea.

Problems

The object of this new section in our planning advice is, where possible, to answer inevitable questions about the risks of sailing in the Red Sea and Gulf of Aden. We won't deny there are risks. Of course there are. But they are ALL low probability. At this point we've no doubt lost those of you with fixed ideas. In which case skip the rest of this section.

The risks fall into four categories. Danger from political terrorism and war. Piracy risk. The usual run of medical and other health risks. Otherwise there is the hurdle of dealing with poverty and unfamiliar cultures.

Even after 11th September 2001, the risks in any category in the Red Sea and Gulf of Aden are almost certainly lower than the risk of being mugged in a score of developed world cities, or of a serious traffic accident anywhere. Frankly, with reasonable planning, care and good manners, you and your property are no more at risk in these waters than in a high season port in Europe.

Terrorism and war

There is only one predictable problem. It is still several years from resolution and is easy to avoid. It's the mess that was Somalia. Don't go near it.

Terrorism

The western media and its obsession with bad news necessarily exaggerate the impact of political Islam on Middle Eastern societies. It is hard to gain a fair perspective. The easiest way is to remember that, though they may be poorer than you, worship in a different way and so on, the people in the countries covered by this guide are like the people where you come from. A small minority are lunatic fringe, infidel-hating fanatics; but remember Timothy McVeigh and Oklahoma City, the IRA, ETA, Corsican Separatism, the Brigade Rosse, etc. A percentage are criminally inclined; but check Yemeni or Sudanese crime statistics against those of Washington, London or Paris. The overwhelming majority of inhabitants will be indifferent to strangers. And a delightfully large minority will be kind, generous, helpful and welcoming.

Bluntly, you have little to worry about. Terrorists have bigger, higher profile targets than cruising yachts. Don't be aggressively 'western'. Respect local people and their culture whatever your private views. Dress and behave accordingly. Learn and use a little Arabic and, for Eritrea, some Tigrinya. Keep a low profile and if you worry about your flag, don't fly it.

Problems

If you still worry, consult your embassy and listen to the daily news on the BBC, VOA, Radio France International, Radio Deutsche Welle and so forth. The cruisers who ran foul of the Yemeni/Eritrea spat over the Hanish Is in the mid–90s could have avoided a lot of grief by listening to the news and talking to their consulate.

War

Many Middle Eastern societies are slowly finding their feet in the long process of post-colonial adjustment. All bar Oman, an oasis of prosperity, are arbitrary creations of post First and Second World War western geopolitics. The forms and principles of a modern, secular state do not easily fit with older, deeply entrenched social structures and beliefs. Unresolved or buried issues periodically surface. The Eritrea/Yemen set-to over the Hanish Is for example, was over juridically sovereign-less territory. The Ottoman Empire had never formally claimed the islands. The British colonial government hadn't either. A formal decision or annexation was needed, no matter that Eritrea chose an unnecessarily macho route to bring things to a head.

Generally the advice on how to keep terrorism in perspective applies also to outbreaks of hostility. Keep in touch with the news. Learn about the societies you are visiting, what potential problems they have and how and where they might erupt. Consult your consulates and embassies. We think, excluding Somalia (and there are hopeful signs even there), that with the UN mediation over Eritrea and Ethiopia and the general step backwards from lunacy after September 11th 2001, the Red Sea and Gulf of Aden coasts are probably at their most settled in years.

Piracy: rumours and risks

Those of you already convinced that the Gulf of Aden is pirate alley are unlikely to be persuaded by anything we write. For the rest, we hope to put things in perspective. During our last visit we were able to sort some truth from fiction. From autumn 1998 to autumn 2001 there were 15 reported incidents. There may have been others but they remain by either unconfirmed by either official or informal sources such as the press and yachtie news bulletins and websites. When we have asked those who claim our total to be a gross underestimate, they are unable or unwilling to provide facts to support their case. Our tally is unlikely to be significantly wide of the mark. Of the 15 certain cases, 7 are confirmed piracy or robbery, of which 3 were within 70 miles of the Somali coast (which yachts are strongly advised to avoid) and 4 in Yemeni waters. Of the 8 remaining, 4 or 5 were contact with pushy Yemeni military patrols, 1 or 2 were importunate fishermen mistaken for robbers, 1 was a case of on board crime and the last a disappearance more likely a tragic mishap than piracy. During this period at least 630 yachts passed N or S bound through the Red Sea and the Gulf of Aden. That makes it a 1/1.5% risk. A problem, yes;

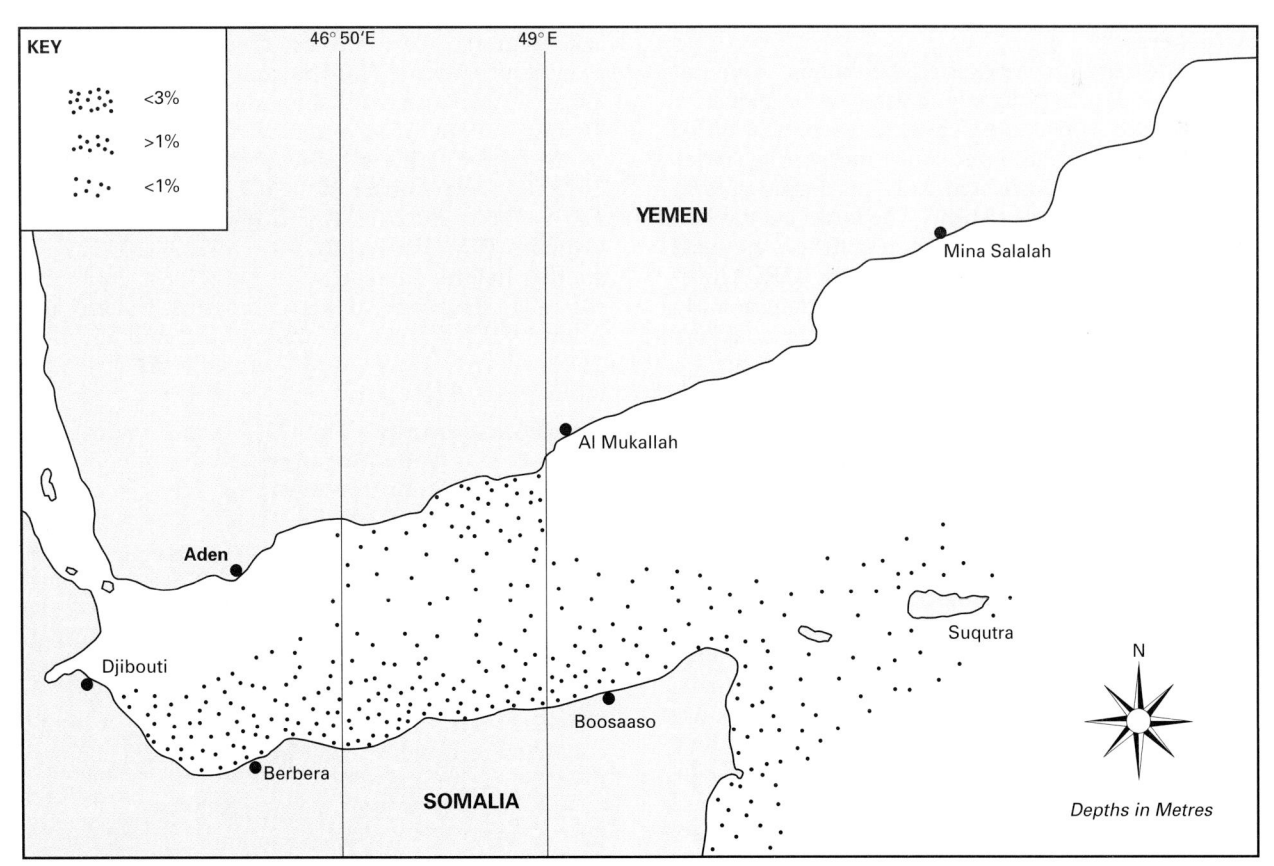

ROBBERY RISK ZONE AND RISK FACTORS (ROBBERIES PER 100 TRANSITS)

Red Sea Pilot

Location/ area	Frequency (USB)	Hours of service (all times Z/GMT)
Indian Ocean/Red Sea/ Diego Garcia	13201	24hr
	11176kHz	1500–0200
	6738kHz	1200–2200
Central and E Mediterranean, Straits of Hormuz & Persian Gulf	23227kHz	0700–1500
	15015kHz	0500–0200
	13244kHz	24hr
	11176kHz	24hr
	6738kHz	1500–0700
	3137kHz	2000–0500

but it should not be exaggerated. Risk zones are shown on page 49. No incidents of piracy, suspected or real, have been reported in Eritrean, Saudi Arabian, Sudanese or Egyptian coastal waters. Some general tips follow but a fuller version can be found in the regularly updated Piracy Advisory on the Red Sea Pilot corrections page of the Imray website www.imray.com

Help
Unless you have organized an escort (see Somalia and Yemen in Prevention and Protection below) you CANNOT EXPECT ANY OFFICIAL HELP. Distress calls about piracy on international distress frequencies are ignored even by nearby ships. The authorities do not or do not wish to hear you. No forces of law and order hear you and, if they do, they are unlikely to be close enough to come to your assistance. YOU MUST BE READY TO HELP YOURSELF.

Attackers
The known attacks are chance encounters - you and a few hoodlums unluckily in the same patch of water at the same time. There is no indication of organized crime, radio frequency monitoring, radar or any other sophisticated aids to detecting and intercepting yachts as victims. The attackers have so far all been in traditional wooden craft not the local Yemeni, long, fast, narrow, usually GRP huri. There is no specific danger time, though early daylight hours have featured twice, and the attackers do not appear to operate at night or in strong weather. The craft had no specific pre-attack occupation. Some appear to be fishing. One was smuggling people. The attackers include both Yemenis and Somalis. There is no consistent pattern save that in every incident the attackers fired warning shots. Some draw close and ask you to heave-to before their warning shots. Others fire shots and close to board immediately. One immobilised its victim by fouling their propellers with a fishing net. They shoot to warn and intimidate, not to kill, injure or disable the boat. Attacks are pushed home fast and hard. Most attackers are apparently nervous. Remember that these are gun culture societies where carrying a gun is a badge of status. The same is true of knives. Knives, guns and warning shots do not usually mean intent to do bodily harm.

Danger zones
If unescorted, avoid the whole coast of Somalia out to 90/100M offshore and especially from 47°30'E to Djibouti waters.

The second problem area is from about 60M E of Aden to near Ras Kalb. Stay at least 20M offshore between 49°E and 46°5'E. If possible transit this area, particularly between 48°50'E and 47°50'E, at night.

Prevention and Protection
Oman The Omani Navy is well armed, very efficient and has a base in Mina Salalah. They actively patrol the coast. There have been no reports of any criminal activity on the Omani coast.
Contacts: Salalah maintains listening watch on 2182kHz and Ch16. Harbourmaster Capt. Ahmed Burham Omar. ☎ (+968) 219500 ext 420, *Fax* (+968) 219253, email AhmedB@Salahport.com
Yemen There is no patrolling of the Yemeni coast. A jointly funded and EU and US backed Yemeni coastguard is to be trained and equipped . . . some time. To help their case for accelerating this programme, the authorities in Aden and Mukalla are anxious that ANY INCIDENT is reported to them, whether or not the threat proves to have been real. They take the matter seriously, although follow up to ascertain facts and punish any Yemenis responsible is usually lukewarm and clogged by red tape unless fired up by major, adverse international publicity. Some yachts have successfully asked for an escort from Mukalla. It wasn't efficient, but may have served its purpose. If you are worried, try asking for an escort in Mukalla or Aden.
Contacts: Aden: Harbourmaster, Captain Ali. Deputy Harbourmaster, Captain Hussein. VHF Ch 16 from about 25M and SSB 2182kHz to about 100M (if the radio is serviceable). ☎ office +9672 202850. Duty Officer ☎ +9672 202262, 202238 *Fax* +9762 206241. An IMO officer in Aden, Captain Roy Facey, ☎/*Fax* +9762 203521, is another helpful contact.
Mukalla: Harbour pilots, Captains Salem and Amin. VHF Ch 16 to about 25M and on SSB 2182kHz to about 60M. Mukalla ☎ 354742; mobile 7951076.
Puntland/Somalia The N Somali (Puntland) coastguard is a Bermuda-registered company and pilots using the British-registered *Celtic Horizon*, a 65m, former stern trawler, which acts as mother ship to several RIBs manned by armed crew. The area covered is from 100M W of Boosasso (roughly 47°30'E), round the Horn of Africa to Eyl, on the Indian Ocean coast. Extremely expensive armed escort can be and has been arranged to convoy yachts through the Gulf of Aden and S Red Sea.
Contacts: The Hart Group (Bermuda) can be contacted in London: ☎ +44 20 7751 0771
email george.simm@talk21.com
In Boosaaso: ☎ +252 5 726121/826005;
☎/*Fax* +252 523 6104, *email* nimrod@brtel.net

There are plans for an international naval force to be established to patrol the Horn of Africa.

Djibouti The French Navy have a presence in Djibouti. Anti-piracy patrols are not part of their brief. They reportedly keep watch on MF 2182kHz and VHF Chs 12 and 16. But distress calls are unlikely to be heard unless the call is made within a short distance of Djibouti. Do not rely on them. You can try to alert them BEFORE your transit to ascertain how best to get help if you need it. You will need to be able to speak French.

Contacts: French Navy (la Marine Française)
☎ +253 351 351 or +253 35 03 48. Ask for the OPO (officier permanent opération).
The Djibouti Port Authority (Port Autonome International de Djibouti), BP 2107, Djibouti can't offer much help but can at least make noises off.
☎ +253 352 331, *email* port@intnet.dj

Eritrea Eritrean forces patrol coastal waters and may stop you and ask to see your papers. Their vessels, uniforms and procedures are somewhat ramshackle. They are sticklers for rules, which they will know and you will not, but civil. There will be a UN maritime presence with some helicopter patrolling as long as UNMEE (the UN mission to Ethiopia and Eritrea) continues. They may respond to a distress call, but don't rely on it.

US Navy Unless US naval forces are actively engaged in the theatre, as over Afghanistan, this is a long shot, but they are known to maintain a LISTENING WATCH ONLY on the schedule shown on the previous page.

In addition there is a USN correspondence frequency, 14467 kHz, watch times unknown, but believed also to be used by keen retired USN personnel. This might work when all else is silent.

Strategy and tactics

1. Inform local port authorities (NOT agents, Immigration, Customs or anyone else) in Salalah, Mukalla, Aden, Djibouti or the Hart Group of your passage plan and ask them to inform your next port of your ETA (see contacts in 'Prevention and protection'). If you are apprehensive, don't be afraid to ask for an escort. Keep in regular touch with your port of departure for as long as you can. Contact your port of arrival as early as possible and regularly thereafter.
2. While transiting the Gulf of Aden, prepare your boat against boarding by robbers. Secure important valuables out of sight, hide any portable GPS or VHF, leave some attractive goodies on display, have some cash 'hidden' where it can easily be found.
3. If you can organise an escort, well and good. Otherwise sail in convoy of at least three or four.
4. Sail no more than 0·5M apart for swift mutual aid but forming a dispersed 'target'.
5. Give positions by bearing and distance from a datum position known only to the boats in company.

Arms or not?

Sometimes a gun works. But be aware of the downsides. You must choose the right weapon – close or long range, single or multiple shot, etc. You must decide whether to license it or keep it aboard unlicensed. You have to decide whether to declare or hide it at each port of entry. Come the hour, you have two choices. Shoot first and hope you scare the blighters off. Shoot second and still hope to scare them away or win any fire-fight. If scare tactics haven't worked, you must use your gun swiftly, accurately and intend to kill. You'll almost certainly be breaking a law. Most countries outlaw all but 'reasonable force' in protecting yourself. Using a gun before anyone uses one on you breaks such a law. Up to you, but we'd advise against.

6. At night run without lights or only deck level port, starboard and stern lights.
7. Agree on how your convoy will maintain contact visually at night. (Illumination of all round white masthead lights for a few minutes hourly works well.)
8. If you have it, use radar actively. Try to identify contacts before they have you visual and steer to stay beyond visual range (>4M).
9. Don't chatter on VHF, don't use Ch 16 and ALWAYS use LO-POWER.
10. Make sure all your group knows of emergency frequencies and contacts for aid.
11. If you are approached, BE FRIENDLY. Smile, offer a welcome. If your contacts aren't pirates, you get off on the right foot. If they are, you've helped keep the temperature low.
12. If you are SURE you are being AGGRESSIVELY AND ACTIVELY PURSUED, IMMEDIATELY broadcast an alert using ALL means (SSB, VHF, SATCOMS and mobile phone if you can get a signal). It's worth a try even if help is not forthcoming.
13. If pursuit closes and fires shots, send a MAYDAY and fire PARACHUTE FLARES. Then SURRENDER. You can try warning shots if you are armed but remember, they may be better armed and meaner minded than you. Be polite. Don't try to keep things back unless they are well hidden. Early, seemingly VALUABLE concessions may satisfy. Your valuables can be replaced.

Follow-up If you are unlucky enough to be a piracy victim, or feel you have been threatened by potential piracy, please, while you can still remember the details, log:

date, time and position of attack or approach

direction of approach and description of suspect craft (including colour, size, means of propulsion, design (local or modern imported), construction material (GRP, wood, etc), any conspicuous features and any name or numbers)

description and number of crew on suspect craft, whether in uniform or not and whether armed and with what description of any contact made (nature of gestures, messages passed, language used, etc)

details of any injuries sustained

details of any damage to your vessel

details of items stolen (description, serial numbers, etc)

details of last direction in which pirates were observed to be moving (approx. course and speed)

Give these details to the authorities in Salalah, Mukalla, Aden, Djibouti or the Hart Group as appropriate (for contact details see above). Please also inform us so we can keep this information up to date, *email* to Morgdav@aol.com (cc Imrays, ilnw@imray.com) and Mike Devonshire at the joint ISAF/IMO sub-committee which targets piracy of pleasure craft. *Email* to piracy@isaf.co.uk

Other cultures, other ways

Wherever you go you will attract curiosity and possibly envy. Unless you never go ashore you'll inevitably come across desperately poor people. Even if you stay aboard there will be the odd encounter. Everybody reacts according to temperament, but for the 99% of cases when those approaching you have no malevolent intent, here are some guidelines.

In anchorages local boats often come alongside uninvited even in the remotest anchorages. This may seem importunate or even threatening. We suggest that you put fenders out and be ready to take lines. In our experience being prepared saves time, aggravation and your topsides.

Officials Personnel may be armed but not in uniform. They will be insistent until you grasp they want to see your papers. Usually someone will speak at least broken English. The more confident they are the more relaxed the encounter. We have met Yemeni, Sudanese, Eritrean and Egyptian officers who have been not only courteous but also charming and, when necessary, helpful. A few words of greeting in Arabic help a lot but a ship's letter of introduction in Arabic (see Appendix IV, photocopiable proforma), complete with details of where you've come from and where you're headed is even better. Together with passports and clearance papers it helps smooth your passage. Officials love bumf. The ones in the Red Sea are no exception. Military and police outposts don't usually have forms for you to fill in. They need to make reports on passing yachts. Make life easier for them.

Fishermen If your visitors are just fishermen they may look pretty roguish. They can be dressed in anything from rags and balaclava helmets to jeans and oilskins. Two things are for sure. They want you to give them something and they have endless patience. If you try to ignore them they may take a long time to go away. Usually they are friendly and, particularly in the Yemen, very noisy. Experience has taught us always to begin by exchanging greetings and civilities, whatever their appearance, for example:

Salaam al haikum (Hello). *Min ain enta?* (Where are you from?)★

Our boat has low freeboard so we sit on the deck to be level with the people we're talking to and fend off with our feet if necessary! A polite exchange of greetings is a preliminary. Your next good move is to offer a little something. If you don't, there'll be a request, maybe for food, drink, cigarettes, clothes or fish hooks. All of this takes place in mime so you can't ignore it and hope it'll go away. Yachts are rightly associated with comparative wealth, so expectations of generosity are high. Never forget Islam is a pointedly hospitable, gift giving and above all alms giving religion. In all the towns you'll see the better off giving to the poor. You may feel poor compared with other boats in your cohort. But compared to local people you're Croesus. They'll be puzzled and disappointed if you turn them down. After all, you're in their back yard and the onus is on you to fit in with local mores.

A pre-emptive query as to whether they have fish (*samak*, in Arabic) can start the interchange on a more businesslike footing. The folk may have little but they do have pride. If they have any fish they will usually offer it in return for what you give them. Take some, even if you don't intend to eat it, and honour will be done. Some may offer fish first and ask for payment in cash, others will barter. Basically, we just give them something we can spare. Some things are always in demand. Sweet biscuits, water, bread, sugar, tea, rice, t-shirts, and (unfortunately) cigarettes, preferably American brands. You may be asked for matches, aspirins, alcohol, bits of rope, diving masks and snorkels. The list can be extended indefinitely. If you give exactly what's asked for they may still ask for more. If you offer a substitute, even a relatively lavish one, they may not look happy with it. All you can do is set your limits and think of the next anchorage and the next poor beggar.

When you provision remember the likelihood of these encounters. Even if you only have a bag of sugar and a smile to give away it saves agonising over what to do, allows you to be gracious and prevents difficulty all round. OK, you may not condone begging on principle, but remember that giving alms is one of the tenets of Muslim belief. Given your style of travelling they will expect you to offer them – even poorer nomads than you – sustenance.

And when they leave? The word for thank you is *shukran* and 'don't mention it' is *afwan*★.

Bargaining By the time you get to the Red Sea you'll probably have some experience at bargaining. For some it's fun. For others it's plain embarrassing. For beginners, whether buying spices or silverware here are a few tips:

1. Learn written and spoken Arabic numerals, at least 1 to 10.★
2. Check a few stalls before buying to get a feel for current prices. (see Appendix III 'Food' for sample prices).

3. If you think the vendor is having you on, turn and walk away. You've got it right if they call you back.
4. Never seem too enthusiastic.
5. Always point out any defects.
6. Always agree on fares in buses and taxis when or before you get in.

There is one item for which you are unlikely ever to bargain. It's bread, called *aysh* in Arabic, and means 'life itself'.

Begging You can't solve the problems of the poor and the maimed, however distressing they are. All you can do is make them no worse and give what you can afford in contexts where you feel happy about it. Take some form of food whenever you go ashore. Give that to beggars, children in particular, instead of money. A few coins for old people and the handicapped are not a bad idea. The giving of alms is a duty for Muslims. You, as a foreigner and heathen, are exempt, but as a visitor adapting a little to local ways is good. For handling 'beggars' in anchorages see above. If you still feel guilty, make donations to the aid charities.

Baksheesh Baksheesh is either a form of tipping or a bribe. If you give cigarettes to a Suez canal pilot, this 'gift' is a form of tip. So is a dollar given to a lad on a beach to keep an eye on your dinghy. It is not charity. We deal with that above. With baksheesh for an agent note it includes tips/bribes given by him to government officials on your behalf. You may disagree with this in principle but in many Arabic countries it is a fact of life, especially in Egypt. By conforming, however reluctantly, life is easier and business more swiftly done. Forms of baksheesh may be requested in all of the countries we deal with except Oman, Jordan and Israel. That said, never offer baksheesh to an official directly, whether he or she is in or out of uniform. Wait to be asked. You are wrong to offer a bribe; it is quite in order for an official to ask for a present.

*For more, see the glossary in Appendix IV.

Medical and health

By the time you reach the Red Sea you will have been cruising for quite a while so we won't go into basics. Pharmacies in Red Sea ports can supply antibiotics, anti-malarials, antifungals, antiseptics, antihistamines and aspirins, usually at lower cost than elsewhere, but check expiry dates. Anything more exotic will probably be difficult. The best place to buy antimalarials is in countries where the disease is a problem.

Always ask a doctor about vaccinations. We deal with these below but by the time you read this the situation may have changed. Attend a course in first aid if you can, it's always a good idea. Clinics and hospitals are few and far between. If you are sailing in a group or if you check in on a radio net you may be lucky enough to make contact with a qualified medic or paramedic, otherwise, you're on your own. A first aid manual, and if you dive, a guide with advice on diving related problems, are essential. Other useful books are in the appendix.

Food and water Many gut related problems can be avoided by making sure your food is carefully prepared and stored. Unless you buy bottled water always treat tap water that you plan to drink wherever you take it aboard. Use chlorine or an equivalent product. A rough rule of thumb that we have always used is 1 drop (0·05ml) of ordinary household bleach (4–6% chlorine) for every litre of water. That's about 1 teaspoon for every 100 litres.

Insects Mosquitoes are the worst pest, both for itchy bites and for diseases. Screens on hatches and ports go a long way towards preventing malaria, dengue fever and several other insect borne diseases. Flies can be a serious nuisance, especially in Port Sudan. Needless to say they are also a health risk. Cockroaches will breed aboard if you let them. Then they are very difficult to get rid of. Never leave cardboard on the boat since they lay their eggs in it. If you see a cockroach, don't rest till you've caught it and squashed it. They are attracted by light and can fly right in through your portholes. Ants get aboard on fruit and vegetables. If the boat gets infested with ants the simplest treatment is a small plastic household 'anthotel' which contains a mixture of sugar and arsenic which they find irresistible. A less poisonous alternative is oil of cloves. Wash all market produce scrupulously before you stow it. We use sea water for this if it's clean, then wipe it off and let the food dry on deck in the shade.

Personal protection

Wear polarised sunglasses. They protect your eyes against the sun, glare and dust and are also indispensable for spotting coral reefs. Take plenty of sunscreen. Wear hats and long sleeves if you are out all day. Humidity is often very low so take moisturising cream too. You will need various pairs of gloves, for diving, sailing and cleaning barnacles off the propeller. The Red Sea is a high growth area. Take shoes suitable for walking in shallow water and across reefs.

Sea creatures

Sharks, moray eels, barracuda and triggerfish The risk of a shark attack is slight. You are statistically more likely to die from a bee sting than a shark attack. Human beings aren't the favourite food of any fish. In some attacks people are bitten and then left alone. Discarded, in effect. They don't taste right. However, you may attract sharks by throwing fish entrails overboard, by wearing glittery jewellery or thrashing about a lot when swimming. You are unlikely to see any Great Whites in the Red Sea and the tiger and bull sharks are uncommon. Amongst dangerous common species in the Red Sea are silky sharks and grey reef sharks. The former can be dangerous in the S Red Sea. The latter are notoriously territorial. A shark that arches its back and points its pectoral fins downwards means trouble. Face sharks that approach you and be prepared to push them away. Moray eels are shy, have poor eyesight and don't like being disturbed. If they bite try not to immediately pull your hand away. That will make it worse because the eel's teeth slant backwards. Big, lone barracuda have been known to attack people, usually in murky water, by mistake. If you disturb a breeding triggerfish it can get aggressive. In all cases common sense and caution will avoid most problems.

Stonefish, stingrays, lionfish, etc. This group, which also includes scorpionfish, surgeonfish, rabbitfish and turkeyfish, will attack if you touch or tread on them accidentally. Their stings are passive defences but can be very toxic. The stonefish and stingray are the most dangerous. People who are allergic can have a violent reaction and may die if they have cardio-respiratory failure. In milder cases pain can last for months. A stonefish antivenin does exist.

Jellyfish Most of these plus some hydroids such as fire coral and stinging plankton can give you a nasty, burning sting. If stung pour lots of vinegar over the sting as soon as possible. Failing vinegar, urine will do. If any bits of the animal are still on the skin remove them with rubber gloves, tongs or forceps. Give CPR in extreme cases. Use an antihistamine cream and/or tablets. Hydrocortisone ointment helps if the skin is very itchy. Scars can last for a long time.

Coral Cuts and scratches will rapidly become infected if you don't clean and disinfect them and keep out of the water. Use an antibiotic cream such as *Bactroban* if the skin does not heal normally. Remember that if you touch any piece of coral it will die, whatever harm it might do to you.

Sea urchins, crown-of-thorns Treat by removing any loose spines. Immerse stings in non-scalding hot water. If you have been stung before you may get a worse reaction the second time. Vinegar, papaya juice or wine (in extremis!) will work as neutralisers and pain killers. Use antibiotics if wounds are infected.

Sea snakes Few sea snakes have mouths that can get hold of anything bigger than your little finger or the edge of your ear, but their venom is very powerful. If bitten, immobilise the limb and apply a pressure bandage. Treat for shock if necessary and get to hospital. Antivenin exists.

Cone shells Never pick up cone-shaped shells, however beautiful. Live cone shells such as *Conus geographus* have an extremely poisonous dart under the tail. Treatment is the same as for sea snake bites.

Ciguatera It is rumoured, falsely, that the Red Sea is completely ciguatera free. The dinoflagellate *Gambierdiscus toxicu*s produces the ciguatera toxin which is endemic in coral reef fish including several species found in the Red Sea. The toxin passes up through the food chain from coral grazing fish to man. Barracuda carry the toxin, as do many large predatory reef fish such as grouper over 2kg. Yellowtail snapper and mahi mahi are usually safe, as are small fish. The symptoms can be severe and include tingling lips, burning pain when cold liquids are touched or drunk, intense itching, gastrointestinal complaints, muscle and joint pain. Symptoms get worse after drinking alcohol but there is no diagnostic laboratory test as yet. The only effective treatment is mannitol, given by injection.

Infectious and parasitic diseases

Malaria The disease is caused by the female Anopheles mosquito. It looks different to others because when it lands its body is angled head down. The Anopheles carries the parasite *Plasmodium falciparum*. The female feeds on blood and transmits the parasite through its bite. The best prevention is to cover up. Long flowing robes worn by Arabs are a multipurpose form of protection. They aren't worn only out of modesty but protect against many insect borne diseases. The ports on the W coast of Yemen are high risk. In Eritrea and Djibouti the damper weather of winter and spring mean that mosquitoes can breed, though in Asmara there's nothing to worry about. In Sudan, take special care whether you call at Suakin or Port Sudan. Coastal Egypt is low risk.

Fit mosquito screens to your ports and hatches and cover up if out at dawn or dusk. Other preventive measures include using aerosol insect repellents on your clothes. The most important areas are ankles and wrists. For skin, the liquids or non-aerosol sprays are best, least sticky and easiest to apply. You can also get very effective mosquito burners, not coils, which give off an inoffensive insecticide. They work with small tablets or a bottle of liquid insecticide. Some are dry cell battery powered, others will work off a cigar lighter and your boat's DC system.

Of course you can also take prophylactic antimalarials. Side effects can be a problem and the antimalarials can mask symptoms. Talk to a pharmacist or a doctor before you buy antimalarials, preferably in the Red Sea. Otherwise consult the MASTA website (see Appendix II). Remember to start taking the pills a week before you expect to go ashore in an urban area and for a month after you've left malarial regions. Symptoms can take some weeks to appear. Chloroquine and proguanil are widely prescribed for non-chloroquine resistant malaria. Malarone is now considered the best for chloroquine-resistant malaria but it's only recommended for trips of 1 month or less. Mefloquine (Lariam), doxycycline (Vibramycin) are alternatives, but have side effects. Drug resistant malaria is common because the parasite responsible mutates.

Whatever you do, there will always be some risk. The most important thing is to carry aboard at least two courses of treatment. In 2002 the recommended treatment was as follows:

For non-chloroquine-resistant malaria:
3 Fansidar tablets taken as a single dose

For chloroquine-resistant malaria:
600mg Quinine, 3 times a day for 3 days plus, after the last dose, 3 Fansidar tablets taken as one dose.

Vaccinations The following jabs are often recommended:

Hepatitis A, one dose followed by a booster 6–12 months later, gives immunity for 10 years. Consider Hepatitis B vaccination at the same time as it is available in a combined vaccination. Gamma globulin injection is an alternative but less effective.

Tetanus-diptheria, one dose with a booster after 6 weeks and another 6 months later, gives immunity for 10 years. Then you will need another booster.

Typhoid, one dose with a booster a month later, gives immunity for 2–3 years.

Make sure that your TB and polio immunisation is current.

Viral or bacterial Meningitis A can be prevented with vaccines which give 3–5 years' protection.

Yellow fever is transmitted by mosquitoes. Jabs are only required if you have stayed in areas where the disease is endemic. We have not been asked for certificates of vaccination while sailing N or S through the Red Sea.

AIDS There is a very serious problem in many parts of Africa and you should take the usual precautions. If you ever go to a hospital or clinic make sure to supply your own syringes, needles and intravenous infusion sets. If you are seriously ill try to fly home. For health related books and websites see the bibliography in Appendix II.

Part III
Pilotage

Introduction

Names British Admiralty latest orthography is used. Where old and new names differ markedly, we keep the older name in brackets, e.g. Mayyun (Perim Island). With changes of one or two letters, e.g. Mersa for Marsa, Gebel for Jebel, or where pronunciation is much the same, e.g. Iri and Er Rih, your native wit will point the way. With inconsistent usage, we choose the most common spelling, e.g. Suakin not Sawakin.

Abbreviations British Admiralty practice is followed with two exceptions. For the nature of the bottom we use lower-case letters. A capital M after figures represents both 'mile' or 'miles' AND magnetic bearings and courses.

Positions Where known, GPS positions are noted on each anchorage sketch. Unless otherwise specified GPS positions are to WGS84. Use with caution and ensure you have corrected to the chart datum. Positions are to two decimal places. No one should be navigating in these waters to accuracies better than 20m. One GPS decimal place (200m), is more prudent.

Plans Sketches are for orientation only. Although accurate enough for eyeball conning, you should not plot positions on them. Use them with up-to-date charts. Key maps are solely for planning and quick reference.

Symbols British Admiralty (and international) conventions are used except as indicated in the 'Key to symbols' on page viii.

Soundings In metres and tenths.

Bearings These are true and from observer to the object observed unless otherwise shown.

Charts Admiralty chart numbers are given for ports and anchorages shown on large scale charts.

1. Southern approaches to the Red Sea

Admiralty charts
Large scale 2896, 3784, 3660, 2950, 3530, 262, 253, 2588, 7, 5
Medium scale 2895, 3661, 5, 100, 253, 2588, 1925
Small scale 3784, 6, 4703, 4704, 4705, 157

US charts
Large scale 62046, 62093, 62095, 62097, 62098, 62302, 62313
Medium scale 62092, 62046,
Small scale 62000, 62024, 62040, 62050, 62070, 62080, 62090, 62091, 62202, 62306, 62310

French charts
Large scale 4792, 6326, 6388, 6947, 7520
Medium scale 6878, 7519
Small scale 6265, 6266, 6947, 6987, 7519

Warning
Although the most recent editions of charts from all the major hydrographic offices are reconciled to WGS84, this CANNOT compensate for shortcomings in the original 19th century surveys. Note also that apart from a 2000/2001 survey including Bab el Mandeb mainly to medium scale, only parts of the Omani coast have been surveyed to ANY WGS datum. Three points of navigational importance follow. First, always navigate, even with GPS, with circles of probable error (CEPs) of at least 2 miles, and at night at least 5 miles. Second, navigating by GPS alone in the Red Sea and Gulf of Aden is stupid. Do not do it. Make daylight landfalls and use hand-bearing compass and radar, if you have it, to establish WGS84-to-chart error factors. Third, treat ALL waypoints with appropriate caution, including ALL those given below, and NEVER assume that the waters between, close to or at waypoints are free of hazard.

Oman

Mina Salalah

Admiralty chart 2896, plan page 69

Approach

There is a conspicuous container terminal with 12 large gantries. Two pairs of port and starboard-hand buoys lead to the port entrance as well as a set of directional lights. Stay in the white sector.

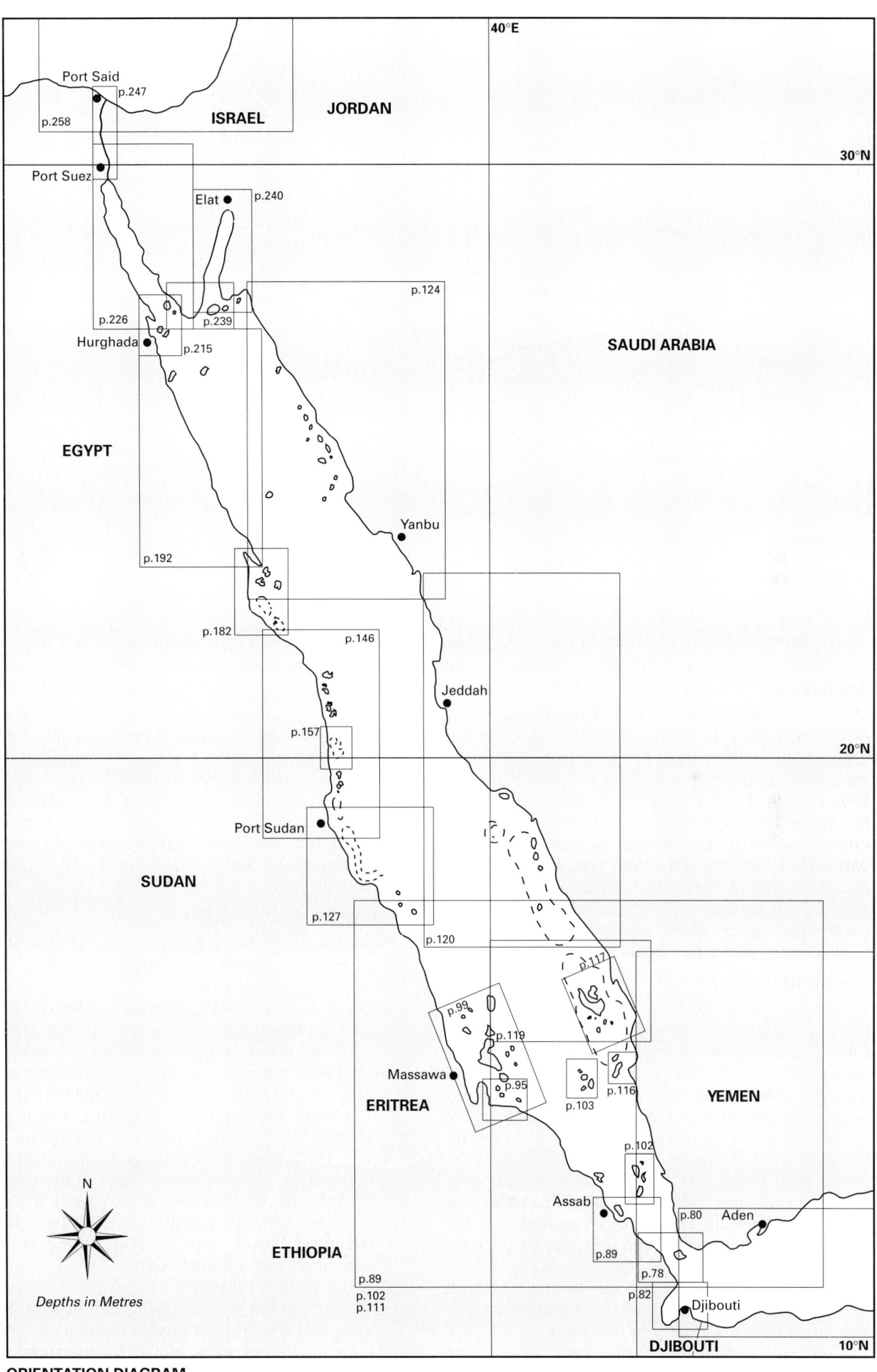

Southern approaches to the Red Sea

ORIENTATION DIAGRAM

Depths in Metres

Red Sea Pilot

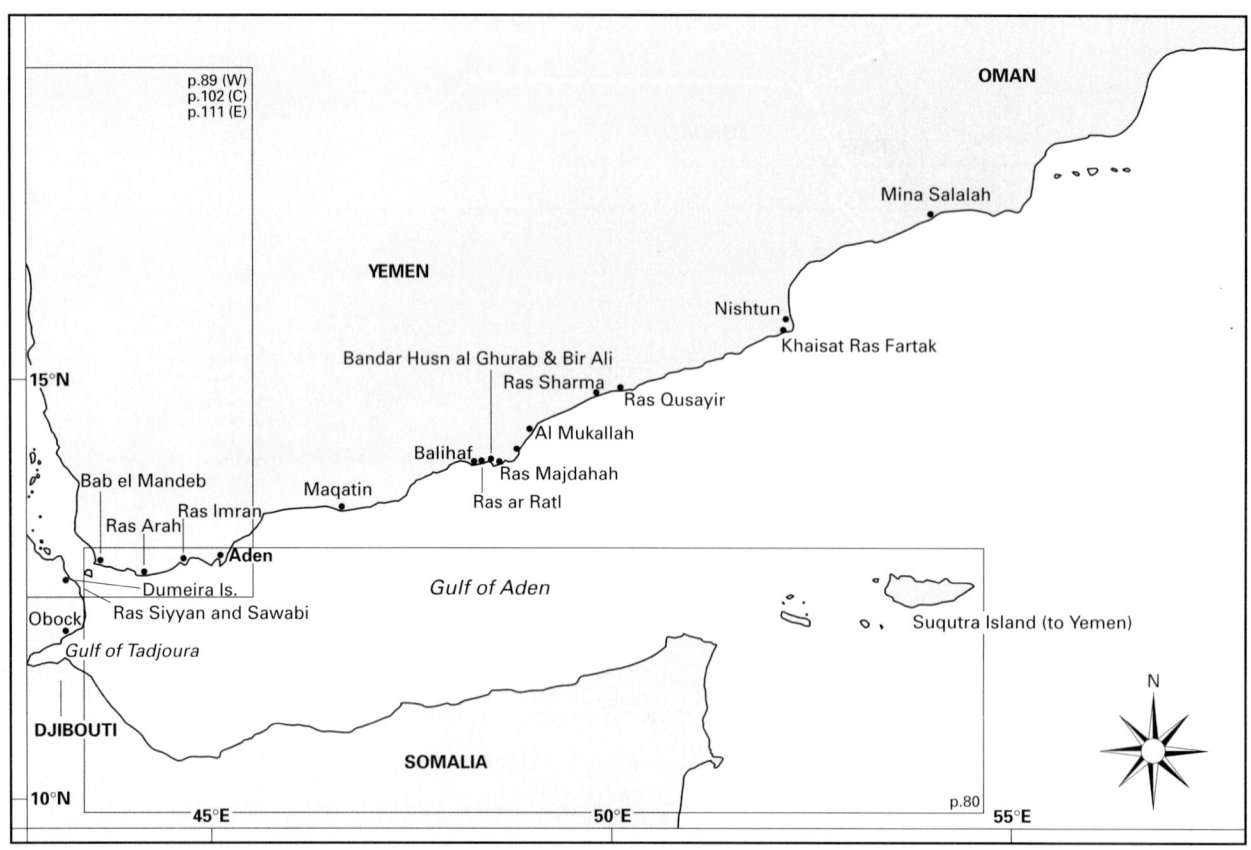

SOUTHERN APPROACHES TO RED SEA EXCLUDING DETAILS OF N SOMALIA AND GULF OF TADJOURA

Anchorage

Yachts anchor in the last but one basin on the S side of the main port, in 4–7m, reasonable holding but rather tight at the height of the season. Pick your spot carefully and be prepared to move if requested. If you get in the way of port operations and you're not aboard they'll do it for you. The Royal Navy of Oman base is at the SW end of the basin. Their patrol boat exits regularly stern first. It needs room to manoeuvre. A clear lane at least 30–40m wide must be left on the E side of the basin. Leave your dinghy tied to any of the iron steps near the wash block clear of any police boats.

Formalities

Call on VHF Ch 16 on approach and hoist a Q flag. Customs will come aboard to do paperwork. In theory, there is no clearance on festival days or public holidays but that said, we were cleared in on a Friday evening. Strictly no visitors are allowed before you are cleared. Customs issues a clearance document and shore passes that allow 8 exits from the port area. Passports are lodged with the port police post at the gate. Call there to get the pass stamped each time you go to town. The officer on duty will tell you the latest time for returning to the port. Explain if you think you'll be out late. Check in at the gate on return. You can apply for an extension of your pass if necessary but if you intend to stay for a long period, apply for a visa on arrival. See Introduction, Country information.

When you want to leave, take your inward port clearance to Customs. If you have signed on crew, go next to see Immigration to process a new crew list and have the new crew's visa stamp cancelled. Then exchange your Customs clearance for your passports and a port clearance at the Police post. Allow plenty of time for all this. If you want to clear out in the evening and leave very early the next day you may have to make a special case but it can usually be arranged.

Facilities and services

Facilities for yachts are improving. Salalah is very well run compared with most Red Sea ports. Haulout is possible here.

Water The wash block's showers and toilets are not very tempting but there is a good laundry area. The tap water is potable for jerry-jugging. To load large amounts, go to the port administration office (outside the main gates on the left) and pay 1 Omani rial (approx US$2·50). You'll get a receipt. Then arrange with port control on VHF Ch 16 where to go alongside at a mutually convenient time to water ship. Laundries are plentiful in the town but most people do hand-washing. Garbage is collected daily from beside the wash block but in 2001 there was no container, just a pile of bags.

Fuel Get diesel and petrol by jerrycan from service stations on the way to town. The nearest is 6M away and the simplest answer is to hire a car. For large quantities only, make arrangements with

Southern approaches to the Red Sea

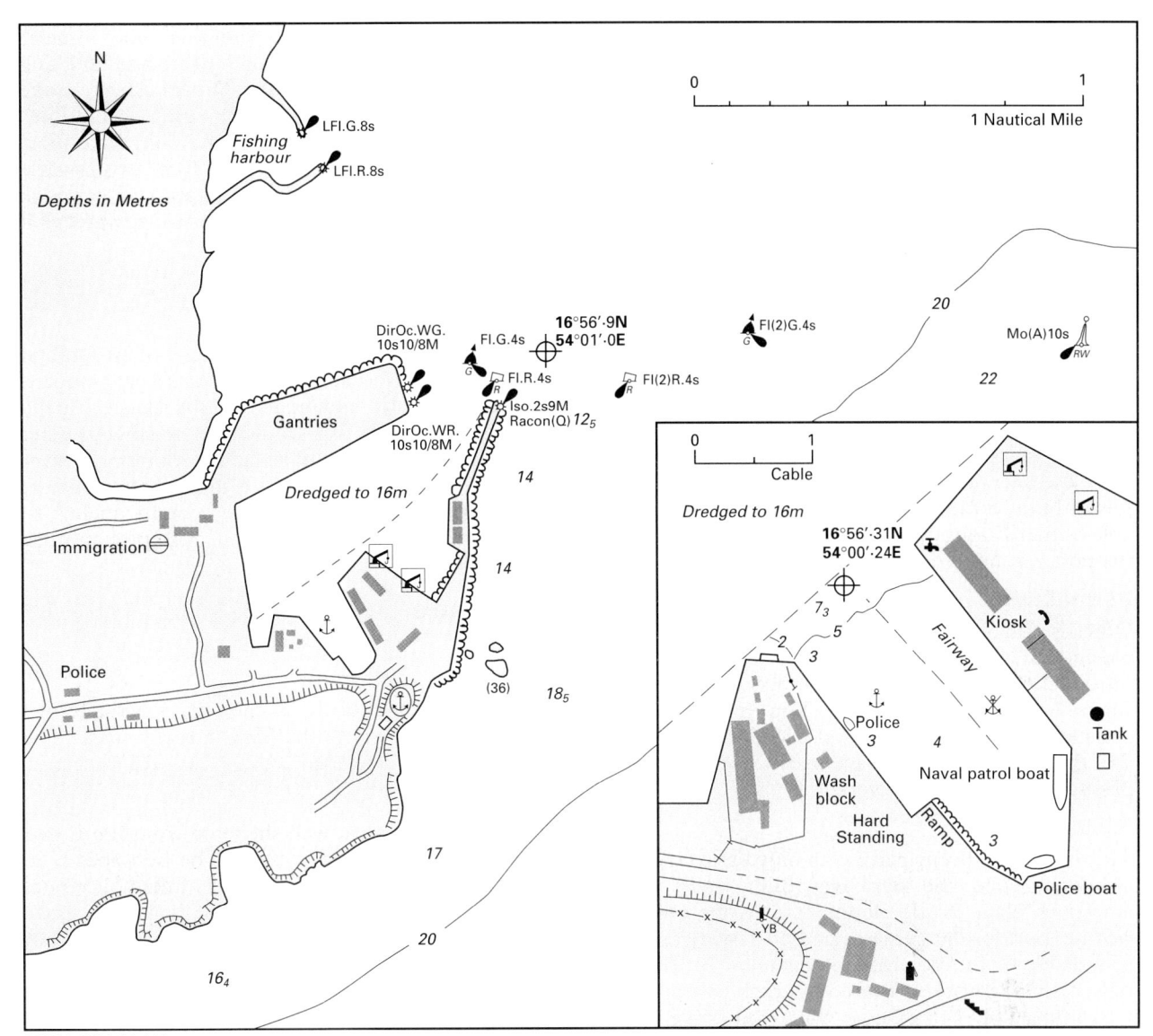

MINA SALALAH

Salalah Port Services for delivery by tanker. Think in tonnes. LPG from the industrial area on the way into town where there are many workshops.

Money Cash is available from ATMs. The nearest is a HSBC near the junction of the port access road and the highway. Salalah is the place to get US$ for the trip up the Red Sea. Try to calculate how much you will need. In 2001 the next ATMs for easy withdrawal of any cash were in Djibouti or Egypt.

Food and stores High quality fresh produce is widely available. There are good supermarkets, e.g. Lulus and Cold Storage (a.k.a. Spinneys), who deliver if you buy enough. Prices are high compared with Yemen. Masoor Indian supermarket on the main shopping street, Almontazah, is also good value.

International Project Services have opened in the port area near the anchorage and can offer some chandlery items and repairs. They hope to offer duty-free stores in the near future.

Communications There is a GSM mobile phone network. Phonecards, on sale from the kiosk in the port and from many outlets in town, are good for both the phones in the port but for international calls use the phone by the kiosk. Cyber-cafés charge about US$2·50 per hour and there are two not far from Lulus. There is a DHL office in Salalah. The PO, in an elaborate low-rise building next door to the Telecoms centre (conspic with a radio mast and telephone painted on a tower), is in the NW part of town and not very easy to find. The PO's Poste Restante holds mail for 2 weeks. An alternative mailbox for Amex cardholders is Zubair Travel Services at the Holiday Inn, PO Box 89, Salalah, ☎ 235581/82, hours 0830–1300 and 1630–1830. The downside is that the Holiday Inn is way out on the far side of town.

Local transport The port is approx 12M from downtown. It is easy to hitch a ride in, but usually you will need a taxi back. A hire car shared with another boat makes modest refuelling, fetching

LPG and major provisioning easier. Hire rates are reasonable if you don't want the latest model.

Note The Gulf of Aden has a merited reputation for opportunist acts of minor piracy. The risks are not as great as rumour makes them. Salalah is the place to prepare for running this modest gauntlet. We give facts, figures and recommendations in 'Piracy: rumours and risks' in Part II, Planning.

S Yemen to Bab el Mandeb

Nishtun
Admiralty chart 3784, plan page 71

Approach
There are no hazards until the entrance where white mooring marker buoys tailed to the W and N walls make the anchorage tight. A rolly alternative is W of the harbour, off the beach. For fuelling you may be able to berth alongside. The light on the hill behind the port was not working in 2001.

Formalities
Officials, full of welcome, will come out. They check papers but speak little English. A ship's letter of introduction is useful. See photocopiable Arabic ship's letter in Appendix 4. You may need to complete some paperwork and exchange passports for shore passes. To leave early ask for your passports back the day before.

Facilities
This is a small, sleepy place with only basic facilities although when the port was built there were ambitious plans for development. It is the only all weather port for the Yemen's poorest province. Fuel is available by drum, payment negotiable in US$ or rials. Drinking water delivered by truck cost approx US25c per 20l. Tell officials when you arrive what you need. Some provisions are available in the village. An agent, Mr Abraham, might introduce himself. There are buses or hire cars for trips inland to the provincial capital, Al Ghaydah, some 50km N, and, if it takes your fancy, beyond.

Khaisat
Admiralty chart 3784, plan page 71

This intriguing anchorage, with its traditional village climbing up the steep valley from the shore, lies behind a small headland W of the immense cliffs of Ras Fartak. It may have been the incense port at Suagros (Ras Fartak) mentioned in the *Periplus of the Erythraean Sea*. Worth exploration in light weather. We didn't try it because of a tight schedule but were sore tempted. Tuck in well for best protection. Use a stern anchor if it's rolly. Khaisat used to be the main port for the province until Nishtun was developed.

Even on this seemingly barren stretch coast, with its bizarre, black rivers of solidified lava rolling down from the coastal hills and stopping abruptly in low cliffs at the sea, there are villages. At night their lights pop up every few miles between Ras Fartak and Al Mukalla. Fishermen from them may come to you and ask for cigarettes, food, drink and anything you can spare. Some just beg. Others offer to trade fish. They wear headcloths and sometimes balaclavas against spray, wind and sun. Like most boats with outboards, these only have two throttle settings, full ahead and stop. An approach can look alarming but once alongside, smiles usually prevail if you smile too.

Ras Qusayir
Plan page 71

The town of Qusayir and the village of Al Qurayn are conspicuous as are the many local boats moored or at anchor. The anchorage N of the islets is usually blocked by dhows. We would only try this in quiet weather or desperation since the bottom is very irregular and Ras Sharma is a much better option. Fit in where you can and use a stern anchor to control roll as necessary. You will probably be visited by the local authorities.

Ras Sharma
Plan page 71

Approach
Jazirat Sharma, S of the anchorage, is actually two islets. The smaller, to the SE is a rocky stack. The larger is pyramid shaped.

Anchorage
This is a lovely area, well sheltered from the E and with excellent holding in sand. The best spot is in the bay just NE of the islets, clear of the police boat mooring. The police post is on the low ridge above the anchorage. They will probably visit you but since not much English is spoken in these parts a ship's letter in Arabic is very useful. Boats from the fishing village of Al Qarn will probably call offering to trade or sell fish. One that visited us even offered to trade a couple of wives, we think, though we weren't clear what the going rate was!

Al Mukalla
Admiralty chart 3784, plan page 73

Approach
Well lit all round the bay from Bandar Burum to Ras Marbat. A waypoint for the final approach from E is 14°29'N 49°10'E though don't forget Rocky Bank. Night entry is possible with care. Note that the BA large-scale plan lat and long are both out by about a cable. There has been reclamation around the coast. This is shown on our plan but not the chart. Khalf Harbour is the new port, intended largely for commercial vessels.

Anchorage
The anchorage is in the old dhow harbour, between the charted wrecks and the sea wall. Anchor fore and aft to counter ground swell. No problem with holding but it can be very rolly. The larger, S'most wreck is the *Maldive Image*. Only the central accommodation structure is left. The N'most wreck

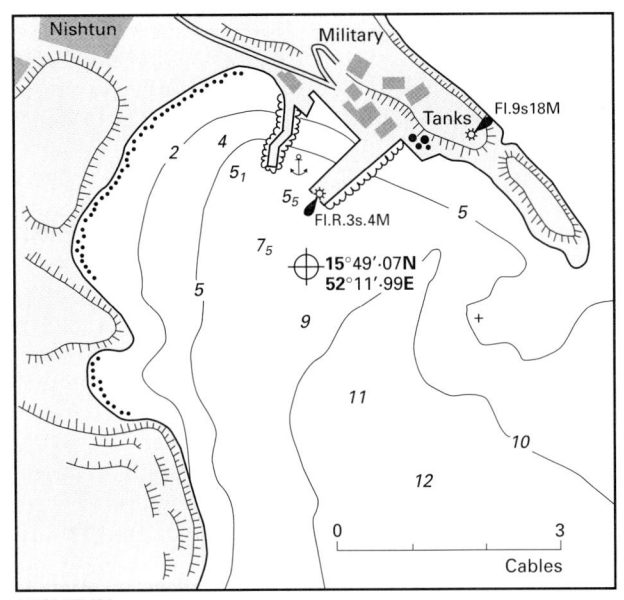

Southern approaches to the Red Sea

NISHTUN

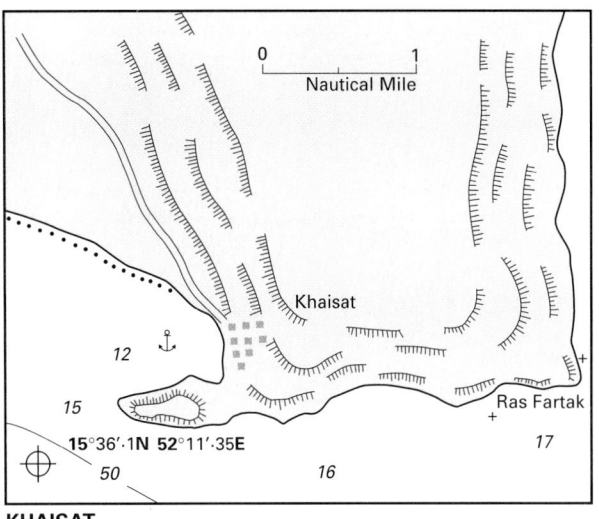

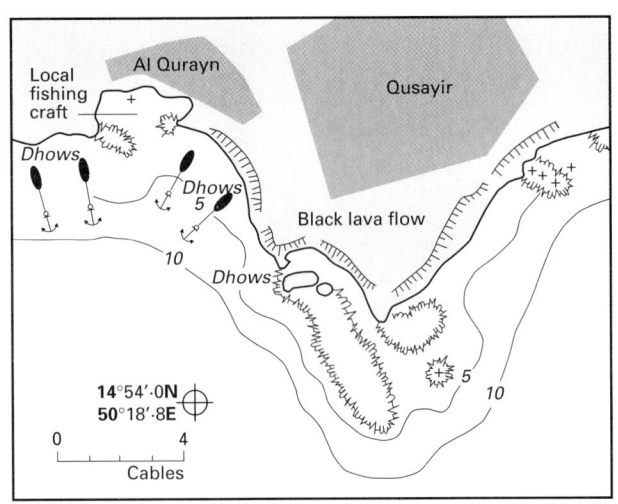

KHAISAT

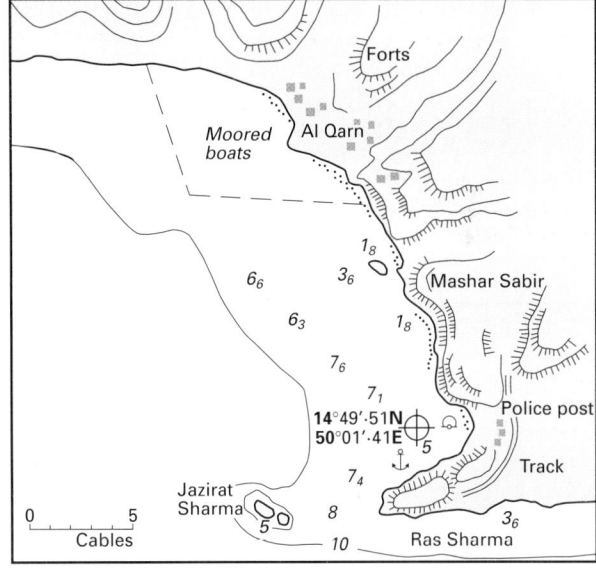

RAS QUSAYIR

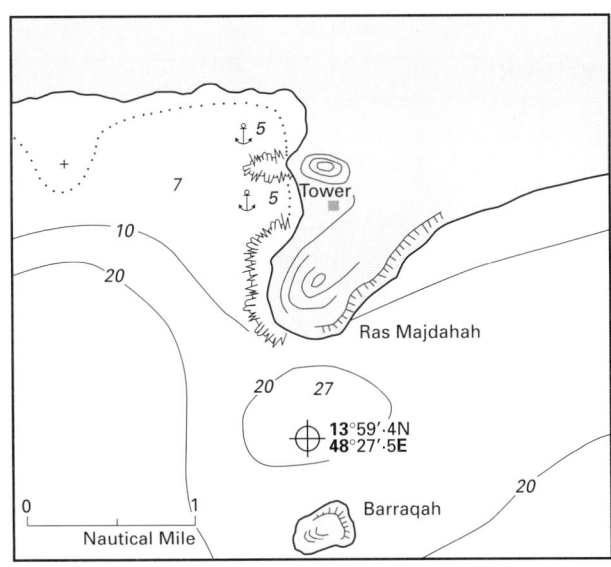

RAS SHARMA

RAS MAJDAHAH

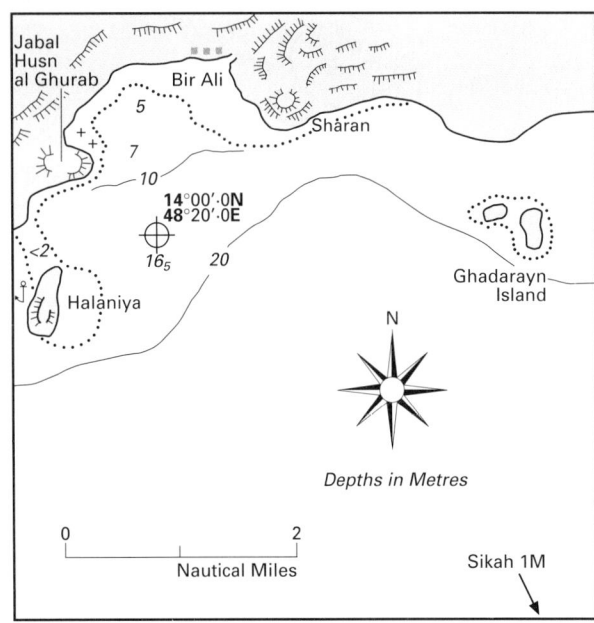

BANDAR HUSN AL GHURAB AND BIR ALI

Depths in Metres

71

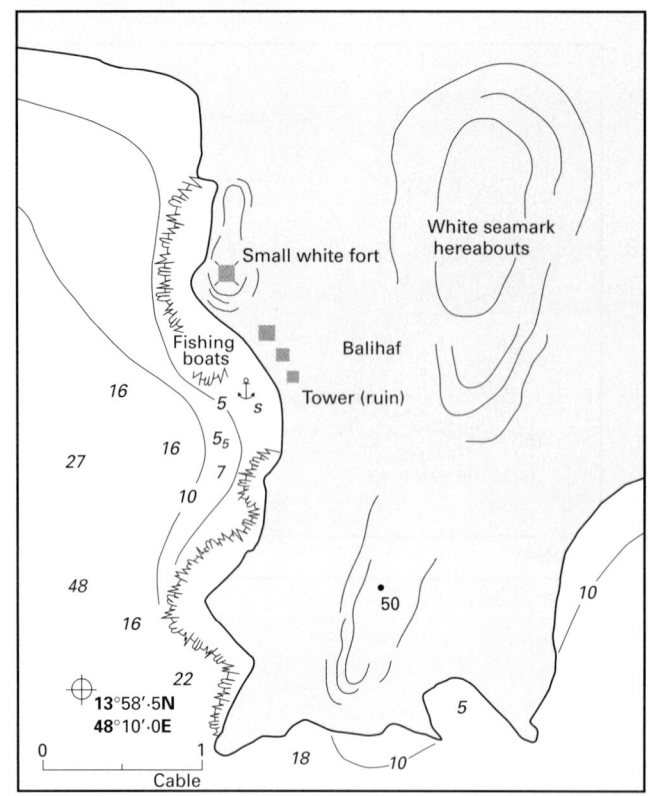

Red Sea Pilot

BALIHAF

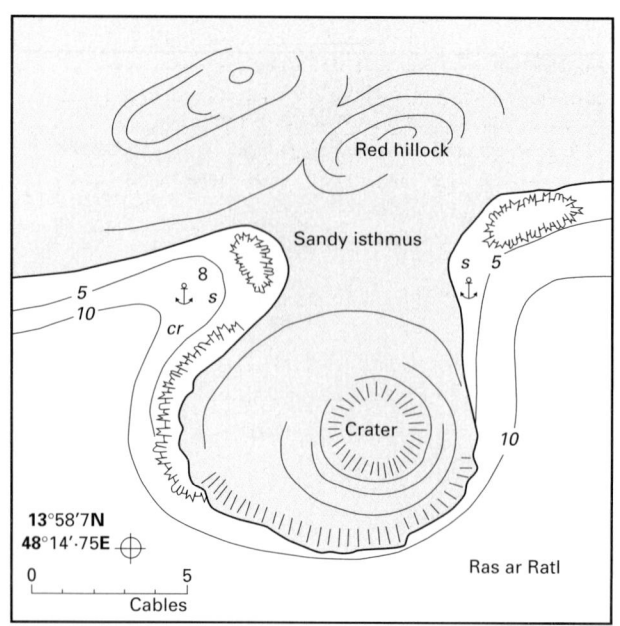

RAS IMRAN

RAS AR RATL

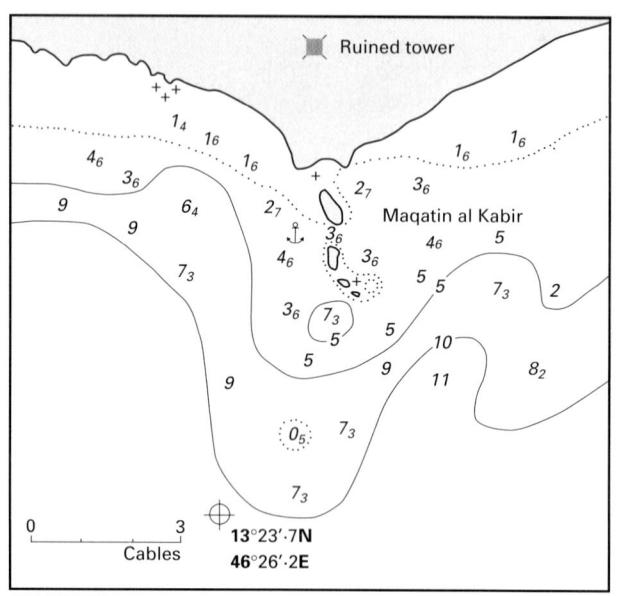

MAQATIN

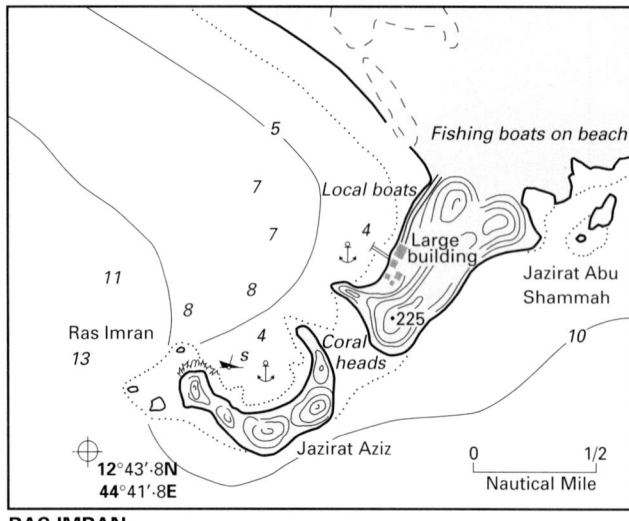

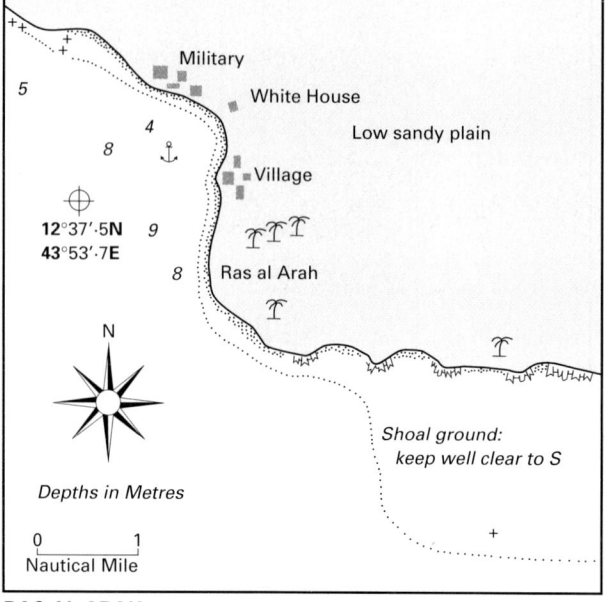

RAS AL ARAH

MINOR ANCHORAGES ON THE YEMEN COAST OF THE GULF OF ADEN BETWEEN BALIHAF AND BAB EL MANDEB

Southern approaches to the Red Sea

OLD AL MUKALLA

- Market
- Internet café
- Ramp
- Moored boats
- M.V Amour
- Wreck being broken up
- M.V Maldive Image (bridge structure only)
- Engine only
- 14°31´·5N
- 49°08´E

Khalf Harbour

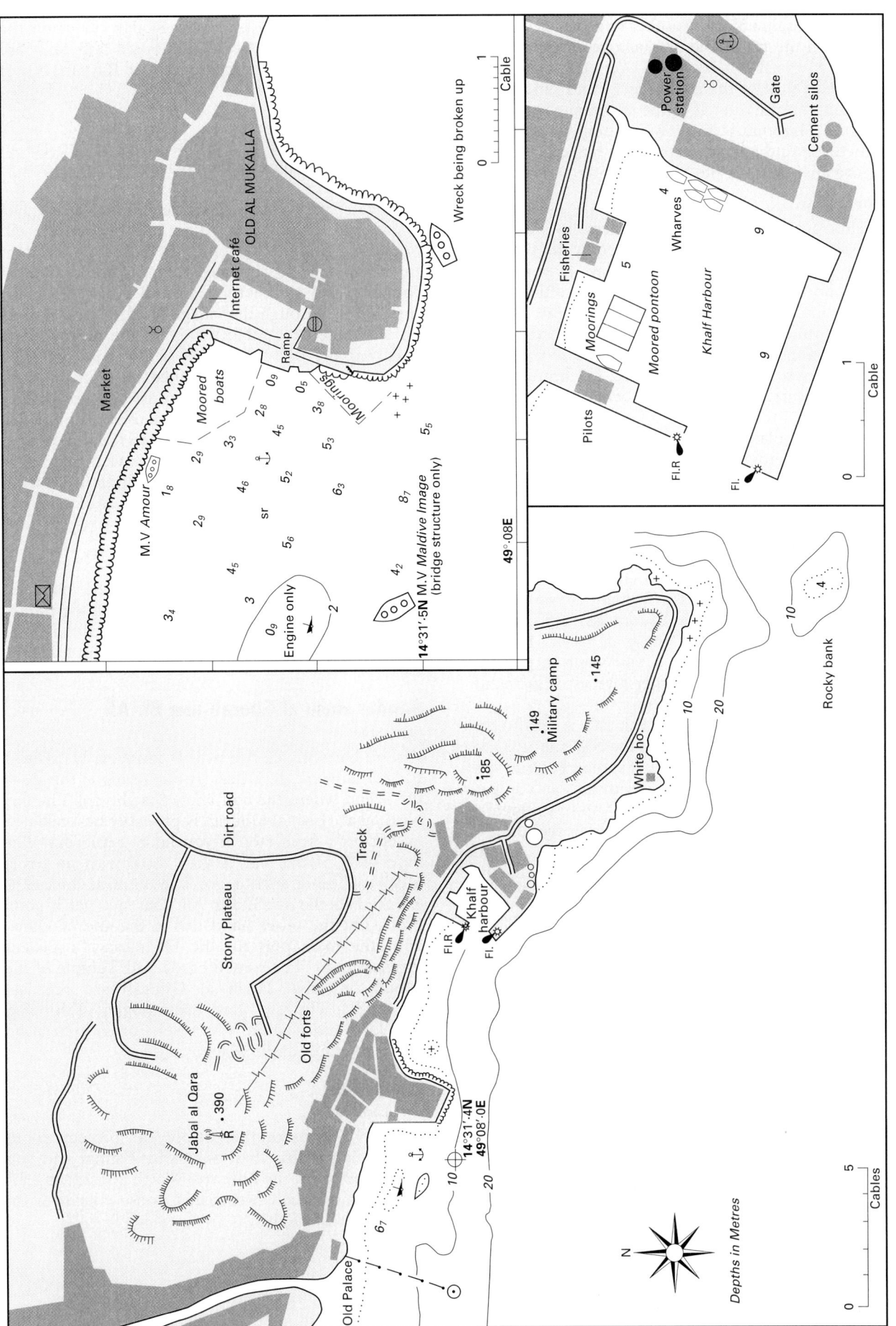

- Pilots
- Fisheries
- Moorings
- Moored pontoon
- Wharves
- Power station
- Cement silos
- Gate
- Fl.
- Fl.R

AL MUKALLA

- Old Palace
- Jabal al Qara •390
- Stony Plateau
- Dirt road
- Track
- Old forts
- Khalf harbour
- Fl.R
- Fl.
- White ho.
- Military camp •145
- •149
- •185
- Rocky bank
- 14°31´·4N 49°08´·0E

Depths in Metres

is only an engine block. Another large wreck is being cut up on the tip of Old Mukalla promontory. The wrecked coaster *Amour* is on the sea wall immediately N of the anchorage. Land dinghies at the ramp on the quay E of the anchorage area.

Khalf Harbour is busy with shipping and it is difficult to find a good berth but it remains the only secure anchorage in the SW Monsoon.

Formalities

Yachts are warmly received. Formalities can be completed at any time though it's polite not to check in or out on Thursday and Friday. If you radio ahead on VHF Ch 16 giving your ETA, officials will come out. You'll need crew lists. Shore passes, usually valid for about 3 days, are issued at the Immigration Office (near the ramp where you leave dinghies) when you first go ashore. Let Immigration know at least 24 hours ahead when you want to leave.

Visas for inland travel are US$80–90. They are cheaper in Aden and embassies abroad. Crew leaving yachts can go to the airport without visas if checked in as ship's crew.

Facilities

Local unofficial agents include Eskander Fadhel who may meet you when you check in. He acts as unofficial interpreter during formalities sometimes. Ask for his prices for fuel, water, gas and laundry and compare them with those in town. He can circumvent hassle for you if you want to pay the modest extra dollar. A weather forecast is available from the Port Captain.

Fuel For a small quantity you can take a jerry to the town's sole service station in the NW suburbs. Be prepared to find that the fuel station has run out of diesel. Local prices are subsidised so if you pay them for yacht fuel, technically you're smuggling and you'll run foul of customs at the dinghy ramp. Life is easier if you use an agent. Alternatively, write a letter of request to the harbourmaster's office stating how much you want. They issue a chit for that amount which satisfies customs. More convenient but more costly is to bunker at the fishing dock in Khalf Harbour. Make prior arrangements with the port manager.

Water Tap water for washing is available for a few rials from restaurants by jerry. Drinking water is sold at about US25c for 20l jerry cans. If you need a lot you can arrange for delivery by tanker to the jetty. There is between 2·6m and 2·9m alongside the old harbour wall at HHW springs. Alternatively, if you decide to bunker in the new port, you can water ship there.

Other services Use a moneychanger on the main street for exchanging US$. The town is good for provisioning, though not as cheap or as varied as Aden. There is a very convenient Internet café about 50m from the dinghy landing point. The PO is about 500m away on the waterfront. DHL and Fedex are also in Mukalla. There is good photo processing by a new Kodak outlet on the waterfront road. Local minibuses and taxis are cheap. The domestic airport is at Rayyan, NE of Mukalla.

General

Mukalla is a very interesting first port of call in Yemen. If you have time, consider a trip inland to visit the Hadramawt or closer inland to the freshwater rock pools. Eskander Fadhel or a travel agent will organise it. The harbourmaster will organise a guard to look after the boat for a few days for about US$20. Locally, for a splendid view of Mukalla, follow the road around Khalf Harbour, turn left up through the village N of the new port to an old camel track that winds up the narrow wadi and over the pass. Turn left at the pass and pick your way towards the old ruined forts on the cliffs above the town. The easternmost sits on an airy peninsula all of its own, connected by a knife-edge ridge. You can push on as we did to the top of Jabal al Qara whence, with some bushwhacking, it is possible walk N of the jabal to return into town from the NW. Allow 4 hours and take plenty of water.

Ras Majdahah

Plan page 71

The passage between Barraqah I and the mainland is clear, deep (27m) and wide. There's good shelter in the NE monsoon behind the headland. The snorkelling is said to be good but watch out for stone fish.

Bandar Husn al Ghurab and Bir Ali

Plan page 71

A sand bar with <2·5m over it connects Halaniya I to the mainland. In E'lies, try anchoring in the lee of the island where the half anchor is shown. The bay N of Jabal Husn al Ghurab is probably best avoided except for a brief stop if you fancy exploring. The inner part of this bay is very shoal with no easily accessible inner anchorage. The outer anchorage is exposed to the swell. Bir Ali does not get a good press. On the other hand this is the site of Qana, once the main port for the Hadramawt's ancient incense trade. The ruined castle and Temple of the Moon on Jabal Husn al Ghurab are the last remnants. The site has been pagan, Ethiopian Christian and Islamic. The results of an excavation are in the museum in Aden.

Ras ar Ratl

Plan page 72

In the NE Monsoon there is only fair protection but there is less traffic than at nearby Balihaf and you are further from the military. Worth considering for a brief stop in light weather. Choose your spot for the anchor carefully.

Southern approaches to the Red Sea

Balihaf
Plan page 72

Ras Al Asida is a good landmark for the approach. Very good protection in the NE monsoon but anchor out far enough to avoid the rocks off the coast, or eyeball your way in through the reefs to tuck in out of the swell. The Aden authorities told us of new prosperity in the area thanks to the oil/LNG terminal a few miles W. There is a local military presence. If you plan stopping, let the authorities in Aden or Mukalla know so they can advise the military. Central government control of the coasts of Abyan and Shabwa Governorates is not absolute. This is where the pirates are most likely to be from!

Maqatin
Plan page 72

Take care to avoid the shoal patch S of the anchorage. There is good shelter from the E but usually a ground swell during NE monsoon. The flood tide sets W.

Aden
Admiralty charts 7, 3660, plan page 76

> '...the sea heaves with the sweep of the coast past the mouth of Aden bay...and half bathed in the light, half black in their own volcanic shadows, the rocks of Aden stand up like dolomites, so jagged and old...There is a feeling of gigantic and naked force about it all and one thinks what it was when these hills were boiling out their stream of fire, hissing them into the sea, and wonders at anything so fragile as man living on these ancient desolations.'
> from Freya Stark, *The Coast of Incense*

Approach

The installations at Little Aden have done nothing to diminish the spectacular approach to Aden. If you are coming from the E, head for the fairway buoy. From the W, pass either side of Jazirat Salil, then use the buoyed channel.

When you are a few miles out, call Aden Port Control on VHF Ch 16 or 2182kHz to give your ETA. If you arrive in daylight they will:
1. Ask for the yacht's name, name of skipper (surname and first name), port of registry, nationality, (gross) register tonnage, last port of call and next port of call.
2. Ask if you know where the anchorage is.
3. Tell you to call again when you near the breakwater and are within sight of the harbourmaster's office.

When you call back, port control will tell you to proceed via the buoyed channel to the anchorage, to check in with customs and immigration as soon as possible after anchoring (open 24 hours) and to visit the harbourmaster's office at your convenience.

If you arrive after dark explain who you are and where you've come from, and they will probably tell you to come in. They may even provide a free escort since technically yachts are not allowed to move within harbour limits at night.

Conspicuous

Aden, with its spectacular backdrop, is hard to miss. The clock tower and the Prince of Wales Pier are good landmarks for the yacht anchorage.

Anchorage

You are expected to anchor N of the Prince of Wales Pier. If you go elsewhere you'll be asked to move unless you have made special arrangements. Holding off the pier is fine though the anchorage can get quite bouncy. The real problem is bunker oil. There is a leak into the harbour from oil that drained into fissured bedrock from ruptured storage tanks, damaged during one of the civil wars. It also leaks from poor maintenance and carelessness at the ABC depot. The best way to clean coated topsides is with kerosene. The harbourmaster is sympathetic and will usually give you permission to anchor for cleaning up on the N side of the fairway, W of the reclaimed area, where marked on the sketch. The only dinghy landing is at the W end of the pier. If you go inside the moored police boats have a long painter and remember about tides. The corner by the sea wall dries out. Outside on the small, mostly immobile harbour tug is better if you're away for long, but it can be hard on dinghy topsides.

Formalities

The Immigration Office always has somebody on duty. It is in the small pink building on the SE corner of the pier area. Multiple copies of crew lists, copies of the main pages of passports and one photo per crew member are required. Bring passports to Immigration every time you go to town to exchange for shore passes that are issued whenever you leave the port area. If you need your passport, e.g. for a bank or embassy visit, explain and you can take it away with you. On your way back to the boat you must retrieve your passport. Ostensibly, without a visa you aren't allowed ashore after midnight and may get fined if you keep a shore pass overnight. Visas are available from Immigration in the port for approx $60 if you want to travel inland. You will need a photograph and a letter signed by the skipper explaining where you are going and for how long. This extra paperwork is all in the interests of security. Customs comes next. The officer on duty is either with Immigration or in the Prince of Wales pier building but paperwork is minimal.

Within 24 hours of arrival you should visit the harbourmaster's office. This is in the tall pale green/grey building high on the promontory SE of the entrance breakwater. Captain Ali, the Aden harbourmaster, is most welcoming and an excellent source of information about cruising along the coast of Yemen. If you intend to stop in the Hanish Islands, now is the time to ask for a permit or letter of authority. This may ensure the Hanish patrols are aware you are coming and may obviate both hassle

Red Sea Pilot

ADEN (INSETS: ADEN AREA AND PRINCE OF WALES PIER)   12°44'.5N   44°56'.9E

76

and improper attempts to extort money. The best time to catch Captain Ali is on weekday mornings, except Friday.

On departure, go first to Immigration with a crew list and passports. If you have a visa and intend to stop at any more ports in Yemen be sure to say so or your visa is automatically cancelled. You will get a clearance document to take to the harbourmaster's office for outward clearance. See the duty officer on the top floor or the harbourmaster himself. If you want to go to the outer anchorage to clean your topsides before leaving this is a good time to make your request. For this, ask Capt Ali in person for permission. You may be visited at this alternative anchorage. If so, explain that you have permission. On the way to the boat call again at Immigration for final clearance. There are no fees though you may be asked, illicitly, for baksheesh. When you have weighed anchor, call the harbourmaster on Ch 16 to say you're leaving.

Facilities

There is no need to use an agent here. Taxi drivers and others will present themselves and offer services but beware, some can be a real nuisance. It's not difficult to find your way around and local people are very friendly.

Fuels The easiest way to obtain diesel is from the Aden Bunkering Company (ABC) since the fuel pump on the pier is only for local boats. No customs clearance is required. Go alongside ABC any morning except Thursday or Friday. It's filthy but has a minimum 3m. Walk to the administration block on the waterfront at the N of the compound. Tell them how much you want and they'll issue a bill to take to the cashier in the building near the entrance gate. You pay, in US$ only, approx 25c per litre, get a chit and go back to the admin block, which issues an order for your fuel. When you've finished go back to the admin block for a receipt. At that point, if you've loaded less than you paid for, you'll get a chit to take to the cashier for a refund.

If you only need a relatively small amount of fuel, write an official request addressed to Police and Customs. Ask a taxi driver to take you to the relevant offices. They will stamp your request and with that authorisation you can get diesel and petrol by jerrycan very cheaply at a nearby service station. Payment is usually in US$ only. Customs may stop you on your return and ask to see your authorisation. For paraffin and LPG seek the help of one of the taxi drivers.

Water No charge for water from the quay taken by jerry, but there's a nominal charge of US$3 to lie alongside to water ship. For block ice from the fish-market arrange delivery with a taxi driver.

Money No credit card or ATM service from any bank in 2001. Moneychangers in Ma'ala offer good rates of exchange. If you need US$, fax or phone your bank at home to arrange a transfer via the Arab Bank in Ma'ala, opposite the Shell service station, or use Western Union in Crater. Allow a week. Note that in Eritrea and Sudan the difficulties and time lags to get US$ are even greater.

Stores Fresh fruit and vegetables are easy to find and cheap. Crater has the best produce market and small supermarkets in Ma'ala and Tawahi have the basics at good prices. Big supermarkets in Khormaksar are overpriced but have special items hard to find elsewhere. Crater and Ma'ala are best for hardware. Eating out is fun and cheap, even if menus are limited. There are several small restaurants near the market in Tawahi, which serve delicious roast chicken, fish and kebabs. Alcohol is expensive and available only in a very few places like the (noisy) Seaman's Club, W of Prince of Wales Pier.

Communications International phone services with metered phones are ubiquitous but overseas calls are quite expensive. There is a GSM mobile phone network, but it was closed down by the government in late 2001 because armed anti-government activists were using it to co-ordinate attacks! There is a good Internet café on the main street in Ma'ala opposite the new Ma'ala Plaza, a modern building decorated with flags. The connection is slow but prices are low. The new PO in Tawahi between the pier and the harbourmaster's office is unreliable. Mail we sent was lost, though our poste restante all arrived. The same is true of the small postal bureau at the pier. Fedex and DHL have offices in Aden, but importing equipment is a bureaucratic nightmare, which even the locals advise against. Minibuses are cheap with a frequent service to Ma'ala and Crater. Pay the driver when you alight. Agree any taxi fare in advance.

Rubbish and laundry There's a skip outside the pier building for rubbish. Laundries in Tawahi have reasonable rates and they sometimes send reps to the pier to tout for business.

General

Aden is dirty and ramshackle but we find it captivating and have always hugely enjoyed being there. It is the best place from which to visit the famous and stunningly beautiful Yemeni capital, Sana'a. For a small fee a military guard will keep an eye on the boat while you are away if nobody on another yacht can do it. There are slow air-conditioned coaches, clapped out and quasi-suicidal service taxis or domestic flights to Sana'a. You can get Egyptian visas in Aden or Sana'a, usually on the same day you apply. Payment must be in rials. For those headed E, visas for India are available in Sana'a. Allow a week unless you really hassle.

Red Sea Pilot

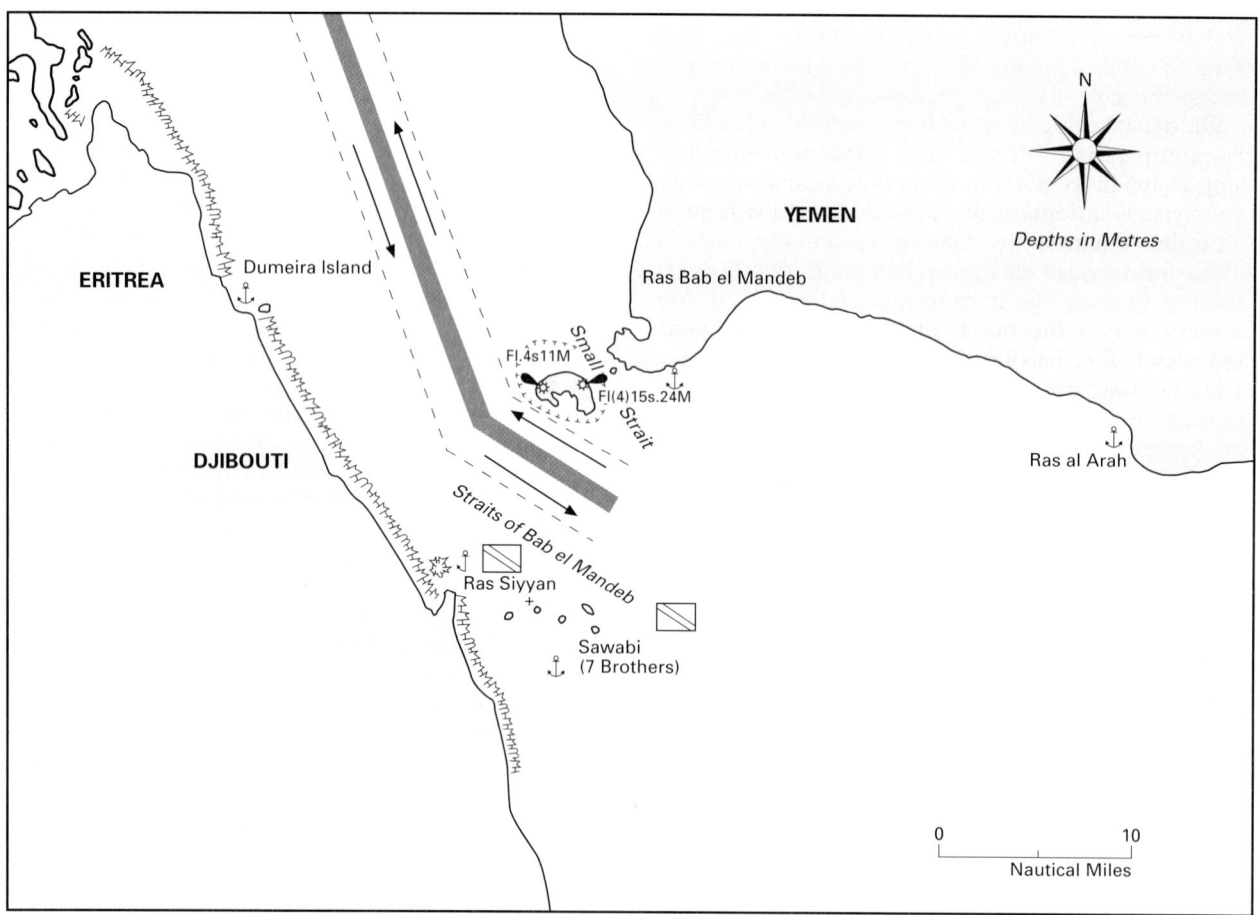

STRAITS OF BAB EL MANDEB

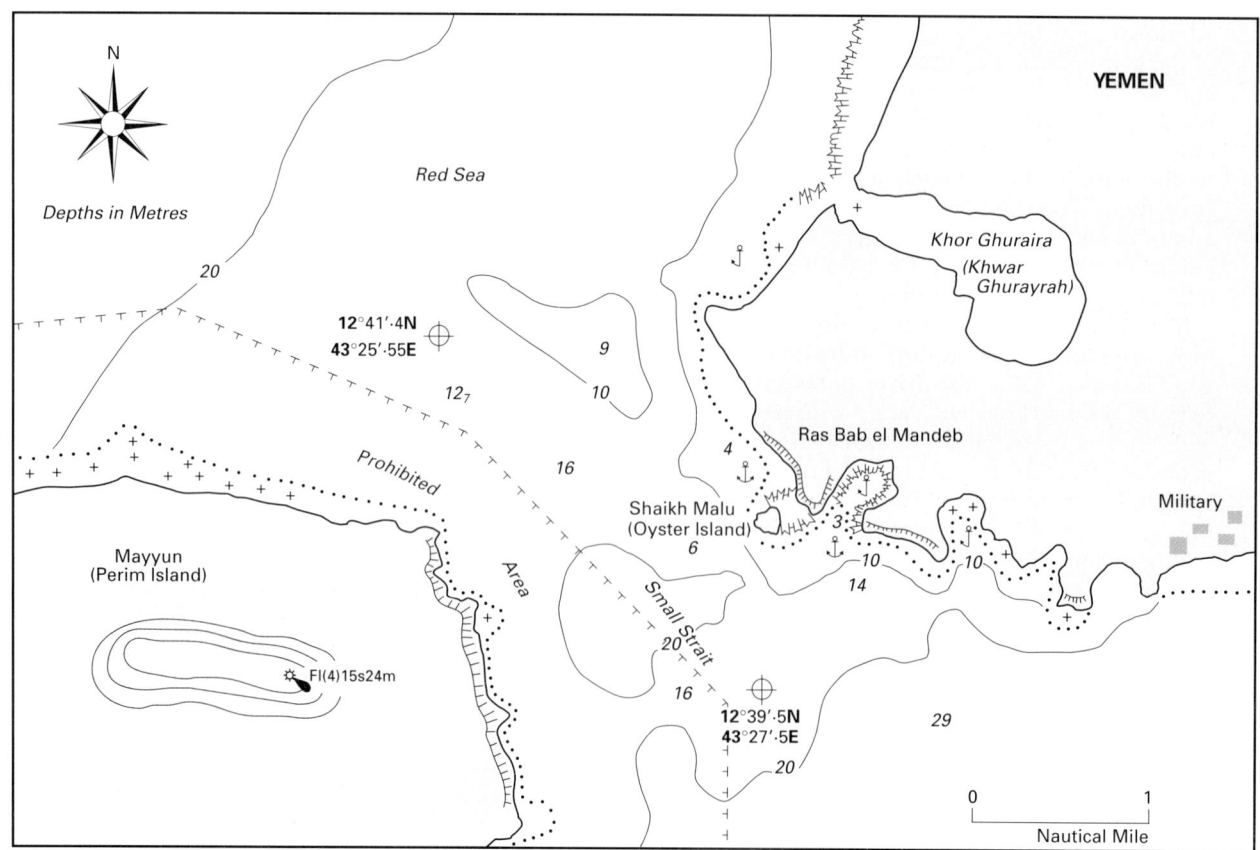

RAS BAB EL MANDEB

78

Southern approaches to the Red Sea

Ras Imran
Plan page 72

Local boats go through the pass between the old crater of Jazirat Aziz and the headland but it looks shoal and we didn't try it. In the approaches to Ras Imran at night, unlit huris (the local narrow-gutted, low freeboard, high prowed fishing boat) at anchor are quite common, so keep a good look out.

The anchorage in the crescent of Jazirat Aziz' crater is a lovely spot with clear water and good holding in sand. Good for any wind from N through SSE, and not bad even in a light NW. Many fishermen will call by. They may try to beg but are happier to sell you fish. You can anchor off the village, but it is noisy, busy and the water is murky. In the SW monsoon there's some shelter between Jazirat al Shammah and the mainland but it is shallow and requires good light.

Ras Al Arah
Plan page 72

Give the shoal area SE of the Ras a good berth. There is a military outpost and you may be visited if you stay long. For a quieter anchorage, try S of the village, abeam a group of rusty tanks (the military sort with guns). Sea breezes can be strong on this section of the coast, although there's less wind offshore. Friendly people in the village where basic supplies are available.

Bab El Mandeb
Admiralty chart 2588, plan page 78

There is an extensive restricted area around Mayyun (Perim I) and yachts entering it may be arrested. The whole area around Perim I and the Yemen mainland shore is a military training area and highly sensitive. The authorities would prefer yachts to use the Large Strait. If you choose Small Strait remember, soldiers only understand straight line courses. Tacking, any sort of zig-zagging, or slowing down is viewed with suspicion. The authorities ask that yachts do NOT anchor around the Ras. If you are desperate (i.e. have a genuine emergency because of gear failure or genuine exhaustion) and plead stress of weather or engine trouble you will normally get permission to stay overnight. A letter from the Aden authorities may ease your way (see above). Co-operation with the authorities, as in the rest of the Red Sea, usually works fine though military patrols have occasionally pressed for baksheesh.

If conditions make continuing at sea an even worse prospect than military harassment, shelter NW or SE of Shaikh Malu (Oyster I), taking the military as they come. Two other possible anchorages are marked with half anchors on the sketch. The anchorage NW of Shaikh Malu can be rolly. In strong SE'lies it's worth trying further N where the other half anchor is shown.

Unless you are going to or coming from Djibouti, Small Strait, E of Mayyun (Perim I), is best for avoiding shipping and for sailing the minimum distance. If you go through either strait in daylight you may be checked by the military. They are often not in uniform.

Currents in the area set approximately in the direction of the wind and tidal influences are slight. Expect currents <2–3kt setting NW in the spring months when most boats sail N, though there are brief spells when the wind is from the opposite direction. This may reduce or reverse currents, but never for long. In fresh SE'lies combined tidal streams and currents can set N–NW <6 knots in both straits. The new tidal atlas in preparation will be worth having, especially if you are southbound. Winds strengthen and veer as they are channelled by the straits.

A new and extended traffic separation scheme is likely to have been introduced not long after this edition was published. Its extent is shown on the diagram on page 21. Note that a correct observance of the Collision Regulations, Rule 10(c) complicates how you get across from Small Strait to the Eritrean shore or vice versa if you are headed S. Be careful how you play this, the application to the IMO for the introduction of the scheme also indicated that the Yemeni authorities will be in charge of enforcement and the apprehension of those who disobey the regulations!

Yemeni islands E of the Horn of Africa and N Somalia

Suqutra (Socotra)

Recent reports are encouraging and the Yemeni authorities do their very best for visitors who make prior arrangements with them. The island is home to the Socotra Biodiversity Project. About 300 plants that grow nowhere else in the world thrive here.

There have been successful circumnavigations but there are no all-weather anchorages. Tidal streams around the coast are very strong. Someone must always stay aboard and should be prepared to move. A fair anchorage in the NE Monsoon is in the S part of Bindar di Sa'ab, 12°33'·5N 53°22'·7E. In the SW Monsoon the most sheltered anchorage is in Bindar di-Lishah, 12°41'N 54°02'E, though that is not saying a lot. The S coast is sparsely inhabited and waterless, though there is a lee during the NE Monsoon, for example behind Ras Kattani (Rhiy Di-Qatanhin), 12°21'·45N 53°32'·37E.

If you are intrigued, you should arrange to visit by going first to Mukalla or Nishtun, inconvenient though it is. Technically Suqutra falls within the same province as Nishtun so it may be wiser to take that route. There may not be long before the island changes. An airstrip to accommodate jets was laid in 1999 and the tourism developers are arriving.

Red Sea Pilot

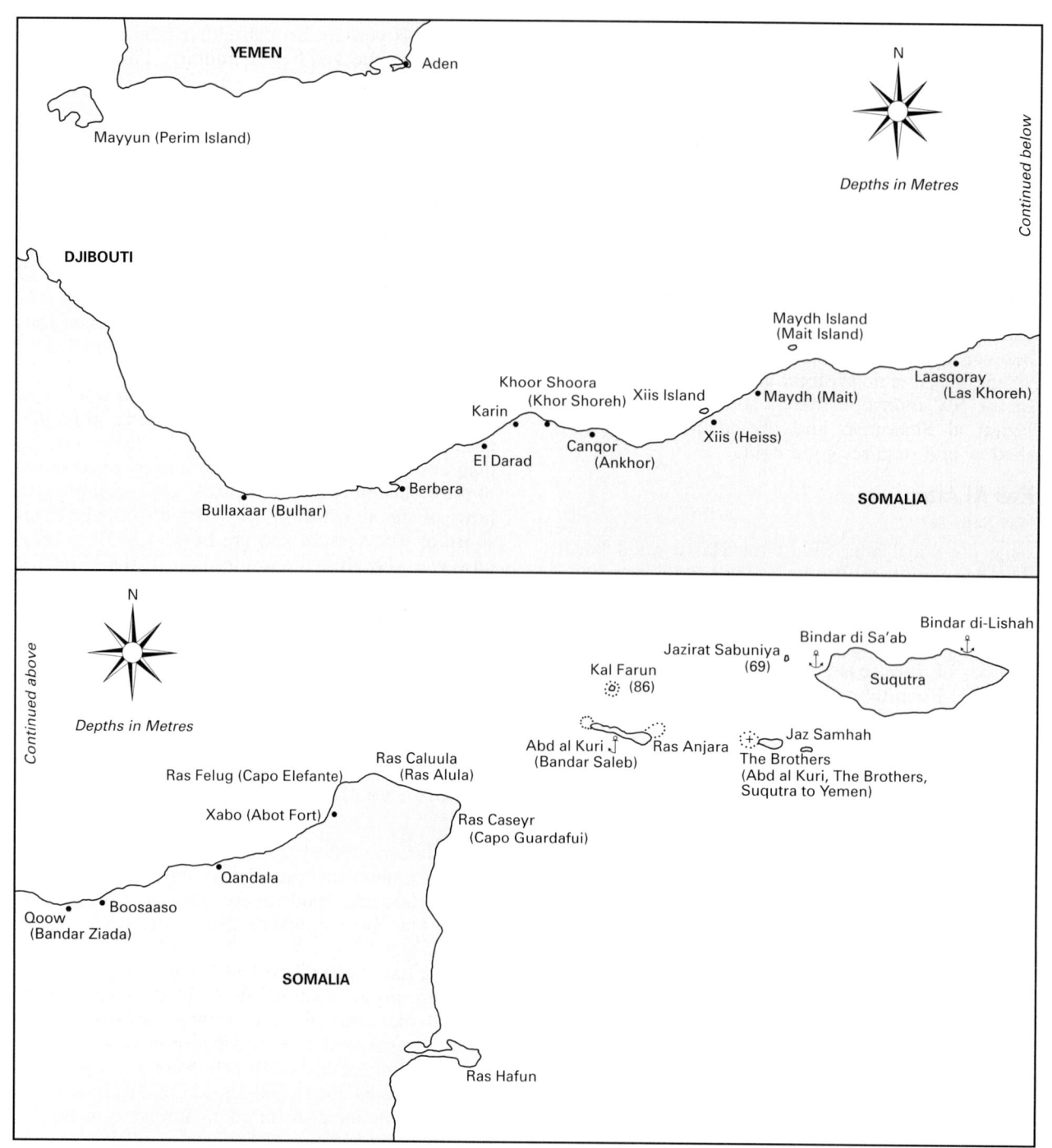

ANCHORAGES ON THE N SOMALIAN COAST AND IN THE ISLANDS OF ABD AL KURI AND THE BROTHERS

The Brothers

The Yemenis call these islands the Sisters. They have a few inhabitants, including a military outpost. Jazirat Samhah, with a high plateau of 700m, is visible from some distance off. Like Suqutra, these offer only vestigial anchorages good for certain weather conditions.

Jazirat Samhah

There is a possible anchorage on the S side where there is some military presence. Worth exploring only if you are desperate and the weather is reasonably settled. Visibility can be very poor and tidal streams strong at the height of either monsoon.

Abd Al Kuri

There are two possible anchorages depending on the monsoon. For NE winds there is Bandar Saleh on the S coast and for the SW monsoon the bay SW of Ras Anjara.

Southern approaches to the Red Sea

Somalia

Again, we give only scanty details of anchorages. In settled times they're worth exploring. At present only desperation would take you here. This is lawless country and anyone getting within 50M of it is taking a big risk. The Admiralty Pilot warns that charts of this area are based on imprecise surveys, and that currents setting onshore may be strong on the approach to Ras Caseyr (Capo Guardafui), the NE extremity of the Horn of Africa. Armed escorts for boats passing through the area can be arranged, but these are very high cost operations. A group of 5 yachts paid US$250,000 for a ten-day escort from Aden to Jeddah in spring 2001! We don't think the risks warrant the outlay but include the details nonetheless.

The company concerned is the Hart Group Ltd, Boosaaso.
Contact Hart Nimrod (Bermuda),
☎ +252 5 726121/826005, ☎/*Fax* +252 523 6104
email nimrod@brtel.net or
george.simm@talk21.com

Caluula
Admiralty chart 2950

This is the first anchorage W of Ras Caseyr (Capo Guardafui). Look for the hospital and the old colonial-style governor's residence. At night the village lights can usually be seen, failing navigational lights, which are unreliable. The chart marks several anchorages, but the best has the light bearing 200° and the fort bearing 090° approx, in about 3m, less than a cable from the shore. Caluula is not a port of entry. If you aren't visited by the authorities but expect to stay long you should go ashore to make your number as a matter of courtesy.

Ras Felug (Capo Elefante)
Admiralty chart 2950

This headland is visible from well offshore and is radar conspicuous. There is an anchorage with protection from winter winds about a cable N of a wreck on the small beach, 5–6m, sand, good holding but a swell usually makes in.

Xabo

There is a wreck on the coast 5½M N of Xabo. A fort in the village is conspicuous. The anchorage in the lagoon is only suitable for boats drawing <1·5m. Hold to the S bank of the pass on a rising tide and anchor near the landing stage used by fishermen. In calm weather you can anchor outside the pass in 10m, sand and rock, to await the tide.

Qandala
Admiralty chart 2950

A navigational light on a house in the village serves as a seamark for approach. It should bear 250° or less to avoid a shoal extending 1M from the shore. When the light, which shows occasionally at night, bears 130°, anchor in 5m a cable off the beach.

Beware of creeks discharging into the sea in the rainy season along this stretch of coast, since they make the water very unclear. Unprotected anchorages on this coast may be dangerous during the SW monsoon.

Boosaaso
Admiralty chart 2950

The town is visible from 5M off. Line up two leading beacons on 150° for the pass through the reef. The most protected anchorage is at the end of the channel, where some shelter is provided by two new jetties, but this is only suitable for boats drawing 2–3m. Otherwise anchor in front of the jetties in 6m, sand, clear of the leading lines. The port is run by the Puntland Ministry for Ports and Transport. Call on VHF Ch 16. The Hart Group run fisheries protection vessel operates out of the port and will assist any yacht in distress.

This is a port of entry; the same regulations apply to both commercial shipping and yachts. There are port dues and health-clearance charges. Immigration will issue transit passes. There is still some UN presence and basic provisions, health care, telecoms, fuel and water are available.

The US sailing directions and others with experience on this coast speak of the dangers of the *karif* (locally known as the *fora*) a strong and sudden SSE wind.

Qoow (Bandar Ziada)

The anchorage here is better sheltered from the SW than Boosaaso, 12M ENE. Anchor in 4m with the mosque E of the village bearing 145°.

Laasqoray
Admiralty chart 2950

Approaching from the E, the watchtower on Ras Laas Macaan will be seen 3M NE of the villages, which are marked by a fish preserving plant. Anchor a cable off the beach in 4m with the ruins on the beach bearing 060° and the small white fort bearing 090°. Swell can make it uncomfortable.

Maydh

Anchor off the beach near the tomb of Sheikh Issakh, 8m, sand. Uncomfortable in strong E or W winds, otherwise fair.

Xiis
Admiralty chart 3530

A rocky islet (50m) marks the anchorage. N of the islet is a rock with less than 2m over it. Anchor in 5m, tucked in behind the causeway (300m long) connecting the islet to the mainland for shelter from E wind and swell.

Canqor

This slight indent in the coast, marked by a white partially ruined fort and a grove of palm trees, gives shelter in light summer winds. In strong SW

Red Sea Pilot

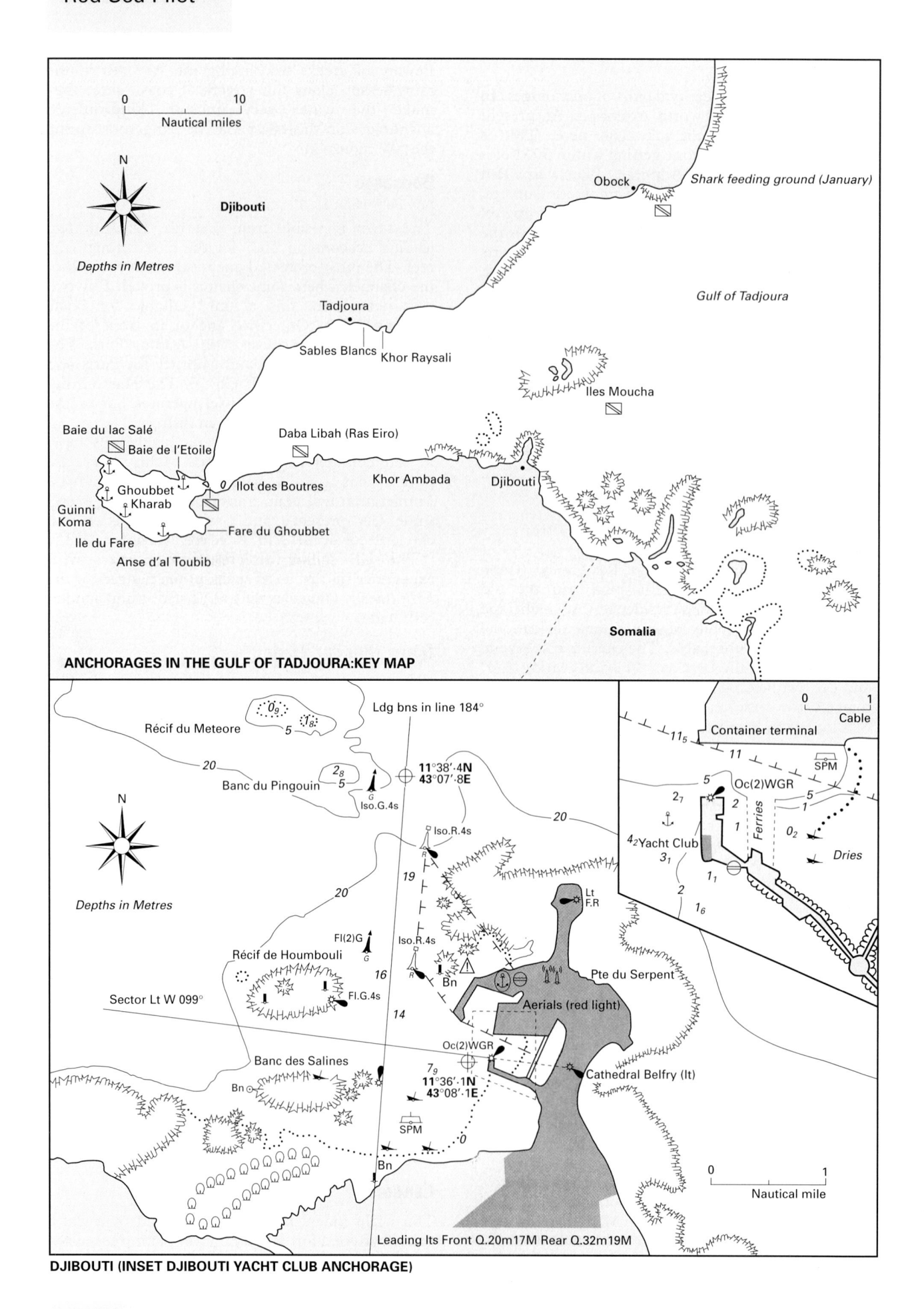

ANCHORAGES IN THE GULF OF TADJOURA: KEY MAP

DJIBOUTI (INSET DJIBOUTI YACHT CLUB ANCHORAGE)

82

weather there is somewhat better shelter in Gubed Rugguuda (Ghubbat Raguda), 2M W of Las Arwein. Both are open to the winter E winds and untenable unless the weather is calm. Anchor close inshore in about 5m.

Khoor Shoora

There is a summer monsoon anchorage in approximately 5m, about a cable from the entrance to the pass, about 3M E of Ras Khansiir. The lagoon is only accessible on a rising tide if you draw under 1m.

Karin

Approach with caution to avoid the 5M long sand and coral bank off the village and anchor in 5m SW of the village.

El Darad

A ruined tower 1M NW of the village serves as a landmark. Anchor in 4m, sand and dead coral, with good shelter from E winds provided by Ras Sudda.

Berbera
Admiralty chart 3530

This is Somalia's premier port and port of entry, with a natural harbour formed by a low sandy spit. One good landmark, 8M SE of Berbera, is Great Gap, a remarkable pass in the irregular mountain range inland. Six peaks show through the gap on a SSW bearing in good light. However, dust storms often obscure visibility. The Admiralty Pilot warns that all lights have been extinguished and all light buoys removed, and that leading marks are difficult to pick out in daylight. There is no clear anchorage. Go to the customs pier in the first instance, unless instructed not to do so by an official launch. Unless you have a special reason for calling at Berbera it is undoubtedly worth avoiding at present. This is a pity because it has an excellent sheltered all-weather anchorage and in better times there was, and one hopes will once again be, good provisioning.

Note The W'most anchorage on the N Somalian coast is at Saylac, but this is not recommended, owing to security risks, unhelpful officials and poor shelter in winter.

Djibouti and the Gulf of Tadjoura

Djibouti
Admiralty chart 262, plan page 82
(French chart 7520 is a good possible alternative.)

Approach

Navigational lights and marks are usually very reliable. The leading light for the main channel is in theory sectoral. Stay in the white sector. Approach from the W between the Recif de Houmbouli and the Banc des Salines. Keep the cathedral belfry, which should be lit, and the light on the end of the government jetty in line bearing 099°. Anchor at the end of the government jetty by the yacht club pontoons. Don't go in too far, it's shoal. Call the harbourmaster on VHF Ch 16 on approach and remember to hoist a courtesy flag.

Dangers

In the early approaches, if you choose to leave the Iles Moucha to the N, note the unlit SPM SE of Banc Maskali. Note that the lit port-hand buoy is well to the W of Banc Mascali. It marks the main channel, which leaves the Iles Moucha to port.

There are several wrecks S of the anchorage.

Note The entire main port area is prohibited to small craft as shown on the plan, though this doesn't apply for checking in by dinghy or if you've arranged to take on fuel or water.

Conspicuous

It is hard to miss Djibouti whether arriving from N or W though, given usual atmospheric conditions, not much starts to resolve itself into detail until you are within 1–2M because the whole town and surrounding area are very low lying.

Anchorage

4–5m, mud, off the Club Nautic (yacht club) in the SE corner of the harbour with good holding and fair shelter, though things can get exciting in a *khamsin*. Leave your dinghy alongside the club's pontoons once you are cleared in.

Formalities

Go by dinghy into the new harbour to visit the *capitainerie* and then immigration. The customs point marked near the yacht club is a security control only. Passports, for which you'll get a receipt, will be held till you leave. To clear out, obtain clearance from the port captain, then go to pay port dues at the port accounts office. Claim passports at immigration on production of your receipt and clearance. Harbour dues amount to US$20–30 for a three-day stay, depending on the size of your boat. Dues are normally payable just before you leave but you can check in and out at the same time, paying in advance, to avoid coming back. Visas, which are obligatory, take half a day to process and cost US$20–30 for 10 days. One-month and 3-month visas are also available.

Facilities

Fuel Diesel is available from Total for approx US30c per litre. Decide how much you need and go to the *capitainerie* for a permit to go alongside. Ask which quay you should use. Then go to town to buy duty stamps from the treasury office (Trésor Public). A taxi will cost about US$3 and will stop at a bank or moneychanger for you. Go to the Total depot outside the port gates, pay in advance and arrange a time to meet the tanker alongside the quay designated by the *capitainerie*.

Water Potable water from the yacht club. If you draw more than 2m you will not be able to get alongside the club pontoon (reserved for members unless you have authorisation) to take on water. You can make arrangements with the *capitainerie* to take on water for a fee but be sure to check for quality before you put any in the tanks. It can be brackish. Delivery is by large diameter hose.

Provisions Excellent supplies, including wine, can be found at the markets and supermarkets. Bargaining is normal in the market. There are several hardware stores. Prices are similar to those you would expect to pay in Europe and come as a surprise after Asia and other Arab countries. Good food at the yacht club. Pay with chits bought at the office, where English is spoken. The manager is Olivier, ☎ +253 35 15 14.

Repairs You can arrange with the port captain to use the slip in the dhow harbour, or you can careen your boat alongside one of the wrecks S of the yacht club. Others have done so successfully. François-Nautique repairs engines and works at the club's pontoons. Spares, at a price, from Marill, in the industrial zone on the way to the airport. Djibouti-Plaisance are the local Volvo-Penta agent. Metalwork by Gambeli, but he is also quite expensive. In general, materials and spares are more expensive than labour.

General

We have had some recent adverse reports of the Djiboutian bureaucracy and prices. There are mixed views about the welcome from the Club Nautic, though everyone likes the showers! A dive company offers services to yachts in transit. Contact Bruno Pardigon of Dolphin Excursions, PO Box 4476. Dankali Expeditions, a branch of the same company has a website, www.dankali.com *email* dankali@hotmail.com, ☎ +253 35 03 13, Fax +253 35 03 80. The tourist office is in Place Ménélik, not far from the market and though short of funds, its staff are very eager.

The port is the centre of economic activity for the country but the presence of the French navy here gives a potentially misleading idea of Djibouti's stability. There was a brief outbreak of violence in December 2000 when the sacked police chief tried a coup d'état. The army supported the President and in the resulting shoot out, with tanks joining in, some 60 civilians may have been killed or injured. The incident was largely subject to a news blackout but normality was quickly restored.

For more, see Djibouti in the 'Country information' section of the Introduction.

Gulf of Tadjoura

These anchorages are taken clockwise from Djibouti to Obock. Most boats that call at Djibouti want to head on N as soon as the weather allows so we give only brief information.

French (SHOM) charts 6894 and 4792 are alternatives to the Admiralty series and possibly offer better coverage.

Iles Moucha
Admiralty chart 262

Approach
Use the pass between Ile Maskali and Ile Moucha. NE of the Maskali light. Depths shoal to 2m in the pass and you need to hold to the N side. A good lookout is necessary. Anchor in 4m, sand, S of Grand Signal Plateau. The area is a designated Territorial Park and is popular for outings from the yacht club. However, the water has been overfished and is murky. Much of the coral is dead or dying.

Ambada
Admiralty chart 253

Avoid swell and excessive depths by heading about 1M W of where the anchor is marked on the BA chart and tuck in behind the reef W of the wadi entrance, 8–10m, sand and coral. If the wind goes NE you can try the narrow channel between the reef and the coast for more shelter.

Daba Libah (Ras Eiro)
Admiralty chart 253

W of the *ras* a tongue of coral extends from the shore on which one can drop a hook and fall back, but only in steady E winds. The bay is deep but possible in good weather. There is a Foreign Legion base ashore who may check you out. Good diving.

Ilot Des Boutres
Admiralty chart 253

Swell can make in in NE'lies but it's a good place to wait for a tide for going into Ghoubbet Kharab.

Ghoubbet Kharab Pass
Admiralty chart 253

This area offers several interesting possibilities once you are inside the pass. There are strong currents in the entrance with up to 7kt in Abou Maya (Petite Passe), which is the safest although only a cable wide. It is easy enough to get in on the flood in winter, but exit is easier at slack high water, before any sea breeze sets in. Slack water lasts only for 5–10 minutes. The wider pass (Grande Passe) in theory has least depths of 3m. This is fiction, the coral has grown greatly and the pass is littered with big bommies. Steep volcanic cliffs line much of the shore. For very experienced divers, the pass is said to provide a wonderful drift dive. Diving in the Ghoubbet itself is rated for whale sharks and manta rays.

Fare du Ghoubbet

This is in the large bay immediately S of the pass. The bay itself is too deep for anchoring so head for the narrow, steep-sided bay down the SE corner. The surroundings are spectacular. A small house overlooks a rude wharf S of the entrance. On the shore N of the entrance there are a couple of large rocks on the reef which projects 50m or so. The entrance itself is barely 100m wide but deep. A wrecked tourist plane is on a hill behind. The anchorage is also deep, the 20m line being within 50m of the shore. It can be gusty. In the right conditions it's more comfortable to drop a hook on the pebbly beach at the end of the bay and fall back.

Anse d'al Toubib

A rocky islet with a conspicuous low wall on top offers good shelter from N through to SE. Anchor in 6–8m about 20m off the reef and fall back. Further out it is very deep.

Ile du Fare (Le Fare du President)

The Fare is a small islet. Anchor in the lee 5–10m, sand, at the end of the beach. Local people occasionally live in the palm grove ashore, and it is a favourite weekend haunt amongst Djiboutians.

Guinni Koma

Admiralty chart 253

Guinni Koma and Baddi Koma are two small islands formed by volcanic cones. The best anchorage, about 1M W of Guinni Koma, is behind a stone dyke which runs out to a rock in mid-bay. There may still be a beacon at the outer end. Give the dyke a wide berth and anchor inside close to the wall in 10m, sand. Take a stern line to the small jetty, if that's still around, to reduce swinging. It is usually safe to leave the boat here if you want to explore Lake Assal. There is also a day anchorage off the spit on the W side of the Guinni Koma, but both wind and current are very unpredictable.

Baie du Lac Sale

Admiralty chart 253

The E swell which makes into the bay means it can be uncomfortable, but off the N shore an islet which runs NW/SE, often connected to the mainland at low water, offers a protected anchorage near its NW end.

Baie de l'Etoile

Admiralty chart 253

Anchor close to the N end of the islet at the entrance, clear SE of the rock marked on the chart, which sometimes dries. 5–10m, sand, good holding and well protected.

Tadjoura

Admiralty chart 253

The anchorage here is poor and can be dangerous in the SW monsoon. Swell makes in during the winter. However, it's a lovely place with friendly people. On approach from the E beware of the shallow bank. There are coral heads about 30m W of the pier. The leading lights were not working recently. You can anchor either in the bay or stern-to with double stern lines to the dolphin if there's room but leave access clear for ferries coming in from the E.

Southern approaches to the Red Sea

Les Sables Blancs

The bay is sheltered from the N and E. Anchor in 10–15m, sand, off the middle of the beach.

Khor Raysali

Admiralty chart 253

Only accessible if you draw less than 2m, in which case you can cross the bar and anchor in 10m.

Obock

Admiralty chart 253, plan page 86

Approach

Approach the harbour from the SE with the leading marks or lights in line bearing 338°. At night these lights are not strong, but the new fisheries pier NW of the light on the E end of the Banc de la Clocheterie is well lit and serves for a night entry. From the ESE, keep the flagstaff near the old Residency building bearing 290° or greater to steer clear of the shoal water S of the Banc de Surcouf. These bearings should take you into the Port du Sud, but note the dangers mentioned below. The channel between the reefs into the Port du Nord-Est, where the more suitable anchorage for yachts is to be found, has a least depth of 4·5m over the shoals in the middle.

Dangers

Shoal water off the S side of Banc du Surcouf may extend further than charted. The E entrance to Port du Nord-Est is difficult and not recommended if you've no local knowledge.

Conspicuous

Towers in the village and the minaret beside the front leading mark should be visible.

Anchorage

5–10m, mud, in Port du Nord-Est but the water is murky. Good protection in N'lies, but there is also some shelter offered from the S by the reefs. Except in a S'ly, if you come in at night it is probably easier to anchor about ½ cable NE of the fisheries pier but the bottom comes up quickly towards the shore.

Formalities

Check in with the police, who are in the building close to the school in the town.

Facilities

Fuel from a service station in the town and water at the fisheries pier. Take care with inflatables, since the pier is encrusted with oysters and there is not much water alongside. Limited provisions are available but when the supply boat arrives twice weekly from Djibouti it's fun to watch.

General

Obock was once the capital of French Somaliland, now Djibouti, but has shrunk since its heyday into a much smaller settlement. Many of the older buildings are now ruins. The French adventurer Henri de Montfried at one time had a house in Obock. There is good diving on the reefs.

Red Sea Pilot

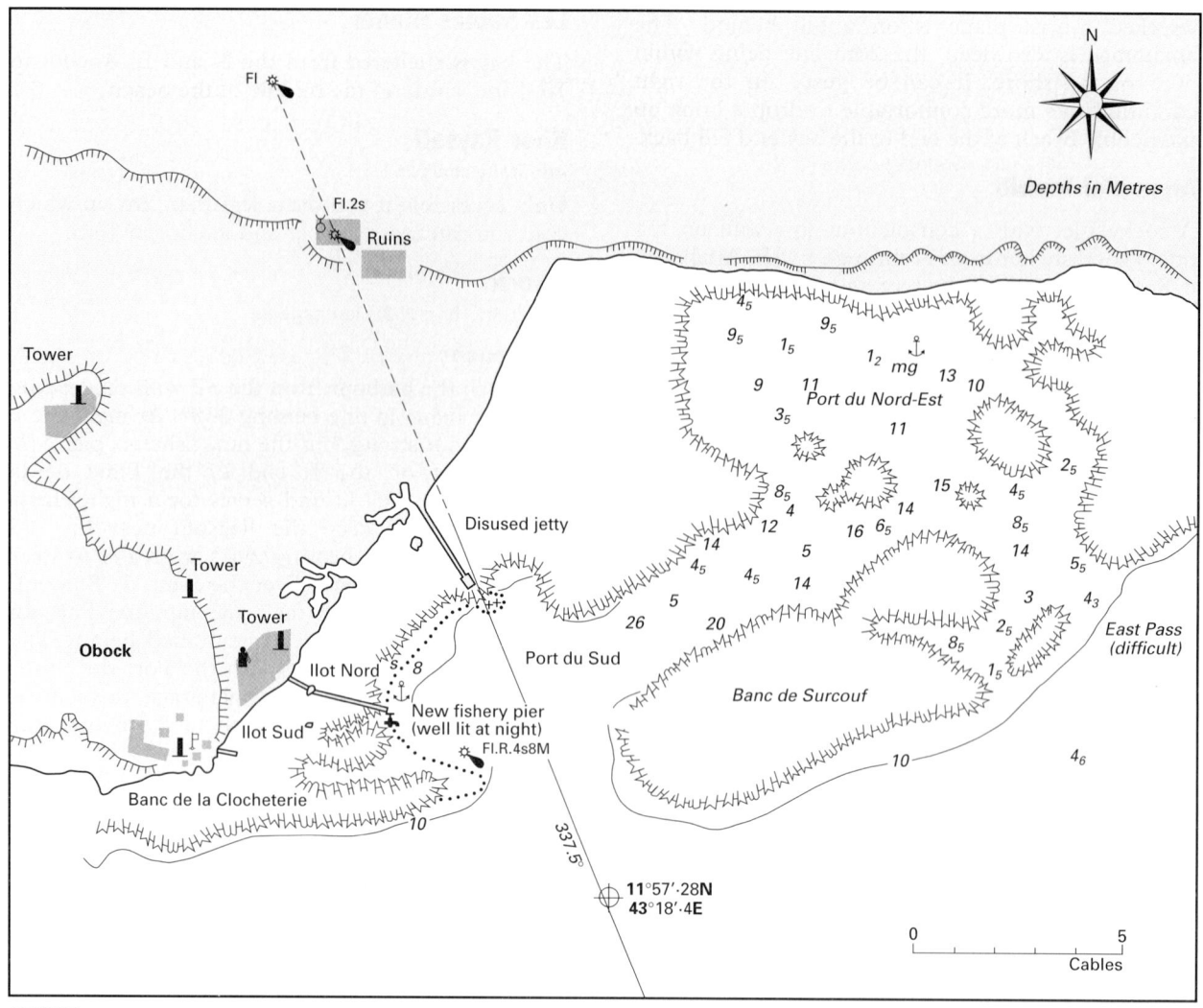

OBOCK

Rhounda Dabali
Admiralty chart 1925, plan page 87
(French chart 6326 is a good alternative for this area.)

Approach
The Admiralty Pilot mentions a pinnacle rock with 0·5m of water over it 2½ cables NW of the island. According to local fishermen there are others which are not easy to pick out since the water over them doesn't change colour. There are strong currents around the islands, with tide-rips.

Conspicuous
The Seven Brothers are brownish in colour, except for Kadda Dabali, which is yellow and has a masonry block on its summit, a beacon near its SE end and a mooring buoy close off the S shore.

Anchorage
2–5m, sand, off the NW corner of the island, near the sand *cay* (which sometimes disappears) NNW of the beach. Note that the 3–4m depths marked further WSW are over reef.

General
A plane occasionally brings divers to a diving base near the lagoon W of Ras Siyyan, 12°29'N 43°18'·5E. There is a possible anchorage at the entrance to the lagoon in settled weather with little swell. However, be warned that all reports since the days of Henri de Montfried onwards are pretty negative. The plus is that the diving in the area is very highly recommended. The two rated dives are in the lagoon pass and over the isolated pinnacle (17m) 5M NW of Ras Siyyan, which has stunning drop-offs. If you luck in with the right conditions, you can find crabs in the mangroves around the lagoon. A possible day anchorage, fair in SE winds, is on the NW corner of Kadda Dabali, 8–10m, coral.

Southern approaches to the Red Sea

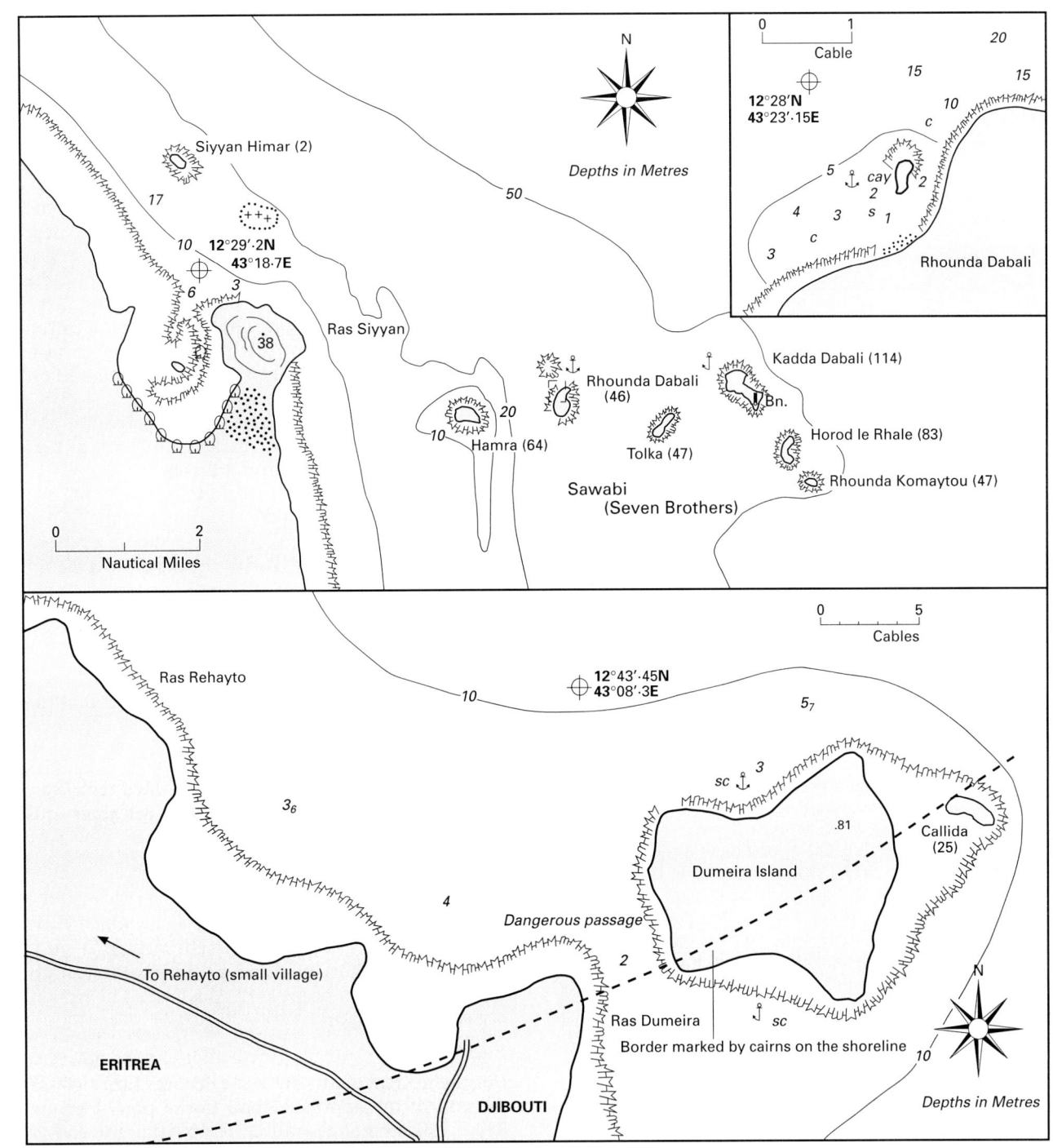

ANCHORAGES BETWEEN DJIBOUTI AND BAB EL MANDEB
TOP RAS SIYYAN AND SAWABI (INSET NW CORNER OF RHOUNDA DABALI)
BOTTOM DUMEIRA ISLAND

Dumeira Island
Admiralty chart 2588, plan above

Dangers
The passes between Dumeira I and the *ras* W'wards and Callida E'wards are shoal and not navigable.

Anchorage
3–5m, sand, amongst coral and rock patches. The best anchorage is on the N side with good holding and protection from S'lies but with strong gusts coming off the hillside. In light NE weather anchor off the reef S of the island, but swell and windward tide can make life uncomfortable.

General
The boundary between Djibouti and Eritrea crosses both the headland and the island and is marked with cairns. You might find the military here but they have been friendly enough in the past. The diving is poor because of sea and currents, but there's good fishing. In mid September Dumeira is a concentration point for migrating raptors.

2. Southern Red Sea

Eritrea

Admiralty charts
Large scale 168, 460, 1926, 2588
Medium scale 81, 171, 453, 460, 1925
Small scale 6, 143, 157, 164, 4704

US charts
Large scale 62121, 62110, 62111,
Medium scale 62110, 62120, 62121, 62130
Small scale 62290

French charts
Large scale 7071
Medium scale 6983, 6984, 7519
Small scale 7099

Warning
Although the most recent editions of charts from all the major hydrographic offices are reconciled to WGS84, this CANNOT compensate for shortcomings in the original 19th century surveys. The recent survey of the area from the Hanish islands to Bab el Mandeb was intended to satisfy IMO requirements for instituting the new traffic separation scheme. It was not a detailed re-survey of all coastlines. However, the Yemeni authorities specifically paid for surveying of charted hazards W of Large Strait to establish that there is a wide, clear passage for small craft as far as Ras Fatuma. There is. Four points of navigational importance follow. First, use the most recent charts properly corrected. Second, always navigate, even with GPS, with circles of probable error (CEPs) of at least 2 miles, and at night at least 5 miles. Third, navigating by GPS alone in the Red Sea is stupid. Do not do it. Make daylight landfalls and use your hand-bearing compass and radar, if you have it, to establish WGS84-to-chart error factors. Fourth, treat ALL waypoints with appropriate caution, including those given below, and NEVER assume that the waters between, close to or at waypoints are free of hazard.

Note South Eritrean waters
a. Military patrols are vigilant in S Eritrean coastal waters. Boats anchoring at Ras Terma and especially in the Rubetino Channel have been moved on but officials are helpful and will usually suggest alternative anchorages in adverse weather.
b. Waters are murky dark green to pale greeny-brown. They disguise hazards even in 2m.
c. Making S against the SE winds and NW set S of Howakil Bay is a trial. At all times tack up as close inshore as you can, seeking where safe to stay INSIDE the line of pale greeny-brown water and never going outside the 30m line where, roughly, the darker greeny stuff begins. Even deeper, rougher water is blue-green. Ideally, stay as close as possible to 10m. The extra bonus if headed S is a slight SE setting counter-current inshore.
d. Much of the Eritrean coast is still mined. Be very cautious going ashore.

Assab
13°00'N 42°46'E
Admiralty chart 1926, plan page 89

Approach
There are strong currents around the entrance to the Rubetino Channel. The channel has a 2m shoal near the E entrance. Hold towards Dercos and you should have 6m. Once inside the isolated reef off the S shore of Dercos it's deep water all the way. Even so, keep a good lookout. The islands are a closed area. The army control them. Permission to visit should be sought in Assab. Yachts have anchored off the town to rest overnight without hassle.

A problem-free approach is to ignore the Rubetino Channel and make the longer and bouncier haul around Fatuma Deset.

The breakwater and N jetty are normally lit, but since the recent war don't bet on it. Unlit SPMs mark the seaward end of a pipeline extending from just N of Ras Caribale. The area between the buoys and the N and S jetties is prohibited and in principle anchoring is forbidden in the port area. In practice this rule seems to apply only to ships.

Conspicuous
Assab is visible from some distance off, especially at night.

Depths and holding
Recently yachts have been asked to Mediterranean moor to the breakwater where there are rings and mooring bollards. Otherwise anchor in 5–8m S of South Jetty.

Formalities
Call Port Control on VHF Ch 6 or Ch 12 on approach. You can call at Assab without a visa, obtain a 2-day shore pass and a permit for transit up the coast and then proceed to Massawa. The permit is free but obligatory.

Facilities
For the nearest bank, PO and telecoms turn right at the end of the approach road to the port. Fuel by jerrycan from a Shell station. Keep left at the end of the approach road then take the road that forks right to the top of the hill. The garage is on the right, about ½M from the port entrance. Good water from the fishing boat jetty near the root of South Jetty, again by jerry. There are small groceries and a market within 10 minutes' walk of the port.

Harena Boatyard on Haleb I. (Lahaleb Deset) has good facilities. Contact Kibrom on VHF Ch 69. The island is controlled by the navy and you will need permission to visit. Make radio contact *before* approaching.

General
Assab was the first Italian toehold, established by the Rubbatino Navigation Co (the chart spelling is

Southern Red Sea

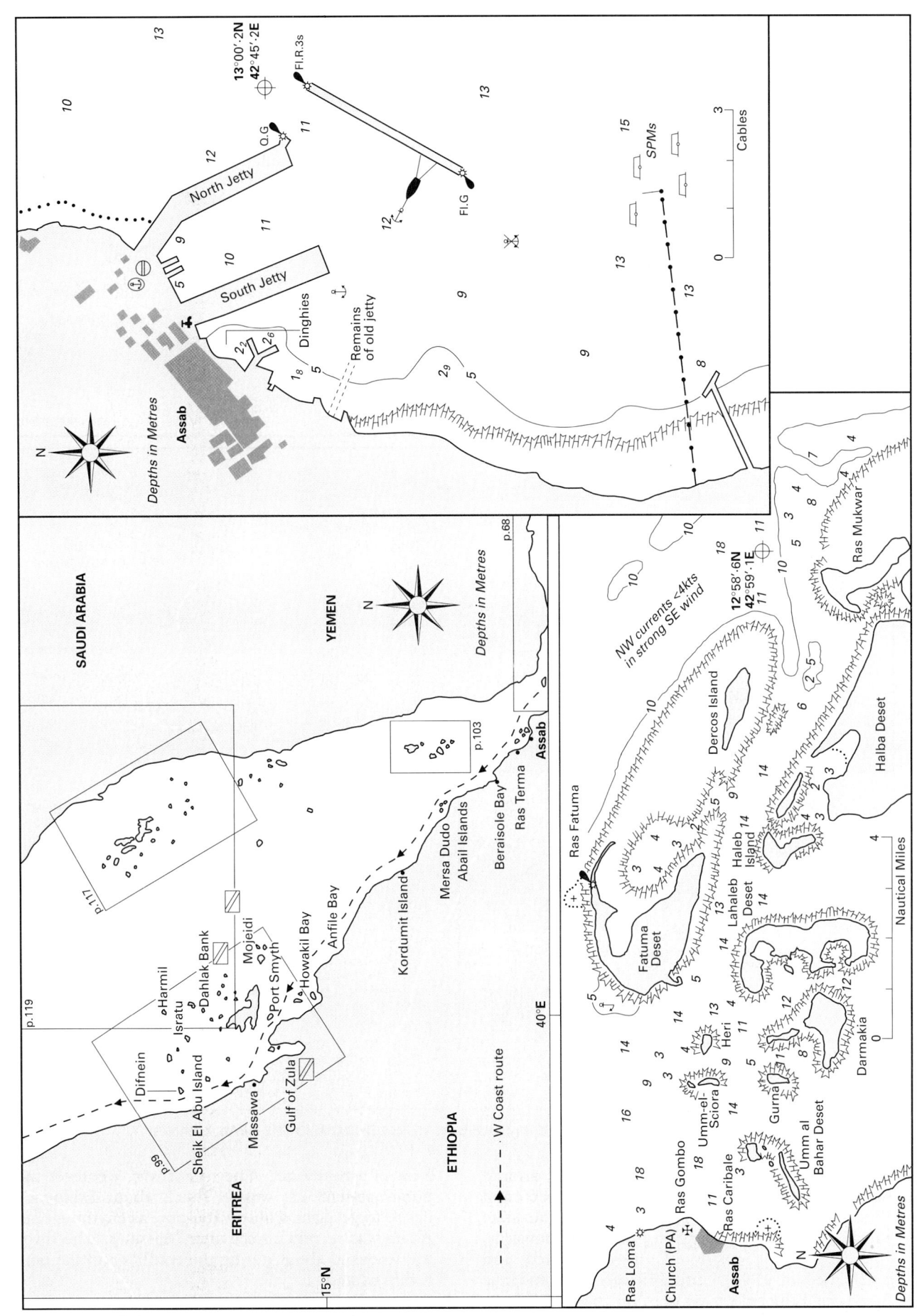

CLOCKWISE FROM TOP LEFT: ERITREA KEY MAP, ASSAB, APPROACHES TO ASSAB

89

Red Sea Pilot

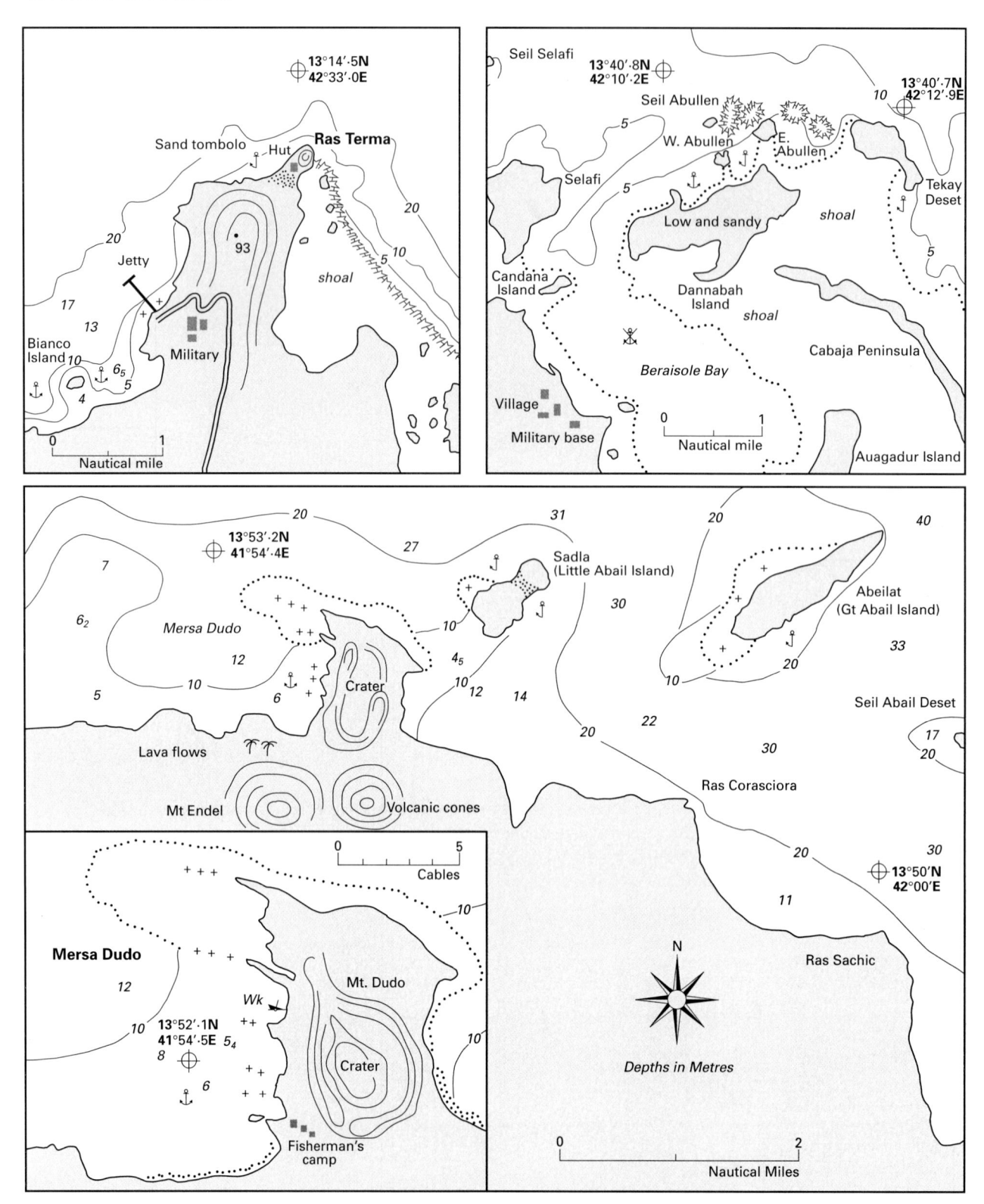

RAS TERMA, DANNABAH ISLAND ANCHORAGE, BERAISOLE BAY, MERSA DUDO AND THE ABAIL ISLANDS

different) by buying parcels of land in and around Assab from the local Afar sultan between 1869 and 1880. In 1882 these were bought by the Italian state which, by creeping expansion from 1885 onwards, established their colony of Eritrea, named and proclaimed in 1896. Until Eritrean independence Assab was Ethiopia's main port. Since then it's been a bit of a backwater. The recent war, whatever the public statements, was as much about Ethiopia's desire to get back a blue water port as anything else. Assab was second choice after Massawa. The town was knocked about during the conflict and the port is dilapidated.

Other anchorages

If you're headed S (or N) and don't want to stop in Assab, surcease from the SE winds can be found, apparently without strong objection, on the bank in the lee of the W end of Fatuma Deset. The water in the Rubetino Channel area is often murky but the best snorkelling is said to be off Fatuma Deset.

An alternative short stop for those headed S (or N), less well sheltered, is in the lee of Sanahor Deset, 13°04'·8N 42°42'·7E, though it is rolly. Note the rocks along the W shore.

Ras Terma
Plan page 90

Approach
There is a large reef with depths of less than 1·5m over it in the middle of Beylul Bahir Selate. The new military camp on Ras Terma is conspicuous. During the ebb tide, give the NE side of the headland a good berth to avoid tide rips which are heavy at springs and would be dangerous in a strong blow.

Anchorage
The military don't like yachts near the jetty. You may be allowed to stay at the head of the bay, E of Bianco It. Otherwise in SE winds, anchor N of the *ras* in 8m in the lee of the low tombolo which joins what was once an islet to Monte Darma. There is a small building on the NW side of the *ras*, which may be a guardhouse, but there was no one around when we were there. Very uncomfortable in N winds.

Beraisole Bahir Selate
Admiralty chart 168, plan page 90

Approach
From a distance the islets in the bay look as though they are a continuation of the range of hills inland.

Dangers
If approaching from the SE take care to clear the shoal area E of Cabija Peninsula. Further N, Fanaadir Rock, 1¼M SW of the S Fanaadir islet, is generally visible and dries 0·3m. There are coral outcrops around the Abullens and an isolated reef S of Karansas Deset.

Anchorage
5–8m N of Dannabah I, excellent holding in sand and good shelter from the SE. The reef shown on the BA chart is not reef but sand. The authorities asked yachts needing shelter to use this anchorage in 2001. They were very courteous and helpful, asking merely to see ships' papers. The inner bay is a restricted area so unfortunately the good anchorage S of Candana I and SE of Selafi I is now off limits. In N'lies you could try tucking in under the lee of the SE end of Tekay Deset where the half anchor is, but it's tight. Fishing dhows use the small bay between E and W Abullen which may get some shelter in N winds from the offshore reefs. The same applies to this bay as to the main SE anchorage, the reef charted on the S shore is not reef. It is possible in N winds that the military would relent and let you anchor inside. You could but try.

Abail Is and Mersa Dudo
Plan page 90

Approach
The passage between Ras Sceraier and the islands is deep and clear with a least charted depth, which we didn't find, of 4·5m between Sadla and Monte Dudo.

Dangers
Shoal ground around the promontory N of Monte Dudo as shown on the sketch-map. It pushes a long way NW.

Conspicuous
The islands stand out as do the extinct volcanic cones of the hilltops and on closer approach the wreck on the beach.

Anchorage
There is a spectacular anchorage in 5–7m, mud and sand, good holding and excellent shelter for SE'lies about 100m off the beach in Marsa Dudo. When the wind dies the swell out to sea works right round the islands and promontory and rolls in, though much attenuated, from the N. More a nuisance than anything. You can also anchor either side of the sand tombolo which joins the two parts of Little Abail. The wind whistles through the gap, but the protection from sea and swell is good. The anchorage S of Abailat (Gt Abail I) has good holding in sand, 9–10m. Reef extends SW of the island.

General
If you've missed the Hanish Is, this is a good substitute in microcosm with wonderful volcanic scenery in amazing colours. Walking ashore is a treat with fantastic views from the crater rim of Monte Dudo. The area of lava flows SW of the anchorage is amazing for the quasi-oasis of palms. The fishermen sometimes camped on the beach are very poor. They beg, but they are no bother.
Note the Bay of Ed and the bays E of it are not sheltered in the prevailing SE winds, which by the afternoon have a lot of E in them. The area offers at best fair weather anchorages and in all circumstances when they'd be feasible, Mersa Dudo is better, with the added advantage of the Abail Is offering shelter in N winds.

Kordumuit Deset (Curdumiat Island)
There is no plan because the island, a high, very conspic, cliffy monolith, is very small and where to go, depending on wind direction, is obvious. There are off-liers on its N side. A wreck lies about 3½M WNW of the island, about 1½M offshore. Anchor in mild SE winds off the slight nick on the NW shore and in calm weather W of the island, in 8–10m sand. Any wind will howl down the steep bluffs and round the island and in strong weather there would be little

Red Sea Pilot

shelter. It will always make more sense to choose Mersa Dudo or Anfile Bay.

Anfile Bay
Admiralty chart 168, plan below

Note The anchorage under Ras Anrata off the village of Tio is more exposed than it looks and has adverse reports. N'bound, pushing the extra 10M to Anfile Bay repays the effort as does making the direct jump to Mersa Dudo if you're S'bound.

Approach
The shoal N of Hant Deset hooks round a long way. Keep to the N of the channel between Handa and Hant Deset for a least depth of 3m and don't turn S to the anchorage too soon. Sha'b Shaks light is extinguished but the settlement at Tio (Thio) has buildings lit in the evening. The beacon on Derebsasa Deset is missing.

Anchorages
The best anchorage for SE'lies is N of Hant Deset in The Harbour. Care is needed on the way in and you'll see 4m or so on the sounder. Good holding in 5m sand. Pleasant walking ashore where there are the ruins of a village and a small cemetery. Fishermen use the anchorage too and will inevitably ask for something. For N winds try the SW end of Derebsasa Deset in 4–5m with good shelter and good holding in moderate to strong winds but beware coral outcrops. An overnight anchorage in SE'lies is in the lee of Estam Aghe, N of the reef tongue that extends SW. This is much used by fishermen. There are other spots, depending on the wind. The north side of Handa Deset feels very exposed though it's sheltered enough if you tuck as far W as feels safe. There are scant signs of the villages charted, though some buildings can be seen on the mainland.

The village of Thio has good local food and has recently been visited by cruisers though the anchorage is exposed.

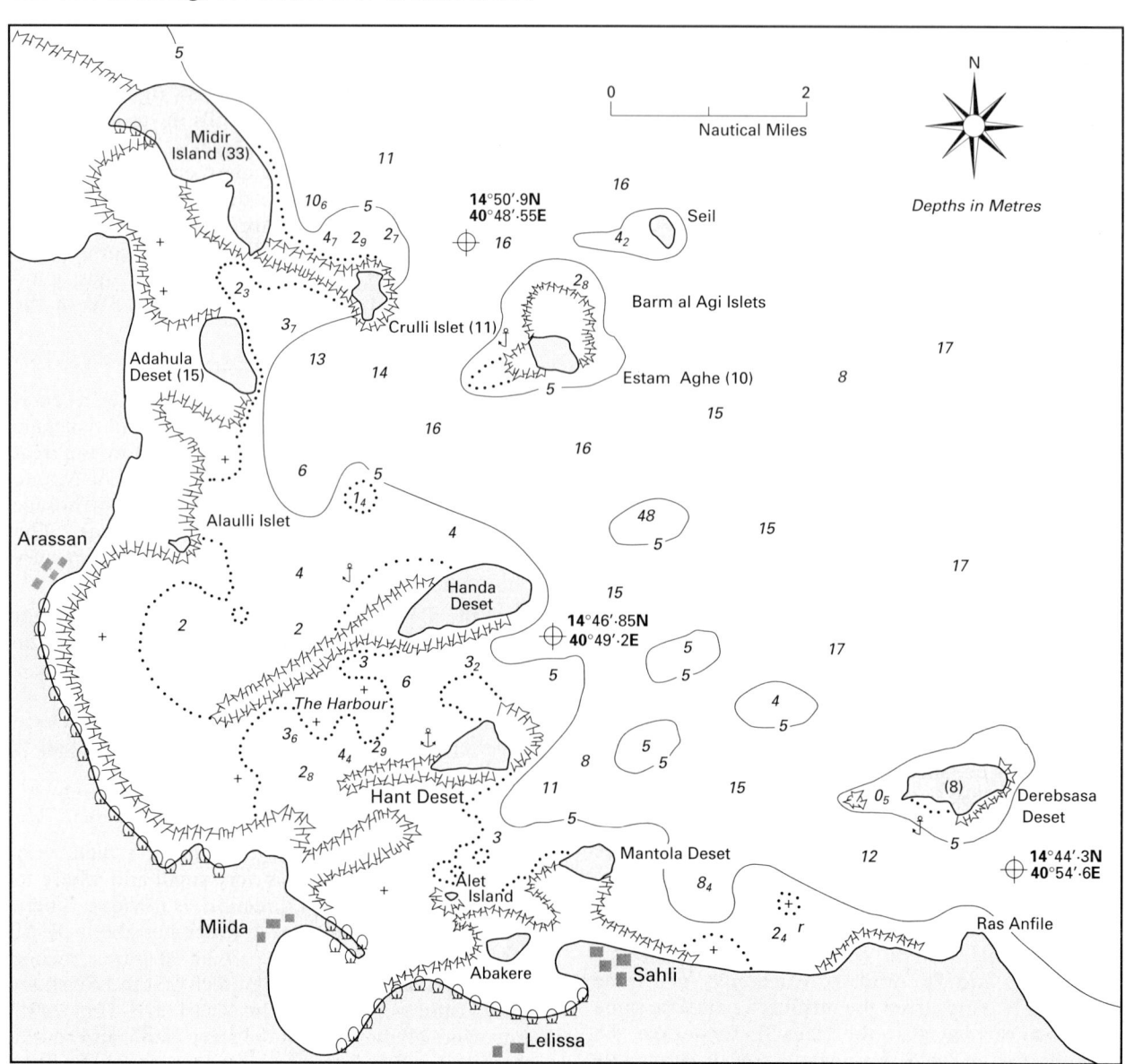

ANFILE BAY

Southern Red Sea

Howakil Bay
Admiralty chart 171, plan page 94

Approach
From seaward the approach is straightforward enough whichever anchorage you've chosen. However, once in the close offing you must keep a careful eye on the echo-sounder. REEFS DO NOT NECESSARILY SHOW. Access to the anchorages is shown on the key-map. The concrete base of the old lighthouse on Umm es Sahrig and the collapsed structure are all that remains. There are confused, lumpy seas SE and S of Howakil Bay in moderate SE'lies.

Anchorages
There are many possibilities between the islands depending on conditions. First, coming from S, there is the headland of Andeba Ye Midir Zerf Chaf for SE'lies but if you're S'bound it's worth pressing on to Umm es Sahrig for very good shelter from the SE under the N side of the W end in 5m, sand. You'll need to probe a bit to find a spot and when you do you'll be very close to the beach! The S side of this spit is fair in N winds.

Further in there is Freedom Anchorage, W of Abbaguba I which is well protected by extensive reef as well as the island, and then Fiddler's Islet, our own minute discovery. The next inlet N'wards, S of Dergamman, is bigger and deeper. Either SW or SE of Adjuz is possible, but with shoals beyond the anchor symbol on the sketch. This is probably the best for a N'ly unless you want to find your way into the good shelter between Debel Ali and Dergamman Kebir on the S of Howakil. Fishermen may call to see you wherever you choose.

Mojeidi
Admiralty chart 171, plan page 95

The island, well offshore, offers reasonable shelter on its W side in fair to moderate conditions and is recommended for diving. Tuck into the bay with the beach. Don't forget that as with all the islands of the Dahlak Bank, holding is not wholly to be trusted.

Port Smyth and Shumma Island
Admiralty chart 168, plan page 95

Approach
The leading marks are hard to pick out. They're a barely distinguishable piles of stones. All the charted buoys and beacons are missing. The light on the S end of the island is working but according to strange rhythms we suspect to be related to movements of shipping in and out of Massawa. It makes a good mark by day.

Dangers
There is a 1·8m patch in the middle of the pass which used to be marked by a beacon. You can go either side with care.

Anchorage
5–7m where marked on the sketch with good holding in winds up to 30kn. There is no shelter from SE winds. In fact E'lies are more common because the island is low. The shelter from sea is good but if there's any strength in the wind, there's a constant swell, the encircling reef being at all times covered. Good walking, wildlife and snorkelling.

Dilemmi (see key-map page 99)
If Shumma doesn't excite you and it's blowing SE there is an anchorage under the lee of Dilemmi on its NW corner, ½M beyond the beacon on Dilemmi's N tip. Anchor in about 10m, mud. Some swell.

Madote (see key-map page 99)
An alternative to Dilemmi. 10m mud and coral. Fairly exposed, but the diving is reported to be excellent.

Dolphin Cove
Admiralty chart 168, plan page 95

The anchorage is in the SE corner of Zula Bahir Selate (Gulf of Zula), with a clear approach and leading beacons in line bearing 084°. The rear one is difficult to see. Anchor in 6–10m, sand and mud, good holding.

Melita Bay
Admiralty chart 168, plan page 95

The bay is on the E side of Zula Bahir Selate and is entered around Ras Nasiracurra, clear of fringing reef and shoal water. You can tuck well in to the head of the bay. There is a fishing village at Nasiracurra. The ruins of the ancient city of Adulis were discovered a little inland from the modern settlement of Zula, on the W side of the Gulf of Zula more or less opposite Melita Bay, by Viscount Annesley's party in 1810. But, as Lionel Casson persuasively argues, the port described in the *Periplus of the Erythraean Sea* could only be Massawa. Adulis prospered as the ancient kingdom of Axum flourished in the 3rd century BCE. And that was its problem because, as trade and population increased, the original location of the settlement, Massawa, suffered from lack of water. Hence the shift of the town 20M S to where the ruins now are. Port operations stayed in Massawa, the only all-weather anchorage, easily accessible by sailing vessel, for miles around except at one point when, for defensive reasons, they moved to Dissei. It is possible to anchor off Ras Malcatto, 15°15'·2N 39°43'·1E, to explore, but it would be much more sensible to do the trip from Massawa.

Dissei Anchorage
Admiralty chart 168, plan page 95

Approach
The island lies SW of the inside route between Assab and Massawa, at the mouth of Zula Bahir Selate. A reef extends 4½M from the N tip of Dissei and there are off-liers SE of the anchorage.

Red Sea Pilot

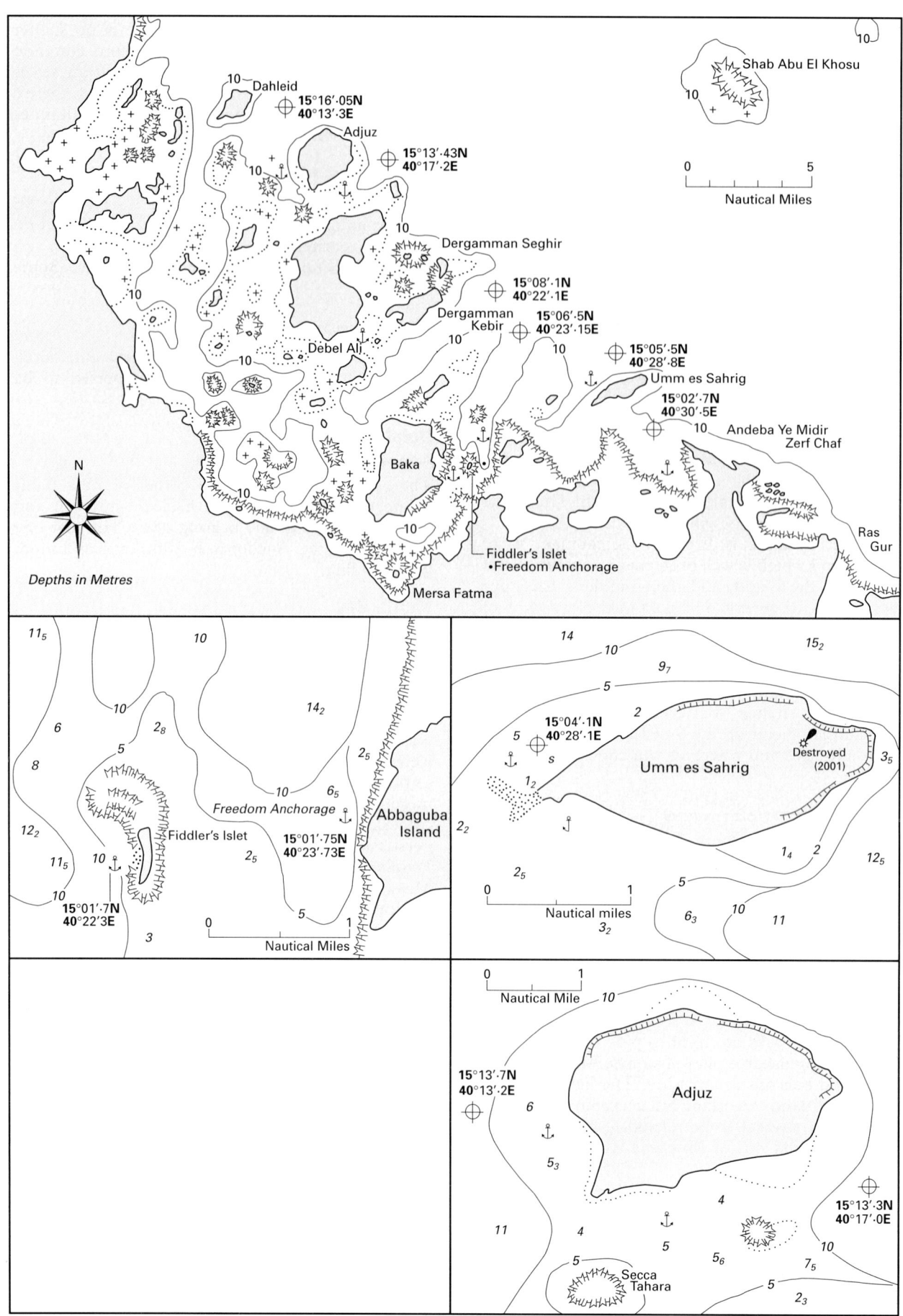

HOWAKIL BAY KEY MAP AND ANCHORAGES IN HOWAKIL BAY

Southern Red Sea

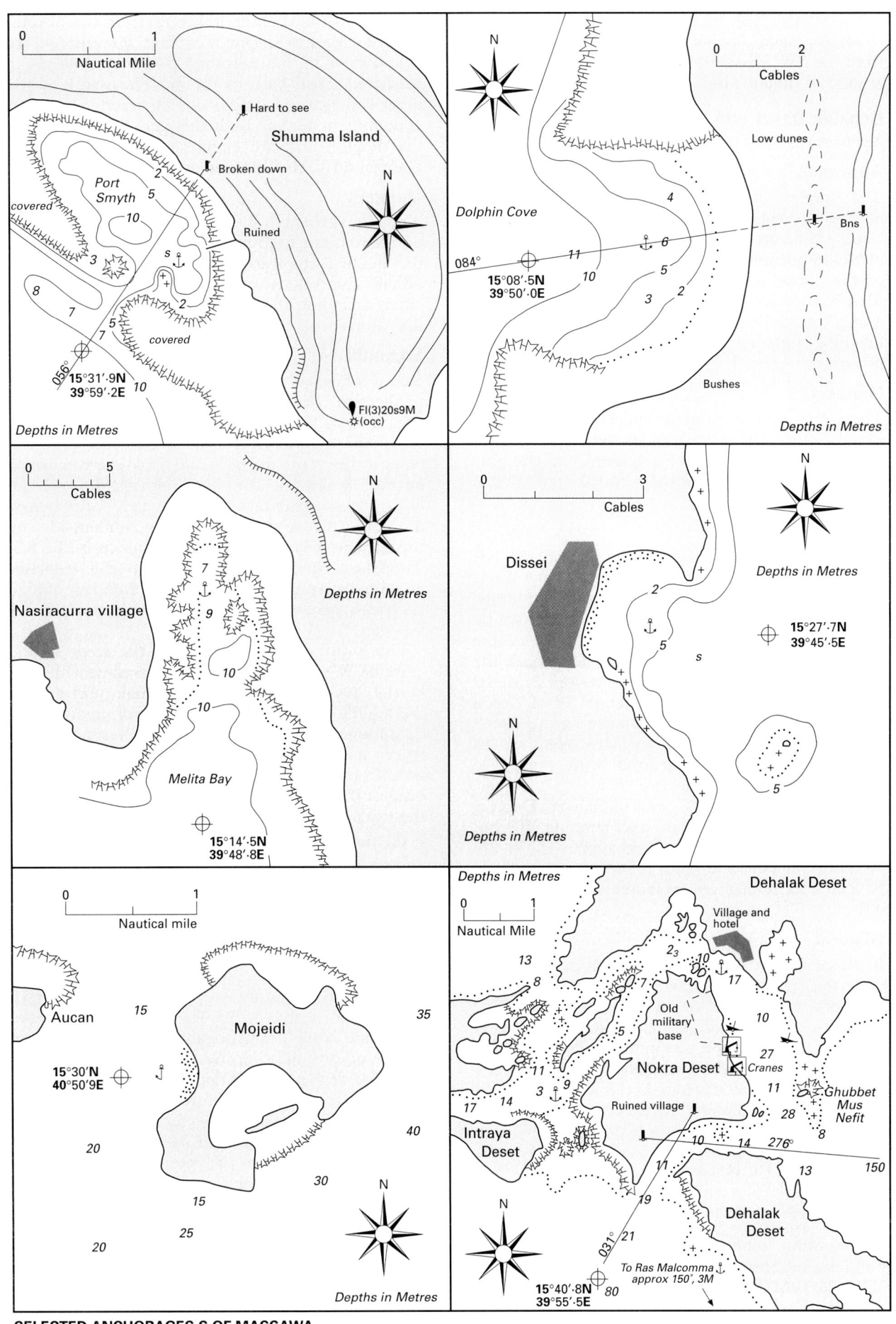

SELECTED ANCHORAGES S OF MASSAWA

The village, with its mosques, is conspicuous on closer approach. Anchor in 4–7m, sand, with shelter from the NE. Dissei is reputed to have good diving around it, though visibility is often poor.

Dehalak Deset and Nokra Deset
Admiralty chart 168, plan page 95

Approach
The simplest entrance to the inner anchorages is between Nokra Deset and Dehalak Deset. Use the leading beacons at the ruined village on the S shore of Nokra bearing 031° and then take a back bearing on the second set of 276° for the lead into Ghubbet Mus Nefit. Least depth 10m in the channel. There is an alternative entrance N of Intraya Deset but it is fiendishly intricate and, given lack of water clarity, not worth the risk.

Dangers
There are many shoal areas and reefs between the islands, even with a lookout in the spreaders you need a constant eye on the sounder. Navigate only in good light. Tidal streams can run at up to 3kts in the channels.

Anchorages
Some possibilities include the channel on the NE side of Nokra Deset, but see the note above about tidal streams. The island itself may still be off limits. It's an old Soviet-equipped naval base, so be cautious about going ashore. The wrecks of the floating docks and cranes more or less mark the dodgy area. Alternatively, anchor between Andebar and Intraya Desets or in the channel W of Nokra Deset. There is a resort where shown on the sketch on Dehalak Deset. Ghubbet Mus Nefit is an interesting day sailing ground with plenty of anchorages along the shore.

An anchorage on the W coast of Dehalak Deset is inshore of Enteara, N of the reef running W from Seil Bayus, off Ras Malcomma in approx 15°38'·2N 39°57'E, 5m sand and mud. It is best entered from N. You'll feel a long way off shore (see key map page 95).

General
In the earliest days of modern European contact with this region, the main port for trade with Ethiopia was here on Dehalak Deset. The description in *The Book of Francisco Rodrigues* (1513) seems to imply the port was in the entrance channel to Ghubbet Mus Nefit on the Nokra Deset shore. Note: If you plan to stay long in this area you should get a permit in Massawa. See below.

Massawa
Admiralty chart 460, plan page 97

Approach
The port is easy to see, day or night. Lights in the area continue to be unreliable, including Assarca Lt but lights on Shumma I, Madote I and Ras Fatuma (Fl(3)26s10M) were working recently. There are no working lights when approaching from the N until close to Massawa. Don't confuse the entrance to Massawa with the entrance to Khor Dakliyat, a restricted naval base to the N. The wreck at the inner end of the leading line is surrounded by an oil containment boom. Both this and the wreck near the dinghy dock are being cleared. Contact Port Control on VHF Ch 16 on approach.

Dangers
None for yachts but NOTE, at low water there is an outboard biting bommie between the anchorage and the landing steps as marked on the sketch. On official charts it is missing. It lies OUTSIDE the charted 2m line and ESE of the end of the pier from the old palace.

Formalities
You will normally be told to go alongside in the port to check in. This may involve tying up to a tug if the port is busy. Immigration and quarantine officials will come to you. If you get no response on VHF enter the harbour and either tie up where you can or anchor in approx 15°36'·8N 39°28'·45E, off the clearly marked salt factory. If you arrive after hours, the authorities may ask you to anchor outside the breakwaters where there is no shelter from the SE. There are occasional exceptions to this rule but arriving at night is not recommended. Once anchored, take the dinghy in to complete formalities ashore.

You must only go ashore via the steps at the extreme W end of the secure port area near the tug berths. You must always use the main port gate in the middle of the secure area. The immigration and quarantine offices are in the white, verandahed, 2 storey stone and timber building set back from the waterfront at the E end of the port area opposite the ornate, if tattered Port Offices. You will complete the usual bumf plus a 'nil declaration' as follows:

I, the master of sailing boat . . . .
declare that I carry:
no passengers
no pets
no stowaways
no drugs.
. . . .(signature and ship's stamp)
. . . .(date)

There is no charge at this stage. Temporary shore passes are issued. If you want to stay more than a couple of days or go up to Asmara you must apply for a visa at the Immigration Office just outside the security gate, on the left before the ruined town hall. To do this you need a chitty from the port Immigration Office to hand over, with your shore pass, in exchange for the visa. Visas valid for 1 month cost US$40 plus, quaintly, a fee of 2·50 nakfa (approx US20c) to pay for the brown card cover for your personal file. You will need one photo per person. Take your passport or shore pass each time you go ashore. It will be checked at the gate.

To clear out, it is better if you give advance notice.

Southern Red Sea

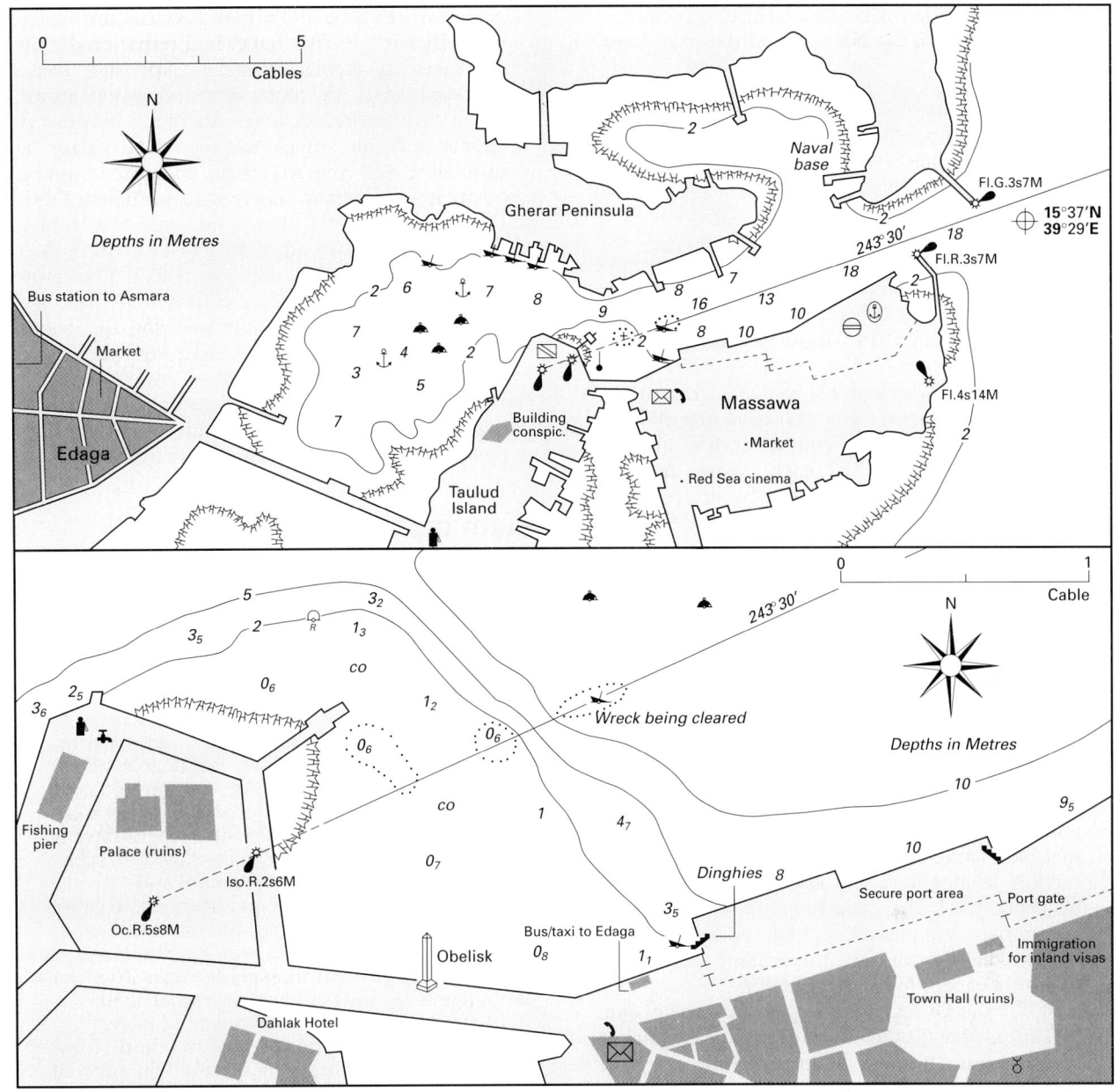

MASSAWA

You usually have to go alongside the wharf where there's a quick check for stowaways. If you ask to stay at anchor you have to take officials to your boat to do the check. Once cleared you must leave immediately. Offices open at 0700 but are usually closed 1300–1600, after 1800, on Saturday afternoon and all day Sunday.

Note The Eritrean authorities are in the process of regulating tourist activity on the coast and offshore islands. Most of the Dahlak Bank has now been designated the Dahlak Marine National Park. Officially, foreign visitors to the offshore islands in the park require a permit from the Ministry of Tourism whose Massawa offices are on Taulud I, on the left side of the main road to the mainland opposite the conspic new building. Charges are US$20 per crew for an unlimited period. Yachts may be excused if they are passing through unless they want to stop between Massawa and Assab. Given the poverty in the country, the charges are fair.

Anchorage

There are 2 mooring buoys at the E entrance to the narrows which lead to the anchorage in Taulud Bay. They usually have trawlers or coasters on them. There is a small red plastic buoy off the pier by the old Governor's Palace with the ruined dome, which lies on roughly the 2m line of extensive shoal ground. It has a partner close to the N shore. We were told they mark the line of an underwater cable. In the anchorage itself there are 3 mooring buoys in approx 15°36'·67N 39°27'·85E, 2 of which have unlit large LASH lighters attached. These limit the space available unless you go further SW into

97

Taulud Bay. Most wrecks have been cleared and lie in a wild jumble on the NE shore of the anchorage.

Facilities

Fuel Diesel is usually available for a reasonable price from the fishing co-operative WNW of the old Governor's palace, by prior arrangement (see below) and with advance payment, at the office in the building behind the pump. Calculate how much you need. No refunds are given. You can either come alongside (least depth 2·5m at the E end) or bring jerries. You can also arrange to bunker in the port if you prefer. Get petrol from a service station. LPG must be brought from Asmara.

Water Potable water and ice are available at the fishing co-operative. You can also buy mineral water in 20-litre jerries from a bottling plant in Edaga for approx US$3 each. Water in large quantities by truck to the port by arrangement with the Fire Department for about US$8 per tonne.

Money US$ can be changed at banks, some shops and restaurants. Black market rates are better. There were no credit card or ATM cash withdrawal facilities in Eritrea in 2000.

You may be met and helped by Mike (Weldemicael Habtezion), who is hoping to set up a yacht service office. In the meantime he tries to meet all yachts after they've checked in. He is a mine of information, a general helper and runs an excellent laundry service. His prices are very good and he will help you to arrange to take on fuel and water at the fishing co-op. For many services rendered he will ask nothing till he has his licence. It's up to you to give what you feel is appropriate, but please don't take Mike for a ride. In 2001 he did not have the backing of the port authority except for doing laundry.

Repairs Try the Government shipyard on the N side of Taulud Bay (ask for Mr Afwerki) or enquire at the Eritrean Shipping Agency outside the port entrance. Unless you carry slings so you can use a mobile crane, haul-out isn't feasible with the local drag-and-chock method on the slipways. There is a machine shop equipped and maintained to very high standards in the road behind the PO not far from the Red Sea Cinema. They can fabricate one-offs for fair prices. Otherwise services are primitive and hardware very difficult to find, though everyone tries to oblige.

Stores Small but quite good groceries on Port (Massawa) I include the Massawa and Bella Vista supermarkets. They have imported Italian food and local alcohol. Locally produced beers and spirits are cheap and good. Wine is also produced with grape juice imported from Italy. It is not so good! There is a small market nearby but a much better market (Edaga in Tigrinya) is across the causeway near the bus station for Asmara. We can recommend the Massawa Restaurant for local cuisine and good value.

Communications The PO's mail services are cheap and efficient. It also handles international calls and faxes but they are relatively expensive. Faxes can sometimes be sent via the shipping agents along the waterfront, where there are photocopy shops and an email service. The latter is unreliable and you may have to go to Asmara. See below for details. There is an inefficient DHL Office on Taulud I almost opposite the Dahlak Hotel, beyond the end of the causeway from Port I and just beyond the new town hall. Deliveries from the US take at least 6 days and there are customs fees for the escort to bring the goods under customs seal from Asmara, unless you go up to collect parcels yourself.

Garbage and laundry There are garbage bins in the port area. Mike offers one laundry service at very reasonable rates. There are others in the town.

Asmara

The capital is well worth a visit. Bus trips are easy to arrange, with spectacular scenery. Take a yellow minibus from opposite the PO, fare about 75c, to the long-distance bus station beyond Edaga. It's about a 10-minute ride. The journey to Asmara takes 4 hours and costs about US$1 single fare. Alternatives are to hire a minibus for approx US$40. The journey is then just 2 hours. Cars can be hired in Asmara at reasonable rates. Hotel prices have gone up but are still very reasonable, all payment must be made in US$. You can find most things in Asmara including *gelataria, pasticceria, pensione, trattoria* and cyber-cafés where rates are cheap but the connection can be slow. Two cyber-cafés are at CTS, beyond the Himbol and Commercial Banks at the S end of the main street and just off the E end of the main street beyond the Telecoms Centre. There is an Egyptian Embassy in Asmara if you want to apply for visas. The local language is Tigrinya but English, rather than Italian, is now the second language of many. Despite the country's poverty the standard of living in the capital is generally higher than on much of the Red Sea coast.

General

Massawa has a long history. As the port of ancient Adulis, it features in the earliest pilot we have, the *Periplus of the Erythraean Sea*. It remained the premier port for trade between Ethiopia and the rest of the world for centuries thereafter. Indeed it could be argued that the independence of Eritrea severed one of the oldest links in maritime trade. For centuries Massawa was an Ottoman fief, graceful relics of which help make Massawa such a charming place to wander round.

Eritrea has the second lowest GDP in the world yet Massawa is a happy, friendly place where the first words you usually hear are, "Welcome to our country, welcome." And the speaker clearly means it whether government official, bank manager or local shopkeeper. Recently, sadly, the government of

Southern Red Sea

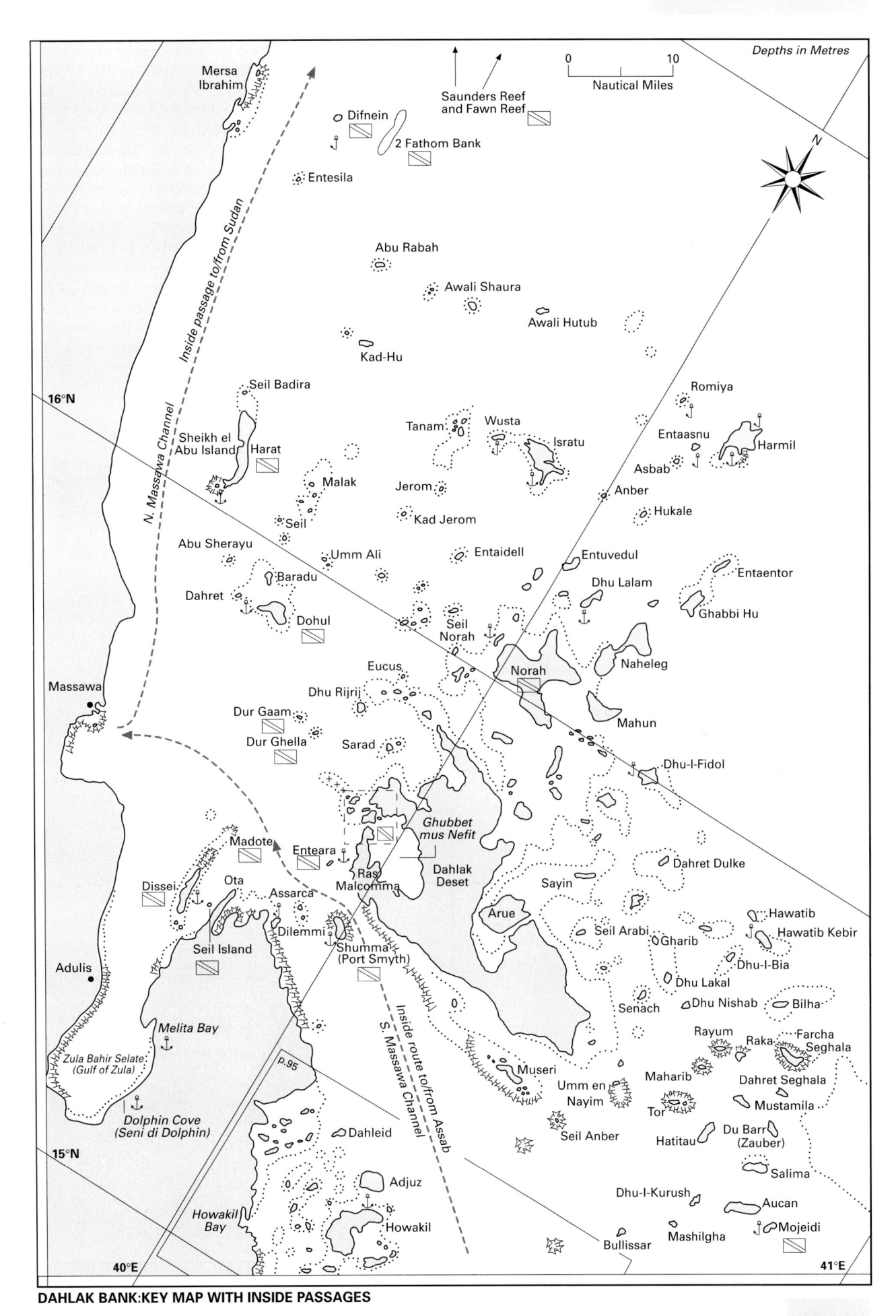

DAHLAK BANK: KEY MAP WITH INSIDE PASSAGES

99

Red Sea Pilot

Eritrea has taken a more authoritarian, intolerant line, so the happy optimism of Eritrea's early days may have a limited life.

Dahlak Bank

Note the chart conventions used on modern metric charts are very confusing. There is NO fringing reef around many of the islands. Instead, they sit on a submerged rock platform on which there are outcrops of reef and a layer of sand. The chart conventions for a rock shelf and a coral reef are very similar and at small scale look almost identical. Often where an island on the Dahlak Bank is surrounded by apparently forbidding dark green reef, what is being shown is the extent of the rock platform, covered with sand and turquoise green water (see entry for Sheikh el Abu below) which may be up to 16m deep! It follows that due caution is necessary if you choose to explore. However, what looks shoal often isn't.

Apart from those S of Massawa already dealt with, there are many possible anchorages on the Dahlak Bank N and E of Massawa. We give detail for a few tried and tested havens. Others have been little visited by yachts and should be tried only by the experienced, even though locals use them regularly. These are marked on the key map for the area (page 000). Remember, water visibility is generally bad and hazards can bite your bum before you've caught on. Second, holding is poor, usually a thin layer of sand over flat rock. The only answer is to lay out a lot of chain. Third, although fish life is rich and snorkelling in spots good, the entire archipelago has strong currents which can make swimming risky. It's safest at slack tide. The good news is that there are plenty of oysters and crayfish. If you intend to spend an extended period exploring the area you ought to have a permit from the Ministry of Tourism in Massawa.

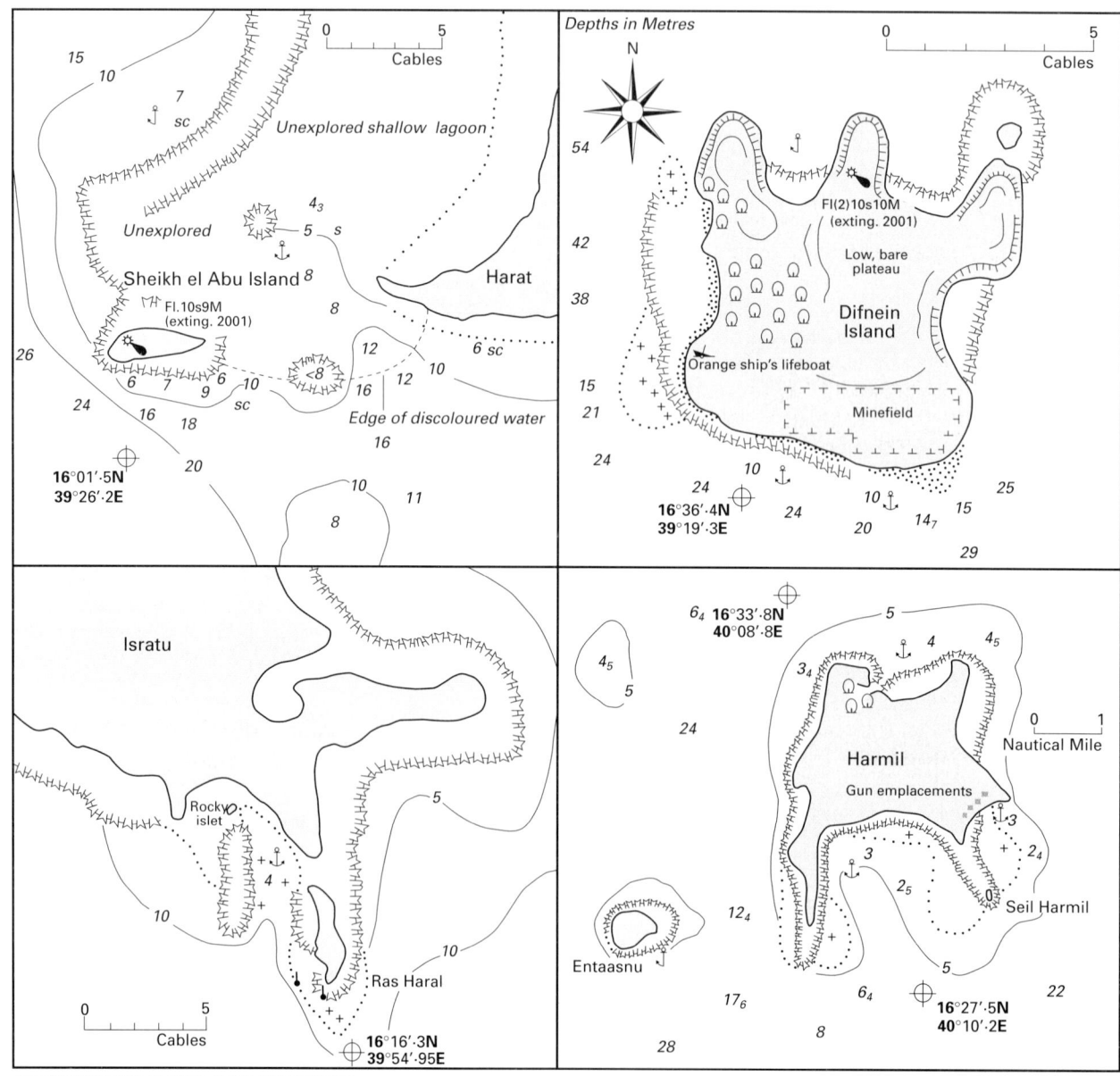

ANCHORAGES N AND E OF MASSAWA

Known anchorages with rough co-ordinates and a crude indication of reliability are as follows from E to W and S to N:

Hawatib Kebir, 15°53'·5N 40°35'·1E, tuck into one of the breaks in the reef on the NW corner, moderate SE only.

Dhu-I-Fidol, 16°00'N 40°16'·2E, beware of the submerged reef running some 9M ESE from the island's E coast. Anchor on the W coast, N of the tongue of reef on the SW corner. Poor holding in thin sand over flat rock so lots of chain. Strong tidal currents.

Seil Norah, 16°02'·9N 39°57'·75E, off the W coast of Seil Norah, a small off-lier of Norah, though the shelf drops off quickly. For fair to moderate weather. In brisker E to SE winds try the other side of Seil Norah, S of Kad Norah 7–8m, sand.

Dhu Lalam, 16°09'·8N 40°03'·9E, you can push right up to the island on its S end and this anchorage offers good shelter in N–NW winds.

Entaasnu, 16°28'·4N 40°05'·9E, and **Romiya**, 16°31'·7N 40°02'·6E. Yachts have found respite from N'lies hanging off the shelves on the SW corners of these islands. Sooner them than us.

Wusta, 16°19'·1N 39°49'·5E. If Isratu feels a bit tricky, try the lagoon between the rock of Fanadir Wusta (which once had a small beacon on its summit and may have still) and the E coast of Wusta in about 3–5m sand and coral. Good for W and NW. Exposed to N through SE.

Sheikh el Abu
Admiralty chart 164, plan page 100

This small island with the lighthouse is on the SW tip of an extensive rock shelf W of the much larger, inhabited Harat I. The best anchorage is in the lagoon between Sheikh el Abu and Harat. Don't be deterred by the colour of the water on the approach from the S. There is an abrupt change from darkish green to bright turquoise green. The former is not shoal but either deep water or coral and rock from 8–16m. The latter is over sand. Depths stay the same as you cross from one colour to another. The sand is over a rock platform and, as is often the case on the Dahlak Bank, may not be thick. Put out plenty of scope. There are coral outcrops in the lagoon but they show up easily in good light. Shelter is better for N'lies than at Difnein. The lagoon looks very shoal in the hook formed by the S end of Harat I. To find shelter in E–SE'lies, you would need to push well in to get a lee from Harat. Alternatively there's a possibility marked by a half anchor N of the reef joining Sheikh el Abu to Harat.

Isratu
Admiralty chart 164, plan p.

There is an off-lier S of Ras Haral to be wary of. Tuck up under the low cliffs in about 4m for the best shelter from all but due S. Scout about among the bommies for a sand patch in which to drop the anchor. You may see some fishermen here but otherwise the only signs of life are the birds. This used to be a smugglers haven during the Eritrean fight for independence. Life is quieter now.

Harmil
Admiralty chart 164, plan page 100

Approach
Seil Harmil (14m) is higher than Harmil itself. There are three anchorages for different wind conditions. Note that the bottom shoals rapidly either side of the entrance to the SW anchorage which hasn't got room for more than two or three boats at most. There is an off-lying rock in the bay on the SE side. Since the island is at most 3m high, there is little protection from the wind but it comes highly recommended by those who have visited it. In winds from the SE the bay on the N coast offers fair shelter, though some swell works round. Unless, that is, you are shoal draft when you can winkle your way into the lagoon, wholly protected by mangroves, off the SW corner of the anchorage.

Difnein
Admiralty chart 164, plan page 100

This is the NW'most island on the Dahlak Bank. Anchor in 10–15m, good holding in sand, close to the S side of the island. The bottom drops away steeply further off. It can be rolly in a NW but with the onset of the sea breeze the wind swings N and things quieten down. An alternative is in the W'most of the two bays on the N coast for S'lies. But we haven't tried it and can't vouch for what you'll find. It looks fairly deep. Don't go ashore. There may be unexploded mines.

Note
There are no reliable anchorages on the mainland coast N of Difnein till you reach Khor Nawarat in Sudan. It is prudent to stand well offshore along this stretch where you pass the Eritrea/Sudan border.

Apart from one deep patch from about 17°10'N to 17°30'N, the bottom is very uneven and you should not be surprised to see the echo-sounder jumping up and down. There is one shoal with only 7m over it in 16°59N 39°13'·5E. Once closing Ras Kasar, unless you are 20M off, there are 11m and 12m shoals and frequent depths of less than 30m.

Red Sea Pilot

Yemeni islands including the Hanish and Zubayr Groups

Admiralty charts
Large scale 453
Medium scale 1925, 164, 81, 453
Small scale 6, 143, 157, 4704

US charts
Large scale 62105
Medium scale 62100, 62110
Small scale 62110

French charts
Medium scale 6945
Small scale 7099

Warning
Although the most recent editions of charts from all the major hydrographic offices are reconciled to WGS84, this CANNOT compensate for shortcomings in the original 19th-century surveys. The only recent survey of the central S Red Sea to WGS84 was to satisfy IMO requirements for the new traffic separation scheme, not detailed resurvey of the coasts. Three points of navigational importance follow. First, always navigate even with GPS, with circles of probable error (CEPs) of at least 2 miles, and at night at least 5 miles. Second, navigating by GPS alone in the Red Sea is stupid. Do not do it. Make daylight landfalls and use hand-bearing compass and radar, if you have it, to establish WGS84-to-chart error factors. Third, treat ALL waypoints with appropriate caution, including those given below and NEVER assume that the waters between, close to or at waypoints are free of hazard.

Hanish Islands

Anchorage possibilities in the group are indicated on the key map. There is then large-scale detail, roughly S to N.

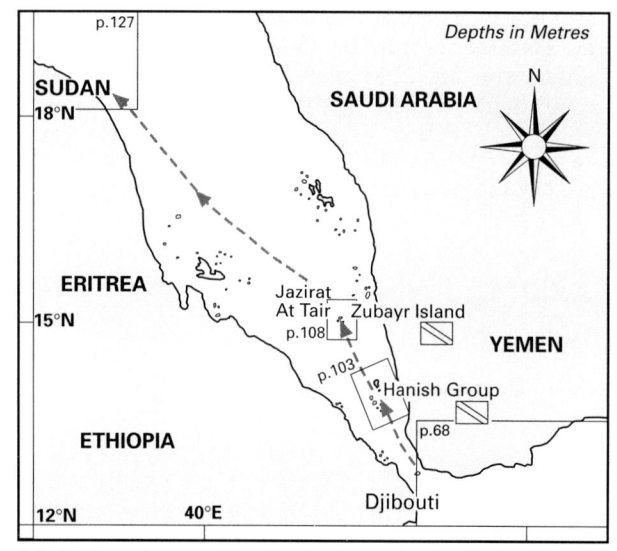

SOUTHERN RED SEA: CENTRAL PART
KEY MAP

Notes
The area is subject to dramatic changes in wind direction in the spring months so anchorages may suddenly become untenable.

Despite the settlement with Eritrea, this is still a fragile zone politically. There is a regular military presence and the Yemeni authorities mount frequent patrols. Some yachts have been mulcted for a 'cruising fee'. To avoid this, ask the Aden harbourmaster (see Aden entry above), for a permit in Arabic to stop in the Hanish.

When approaching and leaving the Hanish Islands, pay careful attention to the new traffic separation scheme (see key map and diagram below left). This is especially true around the Abu Ali Islands to the NE of Jabal Zuqar, where ships in the scheme have very little room to manoeuvre.

Suyul Hanish
Admiralty chart 453, plan page 103

Anchorages
W anchorage 13°37'·05N 42°43'·75E, 6–7m, sand. Good shelter in a SE'ly.

S anchorage 13°36'·2N 42°43'·9E, 9m, sand and coral, indifferent holding. For N'lies only with gusts off the hills.

E anchorage 13°37'N 42°44'·4E, 10m, sand and coral. Sheltered in NW'lies.

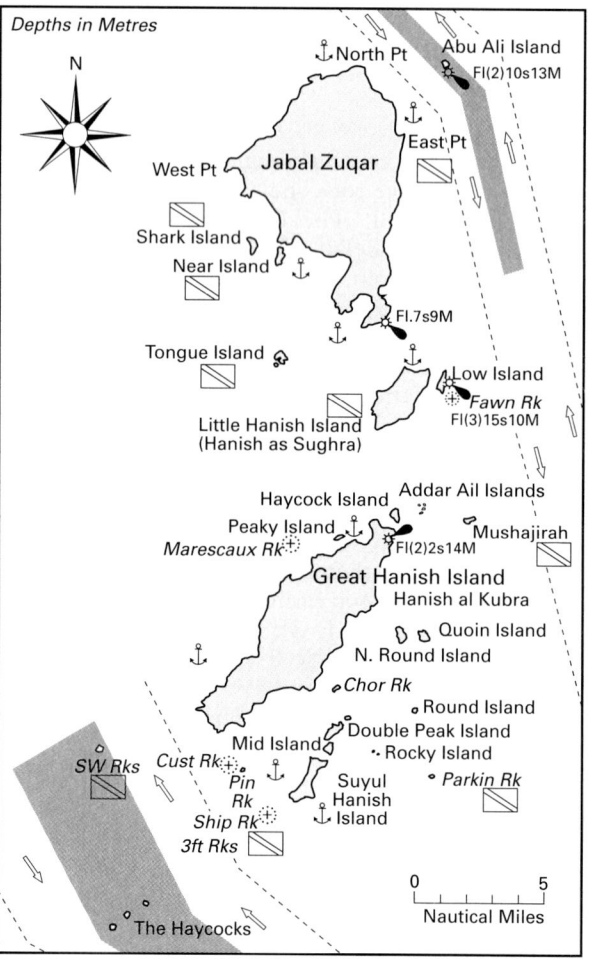

JABAL ZUQAR AND HANISH ISLANDS KEY MAP

102

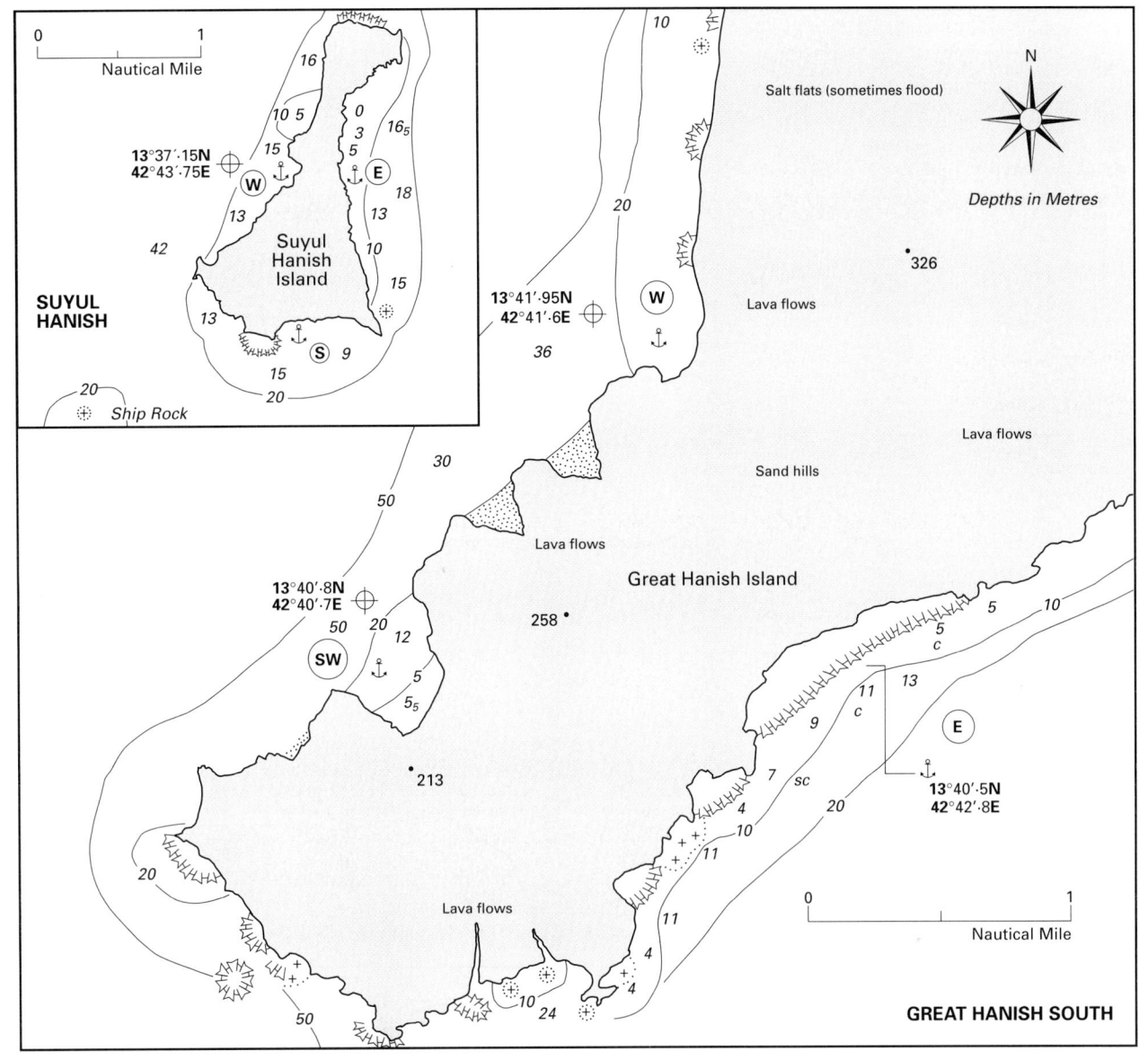

Southern Red Sea

SOUTHERN HANISH ISLANDS ANCHORAGES

Hanish Al Kubra (Great Hanish I)

Admiralty chart 453, plans page 104

Note Regular military patrols are reported here.

Anchorages

SW anchorage 13°40'·8N 42°40'·7E, 5m, sand. There's room for about 2 boats in the S part of the bay, clear of the fringing reef. Protection from the SE is good, but it can be gusty.

W anchorage 13°41'·95N 42°42'·0E, 4–7m, silty sand. This shallow bay offers reasonable protection in light to moderate weather off the volcanic scoria and lava flows.

E anchorage 13°40'·5N 42°42'·8E, 12m sand. The whole large bay has a band of sand roughly between the 20m line and the edge of the reef. In the summer months when the wind is usually NW, almost anywhere in the bay will serve.

N anchorage 13°46'·9N 42°46'·5E, 5–7m, sand and coral. Holding is not good everywhere in the bay. From the SE approach the anchorage between the N tip of Hanish al Kubra and Haycock I. The middle of the channel is deep but narrow. Be wary of the rock a cable off the SW shore of Haycock I, roughly level with the NW tip of Hanish al Kubra. The W shore of the passage has several off-lying rocks close to. There is a light on a column on the NE tip of Hanish al Kubra but it's not reliable. The cliffs to the NE of the anchorage are an extraordinary contrasting gold and black. There is an alternative anchorage nearer to Peaky I.

If you're lucky you may still see antelope and gazelles on the island. Turtles are more common. There is good walking and diving, especially in the N.

Red Sea Pilot

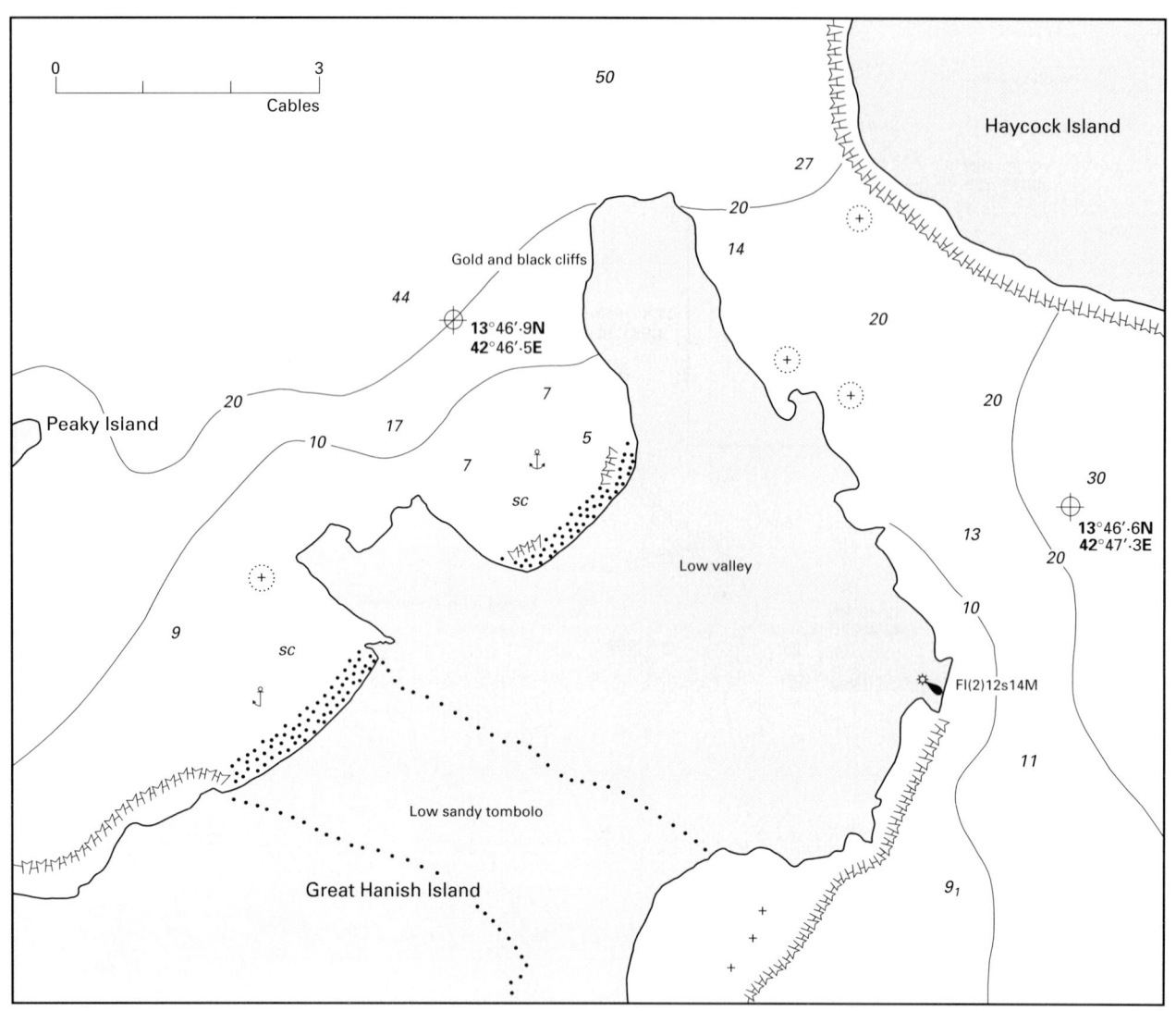

GREAT HANISH NORTH

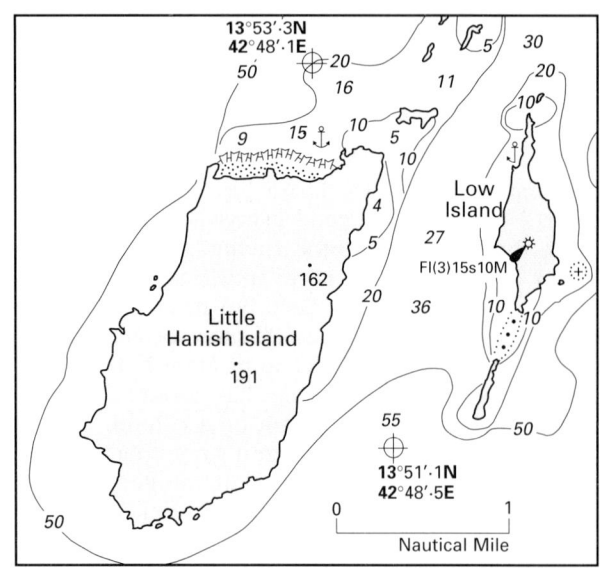

LITTLE HANISH AND LOW ISLAND

Hanish As Sughra (Little Hanish I)
Admiralty chart 453

There is a deep channel free of dangers between Haycock I and Hanish as Sughra. Approach is best from the W because of the rocks and rocky islets lying on a shallow flat NE'wards from the NE tip of the island. Anchor in 14–15m, coral and sand. Good protection in S'lies with good holding if you pick your spot well, though a second anchor may be advisable in strong wind. This anchorage is preferable to the one off Low I but there are better options off Gt Hanish.

Low Island
Admiralty chart 453

Approach from SW past the chain of rocks and narrow islet S of Low I, staying in deep water until the anchorage lies abeam to the E. Fawn Rock, almost awash, lies off the E side.

The light on the highest point of the island is not reliable. There's protection from SE and S winds in the indent off the NW coast in about 15m, though it may be possible to nose further in to shallower

104

Southern Red Sea

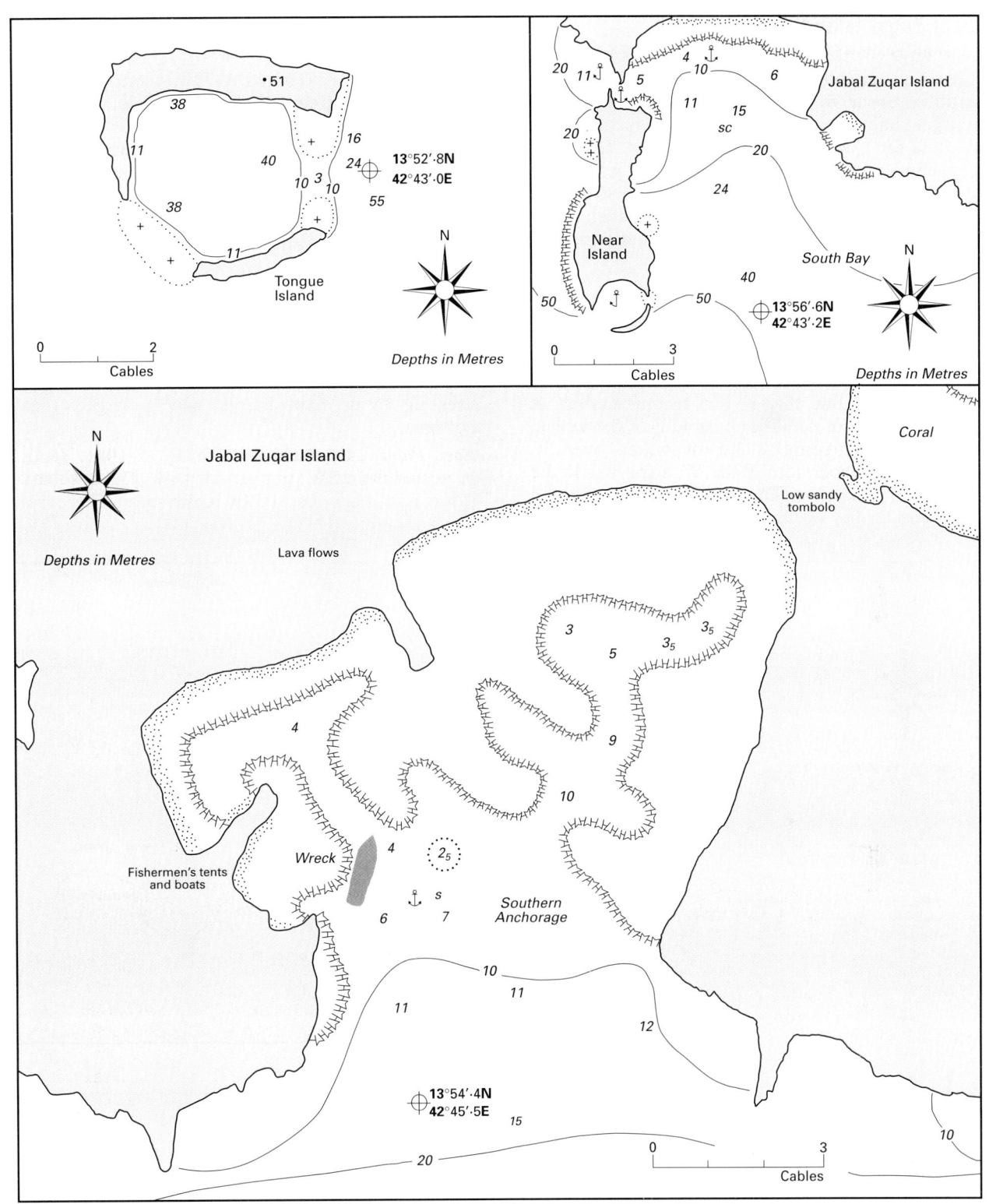

TONGUE ISLAND AND JABAL ZUQAR SOUTH

water. Holding is reasonable in moderate weather. If it's blowing hard, head for Gt Hanish or Jabal Zuqar.

Tongue Island

Admiralty chart 453

The entrance to this lagoon anchorage in an old volcanic crater is through a gap a cable wide on the E side with just 3m depth on the threshold. The main island is semicircular and 51m high, connected in the SW by shoals to a smaller islet which almost completes the circle. Enter in good light only and in light weather. Anchor in 10m, sand, close up against the inner crater wall where there's a narrow shelf. The rest of the crater lagoon is very deep.

Red Sea Pilot

Jabal Zuqar Island
Admiralty charts 453, 1925, plans page 105 and below

Buildings, occasionally occupied by the military are on the shore in the NE corner of South Bay. You may get your papers checked and soldiers are known to press for baksheesh in the form of a 'fee' for a 'cruising permit'. The island is otherwise uninhabited save by nomadic fishermen who may offer to trade. There are at least five anchorages around Jabal Zuqar, which offers good walking and snorkelling.

S anchorage 13°54'·4N 42°45'·5E, 6–8m, sand. There is good protection from N'lies and the holding is good. In good visibility the reefs can easily be seen. On a clear night with no moon in pre-GPS days we were able to enter this anchorage and anchor safely, but this is not recommended! A detached patch of coral lies a cable E of the wreck, though it has some 2–3m of water over it. Penetration deeper into E or W arms would be possible in good light. The almost intact wreck of a small coaster lies on the reef on the W side of the bay. There is a light on a tower (10m) with daymark near the SE tip of Jabal Zuqar.

South Bay W 13°57'N 42°43'·3E, 4–6m, coral and sand. Good holding. Excellent protection in N'lies, fair protection in light SW'lies. Good beachcombing. There is also a small, rough jetty and quarters for the military ashore further over on the E side of the main beach of South Bay.

Near Island 13°57'·5N 42°42'·8E, 5m, sand and coral. Well protected in N'lies, but can be rolly. Two other anchorages are marked on the sketch-map, but these are only really suitable as day anchorages in light weather if you can find suitable depth. The anchorage in the open crater at the S end of Near I is spectacular. It is also possible to anchor off Shark I, which, for divers, apparently lives up to its name. Both have a largely rock bottom.

East Point 14°02'·1N 42°47'·5E, 5–10m, sand, protected from SW, but can be rolly. The headland is low and rocky, backed by white sand. There is a

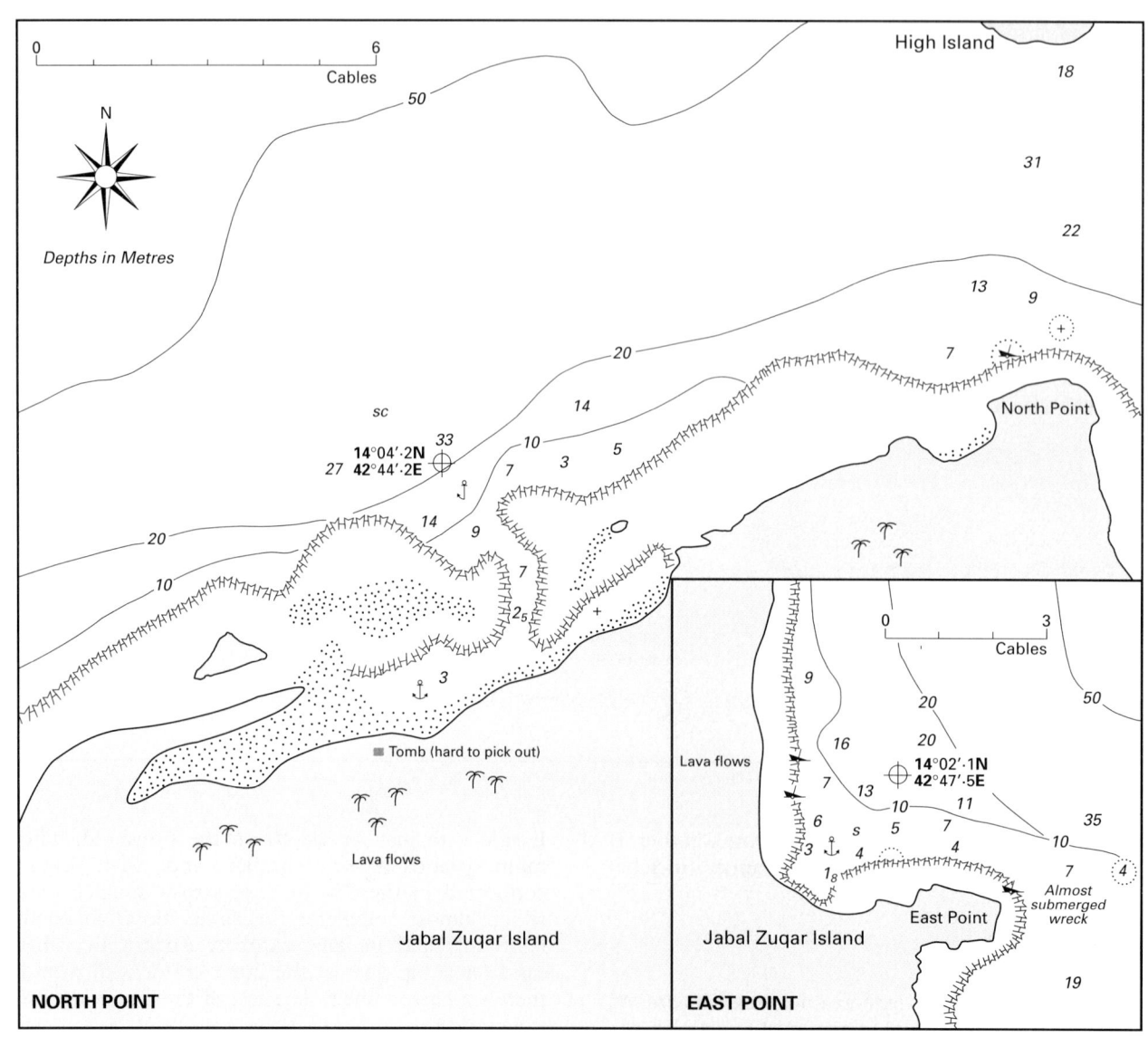

JABAL ZUQAR NORTH

106

Southern Red Sea

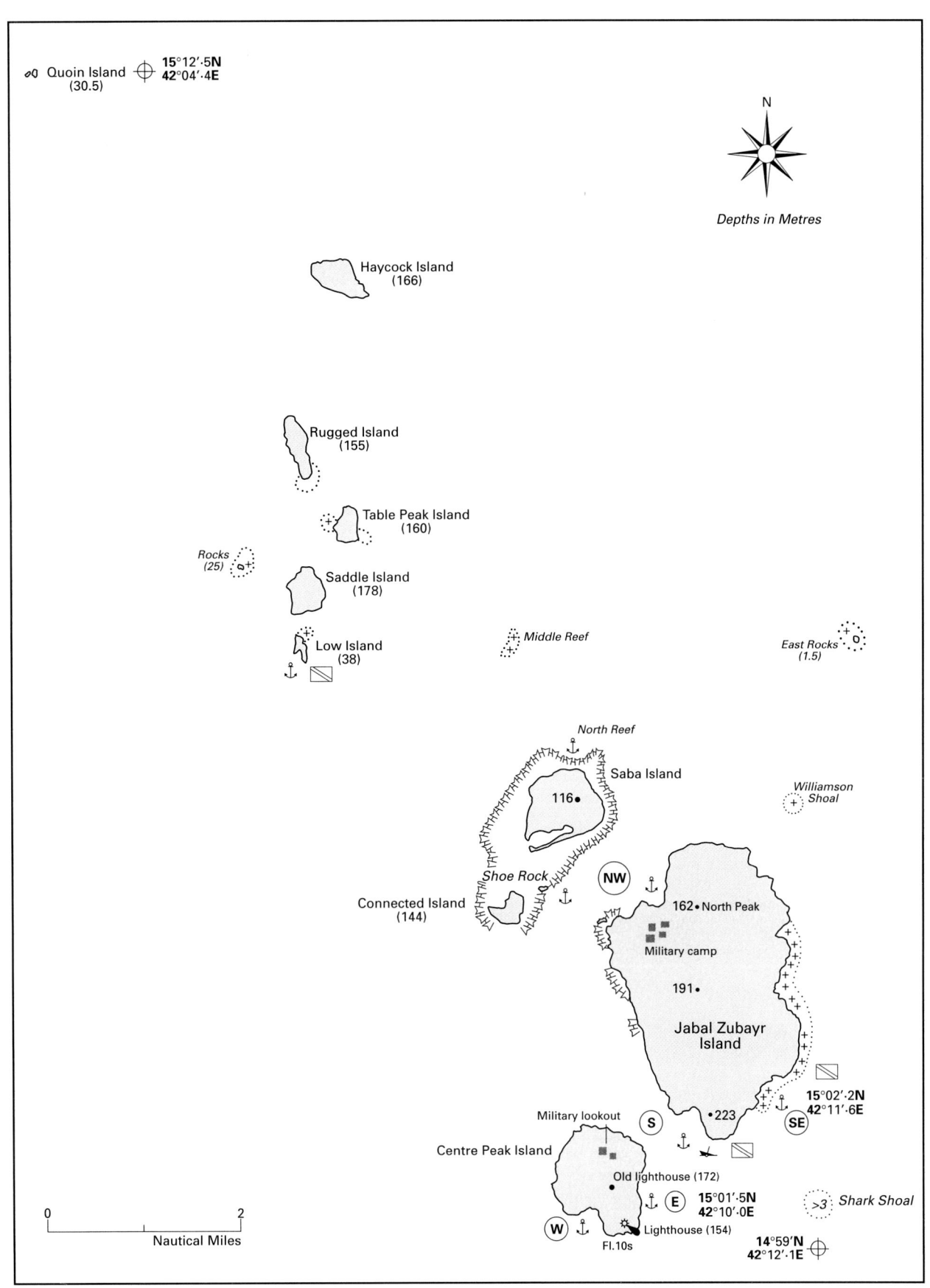

Quoin Island (30.5) — 15°12'.5N 42°04'.4E

Depths in Metres

Haycock Island (166)

Rugged Island (155)

Table Peak Island (160)

Rocks (25)

Saddle Island (178)

Low Island (38)

Middle Reef

East Rocks (1.5)

North Reef

Saba Island 116

Williamson Shoal

Shoe Rock

Connected Island (144)

(NW)

162 • North Peak

Military camp

191 •

Jabal Zubayr Island

15°02'.2N 42°11'.6E

(SE)

Military lookout

(S) • 223

Centre Peak Island

Old lighthouse (172)

(W) (E) 15°01'.5N 42°10'.0E

Lighthouse (154)
Fl.10s

Shark Shoal >3

14°59'N 42°12'.1E

0 — 2 Nautical Miles

ZUBAYR GROUP KEY MAP

107

Red Sea Pilot

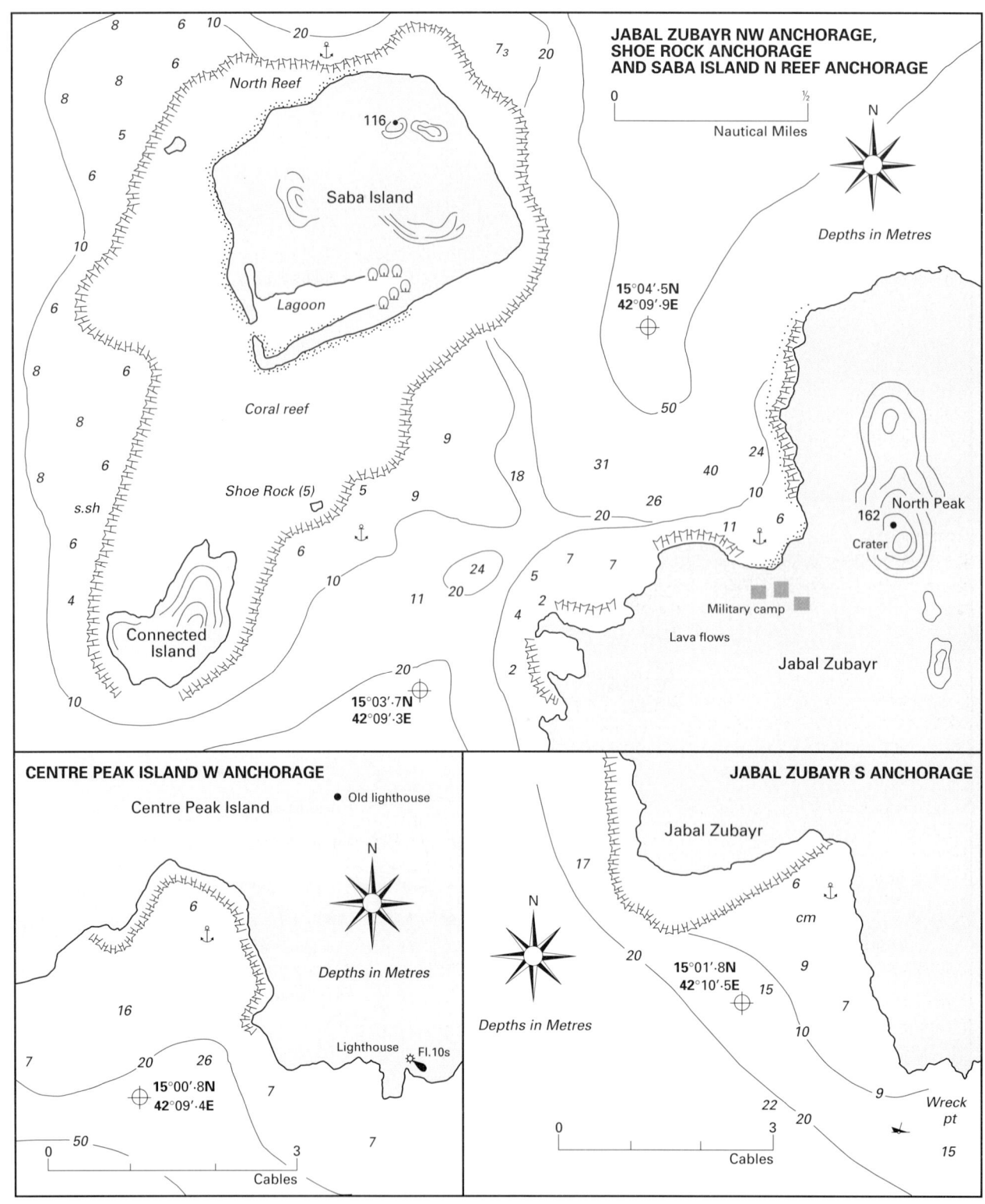

ZUBAYR ISLANDS

shoal patch with about 4m over it E of East Point and a patch of shoal water extending into the anchorage from the fringing reef S of the anchorage. Two wrecks lie on the fringing reef N of East Point. A third wreck, almost submerged, lies on the reef off East Point itself.

Between here and the anchorage at North Point there's good holding in fine sand in 3–7m with Abu Ali light bearing 060°. The bottom is largely coral so, as with many of these anchorages, it's a good idea to have a look at your anchor once you think you are secure.

When in this area be careful to stay clear of the southbound lane of the new traffic separation scheme which, especially around the NE end of Jabal Zuqar, is close offshore.

North Point 14°04'·2N 42°44'·2E. A low sand cay lies on the fringing reef N of a sandspit 4 cables SW of the entrance. The sandbanks on the reefs on each side of the entrance may dry in summer. The best anchorage, in a lagoon, is entered through a narrow dogleg pass in the reef. A dark square tomb is located 1yM WSW of North Point and 3 cables SW of the entrance to the anchorage, but is hard to see. Some vegetation is visible on North Point, remarkable because the island is largely barren. The pass through the reef must be eyeballed from the rigging on the way in. Tidal streams are reported to run at up to 2·5kt across the entrance, flowing SW on a rising tide and NE on a falling tide. Tidal currents in the pass flow along its line. Minimum depth in the pass is 2·5m, but be conservative. In fresh to strong winds the pass is dangerous.

There is 2–3m, sand, good holding with excellent protection from the S once you are in the lagoon.

Jazair Az Zubayr (Zubayr group)
Admiralty chart 453, key map page 107

Approach
From the S, look out for Shark Shoal, which lies about 2½M SSE of the S tip of Jabal Zubayr I. The charted depth is 13m. A recent visitor reports that it's about 3m! Its position is 14°59'·5N 42°12'E. Approaching from the NW, the first in the group is Quoin It, wedge-shaped and 30m high, but not easily visible at night. We describe two anchorages off Centre Peak I, three off Jabal Zubayr, three off Saba I and one off Low I.

Dangers
Shark Shoal, as mentioned above, is in 14°59'·5N 42°12'E. East Rocks, 1·5m are 3M NE of Saba I. They are steep-to up to a cable off, but often break. The islands between Haycock and Saba have shoal water around them, and Middle Reef, 1½M NNW of Saba I, has less than 2m over it.

The wind here is particularly prone to sudden changes in direction, so always be prepared to move.

General
The fish life around these islands gets rave reviews. The coral isn't up to much, a lot of it dead or dying. But there are manta rays, sharks, whale sharks you can swim with if that's what floats your boat and lots to catch and eat.

Centre Peak Island
Admiralty chart 453, plan page 108

The island often has a military presence but there are two possible anchorages if you need them. The first, West Anchorage, in a bay off the SW coast, is rather deep. Holding is good in mud in 15m or so, or less reliable in 6m sand and coral close in, but it can be rolly and gusty. Note that depths drop off very quickly so you need to tuck in. The light on the SE tip of the island is unreliable. The remains of a disused lighthouse are on the highest point (172m). The second, East Anchorage, in 15°01'·5N 42°10'E, is in a small bay off the E side at the end of a lava flow. It's shown on the key map but is not individually illustrated. There is a small pier here, which used to serve the old lighthouse mentioned above. Anchor in about 10m, sand and coral, with shelter from NW–SW.

Jabal Zubayr Island
Admiralty chart 453, plan page 108

The wreck of MV *Star of Shaddia* lies close to the S tip of Zubayr I. The cone-shaped peak (223m) of the island's mountainous ridge is at the S. We mark three anchorages on the key map.

S anchorage 15°01'·8N 42°10'·5E, 6m, coral and mud. Tuck well up so you get some extra shelter from the reef flat.

NW anchorage 15°04'·2N 42°10'·1E, 5–10m, sand and with reasonable holding if you pick your spot well but can be rolly. North Peak (162m), which appears square from seaward but has a crater at the summit, is ¼M inland of the anchorage. There is a military outpost; if they let you go ashore the scramble up to the crater rewards the effort. Local boats shelter here and the fishermen are friendly. Bird life is interesting, with flamingos and boobies.

SE anchorage 15°02'·2N 42°11'·5E, 6–10m, sand, with good holding in the little bay.

Saba Island
Admiralty chart 453, plan page 108

Reef joins Saba I and Connected I (144m) with Shoe Rock (5m) between them. Saba I has some stunted bushes and its lagoon is fringed with mangroves. The lagoon can be explored by dinghy. Flamingos and sea eagles are common. There are a couple of anchorages suitable in light weather only. It is nearly always rolly.

Shoe Rock 15°04'·3N 42°09'·1E, 5–10m, sand.

N reef 15°05'·65N 42°09'·0E, 7–10m. This is really a day anchorage for diving. The bottom drops off very quickly and you have to tuck right in leaving no room for wind shifts.

Low Island
Admiralty chart 453, plan page 110

The anchorage, in 8–10m, sand, is in up against the wall of the extinct crater on the S side of the island. The crater itself is deep.

Jazirat At Tair
Admiralty chart 453, plan page 110

The island has two low volcanic cones at present quiescent. It is a good landmark. Jazirat At Tair light (round metal tower, 20m) is about half way down the W side and in consequence the light is obscured from NE through SE. A steep rocky yellow bluff on the SE side of the island marks the main anchorage in N winds but there are others, none comfortable.

Anchorage A (see sketch) has 5–10m, sand and volcanic rubble, with good holding and protection from the N, but it can be rolly. You need to be

Red Sea Pilot

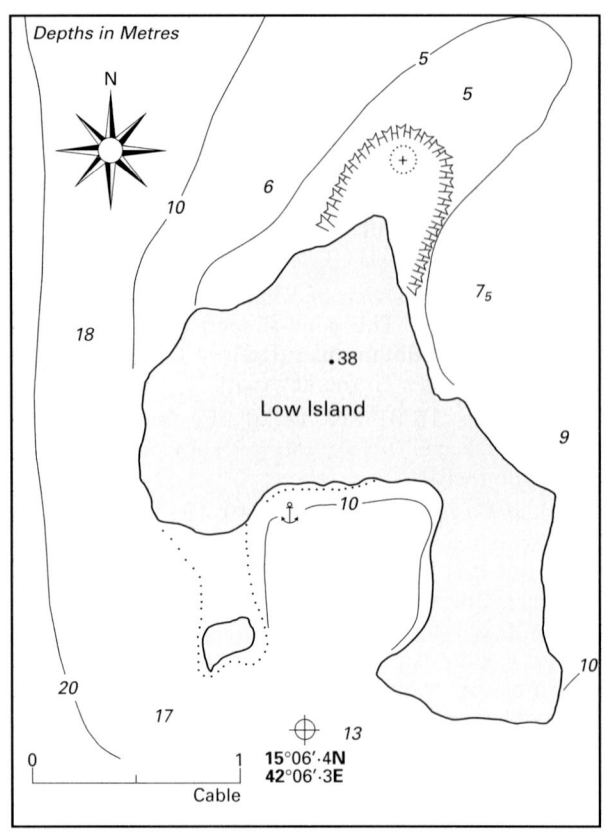

LOW ISLAND, ZUBAYR GROUP

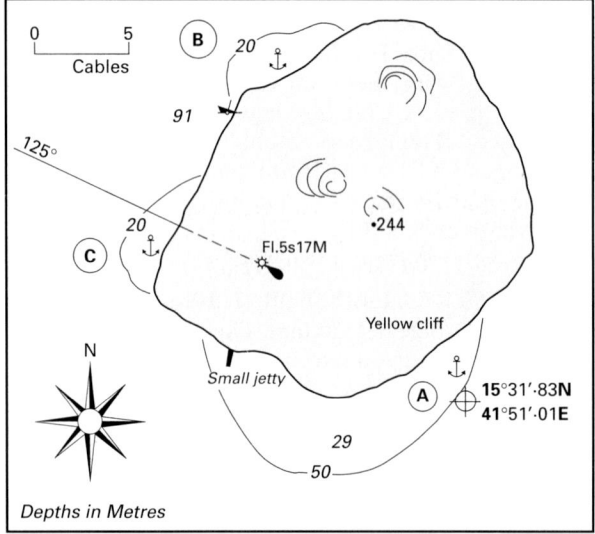

JAZIRAT AT TAIR

Anchorage C has the light bearing 125°. It is near a landing point also sometimes used by the lighthouse keepers. A rude jetty has been reported to have been built on the SW coast and it may now be the landing point. Occasionally you can smell the sulphur from the volcano. There are sharks in the area.

50–100m from the beach to find suitable depths, at which point you'll probably be able to hear your voice echoing off the cliffs.

Anchorage B has 7–10m, sand and coral with moderate holding and good shelter in S'lies off the sandy beach marked by black pumice stone outcrops. Supply boats for the lighthouse land here in S winds as do traditional dhows. Felix Normen writes of a 'little creek', presumably in the coral not the land, in which sand clearings offer anchorage in 7–10m. If you're used to the smaller Mediterranean *calles* and *calanques* you'll know what to look for; and it'll be very tight!

110

Southern Red Sea

Yemen Red Sea coast to the Farasan islands

Admiralty charts
Large scale 542, 1955
Medium scale 542, 548, 1925, 2588
Small scale 6, 143, 157, 4704

US charts
Large scale 62285, 62288, 62292
Medium scale 62285, 62295, 62271
Small scale 62290, 62270

French charts
Large scale 7111
Medium scale 6982, 7517
Small scale 7099

Note
If you are headed S hold into the shore S of Al Mukha to find a slight lee. This reduces a bit the seas you face. The same technique on the Eritrean shore will keep you out of the fiercer N-setting currents offshore and here as there you might find a counter-current in your favour.

Ras Dhubab
Admiralty chart 2588, plan below

This is the first anchorage on the Yemeni coast if you are heading N from Bab el Mandeb. Avoid Chiltern Shoal, S of Ras Dhubab, in rough weather as it makes the sea stand up. The anchorage is for S'lies only in an indent in the fringing reef N of Ras

Warning
Although the most recent editions of charts from all the major hydrographic offices are reconciled to WGS84, this CANNOT compensate for shortcomings in the original 19th-century surveys. The survey of the area from N of the Hanish Islands to Bab el Mandeb, done to WGS84 in 2000–2001, was intended to satisfy IMO requirements for instituting the new separation scheme, not to provide new detailed data of the coast. Three points of navigational importance follow. First, always navigate, even with GPS, with circles of probable error (CEPs) of at least 2 miles, and at night at least 5 miles. Second, navigating by GPS alone in the Red Sea is stupid. Do not do it. Make daylight landfalls and use hand-bearing compass and radar, if you have it, to establish WGS84-to-chart error factors. Third, treat ALL waypoints with appropriate caution, including those given below, and NEVER assume that the waters between, close to or at waypoints are free of hazard.

Dhubab in an opening in the reef. Give the reef a careful berth when approaching from the S. 4m, sand, at the S end of the channel in the reef, with good holding. There are two forts in the vicinity and a small military outpost. Reception for yachts varies.

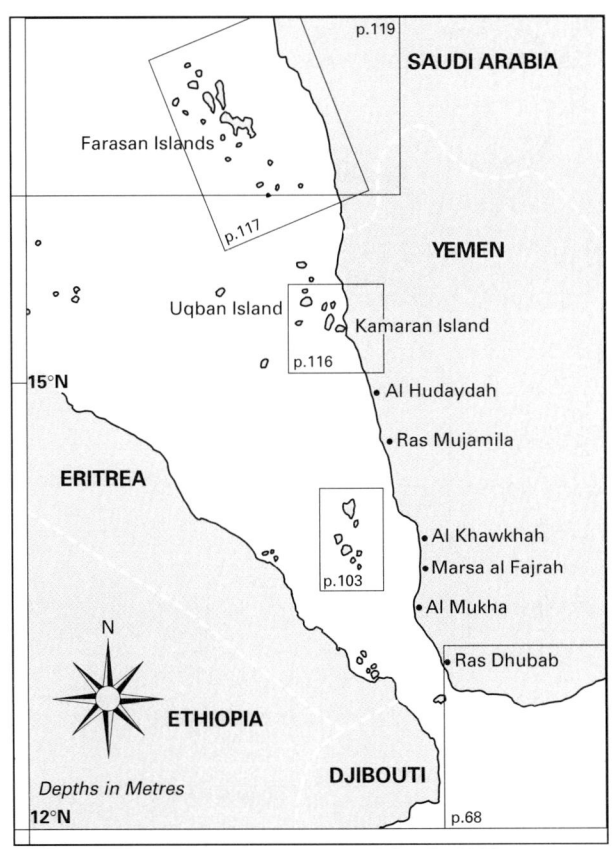

YEMEN COAST TO FARASAN ISLANDS KEY MAP

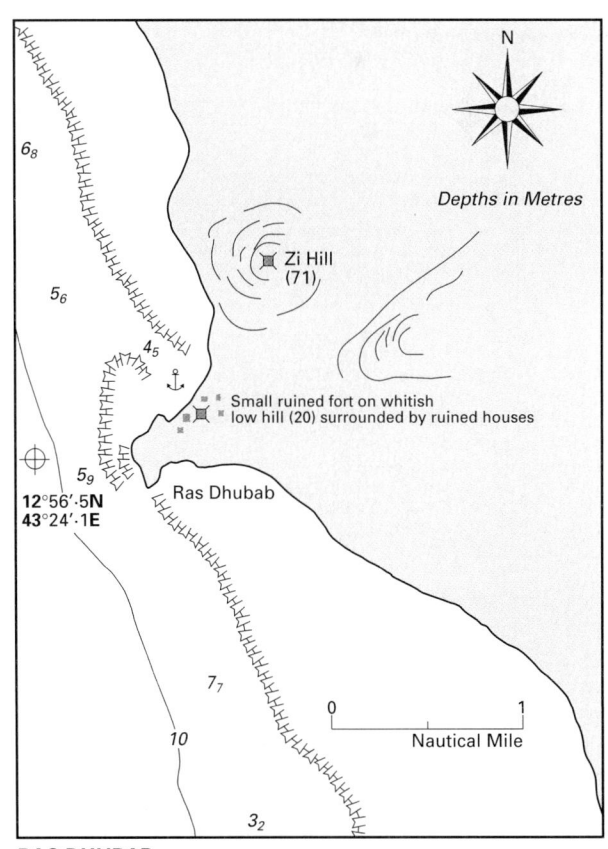

RAS DHUBAB

Red Sea Pilot

Al Mukha
Admiralty chart 1955, plan below

Approach
If you are coming from the S the safest route is to stay outside the 20m line till you are abeam the fairway buoy. In fair weather you might cross the shoals to the S but it's not recommended in any kind of sea. From the fairway buoy there is a buoyed dredged channel. The entry to the port can be difficult in strong gusty SE winds, when there is often poor visibility due to dust storms. Leading beacons, in line 137°, mark the fairway. At night the floodlights on the mole and the lights of the power station 3M N are usually visible up to 10M off but navigational lights are unreliable. Call the Al Mukha pilot on VHF Ch 12 or 16 on approach. He speaks good English and has a reputation for being helpful.

Dangers
Dhows anchor on the sandbank W of the town. It has 2–3m over it and causes waves to break in heavy weather when the dhows restrict movement in the harbour by laying multiple lines and mooring up together. Yachts that have visited Al Mukha recommended anchoring here only in an emergency.

Anchorage
3m, good holding, a cable off the jetty, but keep

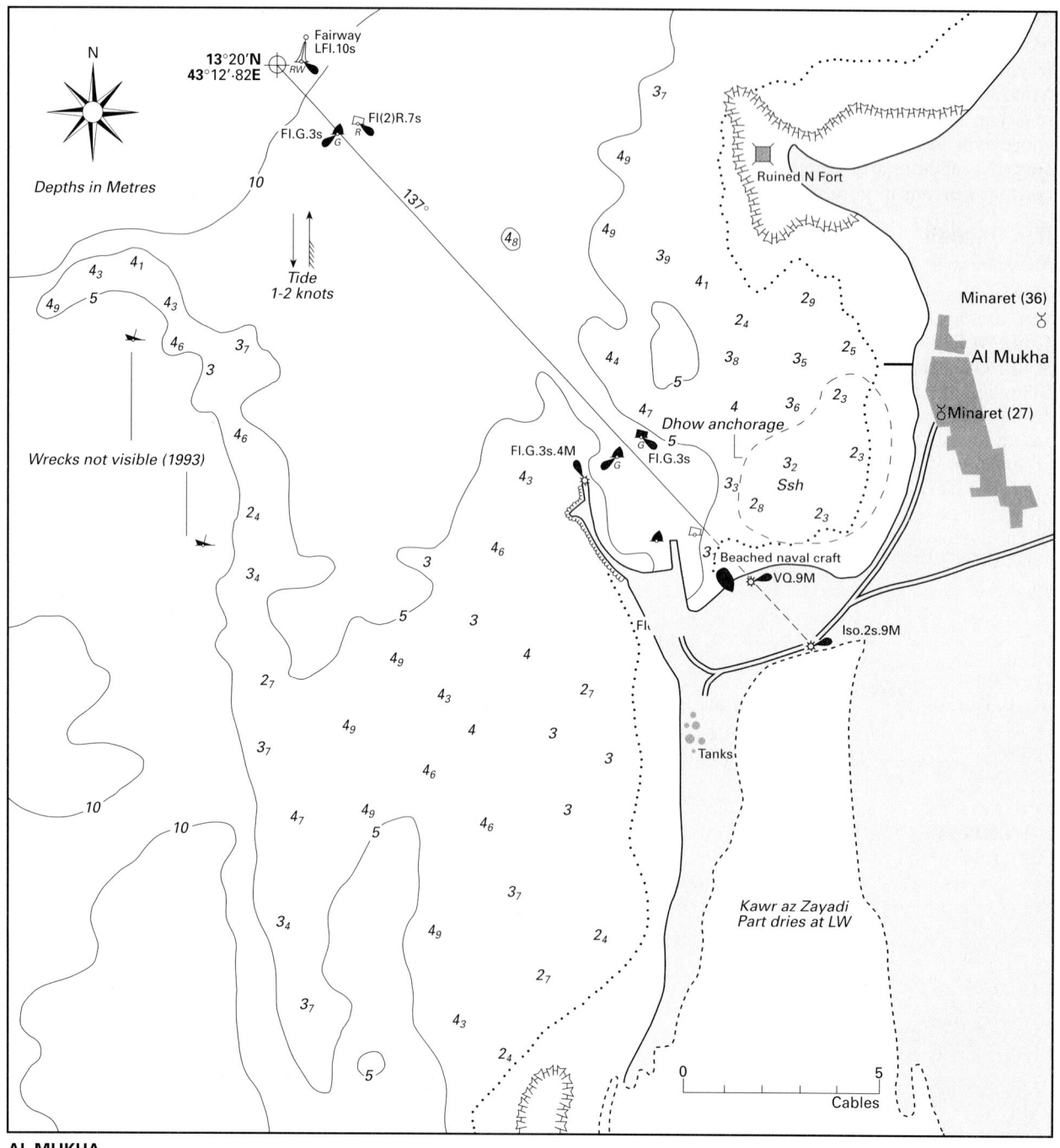

AL MUKHA

112

clear of the leading line and the buoy's mooring chains. An alternative is to anchor with the dhows, but this is only suitable for shallow-draught boats in light weather. There is no shelter except from the S and SE. Even in light S winds the anchorage can be rolly. A sea rapidly builds up in any other conditions, given the long fetch. The harbourmaster will usually be on the jetty to give instructions on VHF.

Formalities
If you want to go ashore officials will probably ask you to tie up at the mole. This is risky because of ship movement and the normal liveliness of the seaway. Ships alongside the mole may take additional berthing lines across to the beached naval vessel. If you do go ashore your passports will usually be held unless you have a visa. Nationals of South Africa or evidence that you've visited the country can be a problem. There is heavy security in the port area. You may be asked for baksheesh.

Facilities
The town is ramshackle and dilapidated but some fresh food, fuel and water can be found. Despite the name, coffee is not easy to find. Take precautions against mosquitoes. Malaria is endemic.

General
Not many boats call here. It was the scene of frequent upsets in its off and on again history as a trading factory for the British East India Company. One of the great surveyors of the Red Sea, Cdr. Elwon, Indian Navy, cut his teeth as a skipper here in charge of the bomb ketch *Thames* in a typical bit of gunboat diplomacy in 1820. Several of the older accounts of sailing up the Red Sea tell of bypassing '...the town of Mocha Yamen, celebrated alike for its breed of Arab horses and its coffee', as Lady Brassey recounts, having sailed past in the *Sunbeam* in the 1870s.

'It is a large white town, full of cupolas and minarets, surrounded with green as far as irrigation extends, and looking like a pearl set in emeralds on the margin of the deep blue sea against a background of red and yellow sand-mountains.'

Even in those days the port was already declining in importance; the British were developing Aden as the region's main port, and a virtual monopoly in coffee beans had been broken when coffee plants were smuggled out to Sri Lanka (then Ceylon) and Java in the early 18th century. By the time W A Robinson sailed past in *Svaap* in the 1930s treatment of single-handed yachtsmen was already unpredictable. Another yacht, the American *Stornaway* foundered on a reef nearby in the 1950s. While her skipper was thrown in jail by the Yemenis, the boat was looted as she lay awash. In spite of these difficulties she completed her circumnavigation, and Carl Petersen, her skipper, was awarded the Blue Water Medal for his achievement in 1953.

Marsa Al Fajrah (Marsa Fujaim or Fejera)
Admiralty chart 1925, plan page 114

Yachts should stand at least 1M off when making in to clear all dangers. The entrance to the anchorage should be passed until it is open to the SE before heading in. Entering in the morning with the sun E'wards may be difficult. The coast is littered with rocks and coral patches for a considerable distance offshore both N and S of the inlet. There may be less water than charted in the approaches. Anchor in 5m well inside the inlet. The nature of the bottom is not known, but there should be good protection from the S.

Al Khawkhah
Admiralty chart 1925, plan page 114

There is a village with palm trees 1M S that is visible from offshore. Approaches are clear except for the sandspit W of the anchorage. Approach in daylight only. Charts of the area are misleading. Holding in the bay is questionable but there is good shelter from S and SE winds. Local dhows and fishing boats use the anchorage. There is a customs and security police post on the road at the head of the bay. Water, fuel and some stores are available.

Ras Mujamila
Admiralty chart 143, plan page 114

Ras Mujamila is only visible within approx 1M of the coast. The best approach is from the NW. There are two anchorages in the entrance to Khor Ghulaifiqa. The S part of the *khor* is formed by a peninsula extending 6½M N from the mainland. The channel between Ras Mujamila and the peninsula has silted up and is not navigable. There is a 2m patch on the S approach to the *ras* outside the 10m line on a bank that extends for 2M E of the point. The best anchorages for SE'lies are either in the small bay which forms the N side of Jazirat Mujamila, or between the two E-tending spits. The first is deeper, depending on how close to the beach you anchor. The second, 5m, sand and mud, good holding, is more protected from the S than it looks. The anchorage on the S side of the southern spit, 5–8m, sand and mud, is for N'lies.

Al Hudaydah and Ras al Katib
Admiralty chart 542, plan page 115

Approach
No lights were working when last visited. Some of the beacons still exist including those on the first leading line but the second lacks any marks whatsoever. In theory the approach channel is dredged to 9·4m, but don't rely on it. A fairway buoy marks the outer end of the approach. There is a jetty on the E end of Ras al Katib and port installations are conspicuous. The water in this area is often discoloured and misleading because of the large amount of sand in suspension.

Red Sea Pilot

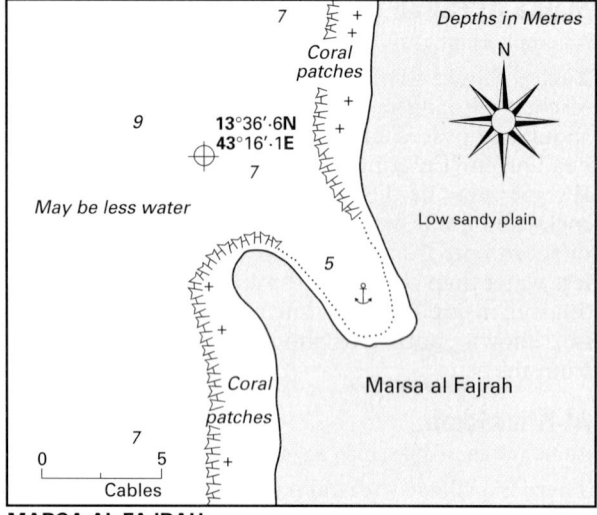

MARSA AL FAJRAH

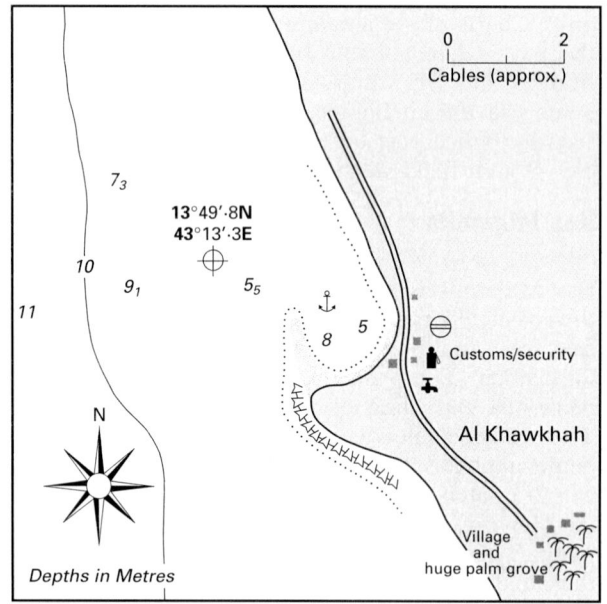

AL KHAWKHAH 13°50'N 43°15'E (very schematic)

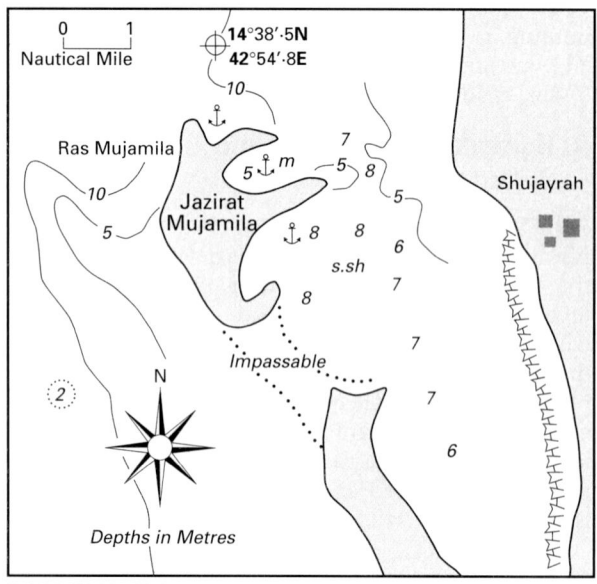

RAS MUJAMILA

Dangers
The channel into the port between islets, sand *cays* and reefs is very narrow in places. Use the large-scale chart of the area if you intend to go into the port but time your arrival for good light. Dredge spoil has built up either side of the channel so you must stick to the leading line. There is a dangerous wreck 2M W of Ras Marsa, 5M N of Ras al Katib.

Anchorage
In the port itself. Yachts used to anchor S of Ras Katib but this area is now occupied by mooring buoys and you're likely to be moved on.

Facilities
It is possible to obtain supplies at Hudaydah. A Yemeni visa makes access to the town easier. There are banks, phone and fax services, shops and PO. There is an airport with international connecting flights, should you need it in an emergency.

General
There was a port here in the early 16th century, but its importance was minor until the Ottoman Turks began to develop it as a rival to Al Mukha. It had a chequered history, and was devastated by recurrent wars. Since 1970 the port has become one of the world's most congested and the city has grown rapidly. It is worth avoiding unless you really need to visit not least because malaria is endemic.

Jazirat Kamaran
Admiralty charts 548, 1955, plan page 116

Kamaran I is really off limits but yachts which have called, after a bit of to-ing and fro-ing, have been allowed to stay. There is a strong military presence here because of the nearby boundary between Saudi Arabia and Yemen. Yachts are not very welcome and you MUST check in at Kamaran itself first.

You will need the large-scale chart. There are overfalls on Arab Shoal on the seaward side of the entrance. Rashshah It, in the approach to the S entrance to Madiq Kamaran, is lit. The colour of the water can often be misleading. There are leading marks, or are supposed to be, S of the As Salif loading jetty, which take you on 060° through the centre of the channel. The salt loading terminal at As Salif itself serves as a good landmark. Approach from the N is littered with islets and shoal water though, if the lights are working, it's better marked than it used to be.

The anchorage at Kamaran town is well sheltered. Tuck down in the SW corner or where directed. Afterwards you might consider going to As Salif. If so, call on VHF Ch 16 first. The port director, Captain Mohammed Ezzi, is very helpful. Once you have permission to land you can obtain fuel, water and provisions. Beware. This is a malarial area. Once you have the necessary clearances, Khor Tuways is a highly recommended alternative especially for wildlife amongst the mangroves.

Southern Red Sea

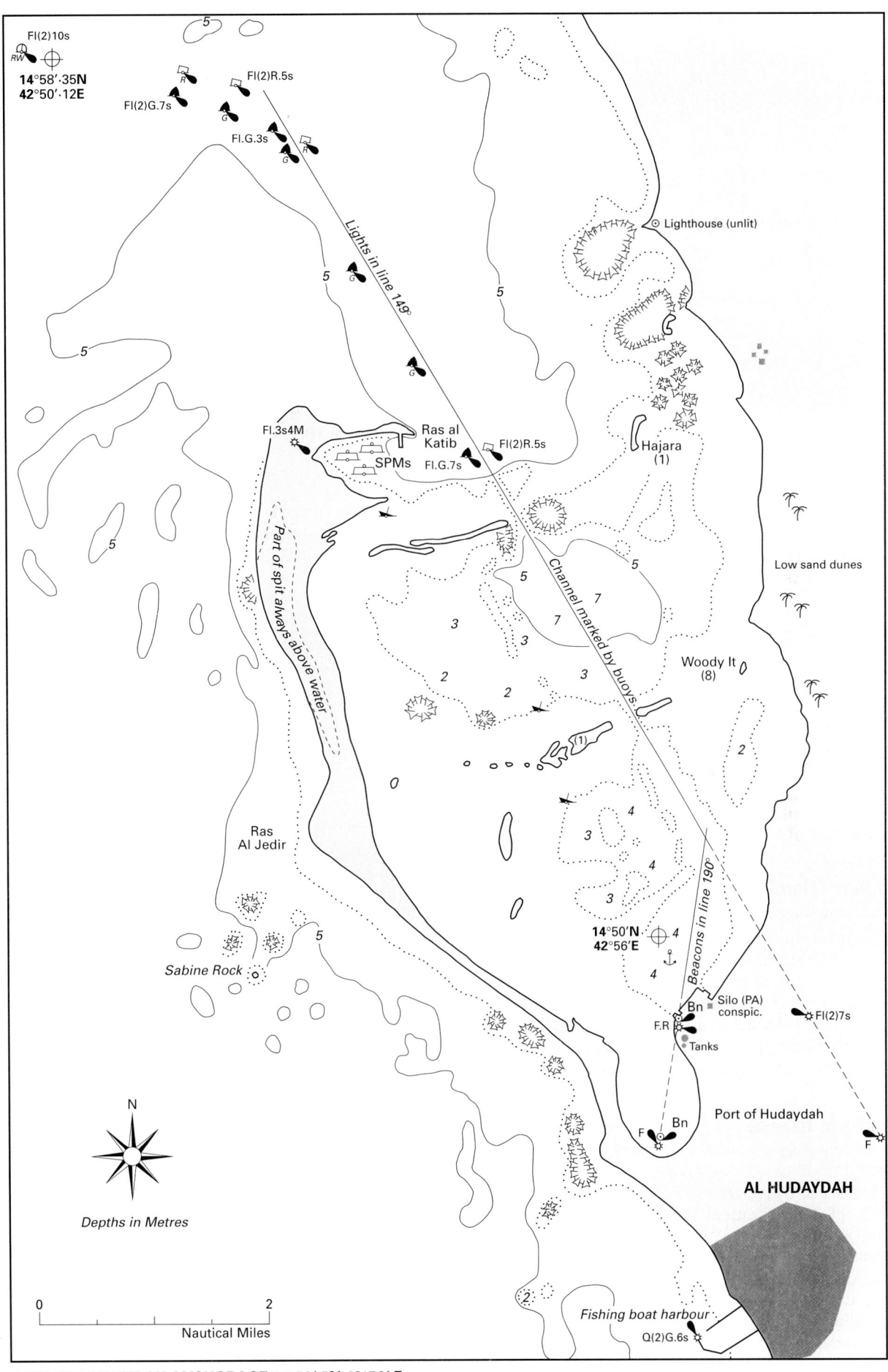

PORT OF HUDAYDAH ANCHORAGE 14°50′.5N 42°56′.E

Red Sea Pilot

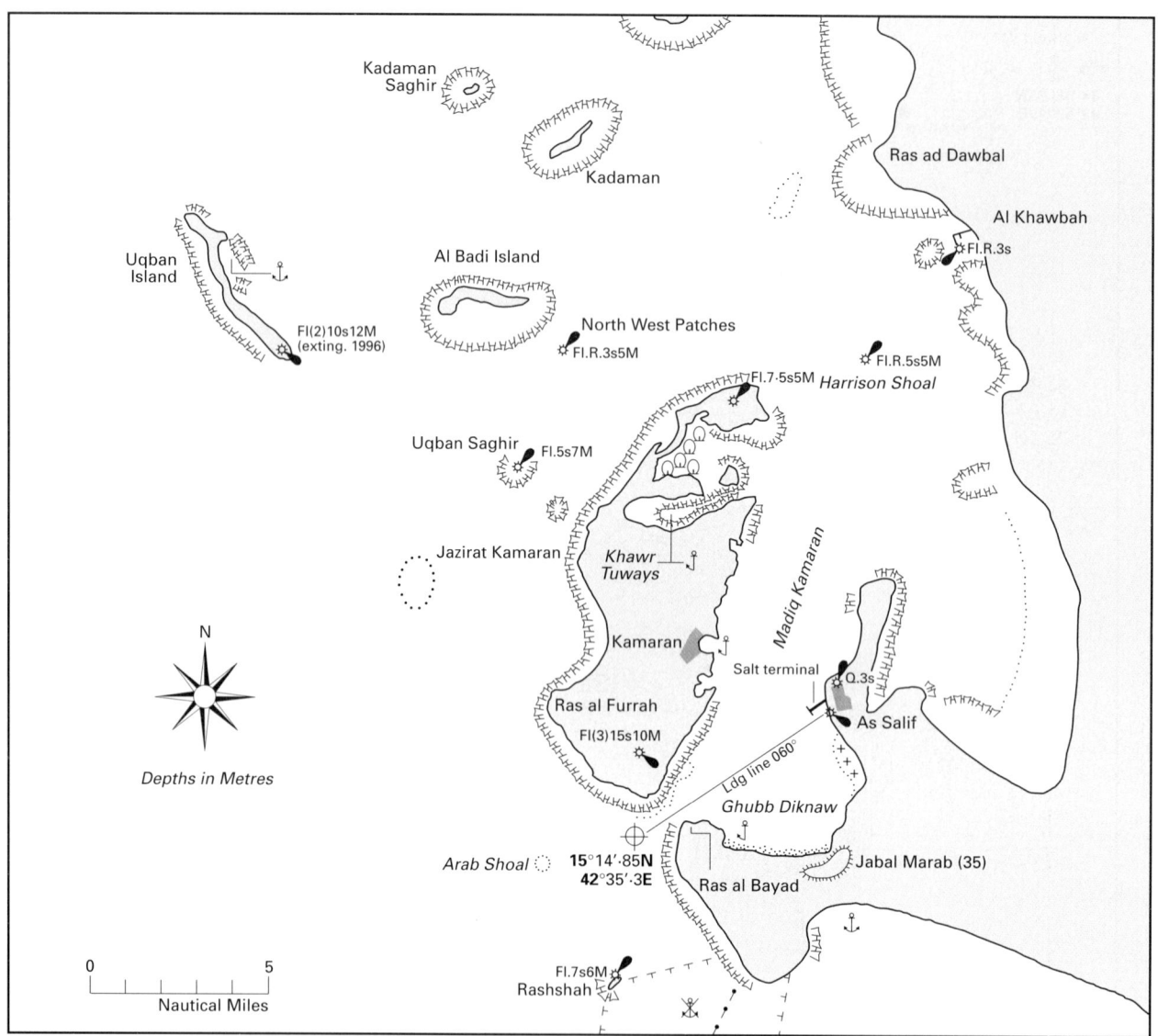

KAMARAN ISLAND AND APPROACHES

Uqban Island

Admiralty chart 548, plan page 118

The reefs show up well in good light. Note the large area of shoal ground off the N side of the island. There is a round tower near the SE end. Depths are irregular but if you push far enough N into the inlet, you'll find reasonable depths and holding. It is well sheltered here from NW'lies, and very pretty. Local boats may visit you, eager to trade. Be ready with fenders.

Farasan Islands

Admiralty charts 15 (see also USDMA 62271), key map page 117

This is a sensitive area and the authorities may warn you off. The Yemeni/Saudi Arabian border is at Oreste Point, 16°23'N 42°45'E. If you manage to continue sailing up this E side of the Red Sea you can take advantage of sheltered waters and S winds last longer than they do on the W side or in the middle of the Red Sea. It is also easier to find anchorages later in the day because you don't have to look towards the afternoon sun. However, although there are reasonably detailed charts they should be used with caution. Eyeball navigation is crucial as GPS positions can differ from charted positions by as much as 2M. Water levels are higher in winter when covered reefs usually show a light green colour. Far more reef is visible in the summer months. However, red or green algae in the water can make differentiation difficult. Most shoals in the deeper parts of the channels are around the 37m (20 fathom) contour, which should be approached with caution.

A recent correspondent from Saudi Arabia tells us that the islands may become a designated World Heritage site. Although barren, they are full of wildlife including gazelles, birds, turtles, and dugongs and are far from being uninhabited. There are several villages on Farasan and Segid I, the main centre being Farasan village. There is a Saudi Coastguard presence in Khor Farasan at the ferry port, at Janabah Bay opposite Kumh I, on Dha al-Fauf I, Ad Dosan I, Wishka I and at Khotib on the

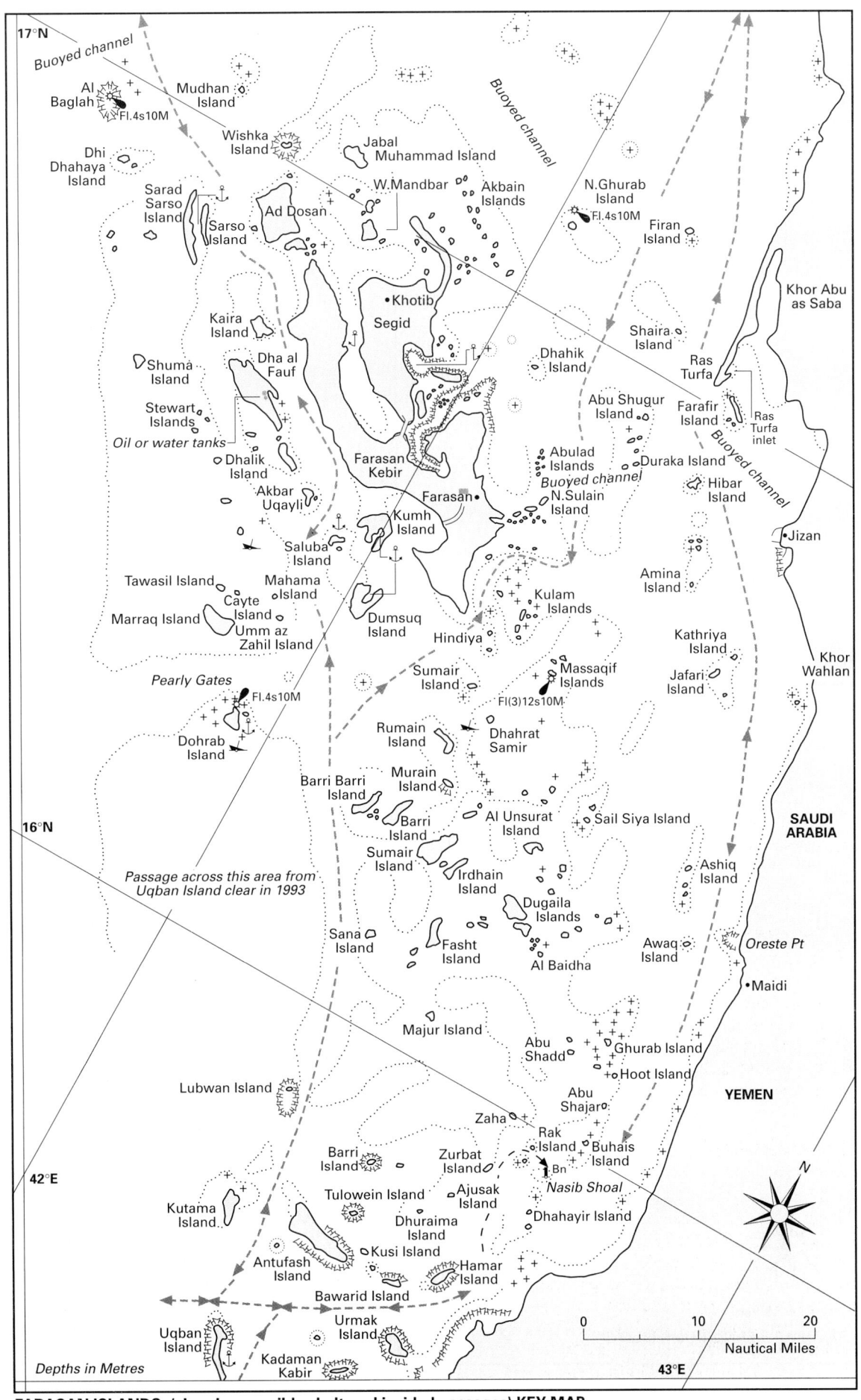

FARASAN ISLANDS (showing possible sheltered insided passages) KEY MAP
Note: Names follow older forms in this section

Red Sea Pilot

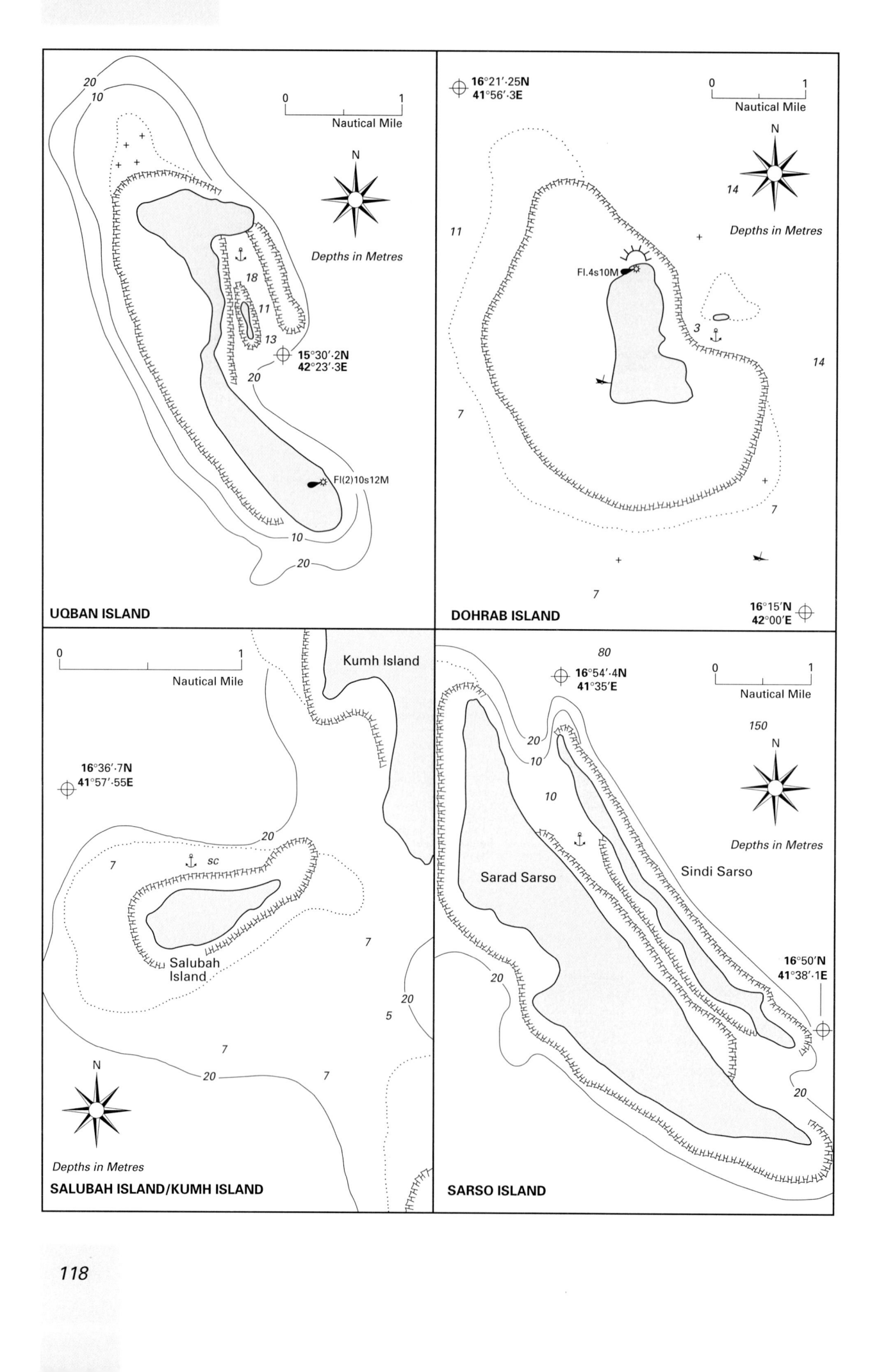

UQBAN ISLAND

DOHRAB ISLAND

SALUBAH ISLAND/KUMH ISLAND

SARSO ISLAND

The Saudi Arabian Coast

N of Segid I. A road bridge connects Farasan and Segid I where shown.

Dohrab Island
Admiralty chart 15, plan page 118

The simplest approach is through the Pearly Gates (see key map, p.) Passage across the bank to the S of the island was also reported to be clear but you would need to keep a very good lookout. The island is low, sandy and surrounded by submerged rocks and reef outcrops, except on its E side. There is a wreck about 1M S of the SE tip. The anchorage is between the island and the islet close off its NE side, in about 3m, sand, with some protection from NW'lies.

Salubah Island
Admiralty chart 15, plan page 118

There is a clear and navigable channel on all sides. The island is high and composed of coral, as is its neighbour to the NW, Akbar Uqayli. Anchor in about 10m, coral with good holding once your anchor is set.

Note There are several other possible anchorages, marked on the area sketch-map (note unusual orientation) with a half-anchor symbol, but these have not been tried.

Sarso Island
Admiralty chart 15, plan page 118

The anchorage is between Sarso (Sarad Sarso) I and Sindi Sarso I on the E edge of Shib Farasan. Depths in the W and N approaches to the islands are irregular and there is shoal water off the NW side of Sarad Sarso The anchorage is towards the S end with good protection in S winds.

3. The Saudi Arabian coast
Admiralty charts
Large scale 12, 16, 64, 326, 327, 328, 2599, 2577, 2658
Medium scale 12, 15, 2659
Small scale 157, 158, 159

KEY MAP A: PORTS AND ANCHORAGES ON THE SAUDIA ARABIAN COAST INSHORE OF THE S FARASAN BANK

Red Sea Pilot

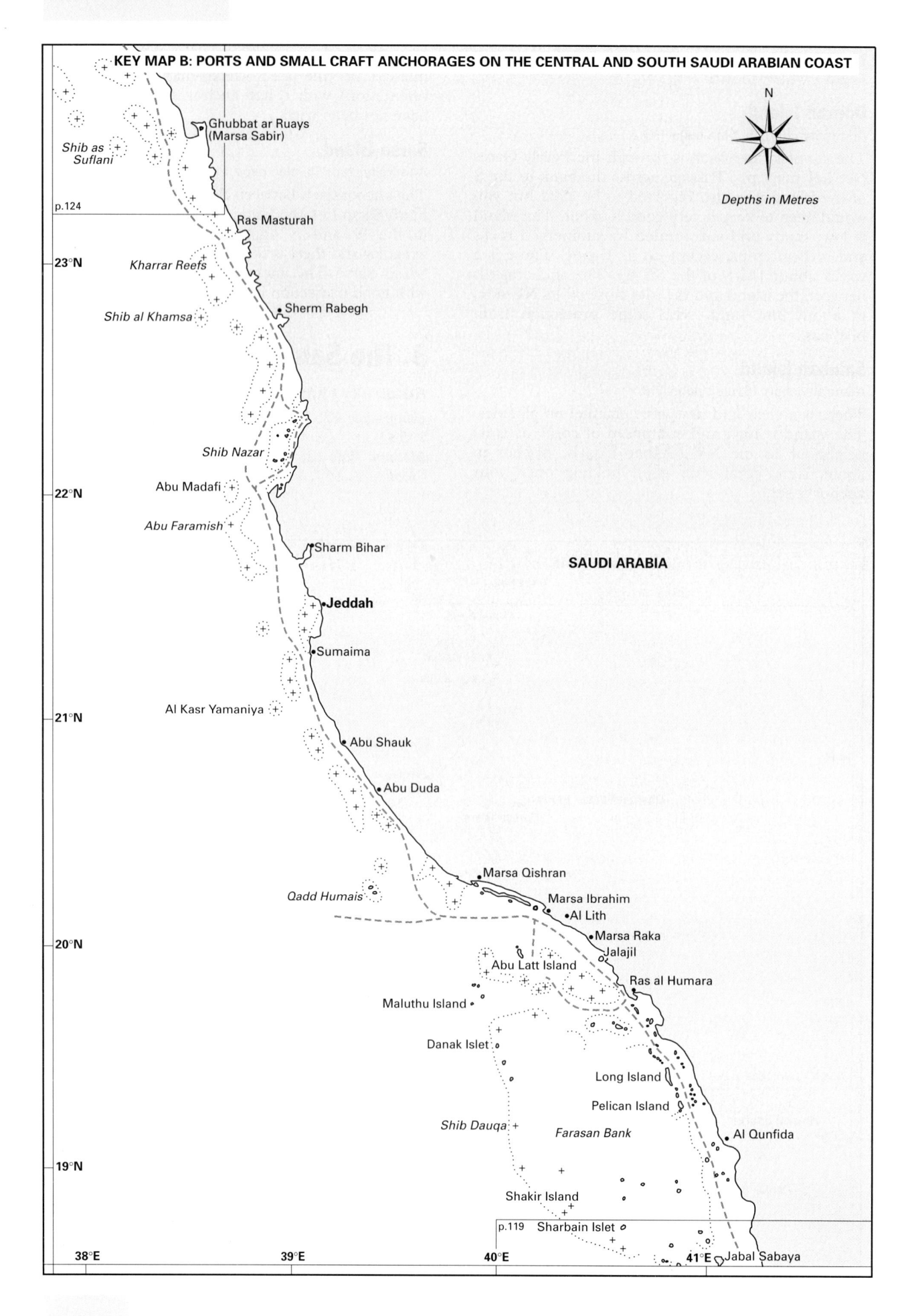

KEY MAP B: PORTS AND SMALL CRAFT ANCHORAGES ON THE CENTRAL AND SOUTH SAUDI ARABIAN COAST

Depths in Metres

- Ghubbat ar Ruays (Marsa Sabir)
- *Shib as Suflani*
- p.124
- Ras Masturah
- 23°N
- *Kharrar Reefs*
- *Shib al Khamsa*
- Sherm Rabegh
- *Shib Nazar*
- 22°N
- *Abu Madafi*
- *Abu Faramish*
- Sharm Bihar
- **Jeddah**
- SAUDI ARABIA
- Sumaima
- 21°N
- Al Kasr Yamaniya
- Abu Shauk
- Abu Duda
- Marsa Qishran
- *Qadd Humais*
- Marsa Ibrahim
- Al Lith
- 20°N
- Marsa Raka
- Jalajil
- Abu Latt Island
- Ras al Humara
- Maluthu Island
- Danak Islet
- Long Island
- Pelican Island
- *Shib Dauqa*
- *Farasan Bank*
- Al Qunfida
- 19°N
- Shakir Island
- p.119 Sharbain Islet
- Jabal Sabaya
- 38°E 39°E 40°E 41°E

120

The Saudi Arabian Coast

US charts
Large scale 62111, 62171, 62172, 62241, 62242, 72276
Medium scale 62115, 62140, 62170
Small scale 62230, 62250, 62270, 62290

French charts
Large scale 6981, 6965
Medium scale 7517, 6981
Small scale 7112, 7113

Key maps pages 120, 124

Warning
Although the most recent editions of charts from all the major hydrographic offices are reconciled to WGS84, this CANNOT compensate for shortcomings in the original 19th-century surveys. Note also that few surveys have been done on this coast to ANY WGS datum. Three points of navigational importance follow. First, always navigate, even with GPS, with circles of probable error (CEPs) of at least 2 miles, and at night at least 5 miles. Second, navigating by GPS alone in this area is stupid. Do not do it. Make daylight landfalls and use your hand-bearing compass and radar, if you have it, to establish WGS84-to-chart error factors. Third, treat ALL waypoints with appropriate caution, including those given below, and NEVER assume that the waters between, close to or at waypoints are free of hazard.

Historically the E side of the Red Sea, outside the offshore reefs, was the recommended route under sail. There is also a well charted inner channel which, with offshore islands, reefs and coastal inlets, offers at least as many refuges as the W coast. One day, surely, we shall be able to explore these waters, hitherto almost unknown to cruising sailors. And we'll be able to find out whether it really is easier to make northing on this coast. Recent news from Saudi Arabia about a new openness to tourism looks promising.

Meanwhile, apart from organised rallies, and mainly with respect to Jeddah, cruisers call only in extremis. Other exceptions are those with Saudi connections, friends or relatives who organise all documentation in advance. Most yachts are not welcome because we are tourists and Saudi Arabia at present excludes tourists absolutely. That said, those in trouble or needing repairs and, naturally, those with Muslim crew are treated sympathetically. For these reasons there is a sketch map only for Jeddah.

Positions in the text are only for orientation on the chart. The idea is to show the most probable entry point from safe water. Name changes are being incorporated into Admiralty publications so we give both forms if there's a possibility of confusion.

Khor Wahlan
16°44'·4N 42°40'E

The inlet is shallow. Anchor in about 4m W of the entrance or a little further S. Harrier Reef, with a small sand *cay* on the E edge, lies about 1M SSW of the entrance. In N winds you might find easier shelter in the lee of Ja'Fari I (16°40'·8N 42°33'·8E).

Jizan
16°53'·2N 42°30'·1E
Admiralty chart 15

Contact the authorities on VHF Ch 16 on approach. There is a buoyed channel and leading lights which take you in on 096·5°. Enter between the S breakwater and the much larger N breakwater with its bulk cement terminal. Pass the basin with its container and Ro-Ro terminal to port and head for the small craft basin. This is in the extreme SW corner of the harbour, S of the container terminal. Be prepared to be directed to the coastguard harbour N of the main harbour instead. This is a malarial area. Take precautions.

Ras Turfa (Ras At Tarfa)
17°00'·5N 42°22'·9E

Anchor in the small inlet on the E side of the point or in N winds under the lee of the point. It is also possible to push up into Khor Abu as Saba and the nearly 10M into Khor Al Ja'afirah for complete shelter, but it would be a long trudge back out.

Between Ras Turfa and the next entry you must find shelter where you may in the lee of reefs and islets.

Widan
17°52'·8N 41°43'·4E

Anchor in the inlet in 5–7m. Sheltered only from SE.

South Khor al Wasm
17°59'·3N 41°36'·9E

Good sheltered anchorage inside the inlet for N'lies but you need to push well N into the inlet to find reasonable depths.

Hasr Islet
18°07'·9N 41°30'E

Anchor in about 10m in an inlet formed by the fringing reef between Hasr and Jabal Its. If the wind is S, try the N side of Hasr It.

Khor Al Birk
18°13'N 41°28'·4E

Anchor in the lee on the E side of the reef extending SW from the entrance, or inside in 6–12m, mud. The BA Pilot speaks of an intricate passage though this may not be so for small craft.

Khor Nahud
18°16'·6N 41°27'·9E

Reef extends for 1M SSE of the N entrance point, but the inlet offers an anchorage in 10–15m in either of the two arms.

WA Robinson sought shelter here in *Svaap* in the 1930s. Somewhat to his regret, he found himself in Khor al Birk instead

Red Sea Pilot

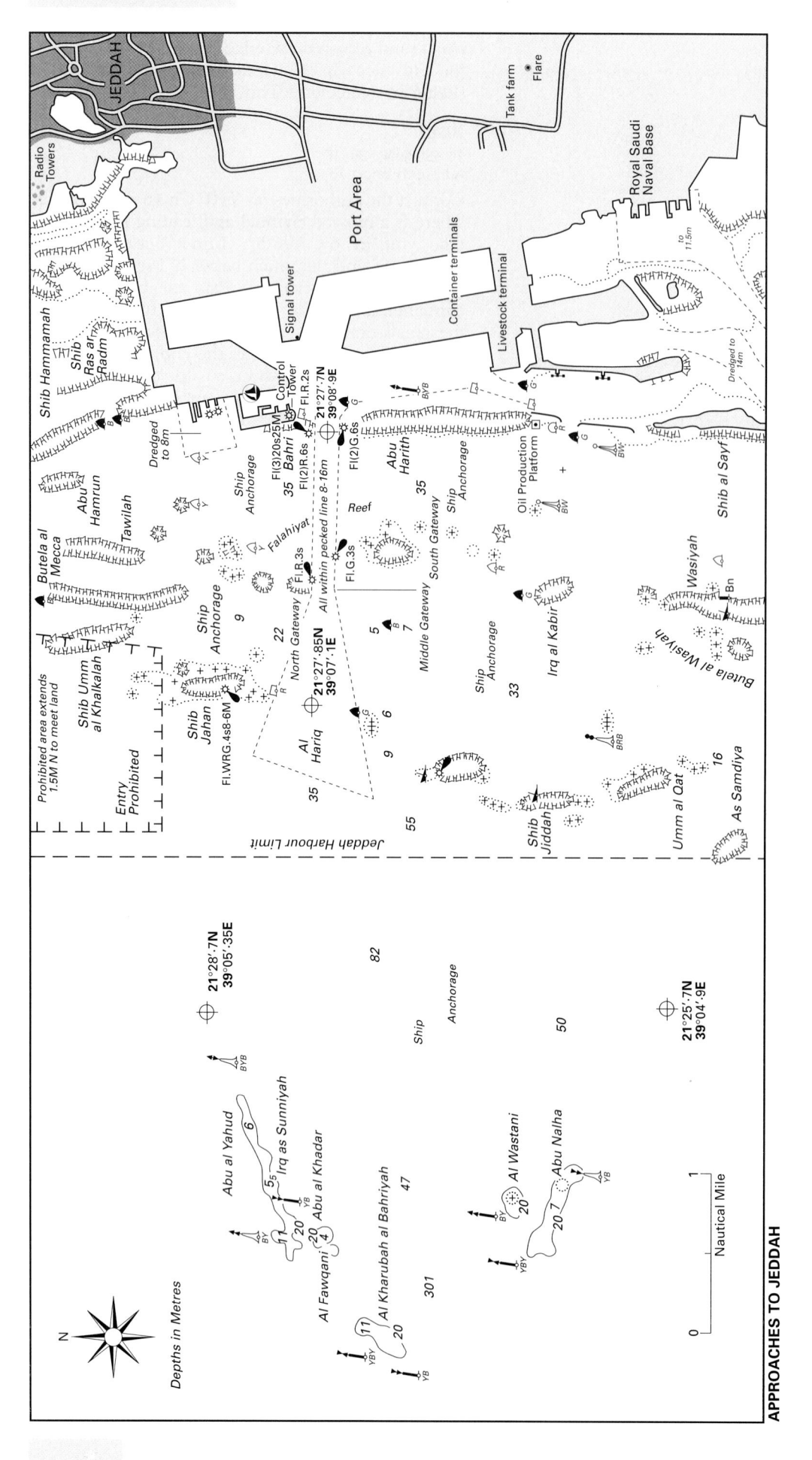

APPROACHES TO JEDDAH

The Saudi Arabian Coast

'...the fantastic Bedouin village of conical brush huts and stone houses that hung beneath a few palms on the barren burnt hillside...' There follows a colourful account of his treatment at the hands of the Emir of El Birk, who kidnapped him and held him to ransom. He complained of discomforts ashore and was allowed to return to *Svaap*. Later he reported he was running out of food. The Emir '...at once sent a whole flock of chickens aboard...'

In spite of the hospitality, he escaped, only to be captured again at Lith.

Khor Amiq
18°26'·9N 41°24'·3E

Anchor in about 11m behind the reef.

Ras Hali
18°36'N 41°14'·6E

Anchor towards the head of the bay N of the *ras*. There might also be a possibility S of the *ras* but you'd have to tuck well in.

Jabal Sabaya
18°33'·9N 41°04'·2E

There is a pass through the reef on the SE side of the island. Anchor in 3m off the remains of a village.

Until Al Qunfida you're again going to be finding shelter where you can behind reefs and islands.

Al Qunfida
19°07'·8N 41°02'·7E

There are several reefs in the approaches and two conspicuous minarets in the town. The N anchorage behind a hook of reef is the main one. The S looks doubtful.

Pelican Island
19°17'·8N 40°53'·3E

This is the S'most of the Fara Is. It has a deep anchorage sheltered from W and SW winds on its NE side and a yacht might be able to tuck into more reasonable depths. For N winds the lagoon and its immediate offing look inviting.

The main channel from here for the next 35–40M offers lots of islands, islets and reefs which may give shelter, especially around Sirrain I, Ras Mahasin and Ghubbet Al Mahasin. The next known option is Ras Al Humara.

Ras Al Humara
19°48'N 40°36'·5E

Anchor in 5–10m about four cables NW of the *ras*. The anchorage is sheltered from the SW by Shib Abu Tuqm, which dries. For N winds the inlets at 19°45'·8N 40°37'·7E and 19°49'N 40°35'·7E look promising. There are also anchorages in the Janabliyat Is 19°46'N 40°35'E.

Jalajil
19°54'·5N 40°31'·5E

There is an anchorage in 10–15m about 1M N of Jalajil, which is a white sandy headland about 8m high. Unless you approach in a long hook N there are off-lying dangers to deal with. For N winds there's a possibility S of the *ras* entered around 19°53'·1N 40°32'·1E.

Marsa Raka
20°01'N 40°24'E

Approach via the Lunka Channel and anchor in 5–7m in the indent in the reef.

Marsa Ibrahim
20°09'N 40°14'E

This inlet is 1½M W of Al Lith and is entered by passing between the three reefs lying off it. Anchor N of the reefs in about 8m for good well protection from N and good holding.

WA Robinson's second kidnapping, described in *Deep Water and Shoal*, took place not far from here. The pilot book of the 1930s was a bit discouraging about Lith, saying of the inhabitants

'In ordinary times piracy and robbing the few pilgrims who attempt to pass through to Mecca are added to their usual means of gaining a livelihood.'

Robinson gives a romantic account of being taken prisoner and riding across the desert bareback on an Arab horse. Once again he managed to escape on *Svaap*, with the help of a dhow captain who was a part-time pirate and part-time slave runner. The story makes present day cruising seem a most prosaic business!

Marsa Qishran
20°14'·1N 39°54'·5E

Anchor in the inlet in about 7m, mud. Marsa Qishran can be entered from the SE in approx 20°10'·6N 40°07'·2E and the narrow inlet wanders about 12M WNW to the anchorage.

Abu Duda
20°33'·8N 39°31'·1E

The area is still unsurveyed but you can probably anchor here with care on the approach. One danger noted by the Admiralty Pilot is Qita Abu Duda, 3M WNW of the entrance, which has approx 2m over it.

Abu Shauk
20°54'·5N 39°18'·7E

Good anchorage here in about 10m inside the inlet.

Sumaima
21°16'N 39°07'E

Use the S entrance, keeping to mid-channel. The N entrance may be shoal. The inlet is deep, but offers shelter from the S. There are four tall radio masts 2M SE of Sumaima which are radar conspicuous.

Jeddah
Admiralty charts 2577, 2599, 2658, plan page 122

Approach
It is forbidden to enter the harbour without authorisation from port control. Yachts usually

Red Sea Pilot

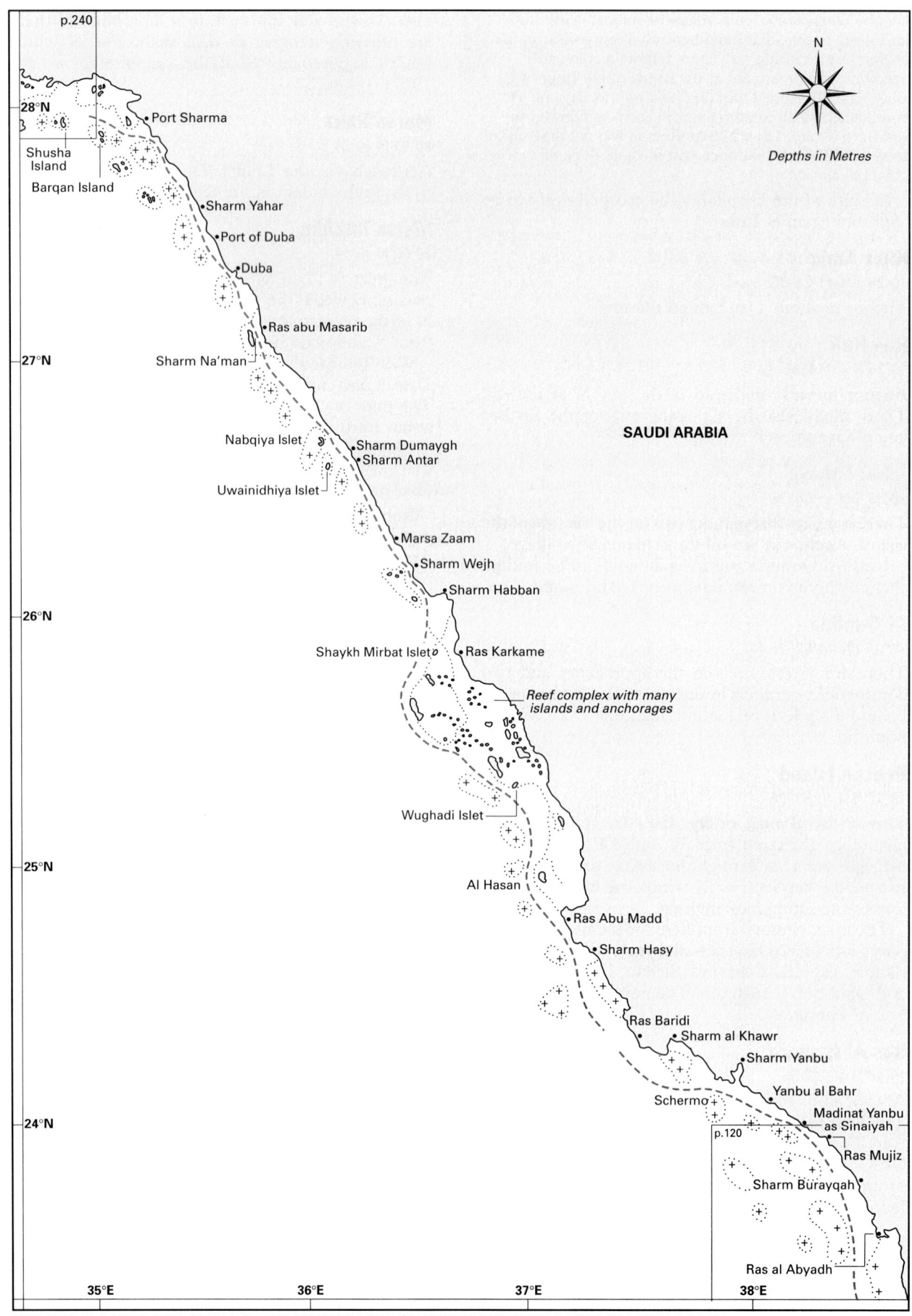

KEY MAP C: PORTS AND SMALL CRAFT ANCHORAGES ON THE NORTHWEST SAUDI ARABIAN COAST

make arrangements in advance if they plan to visit Jeddah. See the notes at the beginning of this section for more general information.

Harbour limits are shown on the sketch. The approach channels are well marked. The best time for entering is about noon, when reefs show clearly as dark green patches in very clear water. The sea breaks on the off-lying dangers N and S of Al Hariq. There is a prohibited area in the N approaches, shown on the sketch.

The port is under continuous development and new lights regularly get added in the approaches. An up to date chart is essential. Two new long distance lights (Fl WRG.14M) have been added in the area of Al Hariq. The lit control tower on the N side of the harbour entrance and the signal tower WSW of the small-craft anchorage make good landmarks.

Anchorage

Approach from N or S via the positions given on the sketch towards the Inner Gateway (21°27'·7N 39°08'·9E). Yachts berth in the Service Harbour (21°28'·25N 39°09'·1E), the basin immediately NNE of the control tower (Fl(3)20s25M) on the N side of the entrance to the main wharf area. It is where the coastguard and pilot boats dock.

Formalities

When 20M out of Jeddah call Jeddah Port Control on Ch 16. They will then probably ask you to go to the main working frequency, Ch 12 and to give your ETA. Movements within the harbour are strictly controlled. You are required to obtain a copy of the *Rules and Regulations Valid for Saudi Arabian Seaports* and carry it on board whilst you are in these waters.

Facilities

Few cruising boats call at Jeddah but those that need help have been courteously received. It is the best place for repairs in the Red Sea but work is expensive. You are not allowed out of the port area without consular assistance or a friend in high places. A yacht which made in without an engine and needing extensive repairs told us that the coastguard was most helpful but that one must use an agent who does nothing whatsoever for US$800 a month!

General

Jeddah contains the 'Tomb of Eve, the mother of mankind', and as the port for Mecca has hundreds of thousands of pilgrims passing through each year. Non-Muslims are not allowed to visit Mecca, or even to see the city from afar. The motorway to the capital at Riyadh has a special bypass, compulsory for non-Muslims, which ensures that no infidel eye sees the holy city. There are US, European and British diplomatic missions in Jeddah, and an international airport. The city has all the modern services and facilities one would expect.

Sharm Abhur (Bihar)
21°42'E 39°04'E

The inlet has a deep, narrow entrance leading to a virtually landlocked anchorage. It is full of wharves and jetties and not likely to be yachtie friendly.

Abu Madafi
22°04'N 38°47'E

Good anchorage sheltered from NW winds off the SE side of the reef. There are two radar-conspicuous wrecks on the NW side. The reef complexes in the channels further W probably offer less exposed alternatives.

Mina Al Qadimah is a closed military area. Between Abu Madafi and Sherm Rabigh you're again left with finding what shelter you can, though there are plenty of possibilities in the area off Ras Hatiba, 22°00'N 38°55'E.

Sherm Rabegh
22°45'N 38°57'E
Admiralty chart 64

Approach either S or N of Shib al Bayda. The entrance to the channel is deep and clear. It is marked by lit beacons and leads NE from the entrance. As with the ports at Yanbu, described briefly below, this is one to avoid except in an emergency. It is a large oil-loading terminal. Notify the authorities on VHF as you approach if you need to go in here. The best anchorage is in 10–20m, sand and coral, at the head of the harbour, in complete protection, but you may be directed elsewhere.

Ghubbat Ar Ruays (Marsa Sabir)
23°34'N 38°33'·5E

There is a good anchorage here, but the Marsa entrance is unsurveyed and the area is littered with reefs.

Yanbu Al Bahr
24°03'·5N 38°02'·5E
Admiralty chart 327

The entrance to the port, which serves Al Madinah, is marked by buoys and beacons which lead to a dredged and beaconed channel. A traffic separation scheme operates in the approaches. If you decide to go in you should make radio contact in advance. Try this one only in extremis. That said, Yanbu is a sheltered, enclosed anchorage. Mina al Malik Fahd (King Fahd Port) at Madinat Yanbu as Sinaiyah, S of Yanbu, is an oil port and, for a yacht, an open roadstead.

Sharm Yanbu
24°08'N 37°55'·5E

There is good anchorage here in 12–15m, hard sand but a yacht that chose to anchor here in 1997 was made unwelcome. There is a coastguard station which will offer help if you are in need.

Red Sea Pilot

Sharm Al Khawr
24°15'N 37°40'·9E

This is a large complex of inlets within which shelter can be found in any wind.

Sharm Hasy
24°36'·5N 37°18'·25E

There is a good anchorage in the inner part of the inlet, although the N half is shallow. This section of coast is reported to be radar conspicuous.

Al Hasani
24°55'·5N 37°06'·4E

Anchor on the E side of the island in 12–15m, sand and coral, fair holding. There is foul ground N and NE of the island. The water is very clear and you can see the bottom in 15m or more.

Wughadi Islet
25°19'·5N 36°58'·8E

There is good anchorage on the bank E of the islet. The reefs and channels N to Mashabih I look intriguing.

Ash Shaykh Mirbat
25°54'N 36°34'E

Anchor SE of this small coral islet. Extensive reef surrounds it stretching nearly 1½M N through E. Currents can run strongly through channels in the vicinity and the water is not always clear enough for good visibility when navigating, even in good light.

Sharm Habban
26°03'·9N 36°33'·7E

This is a narrow inlet with a detached reef, visible in good light, near the middle of its entrance. Pass NW of the reef, taking care to stand clear of the projecting fringing reef on the S entrance point. Anchor in 7–10m, sand and mud. Sharm Habban is reported to lie 2M further S than charted. The above position has been corrected to WGS84.

Sharm Wejh
26°13'N 36°27'E

The town of Wejh is prominent and there is a radio tower (75m) near the airport, about 3M ENE of the town, which also has an aero light. Holding is dodgy in 5–10m, sand, and there's a swell in NW winds.

Anchorages mentioned in the BA Pilot between Sharm Wejh and Sharm Dumaygh include Marsa Zaam 26°17'N 36°24'E, Sharm Antar 26°35'N 36°13'E, and Uwainidhiya Islet 26°35'N 36°05'E. With all of them you'll be finding your way.

Sharm Dumaygh
26°39'N 36°11'E

Take care to avoid the coral heads near the middle of the inlet. Anchor in the NW part of the inlet in about 15m, sand and coral. Good holding and very well protected landlocked anchorage.

Again, details of anchorages N to Duba are scant. Places mentioned in the BA Pilot are Nabqiya Islet 26°44'N 36°03'E, Marsa Zubaydah 26°51'·7N 36°01'·5E, and Sharm Naman 27°08'N 35°45'E.

Port of Duba
27°33'·8N 35°31'·3E

A new port is in operation. There has been mention of plans for a coastguard station and yachting facilities but nothing has transpired as far as we know.

Sharm Yahar (Al Harr)
27°36'·5N 35°30'E

The entrance is difficult to identify, but about two cables inside there is good, well sheltered anchorage in 10–15m.

Port Sharma
27°55'·5N 35°16'E

The port has been extensively developed of late, and has a beaconed fairway 8M long from Yuba I into the harbour. Radio ahead and ask for instructions about anchorage if you need to go in. The area round Burqan I has oil platforms. Again, this is a port of refuge only for emergencies.

Al Khurayba, 4M N of Sharma is thought by some to be Leuke Kome (White Village), the main Nabataean port mentioned in the *Periplus of the Erythraean Sea*, though others, on rather slender grounds, favour Yanbu.

4. The Sudan coast S of Port Sudan

Admiralty charts
Large scale 81, 82, 675, 3492, 3722
Medium scale 675, 81, 82
Small scale 158

US charts
Large scale 62111, 62144, 62142
Medium scale 62143
Small scale 62270, 62230, 62250

French charts
Small scale 7099, 7112

Routes through the Suakin group

(See Key map, Anchorages and routes in South Sudan, below.)

There are broadly three ways N from the latitude of Khor Nawarat to that of Port Sudan. You can hold to the coast, go through the middle or go out round the E edge. Which you choose is likely to hinge on your enthusiasm for diving or fishing and your tolerance for clinging onto iffy anchorages on the edges of reefs. One word of additional caution. DO NOT use small-scale charts for navigating in or close to this area. Above all do not use BA138 (now

The Sudan Coast S of Port Sudan

Warning
Although the most recent editions of charts from all the major hydrographic offices are reconciled to WGS84, this CANNOT compensate for shortcomings in the original 19th-century surveys. In the S of the Suakin Group the only survey ever done was between 1830 and 1834. Note also that NO full survey has ever been done on this coast to ANY WGS datum. Three points of navigational importance follow. First, always navigate, even with GPS, with circles of probable error (CEPs) of at least 2 miles, and at night at least 5 miles. Second, navigating by GPS alone in this area is stupid. Do not do it. Make daylight landfalls and use your handbearing compass and radar, if you have it, to establish WGS84-to-chart error factors. Third, treat ALL waypoints with appropriate caution, including those given below, and NEVER assume that the waters between, close to or at waypoints are free of hazard.

superseded by BA158) in this way. Longitudes on BA138 are displaced W by up to 2M from WGS84.

The coastal route

This offers cruising in flat water and runs from Khor Nawarat, the first reliable anchorage on the Sudan coast, via the Shubuk Channel and the Inner Channel, to Suakin and as many *marsas* as time and inclination suggest. Sketch-maps and descriptions of most of the anchorages are included.

The middle route

This is technically the Inner Channel with cruising variations. From Khor Nawarat or thereabouts head for the Karb Its, Dar Ah Teras, Talla Talla Saqir or Kebir, Harorayeet (Two Its), perhaps Dhanab Al Qirsh (Green Reef), and on between the Eitwid Reefs and Shab el Shubuk to the Inner Channel, Suakin and N to Port Sudan.

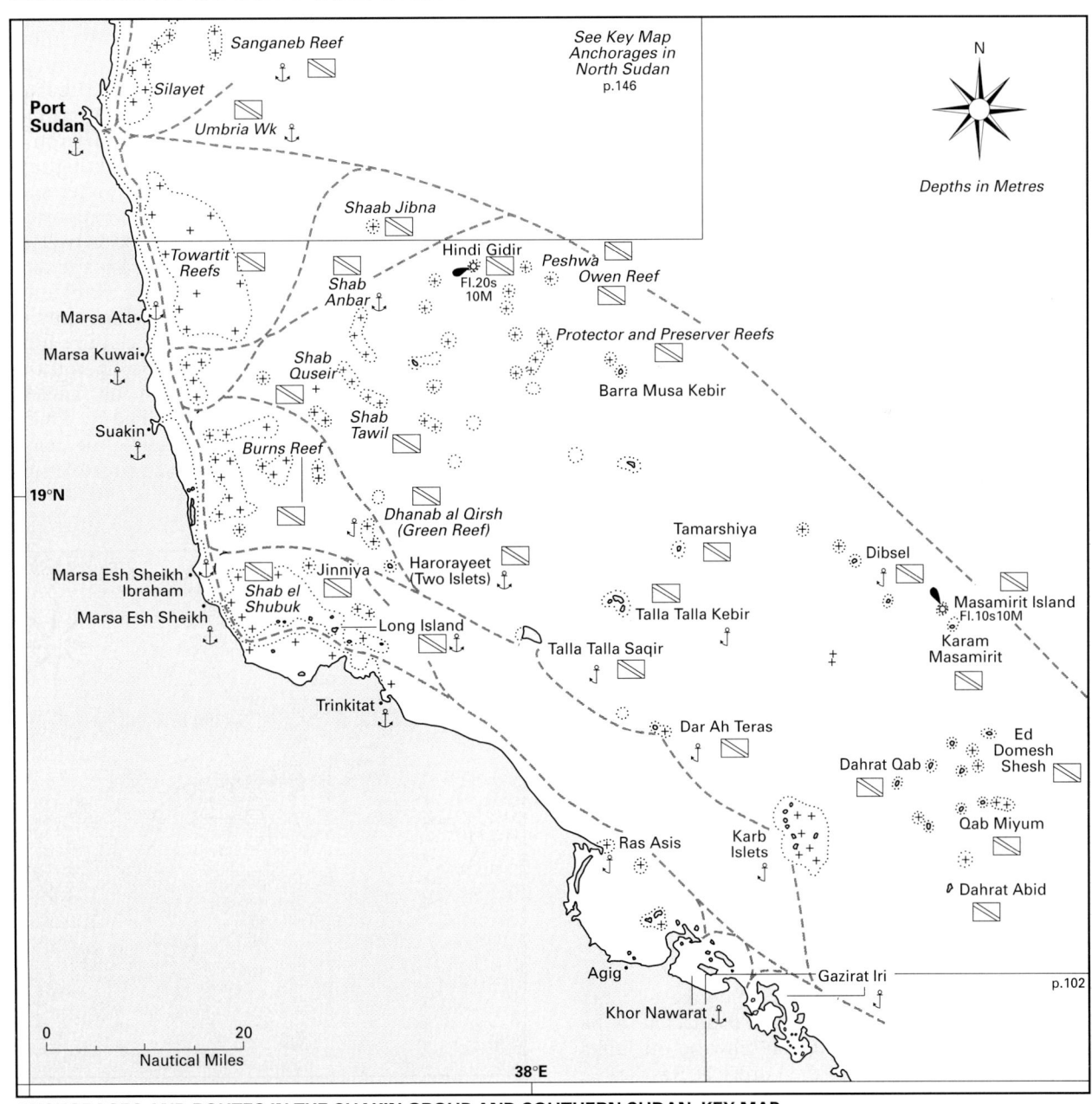

ANCHORAGES AND ROUTES IN THE SUAKIN GROUP AND SOUTHERN SUDAN: KEY MAP

127

Red Sea Pilot

The outside edge

One can anchor-hop (if not in ideal anchorages, at least in 20M chunks) from Dahrat Abid to Port Sudan. This takes in Dahrat Abid, Qab Miyum, Ed Domesh Shesh, Karam Masamirit, Dibsel, Barra Musa Kebir, Owen Reef, Peshwa, Hindi Gidir, and Shab Jibna, but needs settled weather.

Variations

These would begin with the first few anchorages of either the middle or the outside route. From the latitude of Dhanab al Qirsh and Barra Musa Saqir N, exploration is the order of the day. Shab Tawil, Shab Quseir (also called Pinaule and rated an excellent dive), Shab Anbar and Burns Reef are possibles, though only Shab Anbar offers all weather, if in a blow exciting, shelter.

Many of the small island anchorages and the reefs of the Suakin group, even if we give no sketch map, may offer some sort of anchorage in their lee. With the islets where to go is comparatively obvious. With the reefs, you have to explore. Plan your itinerary to give plenty of time for scouting anchorages with your dinghy if possible. Above all, organise yourself so that if a midnight move is necessary you can get up and go in safety. Nowadays this means sensible use of your GPS and its waypoint-saving facility as you make your way in. If, once you're anchored, you make a small chartlet of where you are and how you got there, it will hugely help you if you have to make a midnight move. And if you're really nice you'll send us a copy.

Be wary of the unpredictable currents common in the area. These can very suddenly make a tenuous anchorage a dangerous one and cause what felt like a safe offing to become a grounding.

Known dive locations S of Port Sudan are: Dahrat Abid, Qab Miyum, Ed Domesh Shesh, Dar Ah Teras, Talla Talla Saqir, Talla Talla Kebir, Dibsel, Tamarshiya, Shab el Shubuk and Long I, Harorayeet (Two Its), Jinniya (Corner Reef), Dhanab al Qirsh, Barra Musa Kebir, Burns Reef, Shab Tawil, Shab Quseir, Shab Anbar, Owen Reef, Peshwa, Hindi Gidir, Shab Jibna, Shab Towartit.

Only some of these offer anything other than day anchorages. There are doubtless more in this enormous area of reefs, shoals and islets. If you're heading S, ask the dive operators in Port Sudan if they have any particular recommendations. These people are the real experts. We only touch on the subject.

As usual, the anchorages in this section are arranged by latitude, S–N.

A. The coastal route

Gazirat Iri

This is the first potential anchorage on the S Sudan coast, formed by a sandy peninsula connected to the mainland by drying reef. It may now be off limits because of its closeness to the border at Ras Kasar.

There are few landmarks but the Admiralty Pilot says there is a beacon 1½M SE of Gazirat Abid, 18°10'N 38°29'E, about 5½M SE of the entrance to Gazirat Iri although we couldn't see it. Stay in 20m or more on the approach from the SE to clear the fringing reef until you are N of the N tip of the peninsula. Once off the entrance to the channel, cross the 3·1m bar over the coastal reef into depths of 5–8m. Then follow the intricate channel SE into the anchorage where you should have about 4m, sand and mud.

The ruins marked on the sketch are thought by some to be the remains of the ancient Ptolemaic settlement of Ptolemeis Theron, Ptolemeis of the Hunts, founded between 270 and 264BCE. The area inland from here N to around Trinkitat is known locally as The Delta. In the past it was extremely fertile and was no doubt once good hunting ground but the beasts once hunted here were elephants which says something about climate change.

These ruins and the ruins of Adulis in the Gulf of Zula are reminders of the extraordinary maritime expansion of Egypt under the dynamic Egypto-Grecian Ptolemaic dynasty founded by Alexander the Great. Whether the ruins really are Theron is debatable. Lionel Casson and GWB Huntingford, editors of the *Periplus,* both argue that Theron is today's Aqiq although there have been supporters for Suakin, Marsa Maqdam and Trinkitat as well. In fact the text makes Gazirat Iri a plausible candidate because chapter 3 reads. . . 'The place (Ptolemeis Theron) has no harbour and offers refuge only to small craft.' In the context of the 3rd century BCE, that means very small indeed, so maybe Theron was here. It is an interesting comment on ancient navigation that Khor Nawarat, described by Elwon and Moresby as the finest anchorage on the coast, seems never to have been considered a possible site.

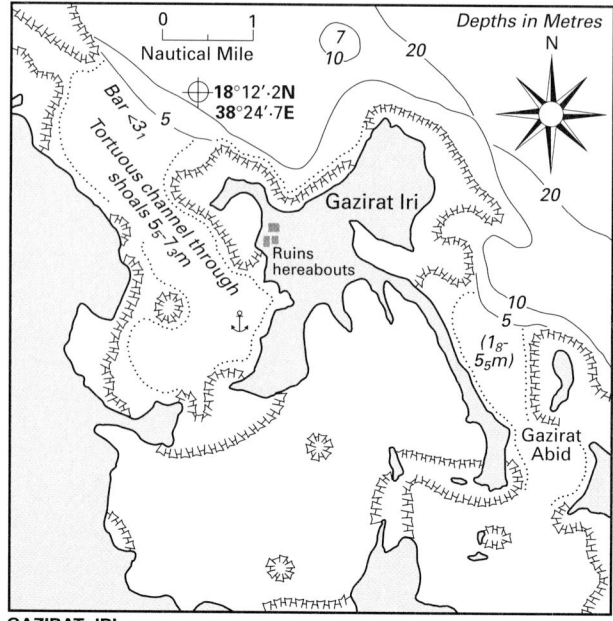

GAZIRAT IRI

The Sudan Coast S of Port Sudan

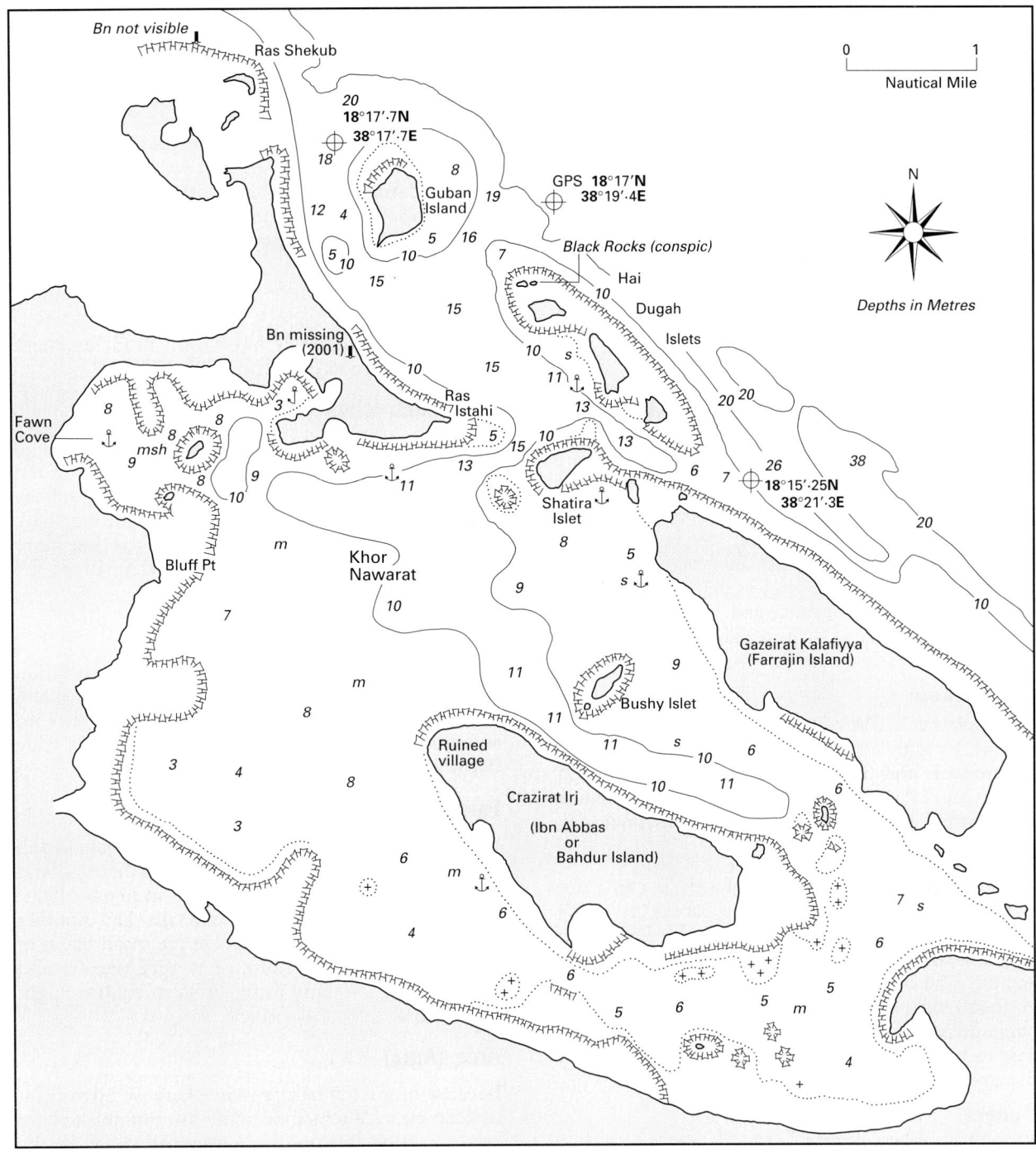

KHOR NAWARAT

Khor Nawarat

Admiralty chart 675

For many yachts this will be the first anchorage since leaving Eritrea or the Yemeni islands in the Hanish Group. Some will come from the Gulf of Aden without stopping since S winds often hold in this far N. This is the area of the shifting convergence zone, a change in wind and even sometimes rain. Ras Kasar, 18°02'N 38°35'E, approx 12M SE, marks the border between Sudan and Eritrea. Keep >5M offshore for 10M S and N of the border if approaching from Eritrea.

Approach

Pre-GPS landfall was tricky in poor visibility due to dust haze on the coastal plain and cloud over the inland mountain ranges. Even now you may not see the pretty featureless and low-lying coast until within 1 or 2M. Strong W currents are common. The entrance to Khor Nawarat is screened by a chain of islets running towards Ras Shekub in the NW. There are three passages, in effect making the *khor* an enclosed harbour, well protected from all quarters. The islets stand on coral and are low and sandy with some scrub and mangrove.

129

The easiest entrance from any direction is between Black Rocks and Guban I. Entrance Shoals, SE of the island, have a least depth of 5·5m over them, but otherwise the channel is deep and clear. From the N another entrance is inshore of Guban I. Guban Patches, about mid-channel, have a least depth of 4·3m; otherwise the channel is clear. The S entrance, through East Passage, S of the Dugah Its, has a least charted depth of 6m and can be bouncy in strong SE winds. Hold to the E shore as you approach the largest Dugah It. The reefs generally show well even in indifferent light, but this is NOT true of the coral spit protruding across to the Hai Dugahs E from Shatira.

Dangers

NE of Khor Nawarat, reef may extend <2M further S and W of the Karb Its than charted. This area is off the sketch on p 000. Inside the *khor* a sharp lookout is needed for isolated coral heads, e.g. to the SW of Shatira I. There is foul ground E of Crazirat Irj (Bahdur or Ibn Abbas I). The approach to the anchorage in Fawn Cove has to be eyeballed through lots of reef. The area in general has silted and depths may be up to 1m less than charted. Take care if you use the S entrance and, again, beware the coral spit extending N from Shatira It into the fairway. We nearly clouted it!

Conspicuous

From seaward, Black Rocks (dark coral boulders) are easily visible even when the light is not so good. In strong E and S winds, a summer phenomenon known locally as a *haboob* and often accompanied by sandstorms, all landmarks can rapidly disappear.

Formalities

None, but it is normal for the military to come out from nearby Aqiq to check your papers. A ship's letter of introduction in Arabic is useful. Be sure to fly the Sudanese courtesy flag on approach and at anchor. The bigger the better. Closer to the border, gunboats on patrol may offer more in the way of harassment. As noted in the previous section, it is wise to keep clear of any international boundaries in the Red Sea.

General

This beautiful but desolate area has good snorkelling (the best is along the seaward fringing reef at the NE end of Farrajin I), fishing, shelling, swimming and windsurfing, and marvellous bird life. It is all but deserted except for the military patrols and a few fishermen, who may want to trade but are generally shy of contact. Camels with riders and the odd 4x4 may occasionally be seen on the mainland. There are plenty of anchorages in Khor Nawarat, to suit all conditions. The following six have been tried recently.

Crazirat Irj (Ibn Abbas or Bahdur I)

There is good holding in 6m, mud, with excellent protection from the N/NE in the lee of the island, which is some 5–6m high, with low cliffs of hard-packed sand. There are the remains of a fishing village on the NW tip. Mosquitoes can be a nuisance.

Gazeirat Kalafiyya (Farrajin I)

As you are coming through the passage S of Ras Istahi, steer well clear of the coral head SW of Shatira I. Anchor in 5m, sand, off the NW tip of the island. Good holding and protection from NE/SE. Snorkelling on the seaward side of the island is outstanding.

Shatira Islet

Anchor S of Shatira It has 6–8m, mud, excellent holding with good protection NE–NW.

Hai Dugah Islets

Anchor in 5–10m (depending on how close you go in), sand, off the W side of the middle islet of the three, N of the fringing reef. Good shelling on the beach but fishermen sometimes leave rotting shark carcasses on the beaches. When we were last there the middle Hai Dugah had been chosen. It can get pretty smelly.

Ras Istahi

11m, with excellent holding and good protection from N, under the *ras*, off the fringing reef. The reef off Ras Istahi may extend further E over the 5m sounding than shown on the sketch. Give it a wide berth by holding towards Shatira It.

Fawn Cove

The entrance to the anchorage must be eyeballed in good light, but the bay provides good holding and all-round protection, in approx 8m, mud and shells. Do not expect to be able to leave quickly! Another anchorage, with at least 3m, is in the small bay E of Fawn Cove but the entrance is very narrow and there is little swinging room. A stern anchor might make sense.

Aqiq (Agig)

Because of the military presence here we advise you to keep clear. We include some information only in case you need to stop or are required to call by the military. This has happened once in the past. Good visibility is required for this approach, as is a close watch on the echo sounder, since shoaling affects this whole area and depths may be less than charted. A rock with 2m over it lies 1M SSE of the S Amarat I. The islets and the mainland coast off Aqiq itself are for the most part fringed with reef extending N from the settlement. There is an isolated shoal patch W of the anchorage.

Ras Asis

The low sandy cape itself is not easily visible from any distance. There is an isolated rock 6M E of the anchorage. Approaching from the N, there are

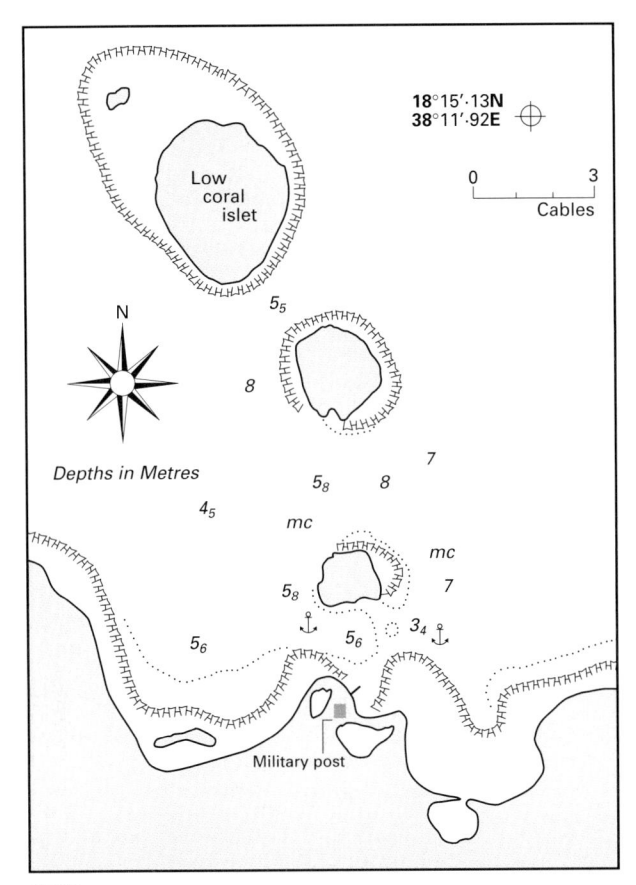

The Sudan Coast S of Port Sudan

approx 5m, sand, with good holding and protection from the N but exposed in E through SW winds. Good for bird life, but can be plagued with jellyfish.

Trinkitat Harbour
Admiralty chart 675, plan below

Approach
The entrance only opens up when you are inside Qita Kanasha, which is always clearly visible and well marked by a beacon (3m) at its N end, square red topmark. The reef S of the entrance breaks in E swell. Strong winds from the S, N or NE can produce sandstorms when sand and dust haze drastically reduce visibility. The shoal ground on either side of the entrance extends further than charted and must be approached with care.

There are now some buildings on the S shore of the harbour conspicuous from seaward. The flagstaff on the SW shore is prominent, though flagstaff is rather a grand name for it.

Anchor in 6–7m, mud in the SE corner. There is good holding with shelter from all but the NE, when a swell makes in but it's more uncomfortable than anything else and a stern anchor might help. There has been silting, particularly at the S end.

Yachts are sometimes turned away from here. It can be a sensitive area and the new buildings may be military. The desert in the area surrounding the harbour floods in the rainy season, October–December. An opening in the SE side leads into a shallow lagoon fun to explore in a dinghy. Bird life is prolific. Local fishermen sometimes have lean-to shacks and tents nearby.

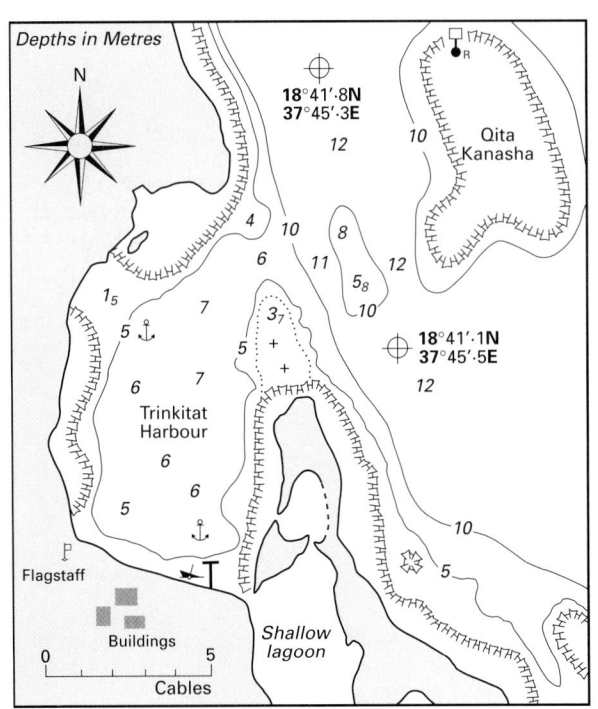

AQIQ

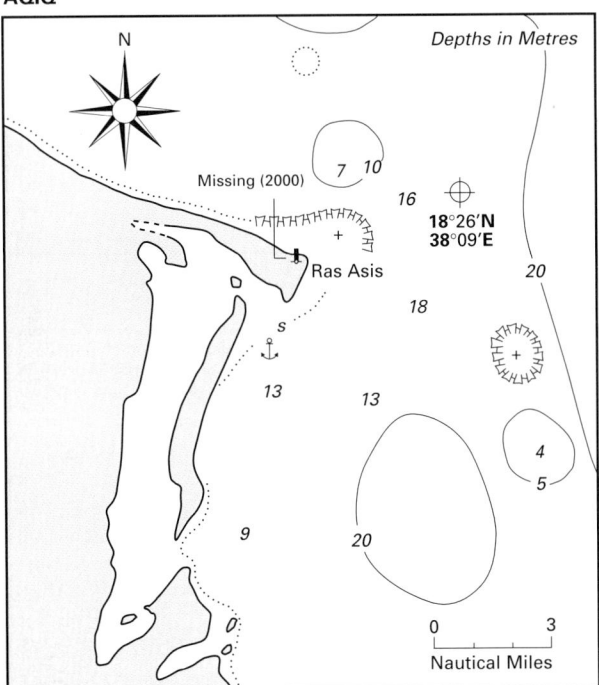

RAS ASIS

irregular depths with shoal patches though we beat down within 1M of the coast without trouble recently. There is a rocky spit with a least depth of 2m extending 2M ENE from the *ras*. The charted beacon on the E headland is missing. Anchor in

TRINKITAT

131

Red Sea Pilot

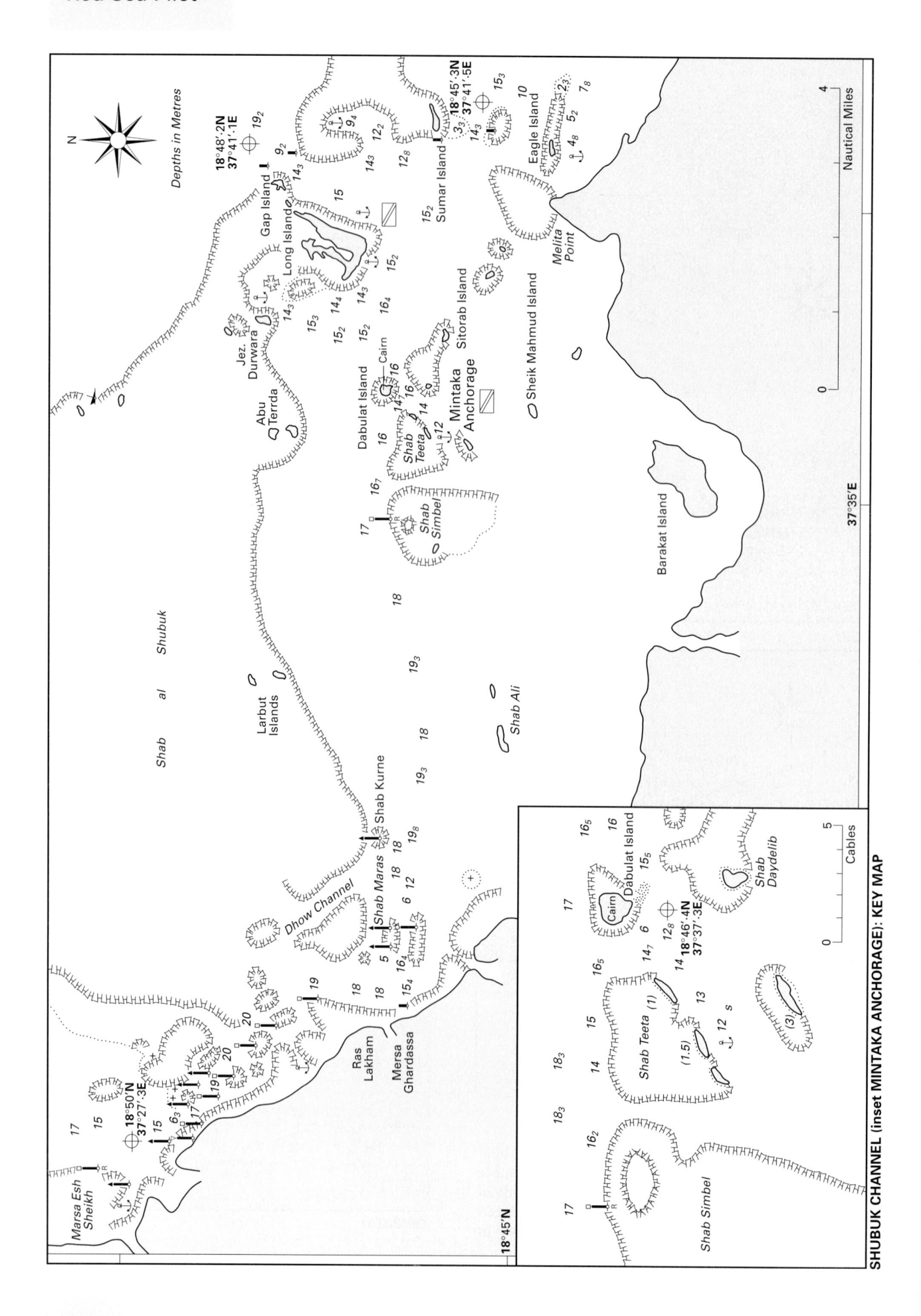

SHUBUK CHANNEL (inset MINTAKA ANCHORAGE): KEY MAP

The Sudan Coast S of Port Sudan

Anchorages in Marsa Maqdam

Marsa Maqdam is the large area of open water in the lee of the E end of Shab Shubuk before you reach the Shubuk Channel. In autumn and early winter, when the wind is often as much E as SE, a swell works its way in through the Suakin Group, but it is very slight. Neither Rambler Shoal nor Fairway Patch is of any concern to a yacht except in strong weather when Rambler Shoal breaks.

Reef Island

Plan page 134

A good anchorage in N'lies, in 3–4m, good holding in sand, under the lee of the low-lying island with the E end of Shab Shubuk offering protection from W through ENE. The 10m depth contour runs quite close to, so you need to be within 100m or so of the reef edge. There might be a slight swell if there's any easting in the wind.

Eagle Island

This is outside the Shubuk Channel proper, S of a small island with nesting ospreys E of Melita Pt, with excellent holding in 5m, fine sand. In calm weather a light swell can make in from seaward through Marsa Maqdam. It is more of a bounce than a roll and not really noticeable. The danger coming from or leaving for the Shubuk Channel is the reef which extends E of Eagle I. The visible reef looks like it ends but shoal ground extends a further 200–300m E and has a least depth of 2·3m, very much less than charted. We nearly found that out the hard way! Keep E of longitude 37°41'·7E on approach or departure.

W of Marsa Maqdam lies the Shubuk Channel, to our minds one of the jewels of the inshore route. Wonderful sheltered anchorages, excellent snorkelling and quiet, excellently beaconed waters for pushing on N.

Shubuk Channel

Admiralty chart 675, key map page 132

The channel provides a well sheltered alternative to the Inner Channel, which is used by heavier traffic. The reefs are marked by many small, low, sandy islets and are intersected by narrow channels. The water is usually clear but it can be sometimes be muddy. All the intricate parts of the Shubuk Channel are well beaconed: red square topmarks, to port heading N to Suakin. The only damaged beacons in 2001 were at the E end of the channel on Sumar I and off Melita Pt.

Passage through the Shubuk Channel requires a sharp lookout, though we recently traversed it S'bound under sail, surveying as we went, with no major alarms. Note that the most recent, metricated versions of BA675 and most electronic charts can be seriously misleading. The official attitude is that if there's doubt because of lack of, or conflicting data, post a sign saying Keep Out. At least the errors are on the safe side.

Approach

The SE entrance is between Sumar I and the beacon that stands on the reefs N of Eagle I, at 18°45'·39N, 37°39'·44E. It is marked by a black cylinder shape with a pole on top to port and to starboard there's a roughly cone-shaped, faded red and white painted pile of cemented-together rocks. The least depth is 9·4m, over an isolated and very visible shoal patch bang in the middle of the entrance. The N channel is between Gap I and the beacon on the tip of the reef N of Sumar I. The least depth is 14m.

Route

The sketch is to scale and bearings are correct but it is NOT for use position plotting. It is for orientation. Do not use autopilot or trundle from GPS waypoint to GPS waypoint. Conventional navigation for position checks is strongly recommended. For planning purposes the following co-ordinates indicate the line of the deep-water track. The trickier bits are well beaconed and, in good light, obvious to the naked eye. The distance from off Melita Pt to the Inner Channel NNE of Marsa Esh Sheikh is approximately 17M.

The following guidelines run from S to N.

From 18°45'·3N 37°41'E with Melita Pt bearing 204°, 1·16M steer approx 285° for 3·5M to
18°46'·3N 37°37'·55E whence steer for about 0·5M in a 'U' via the obvious channel S of Dabulat I to
18°46'·6N 37°37'E whence steer approx 279°, 1·3M with the Shab Simbel Bn (port) roughly due S to
18°46'·8N 37°35'·7E whence steer approx 243°, 1M to
18°46'·37N 37°34'·8E whence steer approx 266°, 4·2M, past Shab Kurne Bn (starboard, black

---

It helps to tick off the beacons as you go. Here is the S to N running order. The numbering is ours, not the Sudan Government's and follows the odd to starboard, even to port convention. For approximate co-ordinates see guidelines.

	Starboard		**Port**
1.	Sumar I: stone cairn	2.	Islet NNE of Melita Pt: pole with no topmark
3.	Dabulat I: stone cairn	4.	Shab Simbel: red square
(5a.	Shab Kurne: black cone, well starboard of the usual route)	6.	Reef S of Shab Maras: white ball
		8.	Reef off Marsa Ghadassa: no topmark
5.	Shab Maras W: black cone	10.	Reef NE of Ras Lakham: red square
7.	Shab Maras E: black cone	12.	Shab Lakham: red square
9.	E reefs SE of El Makglas: black cone	14.	Shab Daala: red square
11.	Mid reefs SE of El Makglas: black cone	16.	E reefs SE of El Makglas: red square
13.	El Makglas: black cone	18.	Mid reefs SE of El Makglas: red square
		20.	El Makglas: red square

Leading line: front: black triangle point up; back: white triangle point down

Red Sea Pilot

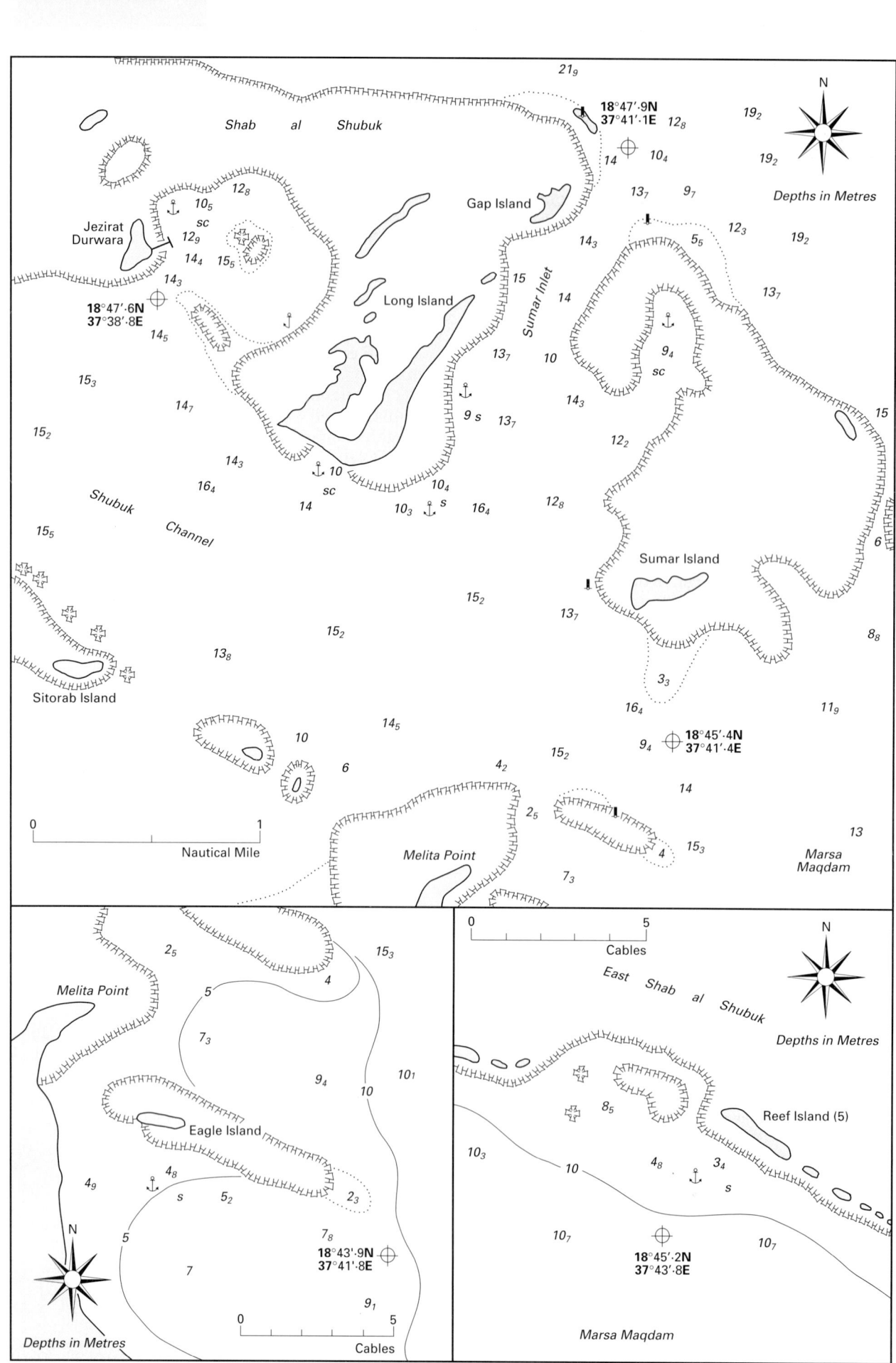

LONG ISLAND ANCHORAGES AND MARSA MAQDAM ANCHORAGES

cone, approx 18°46'·6N 37°31'·3E) approx 0·5M to starboard to
18°46'·2N 37°30'·4E. Near this point you may cross a shoal with 6m over it. Ahead to port will be a bn (white ball, 18°46'·1N 37°30'·3E) and to starboard on Shab Maras two starboard bns (black cones, 18°46'·3N 37°30'·3E and 18°46'·3N 37°29·8E). DO NOT TURN NORTH IMMEDIATELY AFTER THE SECOND STBD BN ON SHAB MARAS. Steer roughly due W, 0·8M towards a bn on the shore (with no topmark at 18°46'·2N 37°29'·1E), to approx
18°46'·35N 37°29'·5E, 0·3M short of the bn with no topmark, whence turn slightly E of N for 1·1M towards a port bn (red square, 18°47'·5N, 37°29'·4E) on the reef NNE of Ras Lakham which you pass and at approx
18°47'·7N, 37°29'·5E alter to approx 325°, 1·5M past port bns on Shab Lakham (red square, 18° 48'·0N 37°29'·1E) and Shab Daala (red square, 18°47'·5N 37°28'·7E) to pass between port (red square, 18°48'·7N 37°28'·2E) and stbd (black cone, 18°48'·7N 37 28'·6E) bns at approx
18°48'·6N 37°28'·6E whence alter to approx 285° for 0·32M to pass between port (red square, 18°48'·6N 37°28'·2E) and stbd (black cone, 18°48'·7N 37°28'·3E) bns at approx
18°48'·7N 37°28'·3E whence alter to approx 318° for 1·1M, passing between port (red square, 18°48'·8N 37°27'·9E) and stbd (black cone, 18°48'·9N 37°28'·0E) bns and across a shoal with least depth approx 5·8m (approx 18°49'·5N 37°27'·4E) to approx
18°49'·6N 37°27'·25E, 15m, where you'll find the leading line (front: black triangle point up, 18° 49'·56N 37° 27'·24E; back: white triangle point down, 18°49'·34N 37°27'·22E) on a back bearing 186° on your port quarter. Steer 006° for approx 1·6M to pass the entrance to Marsa Esh Sheikh (Marsa Sheikh Sad) to approx
18°51'·1N 37°27'·2E whence alter to approx 344° across a very uneven bottom (depths >6m depending on track) to follow the inner channel N'wards.

If you are headed S, reverse the order of the above.

All beacons were in place in 2001. Indeed, the beaconage throughout all of Sudan's coastal waters is at present in very good order. As noted, the reefs show well in good light, and the kink around Dabulat I is not as difficult as it appears, just don't cut the corner. If you're twitched, go N of Dabulat I, HMS *Dahlia* did in the 1930s and reported deep water. Remember, the reefs as shown on the latest BA675 and its electronic equivalents are very misleading, especially around Dabulat I. There are many potential anchorages to be explored in this area, which is said to have some of the best snorkelling in the Red Sea. Some of the better ones have been sketched, others are marked on the key map page 132.

The Sudan Coast S of Port Sudan

Anchorages in the Shubuk Channel

Long Island
Admiralty chart 675, plan page 134

There are good anchorages in sand, about 10m, on the S and E sides, as shown in the sketch but the one with a beach at its head is confined and rolly in a NW'ly. The snorkelling on the SE side is excellent. Unlike in much of the Red Sea the reefs at the E end of the Shubuk Channel, perhaps because of the naturally high average ambient sea water temperatures, are not suffering from global warming and are in good condition with abundant fish life. The lagoon between the two 'arms' of Long I has flamingos and other waders in season, but is only accessible on foot or by dinghy.

Two interesting alternatives nearby are in the lagoon E of Jezirat Durwara. There is an isolated reef just inside the entrance, obvious in good light. For NW'lies anchor in approx 18°48'N 37°38'·5E, NNE of the stone jetty that is almost awash in 10–12m, sand. In SE'lies, try the SE end of the lagoon, though it may be deep at <15m.

Note The new, metricated BA675 and its electronic clones do not show this anchorage at all.

Sumar Inlet
Admiralty chart 675, plan page 134

There is an indent in the reef on the E side of the channel, roughly due N of the beacon W of Sumar I. The anchorage, over sand and coral, offers good shelter in NE'lies.

Mintaka anchorage
Admiralty chart 675, see key map page 132

Approach through the channel SW of Dabulat I which has constant depths of around 12m. The anchorage is very peaceful with good protection NW and N from Shab Teeta and from W through SW and E through SSE from other reefs. There is excellent holding in sand. Good snorkelling. The anchorage is named after the boat of our friends and Red Sea companions Manfred and Petra, who found it.

Marsa Esh Sheikh (Marsa Sheikh Sad)
Admiralty chart 675, plan page 136

Coming from the N, you'll see a beacon on the fringing reef ½M N of the entrance. Don't mistake it for the beacon at the entrance. From the S the approach is more confusing, and best made when the sun is in the E. The beacon you want marks the reef to the N of the entrance but there are three large coral heads partially blocking the fairway. The safest approach is from the NE where the reef to the N is steep to. In effect make a curving approach to the N, then WSW. You should have a least depth of 8m in the channel. Anchor in 8m, mud with good shelter.

General

This may be the first *marsa* anchorage that yachts coming from the S encounter. These inlets in the

Red Sea Pilot

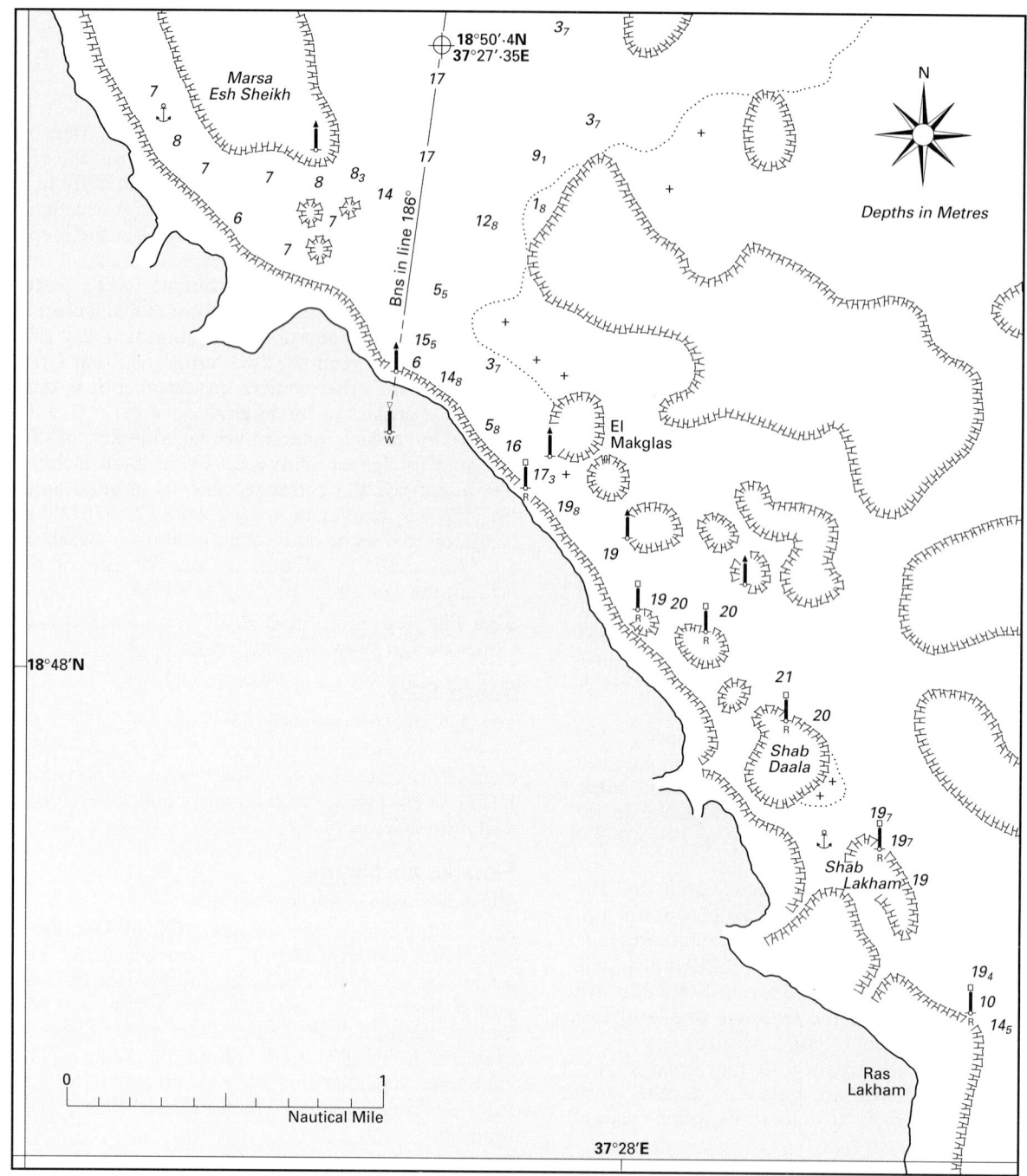

NORTH EXIT TO THE SHUBUK CHANNEL AND MARSA ESH SHEIKH

fringing reef vary in depth, width, and penetration inland, but provide remarkable shelter on the coast from here N. In this particular case the inlet extends for 2M NNW, but shoals rapidly, and is only navigable by yachts for about ½M. Look out for shore birds and camels on the mud flats and coastal plain.

GPS positions plotted on the BA/NIMA charts in this area are too far N by about 100m. This confirms recent reports that GPS WGS84 positions in the vicinity and onwards N as far as Sanganeb Reef can be unreliable.

The Inner Channel from Marsa Esh Sheikh Ibrahim to Suakin

Marsa Esh Sheikh Ibrahim

Admiralty chart 675, plan page 138.

Note on some paper and electronic charts Marsa Sheikh Ibrahim is 1M too far N.

The entrance beacons are fine and there is a beacon marking the channel. Shoal ground in the N arm begins sooner than the BA/NIMA plans show. In general we found <1m less than charted depths throughout. The sand *cay* N of the entrance channel has gone. The charted line of the coast N of Marsa

The Sudan Coast S of Port Sudan

Esh Sheikh Ibrahim is also badly wrong, there being a promontory between the waters N of Marsa Esh Sheikh Ibrahim and Marsa Heidub (Aydob).

Dangers
Shoal water extends from both the beaconed entrance points on either side (see sketch-map), so it does not pay to cut corners. Nets are sometimes set both in the entrance channel and across the anchorage. These are normally marked by buoys.

Anchorage
8m, mud with good holding.

General
Often the most conspicuous feature of the coast hereabouts is the traffic on the coast road, which usually raises clouds of dust. A trawler regularly lies at anchor in the *marsa*, with some lean-to tents close by. Fishermen may be a nuisance and will approach yachts at anchor and demand cigarettes, etc. They may have scant regard for your topsides and are not always courteous. Experiences vary but throughout the Red Sea it is best to cultivate a firm, tactful but friendly reaction when you are asked for gifts of any kind. Often, but not always, fishermen will offer whatever they have by way of catch in return for your generosity. There is good snorkelling and fishing on the reefs.

Note Between Marsa Esh Sheikh Ibrahim and Suakin the match between GPS positions, WGS84 and chart BA81 is not always trustworthy. The coastal fringing reef does not have a steep drop-off and within 100m of it you will find varying depths between 10m and 40m.

There are no really good anchorages on this stretch. Marsa Heidub (18°57'N 37°24'·5E) is 11M S of Suakin and WSW of SW Islet and sometimes used by dhows. The access channel is narrow and depths uncertain. We weren't tempted.

You'll see a puzzling beacon 3M N of Marsa Esh Sheikh Ibrahim and about 1M S of the entrance to Heidub. It marks something inside the reef complex and appears irrelevant to navigation in the main channel. Marsa Entebil (19°01'·5N 37°24'·5E), 4M on up the Inner Channel towards Suakin, opposite Qad Eitwid reef is only an indent. There are two more indents in the coastal reef within 2M N of Marsa Entebil, at 19°02'·5N 37°24'E and 19°04'N 37°24'E. The one further N is marked by a beacon. Both offer fair shelter if you tuck in far enough but the bottom is very uneven. Unless you're really caught out, it's better to push onto Suakin.

Suakin
Admiralty chart 81, plan page 138

Approach
Outer port The Inner Channel leads to Suakin from N and S. From seaward a channel leads from Hindi Gidir (Fl.20s10M 19°23'N 37°55'E), N of Shab Anbar (19°19'·5N 37°42'E) and avoiding Shab al Hareeq (2m, 19°11'·8N 37°31'·3E) to 19°09'N 37°24'·9E, about 1M S of the light on the S of Fikheeb whence Suakin is 260° 3M. Currents, usually setting W or NW, can run strongly in the S approaches to Suakin. The fringing reef both S and N of the entrance dries in patches in summer. The entrance to the harbour is well beaconed but note that there is an old beacon on the fringing reef N of the entrance, N of the newer beacon marked on the sketch.

Inner harbour You pass the new port area and main wharves on the approach to Condenser I. There are two minarets on Old Suakin. Keep the SE'most (the one appearing to the left of your visual field) on 215° as you approach the inner harbour then turn onto approx 181° on the conspicuous chimney of the old cotton factory S of the anchorage. Transit the narrows well to the W side to avoid a reef and then a sandspit on the E side. But don't stray too far W. There is fringing reef fronting the shore of Old Suakin just in front of the ruins of an old, balconied hotel.

Dangers
The fringing reef extends to seaward and into the fairway from the outer port-hand beacon at the entrance to the harbour. The entrance to the NW arm of the harbour is partially obstructed by wrecks, some sunk. The authorities do not like you to anchor here. You may be allowed to wait here till daylight since entry to the anchorage is tricky without good light. There is also a patch with 0·5m over it in the inner harbour.

Conspicuous
The tower marked on Graham Pt is a pale blue mosque with a radar antenna and lookout on top and is very conspicuous.

Anchorage
3–8m, mud, good holding with all-round protection S of old Suakin. There is less swinging room than you might think and the bottom may be more obstructed than shown. One boat, anchored toward the E side of the anchorage, found what they thought was an underwater obstruction when swinging during a thunderstorm.

Formalities and agents
This is not a port of entry, but there are immigration officials in the port and shore passes can be obtained. Call on Ch 14 on approach. You may not receive a reply but it's all right to head for the anchorage.

You can do your own clearance after anchoring by going to the port office at the ferry pier. You will need 4 crewlists, all passports and one photo per crew member. Abu Mohammed, once a yacht agent, now works in the port office. He may still offer to work for you but there will be delays and you will have to pay his fee of approx US$25. He can sometimes be raised on VHF Ch 6. The basic charges in 2002 were:

Shore passes: US$28 each for the first week payable on arrival.

Red Sea Pilot

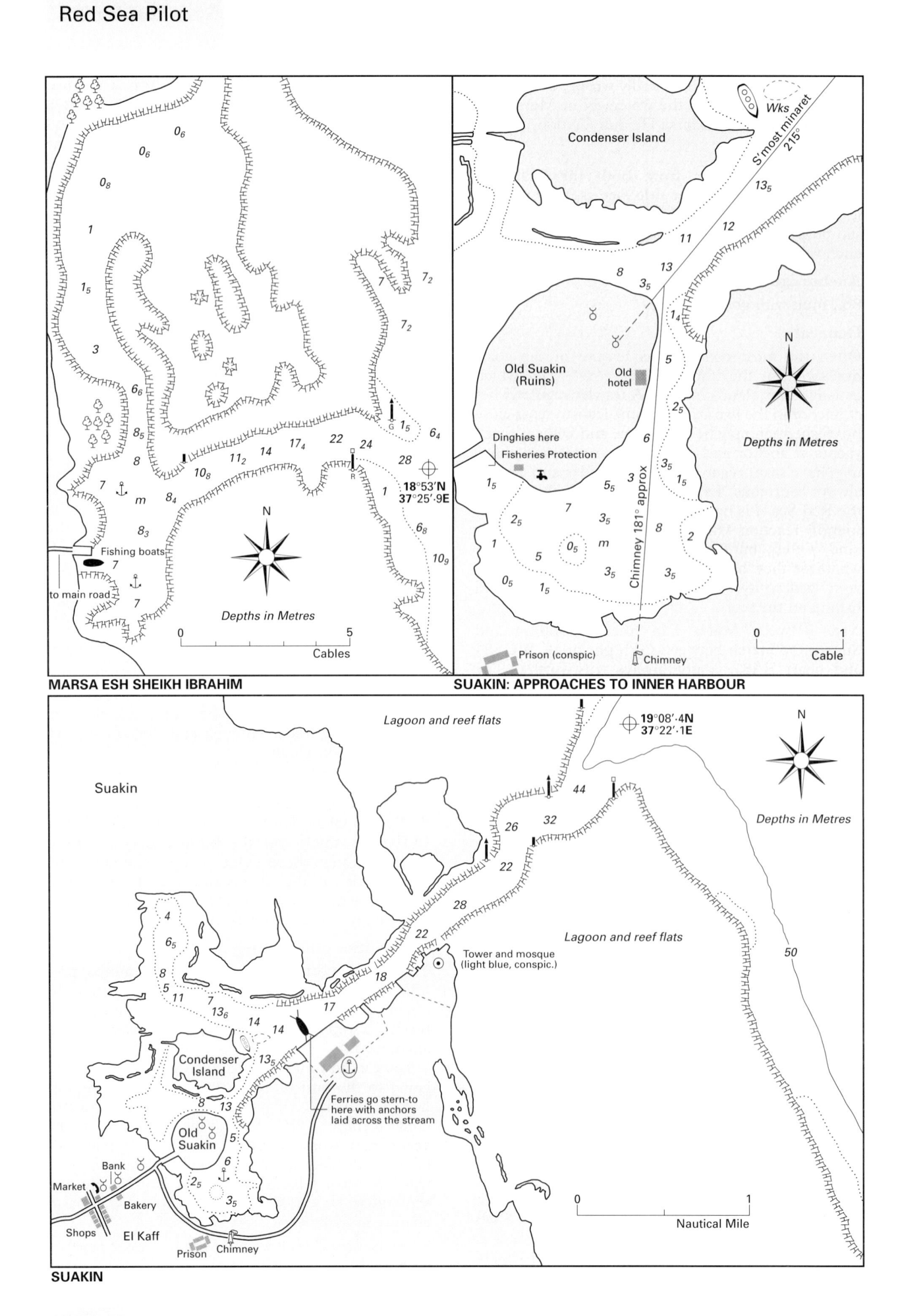

MARSA ESH SHEIKH IBRAHIM

SUAKIN: APPROACHES TO INNER HARBOUR

SUAKIN

138

The Sudan Coast S of Port Sudan

Other dues are US$17·50 per boat plus US 10c per ton and US$5 per head.

When you clear out you should get a cruising permit valid along the Sudan coast. If you clear into Port Sudan the dues are about double this and agents are the norm. Shore passes probably cost about the same but in Suakin you may be able to get a discount for longer stays. On the whole Suakin makes a better stop than Port Sudan.

Facilities

The dinghy landing is at the E end of the causeway joining Old Suakin to El Kaff. Watch out for the sharp metal stakes if you have an inflatable. Don't leave anything tempting in your tender.

Money There is a small bank at the town end of the causeway, which will change US$. Travellers' cheques are acceptable as long as they are NOT issued by a US bank. There were no banks that took credit cards either here or in Port Sudan in 2001. If you find him, Mohammed may be able to help you with exchange if you first go ashore when the bank is shut.

Fuel Diesel deliveries can be arranged either in your own jerry cans or direct from drums to the boat. Make arrangements well in advance of when you want to leave. In 2001 diesel cost US40c per litre and petrol US65c per litre.

Water Well water is delivered by donkey cart and can be taken aboard by jerry from a tank beside the fisheries protection base, which is in a shack on Old Suakin I, SE of the root of the causeway (see sketch). The tank is marked with Abu Mohammed's name. If you need a large quantity, say over 200 litres, arrange to go alongside at the port where water is delivered by hose at approx US$10 per tonne. Fendering is difficult for boats with low topsides because of the height of the jetty.

Stores and other services The market sells fresh produce and bread and there are shops with a surprising range of provisions, given their unlikely frontages, in the market on the opposite side of the crossroads. There are now 'telecommunications centres' in Suakin but it is usually difficult to get an overseas call through. Laundry services are available for US30c per piece. Diesel mechanic, gas refills, etc. can be arranged by an agent or the harbourmaster. Leave your garbage with staff at the 'Tourist Office' near the gates of Old Suakin by the dinghy landing point, but be aware that it will be picked over.

General

The people here are extremely poor but friendly. There is some traditional boat building still done in the anchorage N of Old Suakin. Much of the town is in ruins or very dilapidated. In El Kaff many people live in poor shacks. This is indicative of the state of affairs in the rest of Sudan. Be careful about taking photos near police stations or military installations. They are considered sensitive areas. Adults may also object to being photographed. Always ask first. If you are ashore the military may ask you for your papers but they are always courteous.

The buildings of Old Suakin, made of coral, are now crumbling and deserted. The settlement has had a chequered history as a trading centre since the 10th century BCE. It was the last slave-trading post in the world, used as such until the end of the Second World War. The ruins are considered a tourist attraction and you'll be asked to pay to see them. Don't be tempted to dive on the wreck in the channel. The military may object. Ask first if you want to do anything unusual.

Trips ashore

Mohammed or other agents can arrange trips inland but permission for all travel is required if you have no visa. The cheapest fare to Port Sudan is about US$1 single by bus. There is also a nearby museum of Suakin's history. Another possibility is a day trip to Sinkat and Er Koweit in the Red Sea Hills which can be rather fun. Expect to pay US$12–15 per head. The lower figure is for a covered pick-up seating about 6. Costs generally depend on the type of transport used and the number of people in the group.

Note The beaconage here N and past Port Sudan can be confusing. Suakin used to be the main port, so the direction of buoyage, following the red-port-inbound rule, was to Suakin. That means that from Suakin to Port Sudan the red (square topmark) is to starboard and it continues to be so the rest of the way up the coast of Sudan.

The Inner Channel from Suakin to Port Sudan

There is one excellent, one fair and several indifferent anchorages on this stretch.

Marsa Littlefoot (19°09'·5N 37°22'E) is just N of the entrance to Suakin. The entrance trends 310° then 320° and hooks N in a backwards 'J'. It's obvious as long as you hold to the edge of the reef. Anchor in 6–7m, sand, tucked well up but with only the reef for shelter from the sea.

Shab Damath and Marsa Kuwai
Plan page 142

These lie off the Inner Channel between Suakin and Port Sudan. The chart shows Shab Damath as a continuous reef fronting Marsa Kuwai. It is not. There is a clear break, 3–4 cables wide N of the beacon with depths of 5·5m in mid-channel (19°14'·79N 37°20'·32E). A swell makes through here in the usual N'ly. Depths are irregular in the S approach and the S end of Shab Damath does not always show. W of Shab Damath the bottom is uneven with two shoal patches with less than 5m, one as little as 3·4m. The reef is beaconed and there is a light towards the N end. The only sheltered anchorage is a continuation N'wards of the channel

Red Sea Pilot

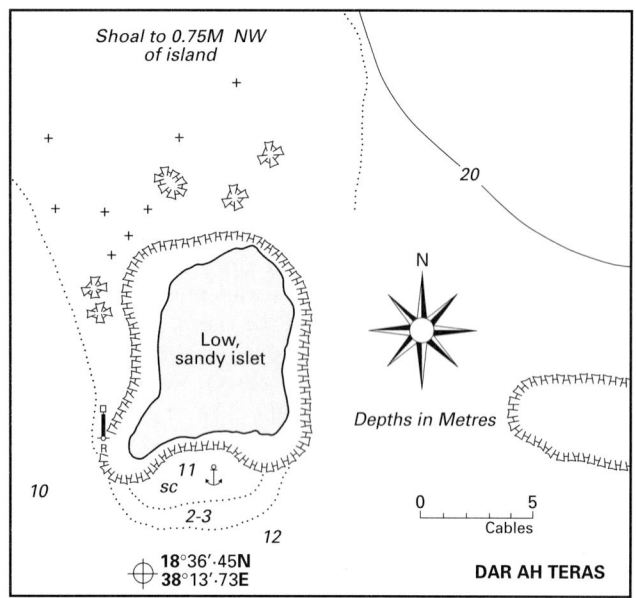

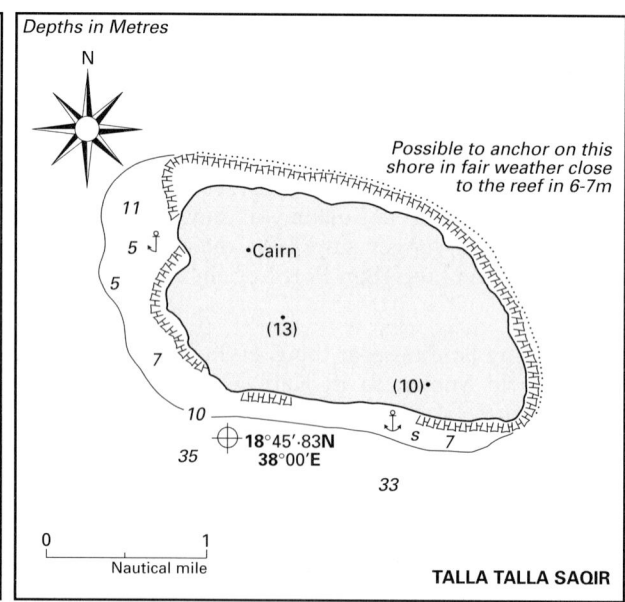

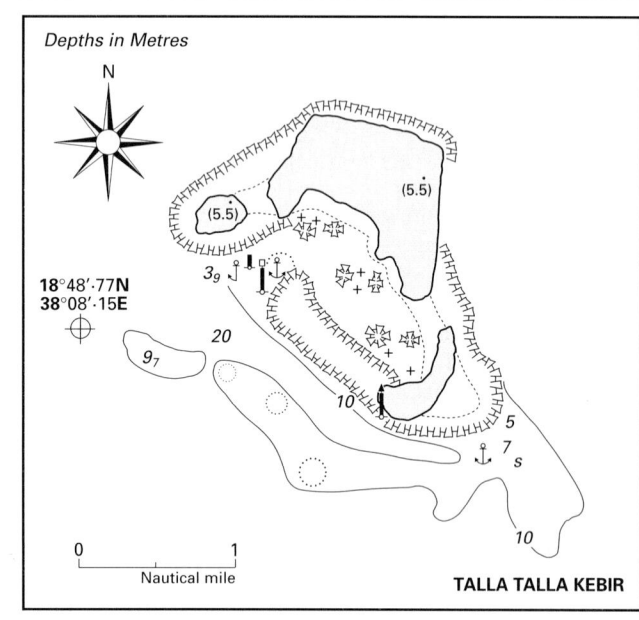

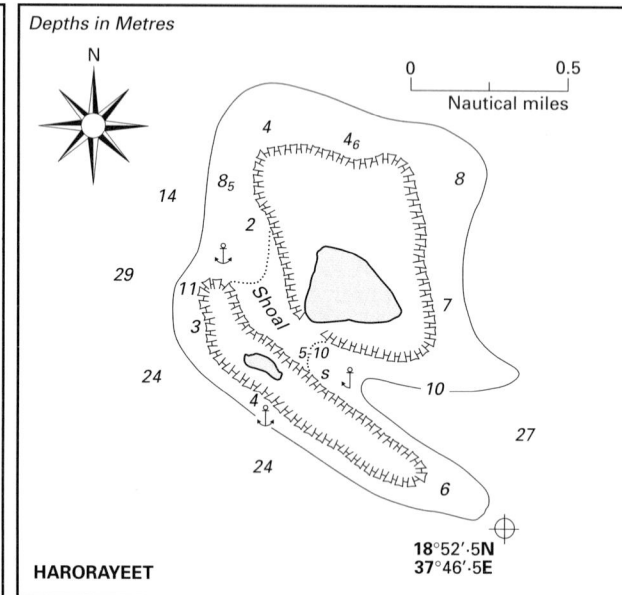

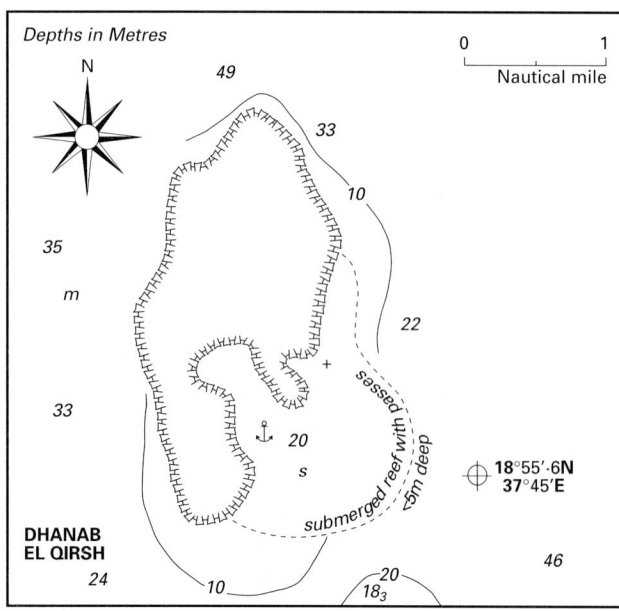

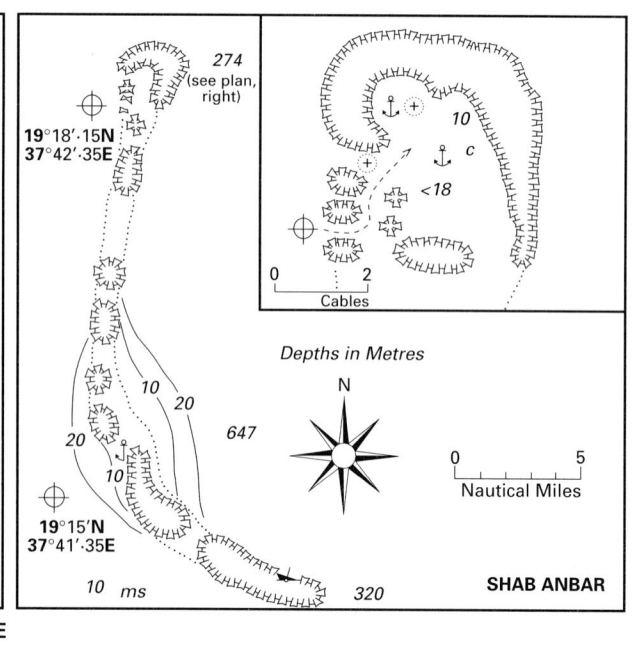

ANCHORAGES IN THE SUAKIN GROUP ON THE MIDDLE ROUTE

The Sudan Coast S of Port Sudan

W of Shab Damath. It is funnel shaped and not large, but with good holding in mud, 5m tucked well up.

There are two new leading lights, in line 270°, on the shore in the N end of Marsa Kuwai in approx 19°15'·33N 37°20'·14E. They mark the deep water passage from seaward N of Hindi Gidir and Shab Anbar, and S of S Towartit Reef into the Inner Channel. The characteristics of the lights are not known. The lower (outer) beacon (W) is on a plinth on the reef, the higher (inner) beacon (W) is on the land.

Marsa Ata
Plan page 142

The reef at the entrance seems to block the obvious approach from the NE, though there is a clear passage (15m) either side of it and there is approximately 3·5m over it. The entrance leads WNW between the beacons once inside the reef patch. The best anchorage is in 10–11m where shown on the bend. The *marsa* stretches about 1M further N with depths slowly shoaling from 10–3·5m. The lagoon to starboard has a very narrow entrance. Further on the main channel branches twice. Each branch leads to an area of shallow mangrove lagoons and channels full of bird and fish life. Marsa Ata is a wonderful spot.

The new duty-free storage area inland of the *marsa* is brightly lit at night.

Between Marsa Ata and Port Sudan there are no good anchorages. There is a conspicuous white single-funnel ferry on the fringing reef in about 19°22'N. Marsa Amid, now called Marsa Bashir (after the President) is a closed area and terminal of a 2000km pipeline built by Chinese convict slave labour. There is an offshore SPLM at approx 19°24'·09N 37°20'·05E. Near the tank farm which marks the end of the pipeline there's a jetty with a tug boat. Keep at least ½–1M offshore in this area. A patrol boat may come out to escort you. The anchorages in North Towartit Reef (19°31'·5N 37°19'·5E) and Towartit Elbow (19°28'·8N 37°18'·1E), which we put in the first edition, are desperate last ploys generally to be avoided. In any sort of wind they are lively, deep and with anchor and chain grabbing coral bottoms.

Note Lights in the area may not be working although the beaconage was excellent and in recent repair in 2001.

B The middle route

Karb Islets
18°25'N 38°25'E
(not illustrated)

This group of six sand and coral islets sits on a coral plateau marked by some pinnacles, with deep water in between. They lie in the Suakin group 20M E of Ras Asis. The sea breaks over the reefs in any swell. Little is known about the area and you need settled weather to explore. The smallest islet in the group is the one to aim for. The best shelter under its lee in approx 18°25'·5N 38°27'·4E.

Dar Ah Teras
Plan page 140

Reef extends 0·75M NW of the islet. Shoal water and a reef with a sandy *cay* lie to the E and SE. There is a beacon with a square red topmark on the SW tip. Cross the belt of reef with 2–3m over it around the S of the island and anchor in about 10m, sand and coral, with shelter from the N if a bit rolly. (One yacht even groped her way to an anchorage here in the dark, but they thought they were somewhere else!)

Talla Talla Saqir
Admiralty chart 675, plan page 140

The island is 15M ENE of Trinkitat Harbour, in the Suakin group. A shallow bank extends up to 6 cables off the W side and most of the shore is fringed with reef. The island is sandy and sloping on the NE side, but has raised coral cliffs on the W and S sides. Anchor in 7m, sand, off the S side, with shelter from the N. For a SE'ly try in 5–10m off the W side though this would probably be rolly. In light weather it is also possible to anchor close up to the reef in 6–7m anywhere on the N or E side of the island. Excellent diving.

Talla Talla Kebir
Admiralty chart 675, plan page 140

This group of three islets lies 10M ENE of Talla Talla Saqir. From a distance the three islets appear to be one. The highest (13m) is the largest, in the NE of the group. W of the S islet there are coral heads with depths of under 2m (see sketch). Three beacons mark the SW tip of this islet. Anchor in 7–8m, sand, off the fringing reef to the SE of the S islet. There is some protection here from the NW, but this is really only a light-weather anchorage. There is another possibility in the S entrance to the lagoon but there is very little swinging room.

Harorayeet (Two Islets)
Admiralty chart 675, plan page 140

Continuing outside the Shubuk Channel, these islets form part of the Suakin group. The reef on which the two islets stand is divided by a narrow channel which is not navigable (see sketch-map). Anchor in the deep indent SE of the larger islet or to the S of the smaller islet, where marked on the sketch. For light to moderate weather only.

Dhanab Al Qirsh (Green Reef)
Admiralty chart 675, plan page 140

This reef lies on the NE side of the fairway of the Inner Channel and is awash on its W and N sides. On its E, SE and S sides it is mostly submerged, but there are a few passes into the lagoon, with <5m of water. Once you're in the lagoon, be wary of coral heads. The anchorage is deep, in about 20m, sand

Red Sea Pilot

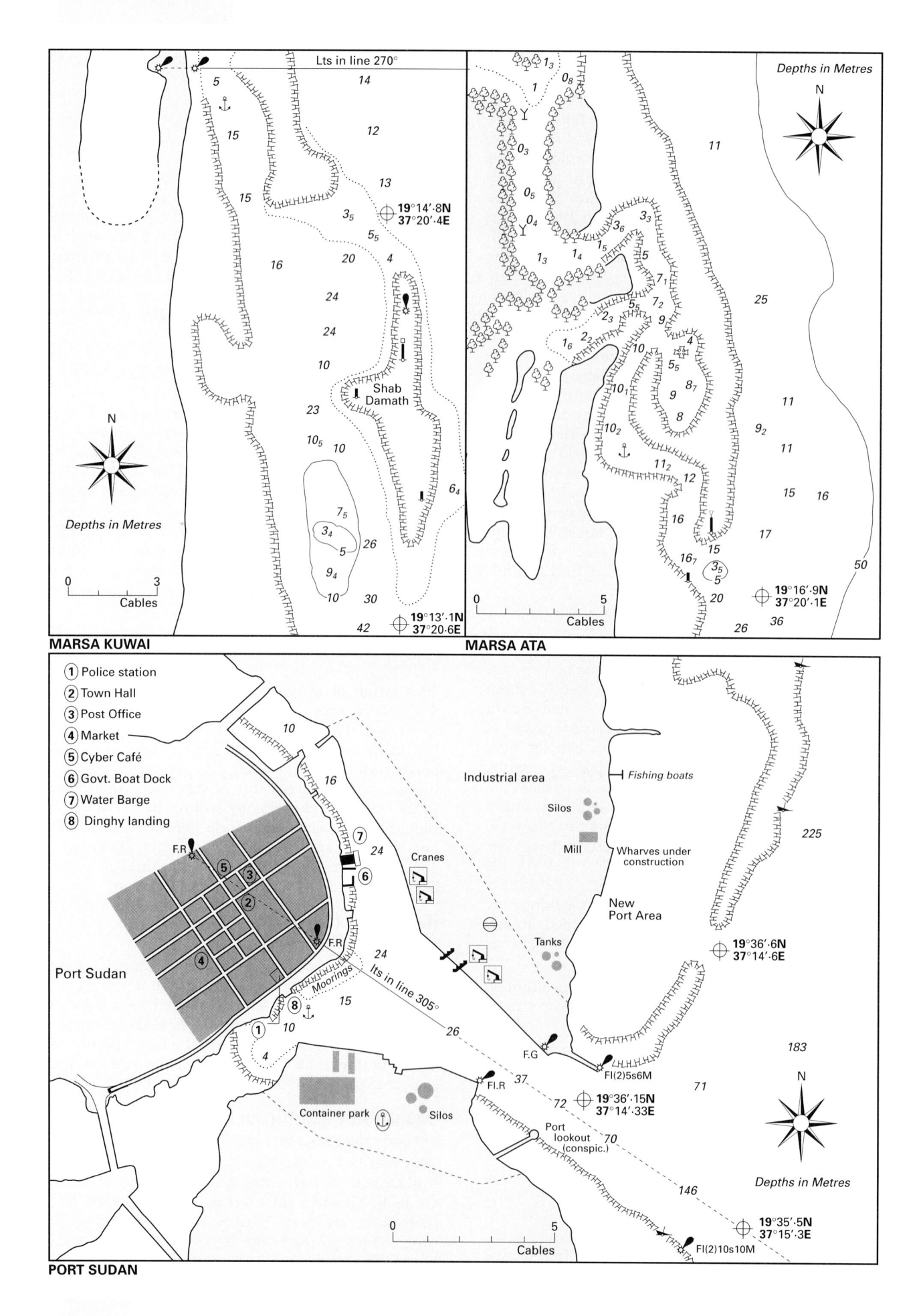

MARSA KUWAI

MARSA ATA

PORT SUDAN

1. Police station
2. Town Hall
3. Post Office
4. Market
5. Cyber Café
6. Govt. Boat Dock
7. Water Barge
8. Dinghy landing

142

The Sudan Coast S of Port Sudan

but fairly sheltered in N winds, if there's not much E in them.

Shab Anbar
Plan page 140

The only really sheltered reef anchorage in the approaches to Suakin from Hindi Gidir. It has a narrow entrance between reef patches (see sketch). Anchor in 10–15m, coral, under the N part of the reef. There is a day anchorage where the half anchor is shown. Good diving.

Port Sudan
Admiralty chart 3492, see plan page 142

Approach
From S (Inner Channel)
The N end of Towartit Reefs lies to starboard 6½M SE of the entrance. The easiest approach is to keep the fringing reef one cable to port. Towartit Elbow beacon, also about 6½M from Port Sudan, marks the point where the reef turns more NW. Towartit Reefs generally show in a moderate breeze and are well beaconed. The N-setting current in the Inner Channel tends to set more NE out over the reefs so watch your position.

From N (Inner Channel)
The approach is deep and clear and well beaconed though note that some of the beacons on the seaward side are not where they're charted. Marsa Gwiyai, to starboard, is the naval base and a closed area. The beacon on S Wingate Reef is less obvious than the ships waiting at anchor. Just before Port Sudan there are three wrecks on the fringing reef and you'll pass the work on the new port where there are several new, silver silos by a large flour mill.

From seaward to the N
Pass S of Sanganeb Reef light (WPT 19°43'N 37°26'·6E), S of Nimra Talata (WPT 19°38'·3N 37°19'·6E) and S of Silayet (Wingate Reefs) South Lt (WPT 19°36'·6N 37°17'·4E).

If you have been visiting Shab Rumi
Day Steer for WPT 19°54'N 37°22'·6E thence to WPT 19°50'N 37°19'·1E and to WPT 19°40'N 37°16'·8E off Marsa Gwiyai.
Night It's safer to pass inside Sanganeb via WPT 19°50'N 37°23'·8E to WPT 19°45'·4N 37°22'·5E to WPT 19°38'·2N 37°20'E whence continue S of Nimra Talata and Silayet South light to Port Sudan.

From seaward to the S
Masamirit It (Fl.10s10M 18°50'N 38°45'E, WPT 18°52'N 38°48'E) is the best initial landmark. It has a light on a metal framework tower on its E side. Steer for WPT 19°26'·3N 38°04'·5E from which Hindi Gidir, the next lit offshore danger (Fl.20s10M), lies some 10M WSW. Shaab Jibna shoal, which breaks, is the next offshore danger S of WPT 19°28'·1N 37°43'·3E. Hindi Gidir on a back bearing of 110° checks you're clear of Shaab Jibna. Head for WPT 19°36'N 37°15'E in clear water, 2 cables N of the leading line 1M from the port entrance.

Final approach to the port
Leading lines with red lights in line bear 305°. The fairway is deep, wide, and clear of dangers. A good position in the channel just S of the N breakwater light is 19°36'·15N 37°14·33E. For approach from the S you will find the leading marks in line conspicuous from 19°35'·5N 37°15'·33E.

Conspicuous
This is a major port. Ships at anchor on the bank SW of Silayet (Wingate Reefs) South light are conspicuous whatever your approach. An oil installation with chimney and flare, about 3M S of the harbour entrance, is a very useful landmark. There are also quite a few more wrecks than charted on the reef N of the entrance. Damma Damma light (red and white chequered framework tower) stands on the fringing reef ½M SE of the entrance. The N breakwater lights are not obvious, so be sure you know where you are if entering at night. A new channel has been broken through the fringing reef to create an extension N of the original port. The lagoon inside is being dredged to create two new bulk loading wharves to handle grain for the large mill N of the existing port, the corrugated steel silos of which are conspicuous.

Anchorage
5–10m, mud, clear of the dredged area of the W spur of the S arm of the harbour (see sketch) if there's room. This can be crowded, though well protected with good holding. You'll get moved if you're in the way of shipping movement, though no one has recently been moved into the N part of the port. The possibility of Mediterranean mooring to the N shore of the anchorage seems to have disappeared with the increase of permanently moored small shipping.

The future of the yacht anchorage is under threat because of the expansion of the port. The container wharf has been extended as our sketch shows. THIS IS NOT YET ON CHARTS. To allow berthing room for ships, the already limited space will all but disappear. There are no proposals as yet for an alternative. We believe the authorities want yachts to go to Suakin instead.

Formalities
Port Control may not respond to a call on VHF but you can head on into the anchorage and agents will come out to you. Frankly life is a lot easier if you use one rather than trying to do clearance yourself. Agents include Capt Halim, El Barbery, El Samkhary and his two sons, Chico and Murtado (Dolphin Travel) the last of whom is learning the ropes. All charge the same fee of US$50. Discounts for groups or rallies are negotiable. Agents will collect 3 crew lists, passports and one photo per person for shore passes which cost US$22 per person in 2000. The price increase in Suakin may soon be reflected here. The agent will arrange for a

Health Official to come out to you to issue quarantine clearance and will later return with shore passes. Always take these ashore with you. There is a check-point at the exit from the dinghy jetty. Customs clearance involves completing a declaration and then taking your agent by dinghy across the port to the Customs offices. No documents are issued at this stage and no fees paid. The Port authority charged a fee of US$32·50 for 5 days in late 2000. Your agent will ask you to pay US$38 to cover bank transaction charges. Fees are payable for each 5-day period, so if you stay 6 days you pay same as if you stay 10 days. Inform your agent at least 24 hours before you want to leave. He will bring a Master's declaration to be completed and will ask for one more crew list. Clearance is obtained by taking him back across the harbour to Customs after paying all your dues. Formalities may be waived if you just anchor overnight and stay aboard.

Health

There is risk of malaria during the wet season. A good pharmacy near PO offers advice. Flies can be a serious pest.

Facilities

The dinghy landing is at an old jetty, W of a moored ex-Dutch vehicle ferry on the N shore of the W arm, just in front of the Red Sea Province Police HQ (sign in English and Arabic conspic). Security is OK. The vile garbage tip is adjacent. Your rubbish will be picked over.

Money Banks exchange foreign currency. The black market rate is not competitive. Note that there are problems in Sudan cashing travellers' cheques issued by US banks. No cash advances on credit cards were available in late 2000. In an emergency you may be able to arrange for an advance via your agent.

Fuel Diesel is available and costs US25c–40c per litre depending on quantity. Visiting yachts almost invariably pay at the higher rate. Petrol can be obtained without bureaucracy at any service station. Gas is easily and cheaply available at the Shell depot, or via your agent. Paraffin and meths (denatured alcohol) are difficult to find.

Water costs US$10 per tonne but the amount is not metered and the charge is to some extent notional. Water quality can vary and it is delivered via a large diameter inflexible hose. Be prepared with reducers. Have lots of fenders ready for berthing. The wharf is not friendly to topsides. Laundry can be arranged at the hotels or via your agent. Charges are about US25c per piece.

Communications Internet access cost $2–$4 per hour depending on cheap rate times. There are several cyber-cafés. The biggest is in the block behind the PO. There is a good GSM phone network and Suditel card phones have been installed in the town. Postage rates are high, especially compared to Eritrea, being similar or higher than those in Europe. DHL is reputedly good. Allow 8 days for deliveries. No duty is payable for a package marked clearly for a yacht in transit but there may be a customs delivery fee.

Stores and other services Provisioning in town is good but expensive for anything imported. The markets have excellent local produce if you can bear the flies, worst in autumn. In the W part of the town there are workshops which can probably fix almost anything if you can find out who to go to. This is a good reason for using an agent. However, facilities for lifting out or docking yachts are close to non-existent, though something may be negotiable with the port authority if you are desperate and prepared to be crudely craned out in slings. There is a poorly stocked duty-free store in the port area with a small chandlery attached. It is a branch of Baaboud Trading and Shipping Agency of Jeddah. Spare parts and charts can be ordered and life-raft servicing arranged. Allow about a week for delivery.

Security within the port area is good, but it has been known for loose items to go missing from decks and dinghies. In the past, these have been returned to boats by the security guards. Good relations with them are worth the effort. A permit may be required for taking photographs. You will also need to apply for a permit for any inland travel; ask your agent where to get one.

General

Port Sudan is a scruffy, run-down place but full of life and offering a warm welcome, beggars and high dues notwithstanding. There is appalling poverty, occasioned by refugees from Sudan's never-ending civil war and border disputes, the regime's international pariah status and rapid population growth. This is Sudan's second largest city with a truly staggering population of 2–2·5 million. Given all this, and although we ended up liking the place, we had mixed feelings after our recent visit. The consensus is that Suakin is preferable because it is half the cost and less squalid. Port Sudan can be visited from there with no trouble.

If you call here you will usually find several Italian dive charter boats that operate in the Red Sea. They are extremely helpful people and we warmly recommend them to divers. Contact details are:

MV *L'Ernesto Leoni*, contact Renato Marchesan and Cristina Pulliero, www.ernestoleoni.com, ☎ +249 12341182

MV *Filicudi*, contact Mary Milanesi and Roberto Innocenti, ☎ +249 12341430
Email info@marybobocruise.com
www.marybobocruise.com

The Sudan coast N of Port Sudan

Admiralty charts
Large and medium scales 82, 3722
Small scale 158

US charts
Large and medium scales 62111, 62142, 62143
Small scale 62250, 62270

French charts
Large and medium scales 6979, 7112
Small scale 7112

See folded sketch charts with this volume

We continue with anchorages from S to N so if you're S'bound, read backwards! Where breaks in the offshore reef make for a fairly clear approach from seaward, relevant details of the channels are included. Positions are for the entrances to anchorages and are accurate to 200m or less, the idea being to get you within eyeball range; then you're on your own. On this section of the Red Sea W coast, it is particularly important to be cautious when transferring GPS-derived positions to charts. With the new BA158 that should not be the problem it was with the old BA138.

Note Marsa Gwiyai is the main Sudan naval base. Both it and Gwiyai Reef are restricted areas and not accessible. The Dutch East India Company's excellent chart atlas by van Keulen in the 18th century has a sketch map of an anchorage in the Port Sudan/Marsa Gwiyai area which was called Port Dradart. It seems most like Marsa Gwiyai in shape although the latitude is given as 19°45'N. If that's so it's interesting that in the end Sudan's main port ended up not at the ancient port of Suakin, nor at the early European choice of Port Dradart but at the previously little known or used Marsa Sheikh Barud, which is what Port Sudan was called before the British created the modern port in the early 20th century.

> **Warning** Although the most recent editions of charts from all the major hydrographic offices are reconciled to WGS84, this CANNOT compensate for shortcomings in the original 19th-century surveys. Note also that NO survey has ever been done on this coast to ANY WGS datum. Three points of navigational importance follow. First, always navigate, even with GPS, with circles of probable error (CEPs) of at least 2 miles, and at night at least 5 miles. Second, navigating by GPS alone in this area is stupid. Do not do it. Make daylight landfalls and use hand-bearing compass and radar, if you have it, to establish WGS84-to-chart error factors. Third, treat ALL waypoints with appropriate caution, including those given below, and NEVER assume that the waters between, close to or at waypoints are free of hazard.

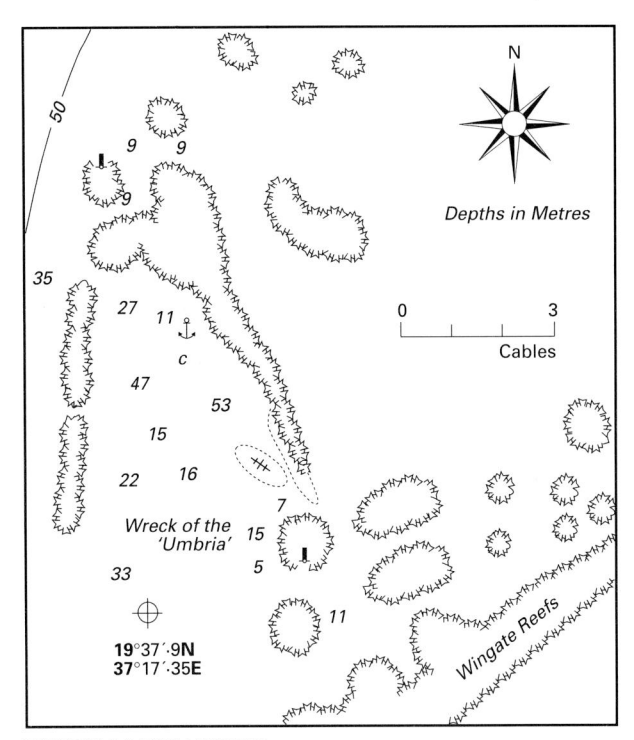

WRECK OF THE UMBRIA

Wreck of the *Umbria*

Approach

The approach from Port Sudan is clear passing E and N of Silayet South (Wingate Reefs South) light. This is really a day anchorage and cannot be recommended except for a short stay in calm weather. In theory anchoring in the area is prohibited, but there is tolerance of short stays for dives. The bottom is very uneven and anchoring, in 10–15m, coral, with reasonable shelter from the N through SE, where marked on the sketch, is a risk. Buoy your anchor and be prepared to dive to clear it. A possible alternative is to pick up a dive mooring a few of which are floating about ½ cable from the wreck. They aren't always free. Weekend changeover periods are best. This is reckoned to be one of the best wreck dives in the Red Sea. The Italian *Umbria* was on the way to Massawa when the Second World War broke out. The skipper, knowing British plans to seize his ship, scuttled it to prevent the cargo of war materiel falling into British hands. Cargo in the holds included munitions, nautical instruments, bombs and Fiat cars.

Sanganeb Reef

Admiralty chart 82, plan page 147

Approach

The reef is mostly steep-to, usually breaks and has a very visible light structure. The entrance is from the WNW with two beacons on the E part of the reef in line bearing 107°. This leads close S of a 3–4m coral head and N of a starboard-hand beacon. Alter to an E'ly course once past this beacon to stay in clear water. Alter S only when you can see the second gap

Red Sea Pilot

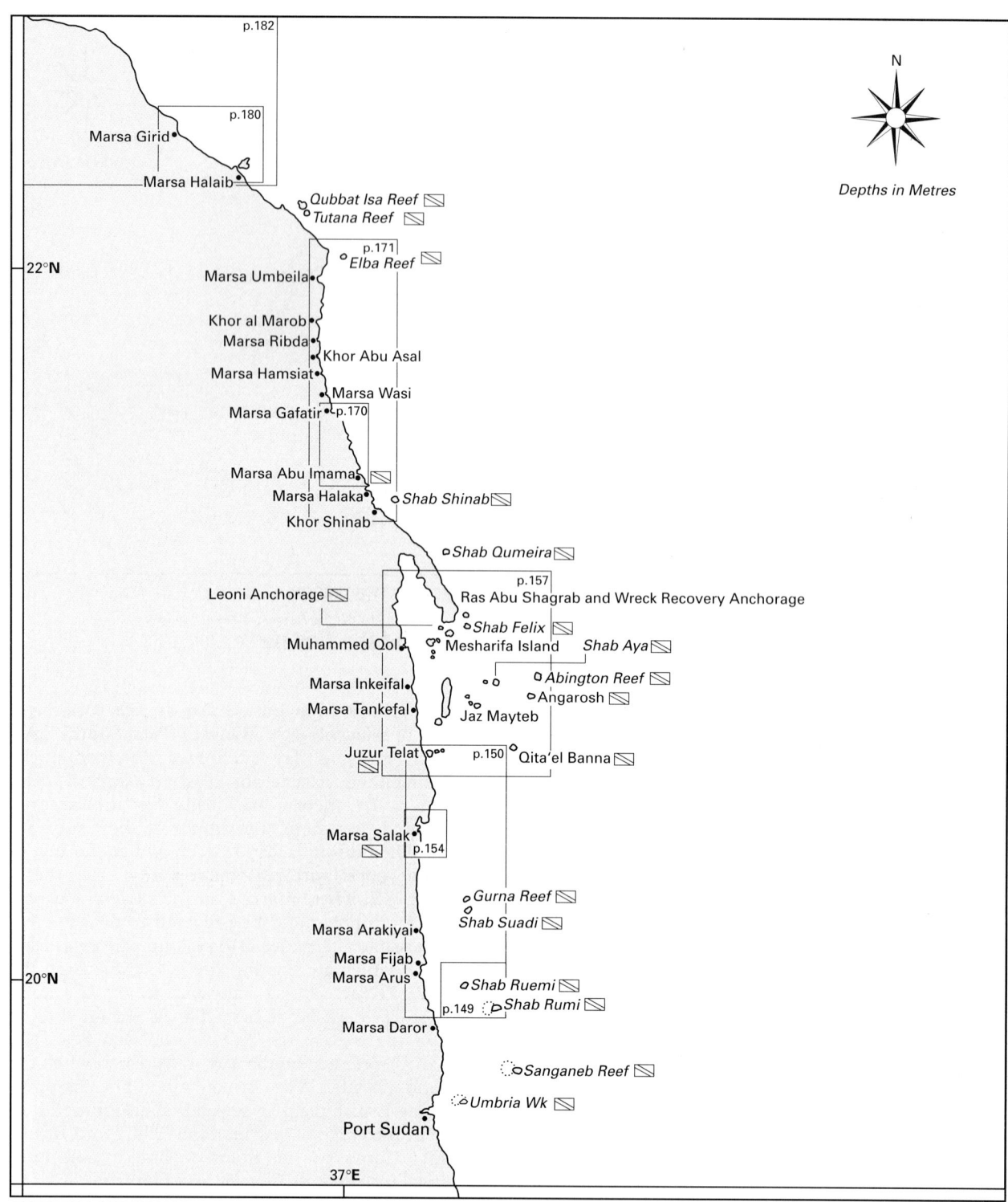

ANCHORAGES IN NORTH SUDAN

in the reef marked by stakes either side. The marks can be confusing, and eyeball navigation, with somebody in the rigging, is essential. The entrance between the stakes is about 50m wide. In the pass there are coral heads with as little as 4m over them. GPS positions in the area are reported to be 100–200m out on the chart.

Dangers
Beacons on the reef may be broken or missing. Depths may be less than charted. There can be strong currents in this area which usually set N and W, but directions can be reversed in winter.

Anchorage
12–15m, sand and coral, in the E part of the N lagoon. The anchorage has room for several boats and is very secure. There are several bommies

The Sudan Coast N of Port Sudan

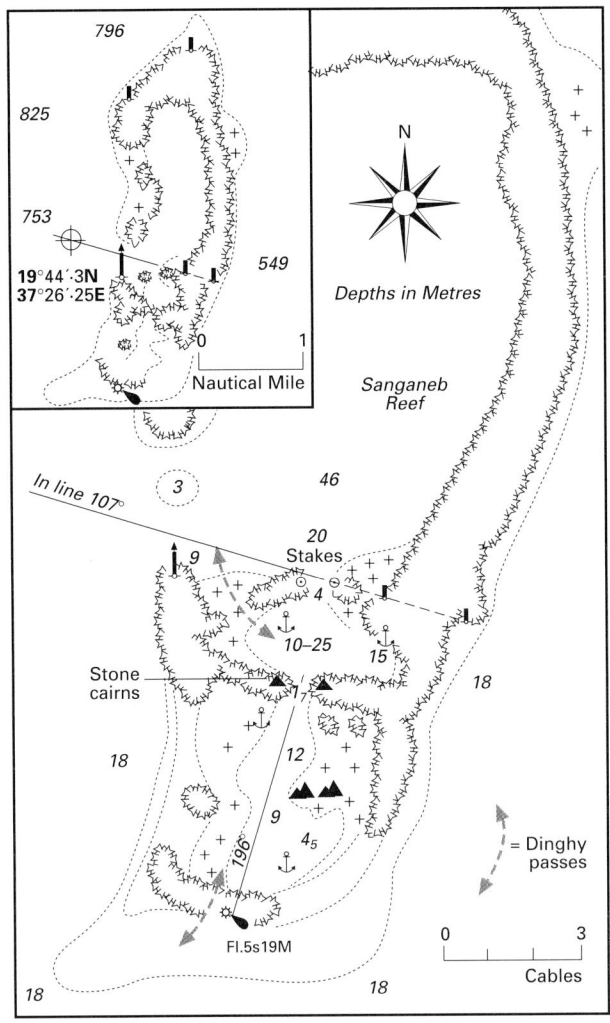

SANGANEB REEF

nearby with only 6m over them, so be careful to anchor clear. In any case buoy your anchor and be prepared to dive to clear your rode. An alternative anchorage is in the NE corner of the lagoon, though the same warning applies. If you're bold and draw less than 1·5m, you could essay the pass into the S lagoon, with the light bearing 196°, where there's a clear sand bottom, 4·5m, in the SE corner. This is a bit exposed in SE winds because the main reef to windward is submerged.

Mooring
If you only want to dive, charter operators recommend dropping back on a long line from the bollards on the end of the lighthouse jetty in the S part of the reef. Anchoring is near impossible here. The dinghy pass into the lagoon to the W of the lighthouse is marked by piles of stones.

General
There is a long wooden pier connecting the S lagoon with the light. On the W side at the lagoon end there is a small platform over the reef, connected to the pier by a gangway. The light, an elegant stone tower with several buildings, stands on an extension to the W of an octagonal stone platform towards the seaward end of the pier. A continuation of the pier, with the battered remains of a cart railway, extends S to the sea. The diving is excellent, both on the outer edge of the reef and inside the lagoon. Sharks, especially hammerheads, frequent the area. One can also visit the lighthouse and meet the keepers.

Marsa Daror
Plan below.

Approach
It is difficult to see the entrance if you look for it in the afternoon, a feature common to many of the anchorages on the coast in this region. The glare on the water caused by the sun in the W is dazzling. Crucially, Marsa Daror is a true reef *marsa*, so there's little above the water or on the featureless, low-lying land to identify it. Except in calm weather, the detached reef immediately SE of the entrance to Marsa Daror breaks, serving as a mark. Coming from the S in good light both channel and offshore reef are easy enough to see. Approach S of the off-lying reef from approx 19°50'·34N 37°16'·57E.

Anchorage
3m, sand, not far from the entrance. Water colour is deceptive, especially after rain. There is shelter from the sea but none from the prevailing wind. Depths may be less than charted as the *marsa* has silted up. The silting causes discolouration of the water, which is usually a cloudy light green and which sometimes makes things look a lot shallower than they are. The anchorage inside the off-lying reef is also sheltered from the sea but the holding none too good on rock and coral. Only tenable in light weather and uncomfortable in any swell.

For orientation and clarification of detail in the Inner Channel from the latitude of Shab Rumi to just N of the Taila Is, see the chart you'll find folded with this volume.

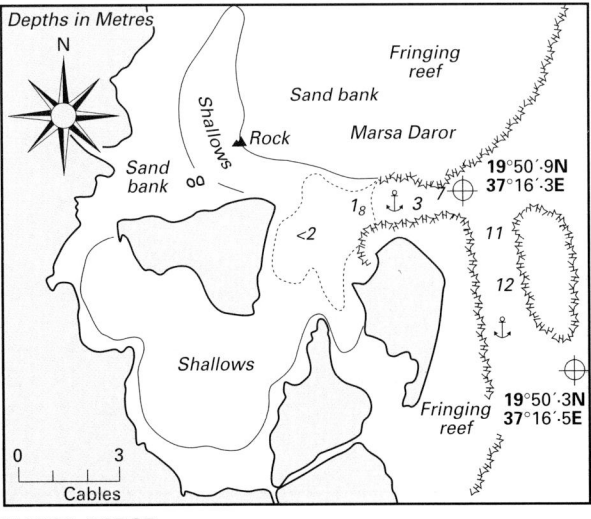

MARSA DAROR

Red Sea Pilot

Shab Rumi

Plan below

Approach

There are two openings through the W side of the reef. A reliable waypoint for the S entrance is 19°56'·42N 37°24'·18E. This channel was blasted through for Cousteau's project in ecologically more naive days. It is only 12–15m wide but clear with about 5m depth. The natural N entrance is wider but more encumbered with bommies which means you have to steer a sinuous course.

The reef complex is orientated NNW/SSE. The SSE tongue extends well S of the beacon.

Anchorage

Very uneven depths in the lagoon make this an iffy anchorage in all but settled weather. The N part of the lagoon can be crowded with dive-charter boats but there are moorings, which you may be able to use at weekends when the charterers go into port. If you do not use a mooring, buoy your anchor and be prepared to dive to get it back. Remember that the lagoon is deep. In light weather your movement on your own cable can wrap it definitively. The lagoon is not peaceful in any moderate to fresh wind.

General

This is something of a diver's Mecca. Jacques Cousteau lived underwater outside the reef for weeks at a time conducting the Precontinent experiments, filming and feeding sharks, and so on. The habitats, known as Conshelf II, are still visible and can easily be found by snorkelling, about ½ cable either side of the S entrance. Nearby is the grave of Alfred Hennebohl, one of Cousteau's co-divers, who died here whilst diving in 1977, aged 50.

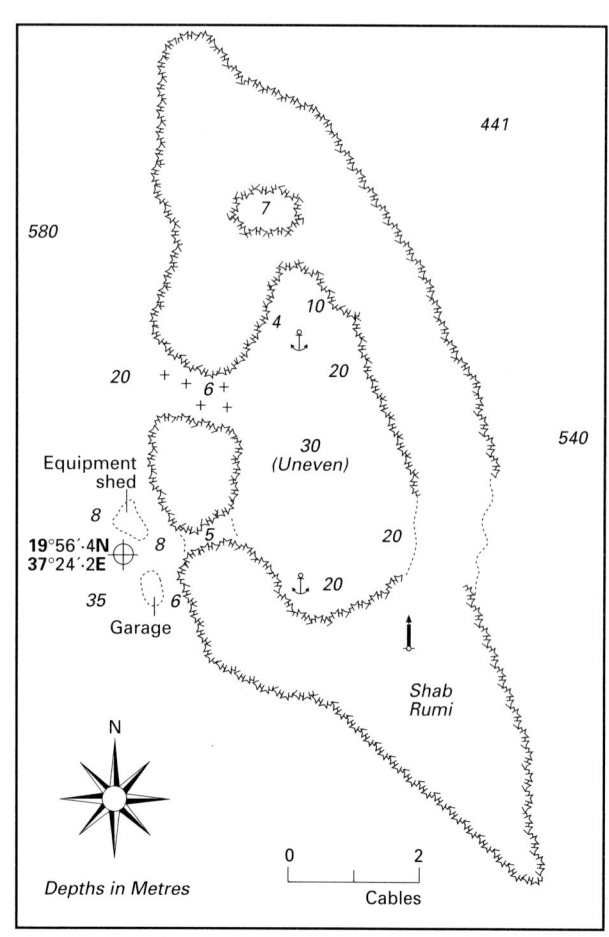

Abu Al Kizan

In the first edition we mentioned the possibility of a direct passage between Shab Rumi and Marsa Fijab. Some friends of ours on *Elsi Arrub*, while anchored with us in Marsa Fijab, said they'd just achieved this very feat. They must have been uncommonly lucky. Abu al Kizan, the huge reef complex between Marsa Fijab and Shab Rumi, is mostly uncharted. It has the form of a vast lagoon 30M long by about 15M wide surrounded by a cordon of reefs of varying width and continuity. The lagoon is deep and mostly clear of dangers likely to be a problem to yachts. Other friends and more recent Red Sea companions Larry and Tracey on *Freedom* tried to repeat *Elsi Arrub's* feat. They had an anxious time, nearly hitting the bricks several times and having to find their way into the lagoon and out again very gingerly. As far as we can see the only feasible passage enters E of Shab Ruemi (next entry) and exits WNW of it into the deep water channel that leads towards Marsa Fijab from seaward.

Shab Ruemi

Plan page 149

Shab Ruemi forms the NE 'wall' of the Abu el Kizan reef complex. There is a pass at 19°59'·38N 37°21'·11E into the main lagoon at the SE end of Shab Ruemi. The anchorage in about 15m, sand and coral, is just S of the 'point' SE of Shab Ruemi's 'claw'. The anchorage in the 'claw' is very deep. The only hazard on the direct route to the anchorage is a large, isolated reef patch in position 20°00'54N 37°19'·81E. It should go without saying that this complex should be penetrated in daylight only in good light conditions and a hand should be aloft as a spotter. With a good lookout it is feasible to exit WNW from Shab Ruemi to the Inner Channel.

Marsa Arus

Plan page 149

Approach

Come in to the tip of fringing reef S of the entrance on a course slightly S of W. Then stay about 100m off the S entrance reef. This will avoid the awkward rocks and shoals in the N approaches. The entrance is narrow, shoaling quickly once abeam of the land. Hold to the middle of the channel until off the spit SSE of the buildings, then favour the S shore to avoid the 1·8m shoal just beyond the point which thrusts S from the N shore.

Dangers

The fringing reef S of the entrance projects considerably further NE than charted and is

SHAB RUMI

The Sudan Coast N of Port Sudan

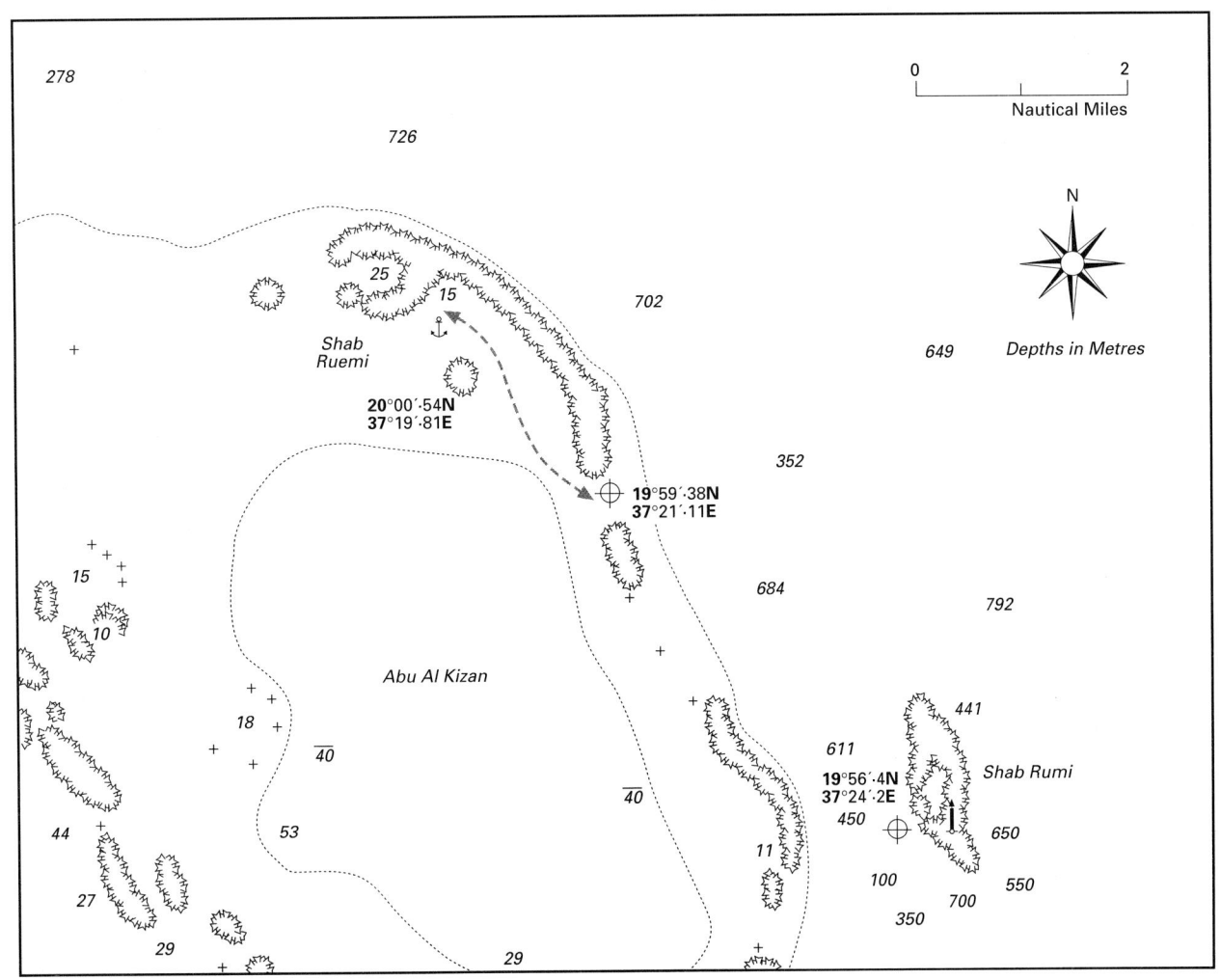

SHAB RUEMI

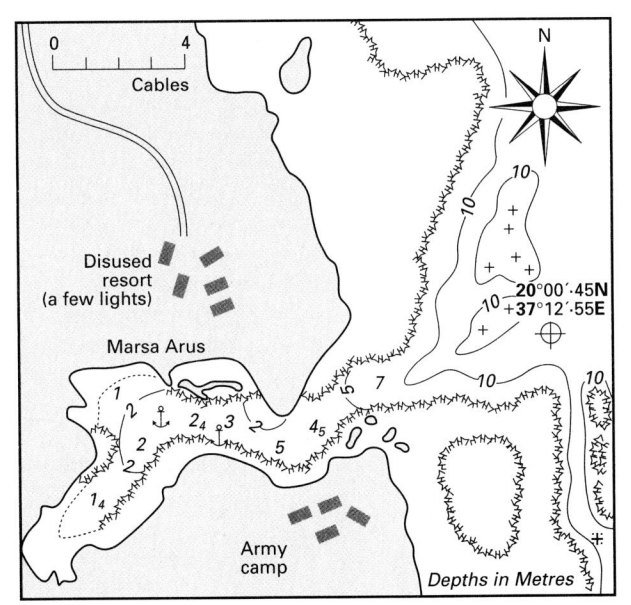

MARSA ARUS

bordered with coral heads. It should be given a wide berth. Note the rocks in the middle of the 'throat' of the approach, which is why you must come in on the S shore. The passage S of the spit is very narrow and the inlet shoals very rapidly further in.

Conspicuous
The remarkably ugly buildings are largely deserted except for a couple of caretakers who can be irksome. This was once, believe it or not, a Club Med diving resort. There is also a small army encampment S of the *marsa*. In every respect Marsa Fijab is to be preferred.

Anchorage
In 3–5m about 2 cables from the entrance, well protected but very confined. If you draw 1·5m or less, you can push further in.

Note Before learning our Red Sea lessons in 1993, we found ourselves still on passage N'wards in this area as darkness fell. An embarrassing error of judgement. We hadn't learned then how invisible *marsa* entrances can be in the afternoon sunlight, and had failed to find Marsa Daror. Fortunately, our fairy godmother, in the shape of Georg, skipper of the Austrian charter boat *Suhail*, saw our masthead light as we nosed into water shallow enough to drop our anchor, all our cable and much else besides, and came out in his inflatable to shepherd us into Marsa Fijab next door.

Red Sea Pilot

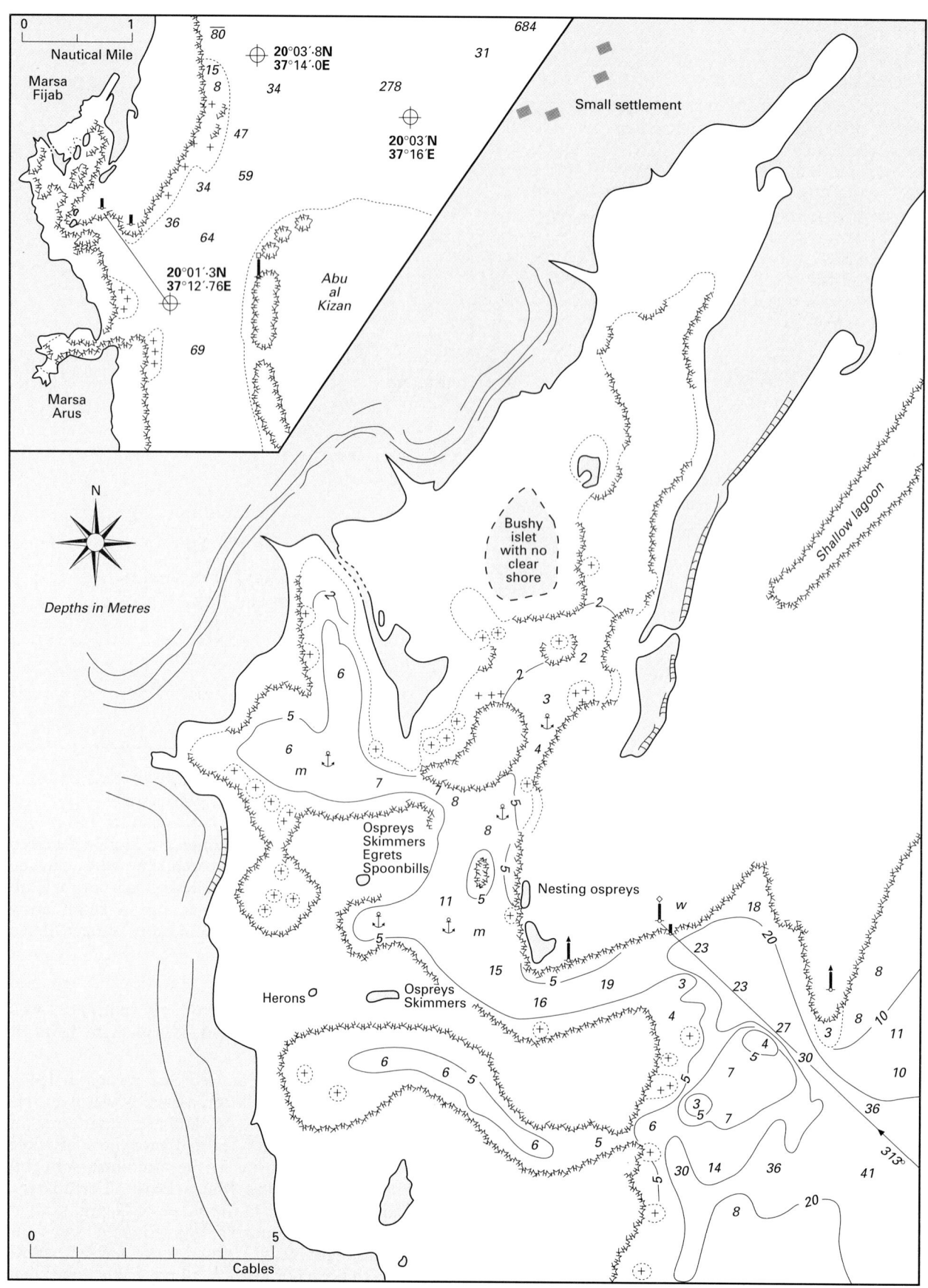

MARSA FIJAB AND APPROACHES

Marsa Fijab
Plan page 150

Approach
A position for the entrance, on the leading line is 20°01'·30N 37°12'·76E. The leading line is marked by a white stake in line with a diamond topmarked beacon and leads into the first leg of the channel on 313°. The turn to port into the channel is obvious, but hold well up to the marks and thereafter favour the N shore. Once past a starboard-hand beacon there is a small *cay* which marks the turning point for the anchorage. Low sandhills on the shore W of the inlet mark the edge of the slightly raised coastal plateau where the coast road runs S to Port Sudan.

Danger
A shoal area with patches of less than 3m extends 50m or more S of the starboard-hand beacon at the entrance.

Anchorage
Several anchorages are marked on the sketch. All have good holding and are well protected. There is a conspicuous shoal patch in the outer anchorage. Access to the NW arm is narrow but the channel is clearly visible in good light with a least depth of 6–7m. The bird life in this area is spectacular, the diving is good, and you will probably see camels being herded along the shore. Sharif, a local nomadic farmer, may paddle out to you on his battered windsurfer hull. He likes to talk, learn more English, French, German and Italian, and will eat anything on offer! He will probably offer to sell you eggs and chickens and arrange camel rides.

Back in the early days of modern Europeans in the Red Sea the Portuguese evidently used Marsa Fijab which they called Bahia de Fuca. There is a recognisable sketch map in an early collection of charts by British East India Company Hydrographer Alexander Dalrymple, which wouldn't be that misleading even now.

Note, again, GPS positions and charted positions often do not agree in this area.

Channel from/to seaward N of Abu Al Kizan
A large channel from seaward to the Inner Channel opens just N of Marsa Fijab. From the S a passage from 20°03'N 37°23'E leading W to 20°03'N 37°14'E will take you clear of all dangers and to within 1½M N of Marsa Fijab. From the N, stay E of 37°30'E until S of 20°40'N whence angle in slowly to 20°10'N 37°18'E, then 20°05'N 37°17'E when the port-hand beacon on Shab Shinnawi (20°05'·3N 37°16'·4E), an isolated S outlier to Shab Suadi will be ½M on your beam. Leave it to STARBOARD. It is an Inner Channel beacon. From there a course leading to 20°02'N 37°13'·5E will take you clear of dangers to the Inner Channel about 1¼M NE of the entrance to Marsa Fijab.

On older charts there is an isolated rock charted at 20°04'·52N 37°16'·45E (BA138) or 20°04'·5N 37°17'·625E (US62250). Electronic charting packages analyse this as an obstruction, shoaler than the surrounding depths, causing seas to break. They do not hazard any guesses as to the depth of water over the shoal. The dive charter operators in Port Sudan do not think this shoal exists. Older charts (both BA138 and US62250) also mislocate Shab Shinnawi, putting it 1M further N than it is. BA138 also puts it 1·3M too far W. This longitude error is a common feature of BA138 in this area. The rock may be a chart compilation lulu related to datum mismatches, positional errors in reports and, in the case of US62250, block correction of the chart to WGS84. But look out for it anyway, just in case!

Reef anchorages
The following reef anchorages are worth looking for in reasonably calm weather, since they are highly recommended for diving, but the positions given are GPS derived and, depending on the vintage of your charts, you may find plotting oddities. We have included all the information that we have, but stress that positions taken from GPS may plot as much as 2M out on the stretch of the coast between Marsa Fijab and Marsa Salak. Yachts sailing here using the old BA138 have found that GPS positions may need moving SW to match charted positions. The larger error seems to be in longitude. The new BA WGS84 chart series should be better. However, since the charts are small scale (1:750,000) reconciliation has been effected by satellite photogrammetry and block datum correction. No resurveying has been done so treat both plotting and any positions you derive with caution when in pilotage waters. Scales have been omitted from the sketches where we lack any precise information. See our sketch chart for an overview.

Note Reef shape, position and beacon placement hereabouts does not well match many official charts. Our version is a 'best fit' between our own observations, the observations of dive charterers we've talked to and official data.

Shab Suadi
Plan plan 152

Approach
The anchorage is at the N end of the reef complex. Charted and GPS positions in the area differ markedly. The beacon on the S end of Shab Suadi is just N of 20°08'·07N 37°17'·25E. From there follow the edge of the reef complex N. Possible anchorages are shown in the sketch. They can be rolly overnight. The attraction of Shab Suadi, though only possible in fair weather, is what's called the Toyota wreck. Inexplicably in the late 1970s the *Blue Bell* hit this reef, was holed, rolled over and sank. It was full of Toyota vehicles. So the fun is diving over the wreck and amongst the one or two vehicles scattered around it. Word has it you can tie your dinghy off on a truck. Note that the wreck is deep, especially at the stern (70–80m), though the bow area is accessible (10–25m).

Red Sea Pilot

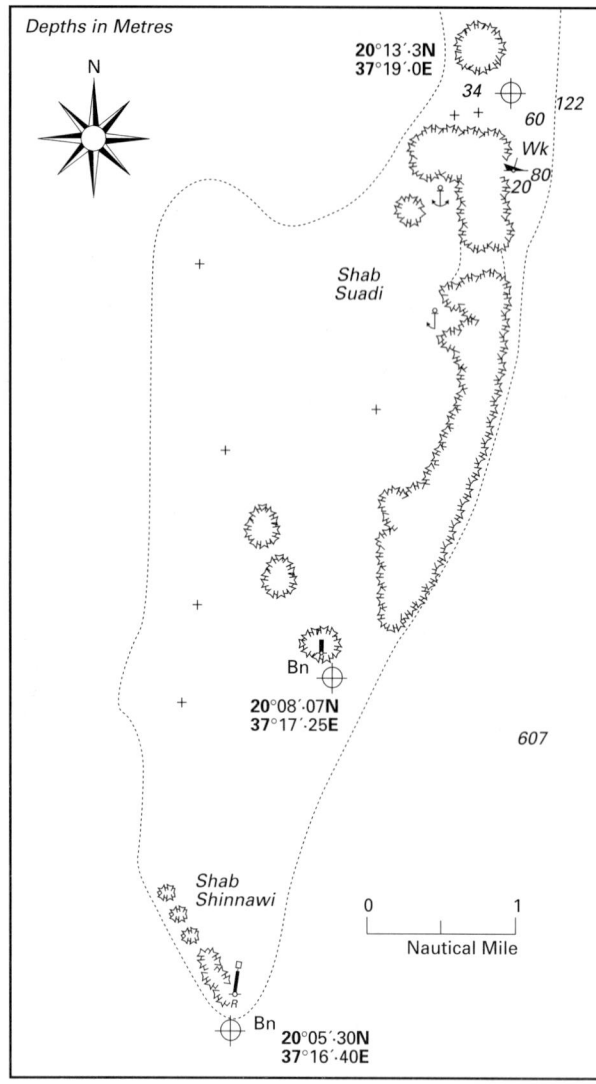

SHAB SUADI

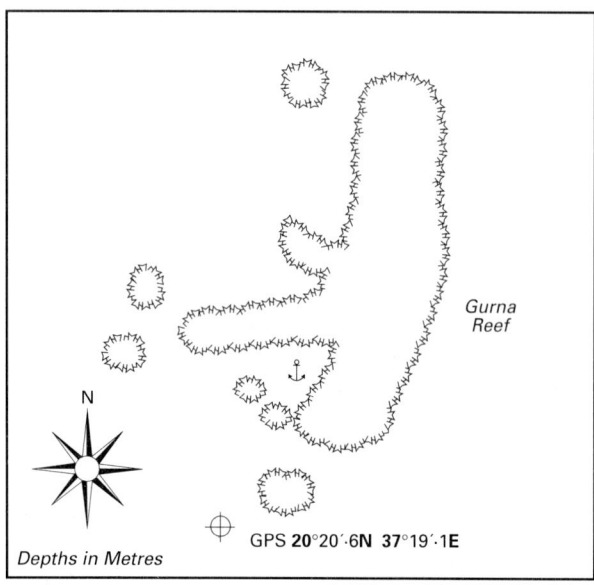

GURNA REEF

Gurna Reef
Plan opposite

Gurna Reef is the isolated, unnamed reef mischarted in 20°19'·6N 37°18'·6E. The correct position is on the plan. This disparity between charted and recorded positions is typical. The reef is 5½M E of the beacon marked on the chart on the fringing reef 6M S of Marsa Salak. The anchorage is the usual sand and coral. You have to search for suitable depths and a clear sand patch. This is for divers only and then only in fair weather. The coral and fish life are highly rated.

Inner channel continued

On your way N from Marsa Fijab you may see Marsa Darra on the chart. This is a shallow, small boat haven, accessed through breaks in the fringing reef, hereabouts very discontinuous. Marsa Darra has a dilapidated jetty by some buildings associated with what were once relatively prosperous salt works and which still produce salt, piles of which you can see glinting inland.

Marsa Aweitir lies 2½M N of Marsa Darra. The entrance is very open and if there's any sort of swell in the offing it runs right in from either side of Shab Shinnawi. The *marsa* looks shallow but we learned locally that this was an effect of rain run-off. It was the rainy season and there were whopping thunderstorms in the hills daily. As a result the *marsa* looked pale green, as if shoal, although apparently there are depths of 15m quite a way in. It is certainly used by fishermen whose GRP boats you will see on the shores. There are fresh water wells not far inland.

Marsa Arakiyai
Plan page 153

Approach

Leave the off-lying reef marked by a starboard-hand beacon to port. The apparent contradiction is because the beacon is actually for the main Inner Channel to Port Sudan. Then alter course SW for the beacons marking the channel through the fringing reef. The port beacon is a 'T' indicating you can leave it to either hand. Leaving it to starboard heads you back S to the Inner Channel through a narrow pass. The channel to the *marsa* entrance leads W, N then W again. On the last leg, past the huts, aim for the mid-point of the gap between the sandspit to the N and the islet to the S. You should have a least depth of about 12m in mid-channel with 8m close to the spit. Once inside the bay, keep a good lookout for bommies, especially SW of the entrance. The coral has grown since the survey and is not always easy to see. There's a pretty clear pool just inside the entrance to the N.

Conspicuous

The buildings show well, but nothing like as well as the old radar truck, now a chicken coop.

Above Mina Salalah, Oman. Cranes conspicuous
Top left Nishtun port
Below Al Mukalla, old dhow harbour and yacht anchorage
Bottom Al Mukalla, Khalf Harbour

Above Ras Fartak, S approaches to Nishtun

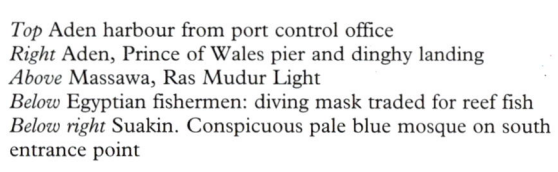

Top Aden harbour from port control office
Right Aden, Prince of Wales pier and dinghy landing
Above Massawa, Ras Mudur Light
Below Egyptian fishermen: diving mask traded for reef fish
Below right Suakin. Conspicuous pale blue mosque on south entrance point

Top Jabal Zuqar, S anchorage
Right Sanganeb reef, preservation notice
Below Ras Imran, approaches

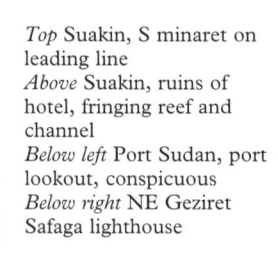

Top Suakin, S minaret on leading line
Above Suakin, ruins of hotel, fringing reef and channel
Below left Port Sudan, port lookout, conspicuous
Below right NE Geziret Safaga lighthouse

Above Marsa Arakiyai, view from anchorage, E across spit
Below Marsa Umbeila, tomb N of entrance
Top right Marsa Abu Imama, view S from North Bay
Right Traditional Yemeni trading craft
Bottom right Marsa Alam, anchorage from police HQ

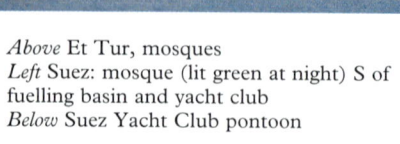

Above Et Tur, mosques
Left Suez: mosque (lit green at night) S of fuelling basin and yacht club
Below Suez Yacht Club pontoon

Top left Marsa Thelemet, 2nd pair of leading marks (with camels)
Top Port Fouad yacht club, palm trees conspicuous
Left Ismailia yacht club during refurbishment
Bottom Al Mukalla, quay E of anchorage

Left Port Fouad, twin minarets conspicuous
Above Fishermen or pirates?
Bottom Oman. Mina Salalah anchorage

The Sudan Coast N of Port Sudan

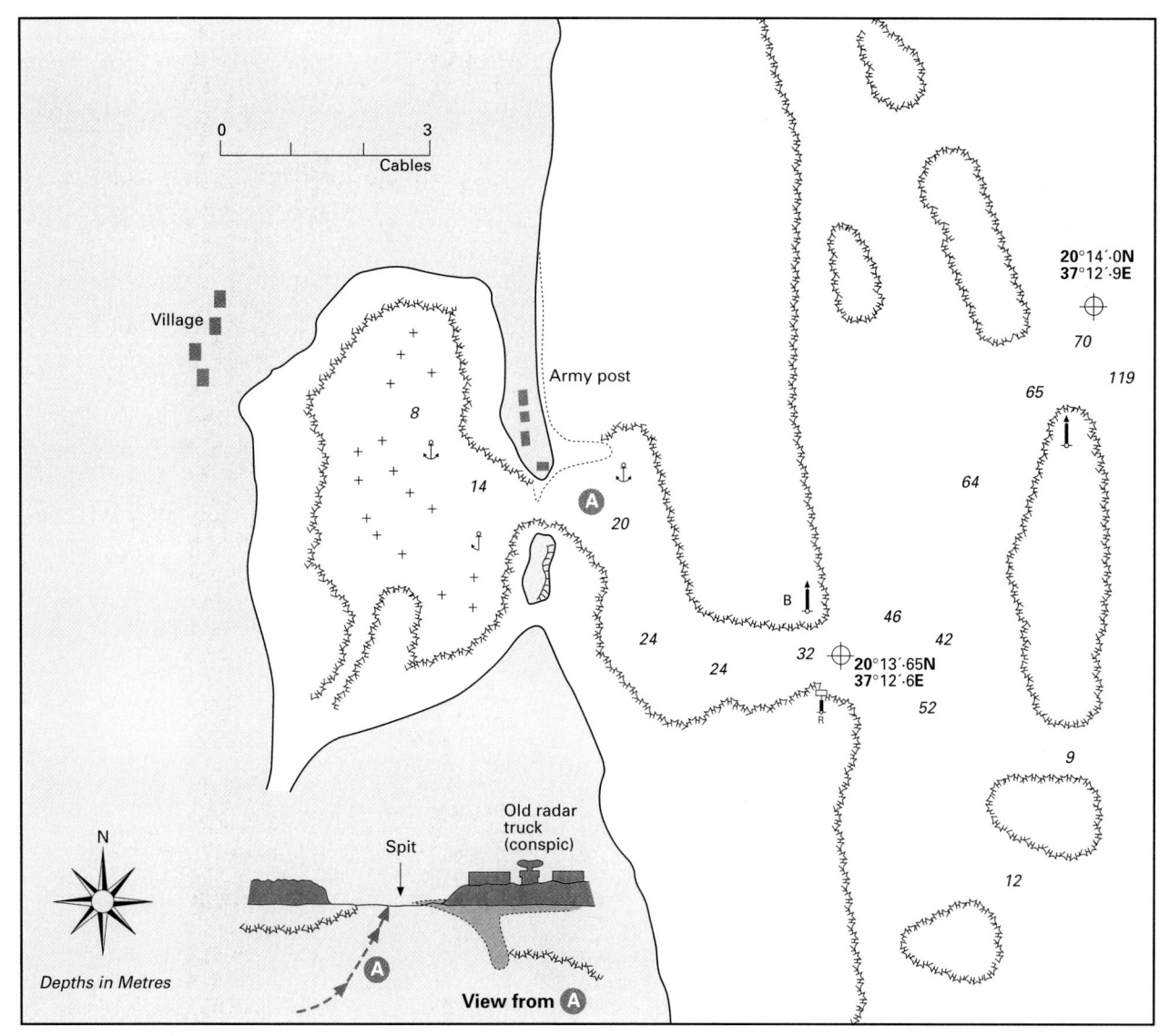

MARSA ARAKIYAI

Anchorage
8–14m, sand with some coral, inside the bay, with the buildings just N of the tower roughly abeam. An alternative anchorage outside the *marsa* is deep with good holding in mud. The *marsa* offers excellent shelter. There is a thriving fishing village inland, its boats pulled up around the spit. The soldiers are friendly and will ask to see your papers. They also like to trade.

General
This was once called Port Salvadora. The name is noted on one of the few remaining copies of the first detailed, scientifically surveyed chart of the Red Sea. The survey was done by Commanders Elwon and Moresby of the Indian Navy but the name marks both the early penetration of the Red Sea by Portuguese mariners and their huge influence on the history of navigation.

Marsa Salak Complex
Plan page 154

Approach
Coming from the S, you will find a beacon on an apparently isolated reef in approx 20°21'·8N 37°10'·6E. This is incorrectly positioned on older charts. BA138 has it at 20°20'·2N 37°12'·3E and US62250 puts it at 20°20'·2N 37°13'·6E. Give it a berth of at least ½M to seaward to avoid the shoals around it, especially to the SE. Fishing boats pass inshore of this reef, but we weren't tempted to try. In about 20°25'N, look for the leading beacons in line on 350°. They will take you W of Shab Salak and the small, mid-channel reef with its 'T' beacon.

If you intend by-passing this area, a passage between 20°24'·75N 37°14'·17E and 20°26'·53N 37°14'·51E passes E of Shab Salak and W of the ill-charted reefs of the barrier reef. From the last waypoint steer to pass the wreck (boiler only) in longitude 37°15'E.

153

Red Sea Pilot

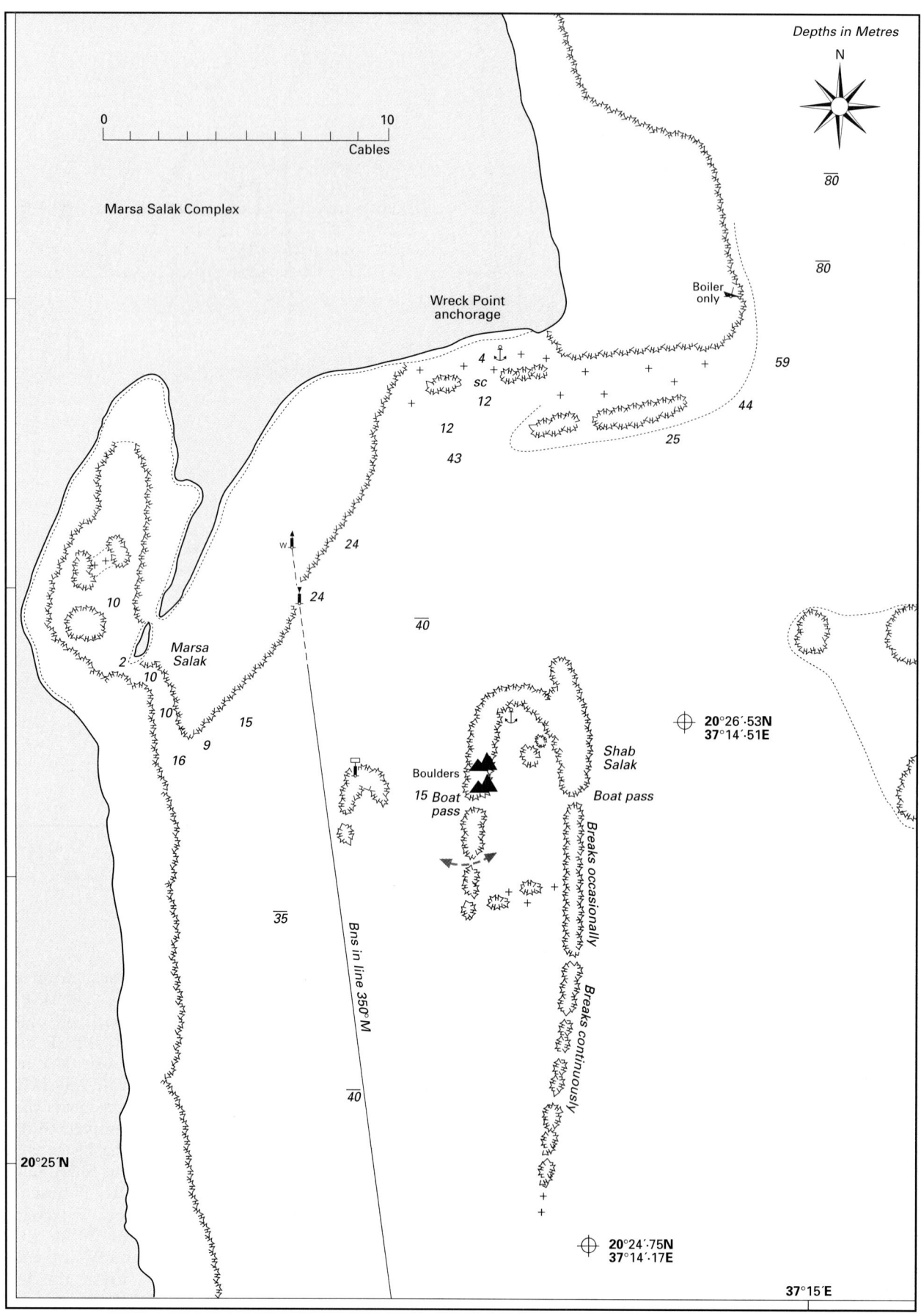

MARSA SALAK COMPLEX

The Sudan Coast N of Port Sudan

Conspicuous
Heading N, the beacons and leading marks are easy to spot. Heading S the boiler, all that remains of the wreck of an old tug on the reef N of the anchorages, stands out.

Anchorage
The anchorages are tricky and need careful eyeballing. Go carefully and use your GPS warily. There are three possibilities, none terribly inviting.

Marsa Salak Enter on approx 350°M from near 20°26'·5N 37°12'·7E. There are depths of 10m in the narrow channel, but only 2m at the neck of the tight, narrow turn round the S of the sandy islet closing the inner *marsa*. Once inside there is a good anchorage in 10m, sand. Only for small, bold boats. Fishermen work from here and will inevitably try to score some goodies.

Shab Salak Boathook shaped, this needs very good visibility. The only safe entry is from the W, about 4–5 cables SE of the T-topmarked beacon, through a clear gap N of two large off-liers at the end of the W tip of the 'boathook'. The opening N of this, near some boulders, is dangerous and for small boats only. The same is true for an entrance on the E side more or less opposite. Fishing boats out of Marsa Salak use both openings but don't be misled. Don't try to approach directly from the S. The 'opening' of the boathook is effectively closed by two large off-liers and some bommies. The E leg or handle of the boathook extends S and slightly W for over 1M to around 20°25'N, though not all of it breaks. The reef itself is dying and 30%–50% is already dead.

Wreck Pt anchorage Pass either W or E of Shab Salak. The leading marks (front: black triangle point down, back: white triangle point up) on 350° take you clear W of the small, beaconed reef (red T topmark) in 20°26'·3N 37°13'·5E. Whence follow the fringing reef E of Marsa Salak. You have to feel your way through a veritable jungle of bommies and off-lying lines of reef to find 4m, sand and coral patches in the lee of the land. Finding a clear space tries the nerves.

Channel from/to seaward E of Marsa Salak
This break in the barrier reef is some 5½M wide, N–S. Coming from the N you need to watch out for Abingdon Reef – the light is often not working, unmarked Angarosh and Merlo Reef (20°50'·65N 37°25'E) and especially the more isolated and also unmarked Qita el Banna (see below for more information). Leaving to seaward, 20°30'N 37°19'E is in the middle of the opening from where any course between NE and SE takes you quickly to the open sea.

Juzur Telat (Taila Is)
Plan page 156

Approach
These are three sandy *cays* on an extended coral platform 12M N of Marsa Salak. The approach up the Inner Channel is clear of danger. You'll pass a doubled starboard hand beacon on the fringing reef at Ras Medar, 24°34'·85N 37°14'·4E. Once within the lee of Juzur Telat, the big reef complex E of the Taila Is, the bottom gets very uneven, though we found no soundings below 12m on a course direct to Marsa Salak from the middle Taila I. The passage between the islands and the mainland is beaconed as shown and has a least depth of 9m.

Dangers
Beware of the isolated coral head S of the W tip of the W'most island.

Conspicuous
There is a grove of trees with a conspicuous palm on the shore of the small *marsa* E of the pass.

Anchorage
6–10m, sand and coral, S of the islands, W of the coral patches, well protected from N'lies and with adequate holding in sand and coral. It pays to hunt around for a good spot since the quality of bottom varies a lot. There are three N wind alternatives. The small *marsa* on the mainland shore is quite well sheltered if you tucked in, though may be more shoal than we show since we only poked our noses in. However, it was given a favourable mention in the 1932 edition of the BA Pilot, which says the holding is good in mud. The half anchors on the sketch mark the other three. Fishermen shelter under the E Taila I and the access looked clear. The anchorage in the 'V' of reef NNE of the middle Taila Is in 20°38'·75N 37°14'·12E comes recommended by one of Port Sudan's most experienced dive charterers. He also recommends the inverted 'V' at 20°39'·15N 37°14'·15E.

In a S wind you can anchor N of the islands. The anchorage shown in approx 20°38'·5N 37°13'·6E has been tried and found OK. The same remarks apply about holding and finding a good spot. Depth unknown.

General
These islands, at one time called Jezair Dabadib, are small but prettily set with good shelling and interesting wildlife. There is clear water and good snorkelling but you may see sharks. This isn't a bad place to wait for a weather break if you're going N.

Channel from/to seaward N of Juzur Telat
The obvious point to make, though it needs making since it seems that with GPS people have thrown navigational good sense to the winds, is that given potential errors, yours and the chart's, night passages through the reefs are stupid. Avoid them. GPS is an aid to navigation, not a substitute for it.

Qita el Banna is the obvious nasty in the outer approaches. Head for 20°40'N 37°25'E to keep clear

Red Sea Pilot

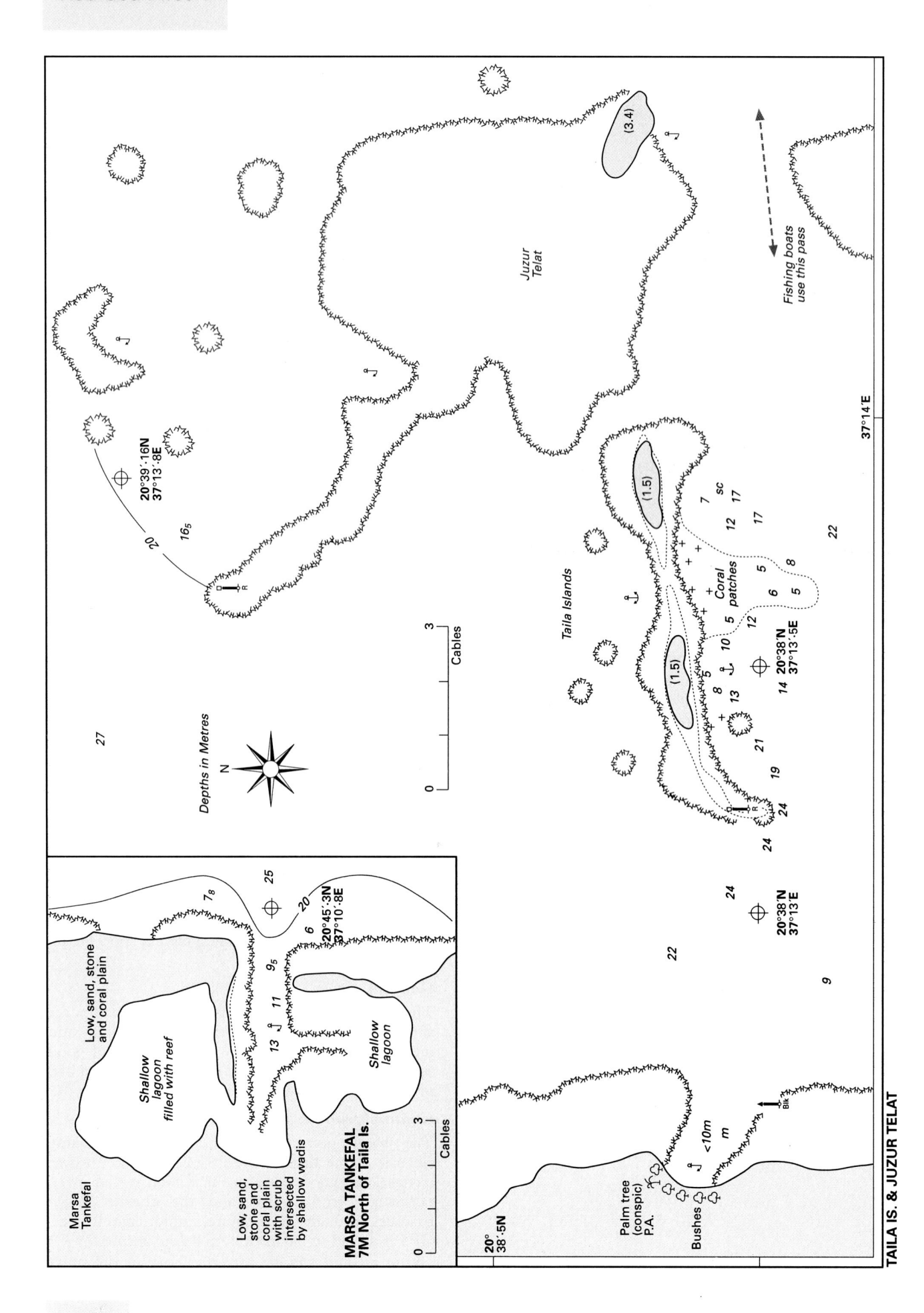

TAILA IS. & JUZUR TELAT

The Sudan Coast N of Port Sudan

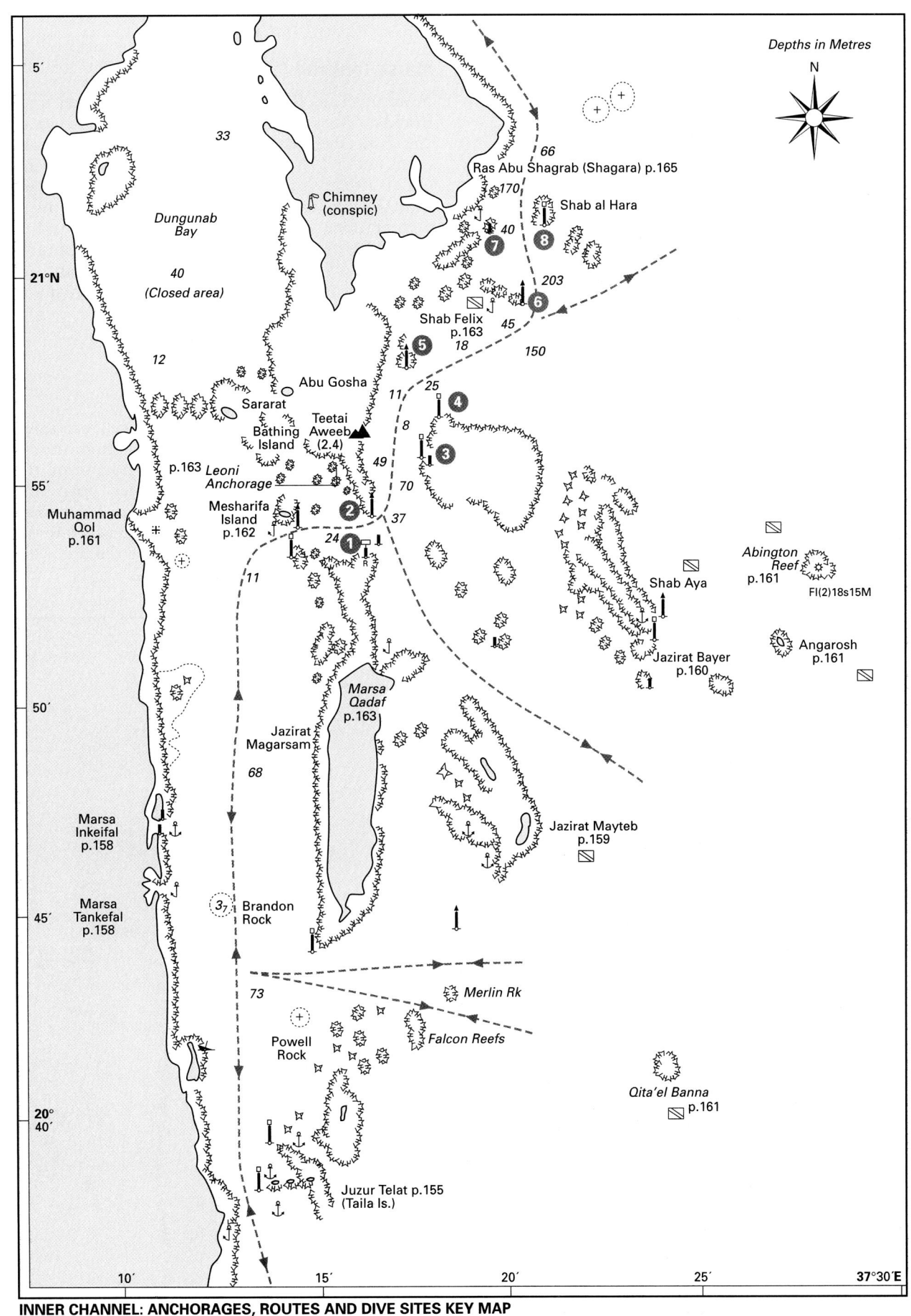

INNER CHANNEL: ANCHORAGES, ROUTES AND DIVE SITES KEY MAP

Depths in Metres

- Ras Abu Shagrab (Shagara) p.165
- Shab al Hara
- Chimney (conspic)
- Dungunab Bay
- 40 (Closed area)
- Shab Felix p.163
- Abu Gosha
- Sararat
- Bathing Island
- Teetai Aweeb (2.4)
- Leoni Anchorage p.163
- Muhammad Qol p.161
- Mesharifa Island p.162
- Abington Reef p.161 Fl(2)18s15M
- Shab Aya
- Jazirat Bayer p.160
- Angarosh p.161
- Marsa Qadaf p.163
- Jazirat Magarsam
- Marsa Inkeifal p.158
- Marsa Tankefal p.158
- Jazirat Mayteb p.159
- Brandon Rock
- Merlin Rk
- Falcon Reefs
- Powell Rock
- Qita'el Banna p.161
- Juzur Telat p.155 (Taila Is.)

157

if you are coming in from seaward to the S. To pass from that position S of Merlin Rock and N of Falcon Reefs, head for 20°43'·35N 37°17'·6E then alter due W until clear of Jazirat (Jaz) Magarsam to the N with the port-hand beacon to STARBOARD. Again, it marks the Inner Channel. Continue due W to clear Powell Rock to the S.

Coming from seaward to the N start at 20°45'N 37°25'E to stay N of Qita el Banna, thence to 20°44'·25N 37°18'·35E to pass between Merlin Rock and the beaconed off-lier SSW of Jaz Mayteb. Thence head towards 20°43'·35N 37°15'E S of the port-hand beacon on the reef S of Jaz Magarsam.

An isolated, uncharted danger with 2m or less water over it was reported to us in approx 20°42'N 37°14'·3E in the SE approaches to Jaz Magarsam. Positional information was not precise and this may have been an unexpected encounter with Powell Rock.

There is an additional route for those coming from seaward which takes you into the Inner Channel N of both Jaz Mayteb and Jaz Magarsam. From 20°50'N 37°20'E it leads NNW to the sharp turn in the Inner Channel track at the S tip of Karair Berer at 20°54'·2N 37°16'·6E. There is one stake on an isolated reef at 20°51'·6N 37°19'·2E. Leave it to starboard going N, port going S.

Inner channel continued

Marsa Tankefal (Little Inkeifal)

From the Taila Is the S approach is clear apart from Brandon Rock, with 3·7m of water over it, about 2M E of the entrance. Further E Jaz Magarsam (Mukawwar I) is a bare, rocky sandstone tableland like an aircraft carrier which shows up well on radar. Little Inkeifal is both more spacious and very much prettier than we'd previously supposed. There is reasonable protection but the reef is encroaching on the inlet and there is little swinging room. An interesting day stop. There are shallow lagoons N and S of the anchorage.

Marsa Inkeifal

Admiralty chart 3722, plan below

Approach

There is a clear approach, with a well beaconed entrance. However, if approaching from the N beware of the shoal that extends S and E from the starboard-hand beacon. Preferably, open the entrance right up before altering course for the run in. This is straightforward as long as you keep to the middle of the channel. A stake marks the tip of the reef on the S side of the entrance.

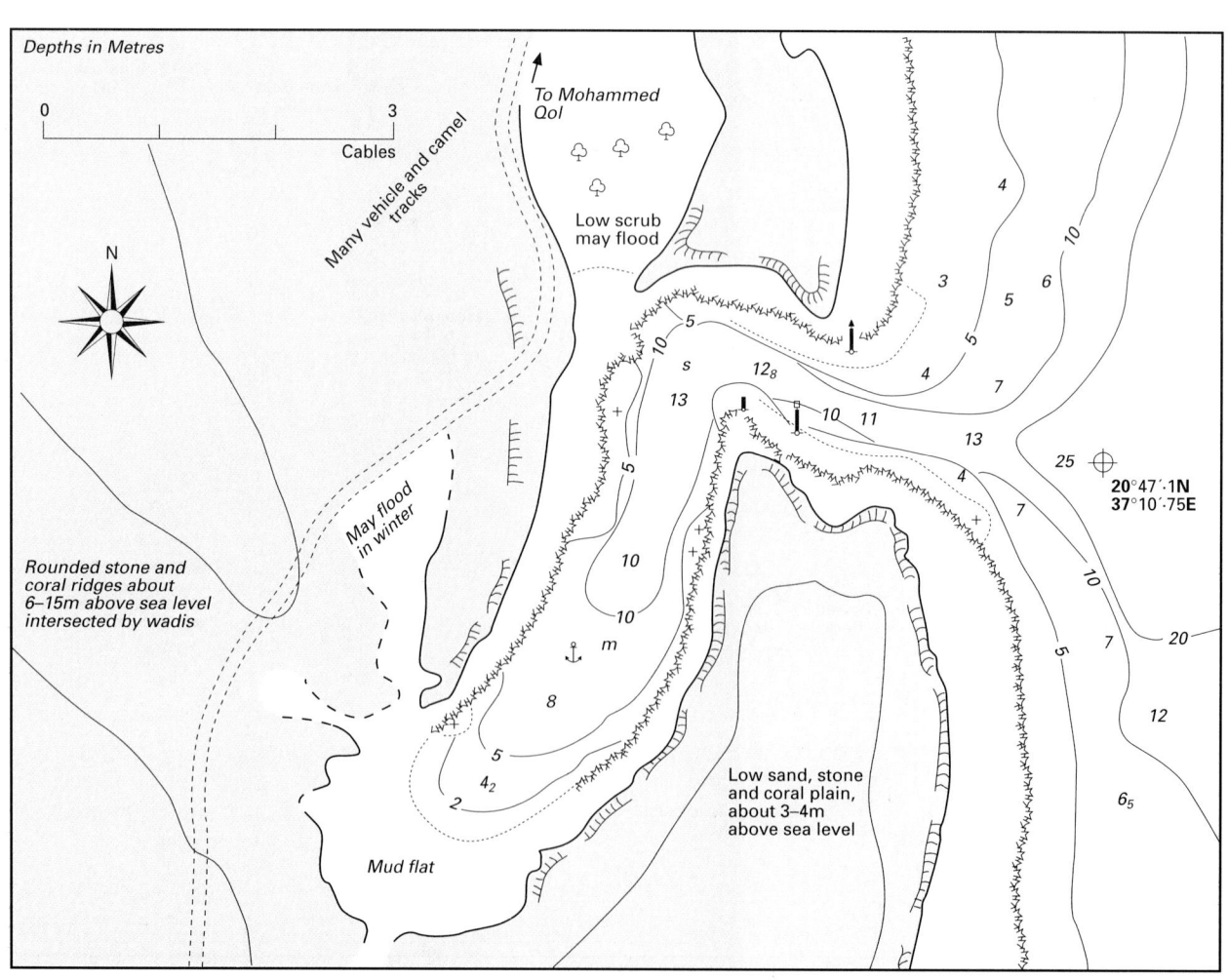

MARSA (KHOR) INKEIFAL

Anchorage

5–10m, mud, at the S end or 10–15m, sand, at the N. Choose whichever suits the conditions, remembering that dramatic reversals of wind direction are not uncommon, so that midway between the shores is best. The low cliffs on the seaward side of the *marsa* add to the sense of security in this excellent anchorage. It seems to us that Dungunab Bay in some way enhances the N wind, making it noticeably fresher hereabouts.

General

There is interesting bird life in the lagoon N of the anchorage, though it is often dry in winter. Boats are sometimes asked to produce their papers but no problems with the authorities have been reported. The entire area makes for good walking and there is a very pleasant coastal walk to Marsa Tankefal (Little Inkeifal) and back. This is a highly recommended anchorage. The times we've been here the difficult thing has been deciding to leave.

Reef anchorages and offshore reef diving

Jazirat Mayteb

Admiralty chart 3722

Approach

From the SW, the approach is clear up the Inner Channel, leaving the beacon on the reef S of Jaz Magarsam to port and the beaconed, isolated reef SSW of Jaz Mayteb to port or starboard depending on which anchorage you've chosen. From the SE, leave the beacon on the isolated reef SSW of Jaz Mayteb starboard. Coming from the N the approach E of Marsa Qadaf looks clear and is beaconed for exit to seaward, but note the unsurveyed stretch you must pass through. Choose good light.

Dangers

The N part of the channel between Jaz Magarsam and Jaz Mayteb is largely unsurveyed, though a

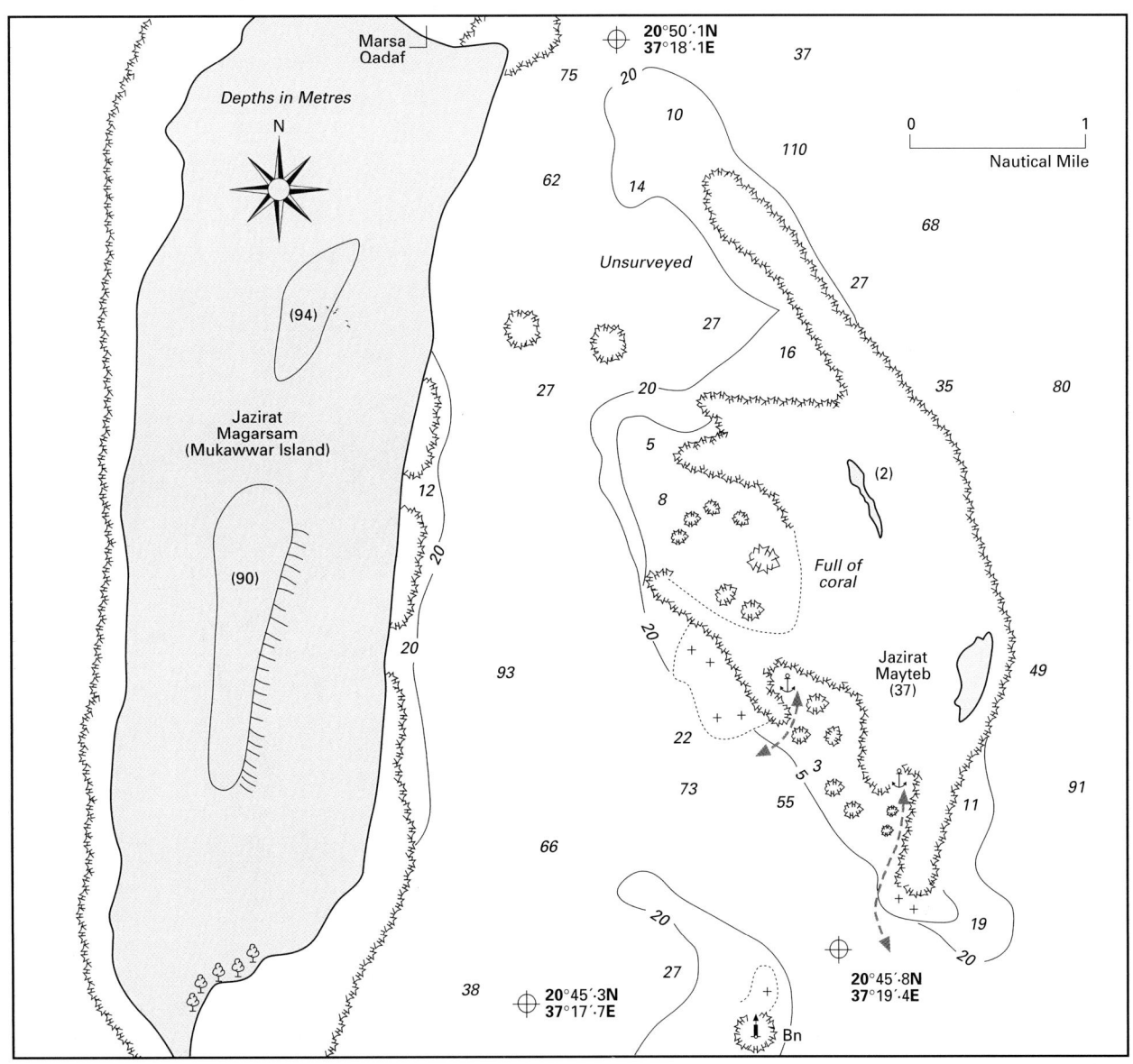

JAZIRAT MAYTEB

Red Sea Pilot

couple of isolated reefs are charted in the N end. W'ly currents can be strong hereabouts.

Anchorages

Two anchorages are shown on the sketch-map. Once you are closing Jaz Mayteb it is a matter of feeling your way in. Good light is needed. There is not much swinging room anywhere and a stern anchor might be wise. Whatever, buoy your hooks. To find the S anchorage, feel your way along the W edge of the finger of reef pushing S from Jaz Mayteb. Be wary of the off-liers off its tip. Once in it is secure and you're well protected from sea. The entry to the N anchorage is round the tip of the arm of reef to the S of the anchorage as shown. Depths decrease rapidly.

Reports from dive-charter boats that have used these anchorages emphasise the excellent diving, with turtles and a wide variety of marine life.

Jazirat Bayer (Shambaya I)
Admiralty chart 3722, plan below

Approach

Approach the anchorages in the channel and lagoon NW of the island via the beaconed entrance in the S. Only those with good eyesight, strong nerves and much bravado would hazard the approach from the N. From the SW, leave the stake on the off-lying reef S of the island to port and pass between the beacons into the channel. Abington Reef, which should be lit but is probably not working, Angarosh, and Merlo Reef (unnamed on the BA and US charts) are in the approaches to the channel from the E.

Dangers

There is a bank with coral outcrops on its S end lying parallel to the marked channel and ending four cables NE of the entrance. There is unsurveyed

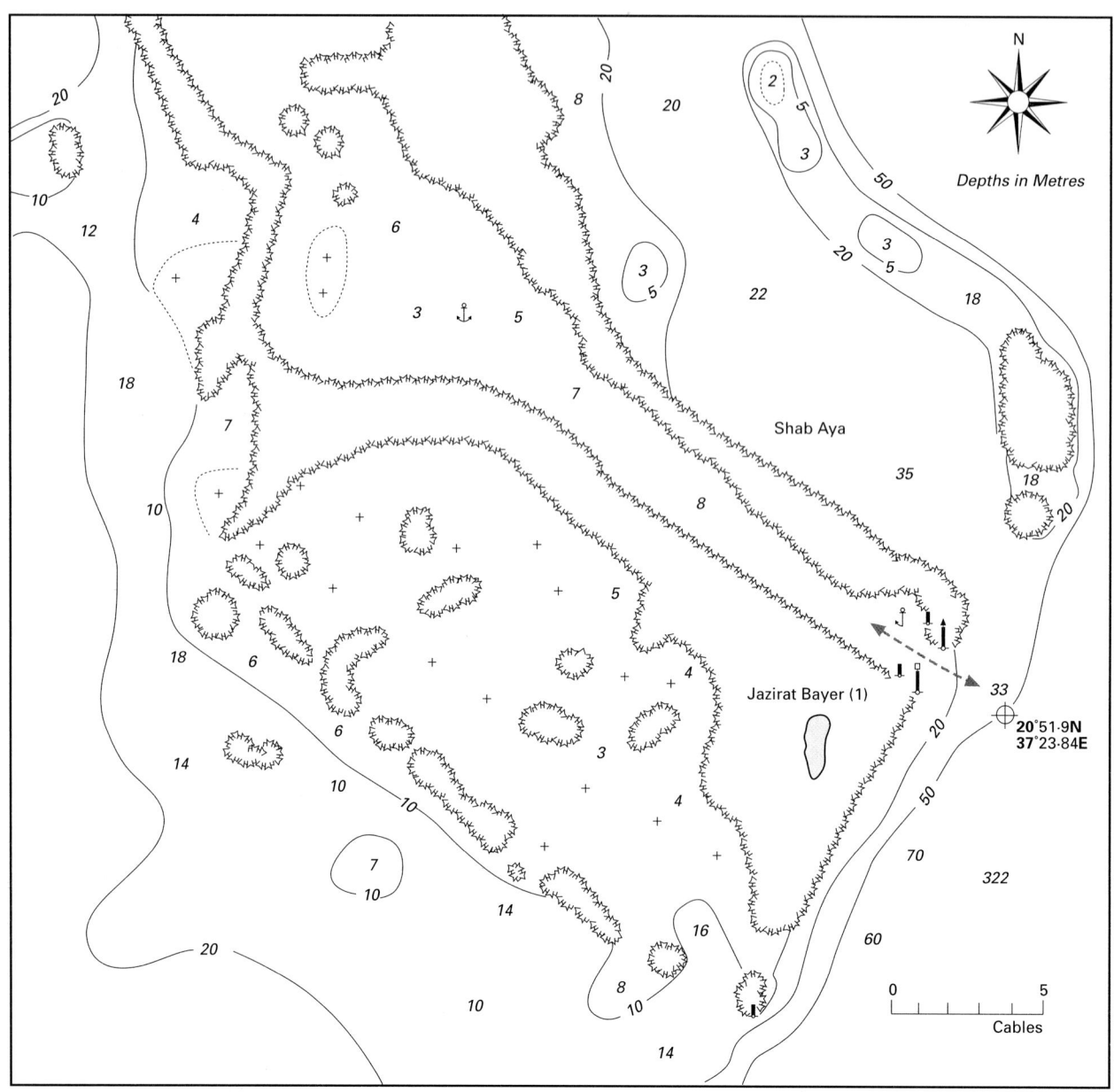

JAZIRAT BAYER AND SHAMBAYA

The Sudan Coast N of Port Sudan

ground to the N and NW of this area, which (the chart warns) is also affected by strong currents.

Anchorage
3–4m, sand or sand and shingle, in the lagoon, with good shelter from prevailing winds, or 7–8m in the channel tucked into the slight bay immediately NW of the entrance.

General
This one is only really for keen divers. It offers an excellent hovering point for good weather at Abington and Angarosh. Marine life here, as in the above anchorage, is reported to be outstanding. There used to be a research structure on the reef just N of the entrance but it has been destroyed. In 1833 the Hon. East India Company's Sloop-of-war *Nautilus* was wrecked on or near Shab Aya, presumably the wreck is still down there somewhere.

Qita El Banna
20°41'·55N 37°23'·3E Key map page 157

The reef is circular and the small 'shelf' at about 15m is pretty vestigial. Falling back on a long rode from a reef hook stuck on the edge of the reef surface is probably the best way to go. You'll need truly quiet weather. That said, this is rated as a great dive with stunning drop-offs all round and memorable fish life including the sharks.

Angarosh
20°51'·8N 37°26'·25E Key map page 157

Once charted as Umm el Burush or Mother of Piastres, Angarosh, which apparently means Mother of Sharks, is rated as one of the world's great shark dives, especially for hammerheads, as well as masses of fish life including vast schools of barracuda. Aligned NW/SE, the small, flat-topped, ovoid 'cake' reef has a sand *cay*. The drop-offs all round are vertiginous except on the slight 'plateau' which lies to the SE. The plateau is deep at 25–45m and, again, falling back on a reef hook on the surface reef is probably the way to go. As with all these offshore, isolated reef dives, Angarosh is only feasible in settled weather.

Abington Reef
20°53'·63N 37°27'E Key map page 157

Marked by a lighthouse structure, which is a good day mark. The light isn't always or even often working. There are depths from 25–35m on a wide shelf looping from SE through NE of the surface reef with isolated shoaling round to SW. The best diving is thought to be on the SE-tending part of the shelf. As with the other offshore reef dives hereabouts, apart from the usual attractions of the coral, the fish life, including occasional manta rays, is the draw. Again this is for the experienced – the dives are deep – and for settled weather using the anchor-on-the-reef-and-drop-back technique. Chain damages reef.

Return to the Inner Channel

Muhammad Qol
Admiralty chart 3722, plan below

Approach
The passage through the reefs E of Muhammad Qol is dealt with below. From the Inner Channel, approach is clear until you are about 1M off, in Rawaya Anchorage. From there follow the two sets of leading marks. The larger, drying reefs in the approaches are beaconed, but there are many shoals not far off the recommended track. You will need good light for this approach, so that the reefs show well.

Dangers
The isolated patches mentioned above are well charted, but extra care is needed at the helm in this approach. There is not much room for error.

Anchorage
5–12m, mud, off the jetty. Good holding.

General
The fishing village is small, can be a bit smelly but is interesting nevertheless. It's possible to buy a few groceries here. US$ cash in small denominations is acceptable if you have no Sudanese currency. There is a military-cum-police presence, and you will probably be visited. If you are approached by officials the boat will probably look like a civilian runabout. The usual presentation of papers, a few civilities and an offer of hospitality to the people of this hospitable culture will satisfy enquiries.

Dungunab Bay
20°57'N 37°12'·5E Key map page 157

This area is unsurveyed, it was always the devil to get into because there's no clear, beaconed passage and you had to pick your way. Luckily temptation is

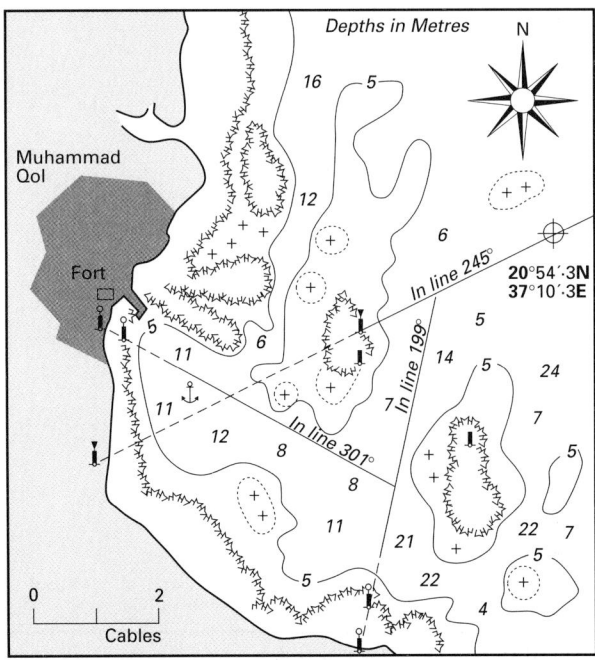

MUHAMMAD QOL

161

Red Sea Pilot

now easily resisted. Entry is prohibited because there's a pearl farm.

Mesharifa Island

Admiralty chart 3722, plan below

Approach

From the S or from Muhammad Qol, approaches are clear. Keep Gad Mesharifa a good ½M to the E until you come level with the beaconed entrance to the channel which leads E and N from here. There are uncharted shoals S of the island. We found one before GPS days, fortunately without hitting it! If Mesharifa bears 020° or greater you will stay in safe water.

Dangers

There is unmarked shoal ground extending SW from the W tip of Mesharifa I. Beware of strong E/W currents.

Conspicuous

The beacon (4m) on the S end of the shoal ground off the SE end of the island is easy to see.

Anchorage

4–5m, close S of the island. The holding is poor in thin, loose sand over rock so you need plenty of scope, which makes all the coral outcrops round about look most unappealing. There is a possible alternative off Bathing Islet.

Note Patrol boats from Muhammad Qol may approach, wanting to see your papers. You may also be bothered by fishing boats from Mohammed Qol come to beg.

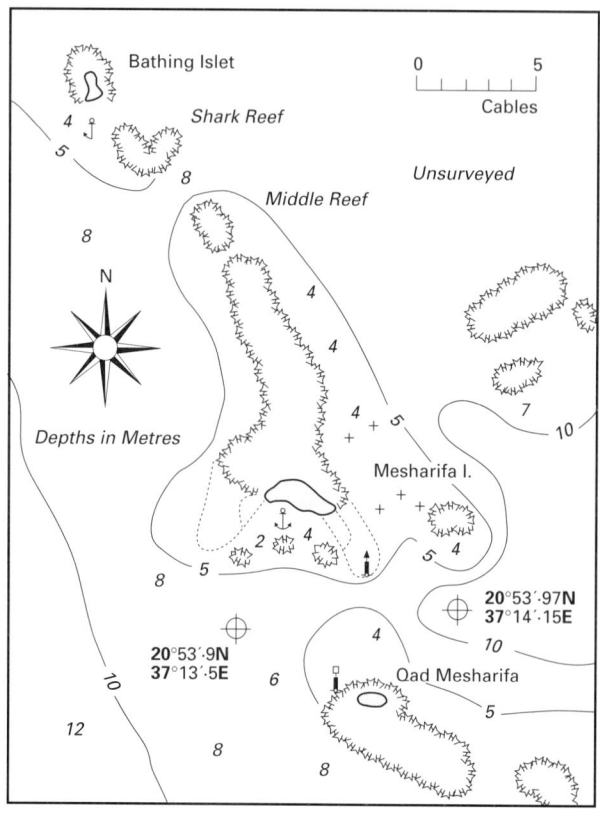

MESHARIFA I.

Directions for the Inner Channel from Mesharifa I

Note The directions are written assuming that you are heading N, and so courses for a northward passage are given first, followed by the reciprocals.

The Channel continues to Ras Abu Shagrab. All beacons have been recently repaired and fitted with new topmarks.

Start of recommended track: 20°53'·95N 37°13'·72E. (There is another approach to the passage from seaward, shown on the key map of the area between Jaz Mayteb and Jaz Bayer, but it requires very good light.)

Course: 084°/264°.

Distance to next turn 2·68M.

Beacons: A port-hand beacon to starboard after 1M, then T and starboard-hand beacons at 2·28M, with an additional stake beacon to starboard 0·3M further E.

Depth: Least charted depth 4m; average 20m.

At 20°54'·19N 37°16'·56E alter to:

Course: 000°/180°.

Distance to next turn 2·45M.

Beacons: Astern to port, Nos.1 and 2; on your port bow there is a conspicuous rock, Teetai Aweeb (2·4m), about 2·5M away.

Depth at course alteration 24m.

Note The alternative approach from E of Jaz Mayteb joins the channel at this point. See sketch-map of the area showing the tracks through the reefs, Juzur Telat to Ras Abu Shagrab.

At 20°56'·65N 37°16'·56E alter to:

Course: 035°/215°.

Distance to next turn 1·42M.

Beacon No.3 will be seen to starboard after changing course. It is a double comprising a stump and a port-hand beacon, and Teetai Aweeb (2·4m), the conspicuous rock, will be seen to port.

Depth at course alteration 20m.

At 20°57'·84N 37°17'·43E alter to:

Course: 055°/235°.

Distance to next turn 3·21M.

Beacons Nos.4 and 5, due NNW and SSE.

Depth at course alteration 12–15m.

If clearing to seaward alter to 063°, or if heading in from seaward steer 243°, aiming to leave beacon No.6 about 1M to starboard.

At 20°59'·72N 37°20'·3E alter to:

Course: 355°/175°.

Distance to run 6M.

Beacons: Stump of beacon No.6 bearing 300°.

Depth at course alteration 70m.

Hold beacon No.7 hard to starboard to clear the coral head S of it and enter the anchorage under Ras Abu Shagrab. Alternatively, leave No.7 to port and subsequently beacon No.8 on Shab al Hara to starboard if continuing N. 6M from this point on 355° takes you clear of all dangers N'wards.

The Sudan Coast N of Port Sudan

If you are approaching from the N, you'll need to run the directions of the passage in reverse. The co-ordinates of the position 6M, 355° from near beacon No.6 are 21°05'·7N 37°19'·7E. A course of 175° from this position leads clear of danger until, after about 2·5M, you are within eyeballing distance of Beacon No.8 on Shab al Hara. We tried this and it works. Do note that if approaching the Inner Channel from the N, be especially careful of the two reefs, 2·25M ENE of Ras Abu Shagrab on which the sea normally breaks.

Marsa Qadaf
20°51'N 30°17'E, see key map page 157

The anchorage marked off the N end of Jaz Magarsam at Marsa Qadaf is something of an unknown quantity, but would bear exploration if a southerly came on, you wanted to get out to Abington or Angarosh and didn't fancy trying either Jaz Mayteb or Shab Aya. It looks as if it has a wide, clear entry, but the head of the bay may be full of hazards. Suck it and see.

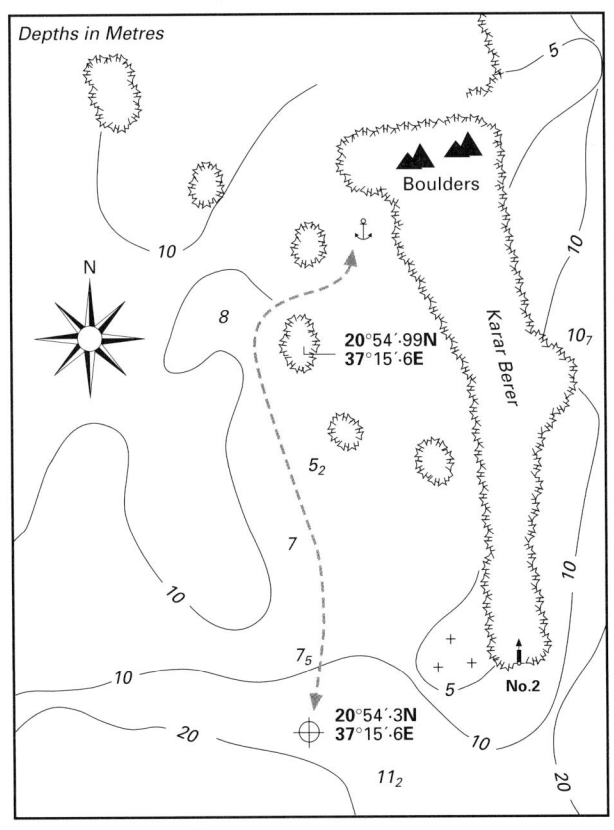

Leoni anchorage
Plan below

An apparently wonderfully sheltered anchorage in behind Karar Berer protected from the N by the vast reefs S of Jaz Dungunab and from the S by the reefs N and E of Jaz Magarsam. Entry requires care and a lookout aloft. Head N from the E end of the deep pool at 20°54'·3N 37°15'·6E across shoal ground with a least depth of 5m until the second large isolated reef at 20°55'N 37°15'·6E. Leave this reef to starboard and pass between it and the next isolated reef at 20°55'·2N 37°15'·61E to tuck under a W-tending hook of reef with a couple of large boulders on it. You'll be in 10m or more over sand and coral. Pick a spot. There is good diving reported on the E face of Karar Berer off Beacon No.2 but be wary of currents.

Shab Felix
Plan below

This reef anchorage is N of the passage from Mesharifa to Ras Abu Shagrab, in the protection of the reef marked by Beacon No.6. Anchor in 8–10m, sand and coral W of the beacon where, now and then, fishing boats take shelter. It may be possible to fiddle your way towards the northern half anchor, the spot commended by Red Sea doyen Felix Normen, after whom this anchorage is named, but you need good light and to like tight spaces. There is good diving at the SE end.

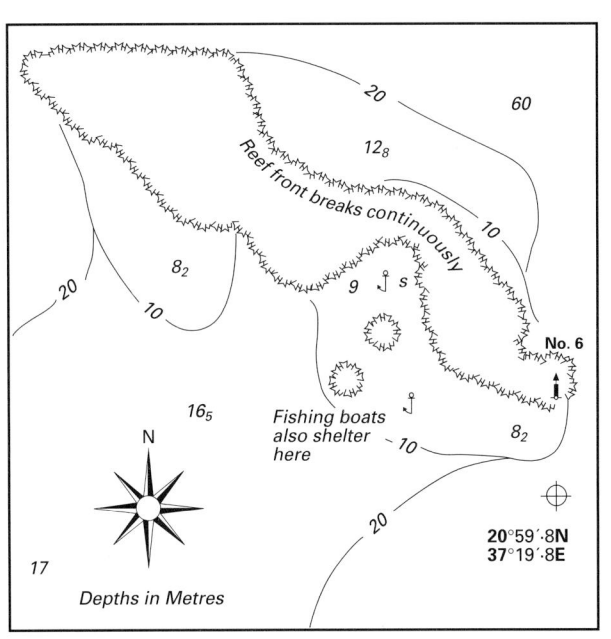

SHAB FELIX

Wreck recovery anchorage
Plan page 164

This anchorage was used over a protracted period by the dive charterers in Port Sudan whilst they recovered the wreck of a large Turkish built *gulet*. It was intended for the dive charter business but stranded on Merlo Reef when headed S for Port Sudan on autopilot. That should be a warning to everyone. The anchorage offered good shelter from N'lies and is relatively easy of access. It is in 10–15m sand and coral and is accessed, from 21°00'·8N 37°18'·7E, via the tongue of deeper water, S of an isolated, uncharted bommie E of the anchorage.

LEONI ANCHORAGE

Red Sea Pilot

Waypoints:
1. 21°01´·36N 37°19´·06E
2. 21°01´·51N 37°18´·94E
3. 21°01´·58N 37°18´·84E
4. 21°01´·63N 37°18´·57E

21°01´·8N 37°18´·8E

Small boats only

Depths <20 with scattered bommies

No.7

unsurveyed

Wreck Recovery Anchorage
21°01´·0N 37°18´·4E

21°00´·8N 37°18´·7E

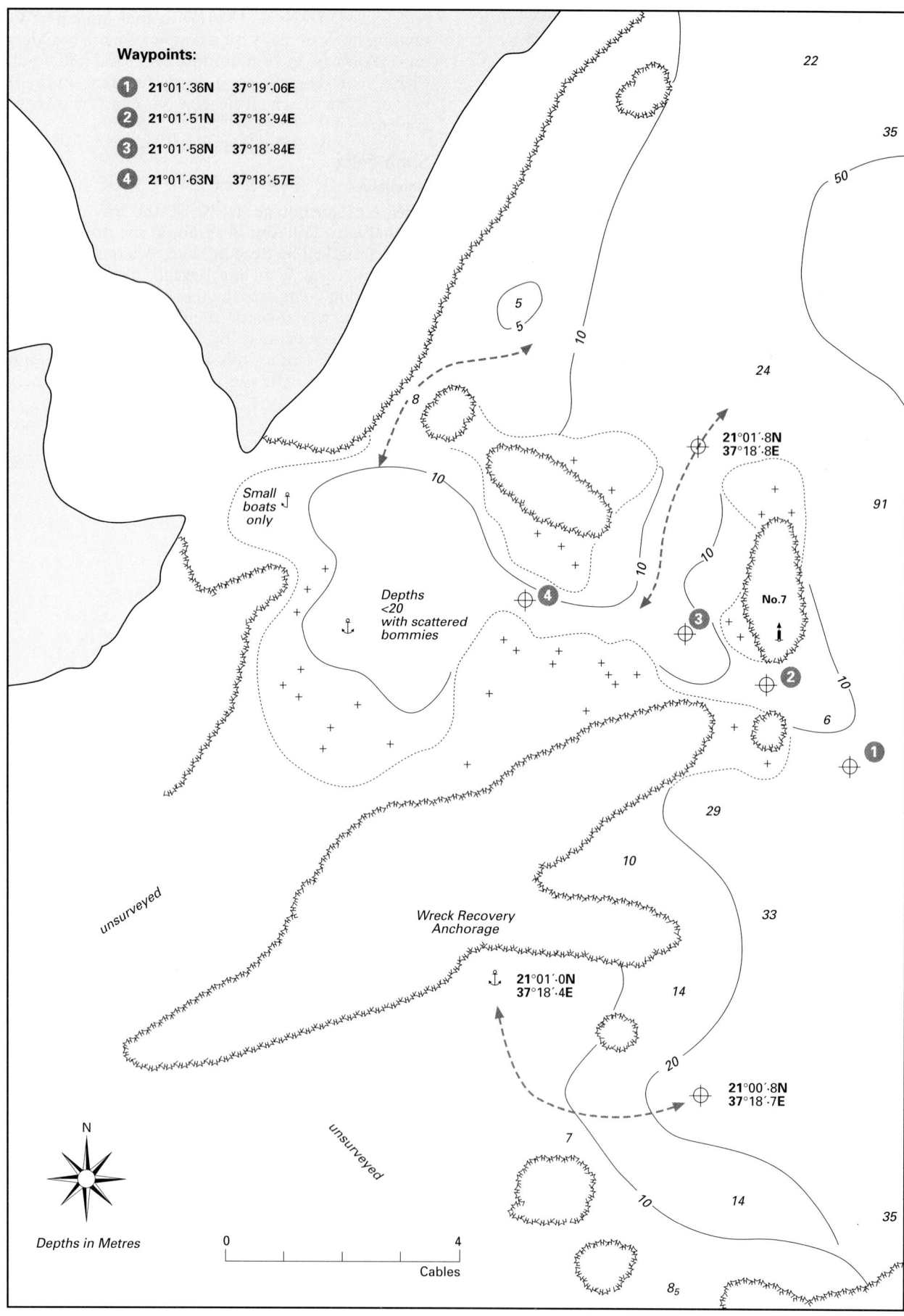

unsurveyed

N

Depths in Metres

0 — 4 Cables

WRECK RECOVERY ANCHORAGE & RAS ABU SHAGRAB ANCHORAGE

164

The Sudan Coast N of Port Sudan

Ras Abu Shagrab
Plan page 164

Approach
The approach is quite difficult and the anchorage is not so rewarding. If the wind dies over night, the swell makes in and you roll about. Still it's better than plugging on N against headwinds as long as you stop when there is still good light. From S, steer approx 320° towards 21°01'·36N 37°19'·06E, S of Beacon No.7, thence 305° to 21°01'·51N 37°18'·94E into the entrance to the anchorage area. The next bit is narrow and hairy in a blow. Steer roughly 300° from the last position to 21°01'·58N 37°18'·84E, then wiggle 280° through 21°01'·63N 37°18'·57E in a least depth of 20m to the anchorage. The narrow pass N of the anchorage is feasible with a least depth of 8m in the middle of the narrow channel. The 5m-deep bommie N of the pass is well clear of the entrance and can be avoided with a good lookout.

From the N the pass from/to 21°01'·8N 37°18'·8E is feasible but more encumbered with reef than charted and needs a sharp look out.

Anchorage
The bottom is very irregular, in sand and coral with depths of up to 20m amid scattered bommies <3m. Be careful where you drop the hook. A good alternative for small boats is in the little bay WNW of the spot marked on the sketch. This is calmer with good holding and is used by fishing boats, but is tight and shallow.

Shab Qumeira
Plan below

Approach
The reef is detached from the mainland and marked at the S end by a yellow beacon with a diamond topmark. From the S pass close W of the beacon, then run parallel to the coast on about 315° until the

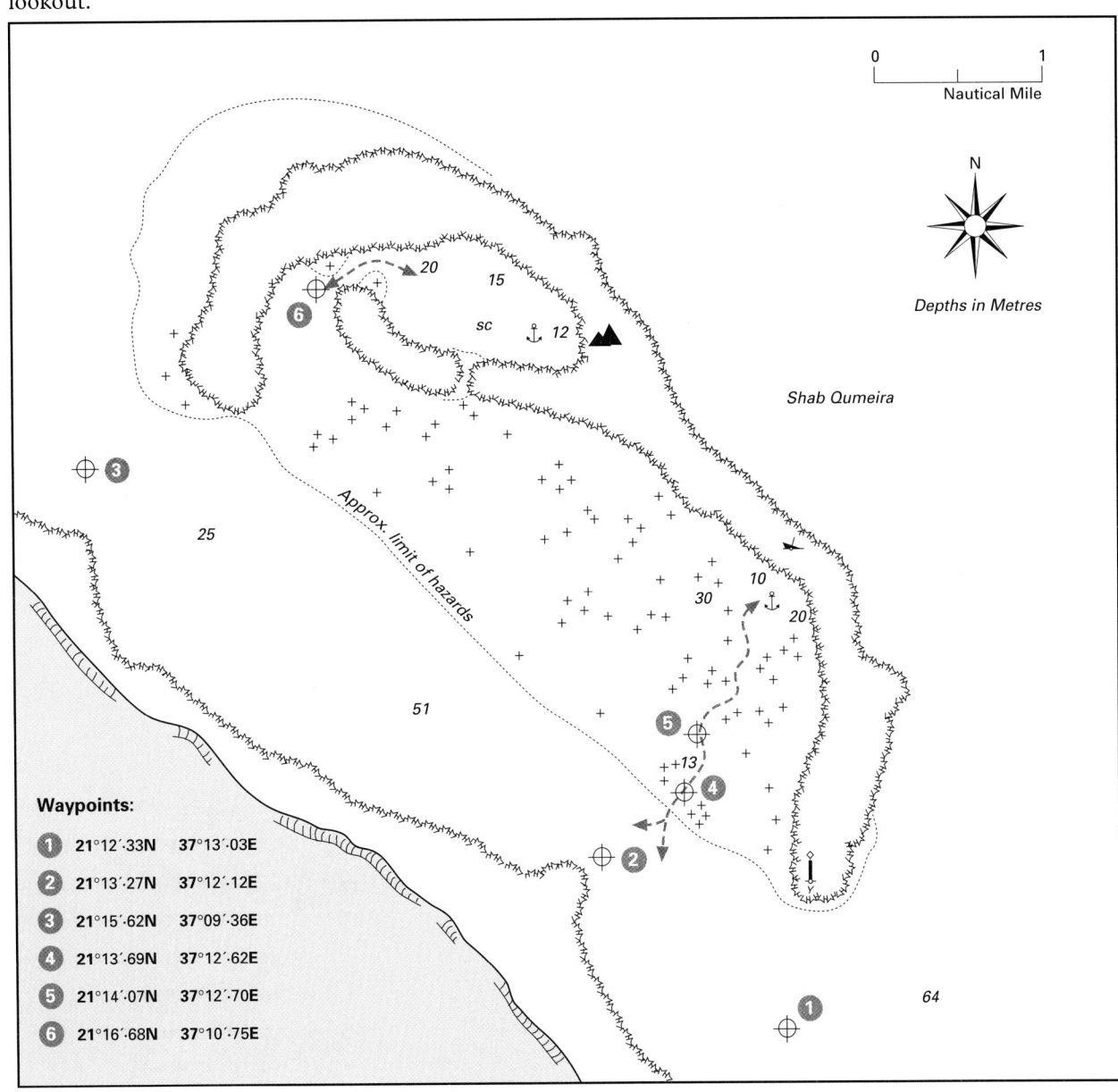

Waypoints:

1	21°12'·33N	37°13'·03E
2	21°13'·27N	37°12'·12E
3	21°15'·62N	37°09'·36E
4	21°13'·69N	37°12'·62E
5	21°14'·07N	37°12'·70E
6	21°16'·68N	37°10'·75E

SHAB QUMEIRA

Red Sea Pilot

wreck bears 020°. The fringing reef off the mainland bulges out into the channel near this point. A GPS waypoint of 21°13'·27N 37°12'·12E is clear to seaward of the bulge by half a cable or so. From there you must steer a sinuous course to anchor near the wreck in 10–20m, sand and coral. However the bottom is very encumbered with coral and access and egress require good light, quick reactions and lots of time. A better spot, though the bottom isn't much of an improvement, is in the lagoon in the NW. Find the NW tip of the reef crescent about 0·5M NW of 21°15'·64N 37°10'·12E and then follow its S edge to an opening between the fringing reef and the inner lagoon barrier at 21°16'·68N 37°10'·75E. There are two bommies in the entrance, one lying off the N side on the 'outside' and the other off the S side on the 'inside'. There is 20m in the entrance. The lagoon shoals to 12m as one steers roughly E at which point one is quite close to the boulder.

Dangers

Scattered bommies rise out of the deep water on the approach to the anchorage, well away from the reef proper. Have someone up in the spreaders as a lookout and zigzag between them. They are easy enough to see in good light as brownish patches of water, alarming only if unexpected.

Note The waters between the numbered waypoints on the sketch MAY NOT BE HAZARD FREE. The waypoints represent known points of safe water. Between them you must pick your own way.

Conspicuous

The wreck on the reef is visible from 4–5M away. It looks like a large block structure. Closer to it's recognisable as an old three-castle tramper of about 2,000 tons. There is a conspicuous rock on the reef WNW of the wreck which, from a distance, looks like a small boat at anchor.

Anchorage

In the NW lagoon 10–12m, as described above, holding is fair in sand and coral and shelter is excellent. Or close to the wreck in 21°14'·88N 37°13'·09E in 10–15m, sand and coral.

For orientation and clarification of detail in the Inner Channel from just S of Khor Shinab to just N of Elba Reef see the folded chart you will find with this volume.

Khor Shinab

Plan below

Approach

There is an islet on the reef (8m) about 3M SE of the entrance to Khor Shinab with a reef 2M E of it and another 1½M NE. These reefs always break. There is also a spur of fringing reef bulging E into the channel in approx 21°20'·5N 37°05'·2E. There are no markers at the break in the reef, which is quite invisible from the SSE by early afternoon, but no problem coming from N. The fringing reef appears to be continuous. If necessary, sail past where you expect the entrance to be, looking for the low spit which ends in a headland that shows dark against the shore. Apparently finding the entrance is no problem on radar no matter what the angle of your approach. The entry tends 240°. Persistence will be rewarded. This is a beautiful place.

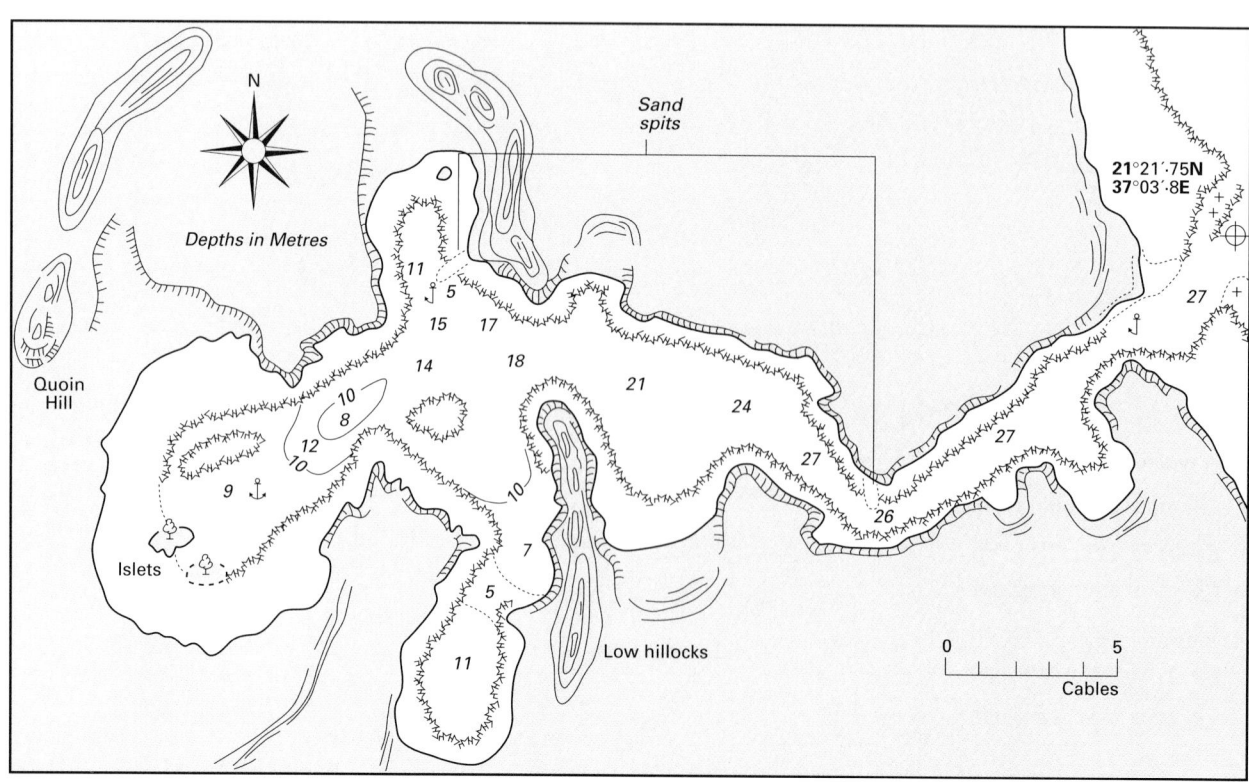

KHOR SHINAB

Dangers

As with many of the *marsas* on this coast, the fringing reef extends further on the N side of the entrance than it does on the S. Here, you'll see several small boulders on it. On the S side there's a shoal patch off the reef tip.

Conspicuous

In good light you might be able to make out Quoin Hill at the head of the inlet. Although the hill is only 59m high, it has a very rugged outline, with two summits.

Anchorage

The *marsa* is big, impressive and relatively easy to negotiate but the anchorages are deep unless you go right in. The best holding is in the last 'bay'. An alternative is 5–10m off the sandspit in the bay in the NW corner. It looks corally but is well protected with good holding. In light weather you can also anchor off the sandy beach just inside the entrance. Put an anchor on the sand and fall back into about 10m.

General

This has to be many yachts' favourite anchorage in the Red Sea. It has desert colours and a sculptured landscape to match one's imaginings. The diving is good and the walking is magnificent. Do climb Quoin Hill. The views in every direction are wonderful. A quoin, by the by, was a wedge, roughly the shape of Quoin Hill, used in the 18th and 19th century navies to control the elevation of the barrels of cannon on their carriages.

Khor Shinab was the rendezvous from which the Honourable Company's Ships *Benares* (Cdr Thomas Elwon IN) and *Palinurus* (Lt Robert Moresby IN) split to begin the first and only full scientific survey of the Red Sea in 1829–1830. Curiously, Khor Shinab's orientation was wrong by some 45° clockwise on large scale plans, an error that wasn't corrected until Lt. Eeles' work in HMS *Grappler* in 1885-86.

The Portuguese were familiar with Khor Shinab before the British arrived and from the 16th to 19th centuries it had the charming soubriquet of Khor Abu Mishmish. According to the marginal notes of a 19th century British Admiralty scribe on the UK Hydrographic Office archive copy of Elwon and Moresby's chart, this name translates even more charmingly as Khor Father Apricot.

Channels to/from seaward: Shab Qumeira to Marsa Abu Imama

There are four ways into the Inner Channel from seaward in the area between Shab Qumeira and Marsa Abu Imama. The most open is between Shab Qumeira and the small reefs SE of Khor Shinab but in winter it places you to leeward of Khor Shinab and to windward of Shab Qumeira. Three alternatives are either N or S of Shab Shinab and N of Shib Halaka. The opening S of Shab Shinab is 2M wide, mid-point at 21°21'·7N 37°06'E. That N of Shab Shinab, mid-point at 21°25'N 37°04'E is also 2M wide. Be wary of the isolated reefs at 21°24'N 37°07'·7E and 21°24'N 37°09'E. The opening N of Shib Halaka is nearly 3M wide but has a reef at 21°29'·6N 37°02'·5E. The opening between this reef and Shib Halaka is only 1M wide, mid-point at 21°28'·8N 37°02'·6E. N of it there's a bit more room (1·5M) with a mid-point at 21°30'·5N 37°03'E.

The passages through the barrier reef N of Marsa Abu Imama are generally risky because of ill-surveyed, isolated reefs from 21°31'N to N of Elba Reef. There are several wide openings, but the uncertain numbers and positions of hazards both in the openings and in the offing make daylight essential. Even with good light you need the greatest care and the sharpest lookout. In any sort of brisk wind you may see the hazards too late to avoid piling up on them. It is worth noting that in a force 7–8 in Foul Bay, the sea breaking on the N flanks of Elba Reef was not visible until we were within half a mile.

Shab Shinab

21°23'N 37°05'E See fold-out chart

This horseshoe-shaped reef is about 2½M NE of the entrance to Khor Shinab and has an opening on its W side into a shallow lagoon. In strong winds the unmarked channels into the lagoon are difficult to see because of breaking seas. Anchor in about 10m, sand, with good holding, in the N part of the lagoon for shelter from N'lies. Fair weather only.

Marsa Halaka

Plan page 168

Approach

Between Khor Shinab and Marsa Halaka there is another bulge in the fringing reef at 21°23'·5N 37°03'E, so it is worth standing well clear. The entrance is open, clear and hard to miss. The opening leads W, turns more SW and then kinks before running on broadly SW to another beautiful anchorage.

Anchorage

The best anchorage is 1·65M from the entrance, through a 50m wide, slightly kinked channel, in the W'most bay near a remarkable rock looking like a bird's head and N of two islands with mangroves. 7–9m sand and mud, very good holding. The inner bay is full of wonderful bird and fish life. Fishermen can usually be seen working the shallows and there are sometimes camel trains on the coastal road, or coast track, which runs across the head of the *marsa*.

Back in Elwon and Moresby's day this was called Khor Dullow. For years the BA had a large-scale plan, surveyed by Cdr Elwon and now withdrawn. It was rather misleading showing a far more confined and boring spot. Khor Dullow, properly Khor Dhu-l'awa, along with the name Khor Delaweb (now Marsa Hamsiat), much contributed to the cartographical confusion hereabouts. Indeed in 1922, when a Sudan Government official

Red Sea Pilot

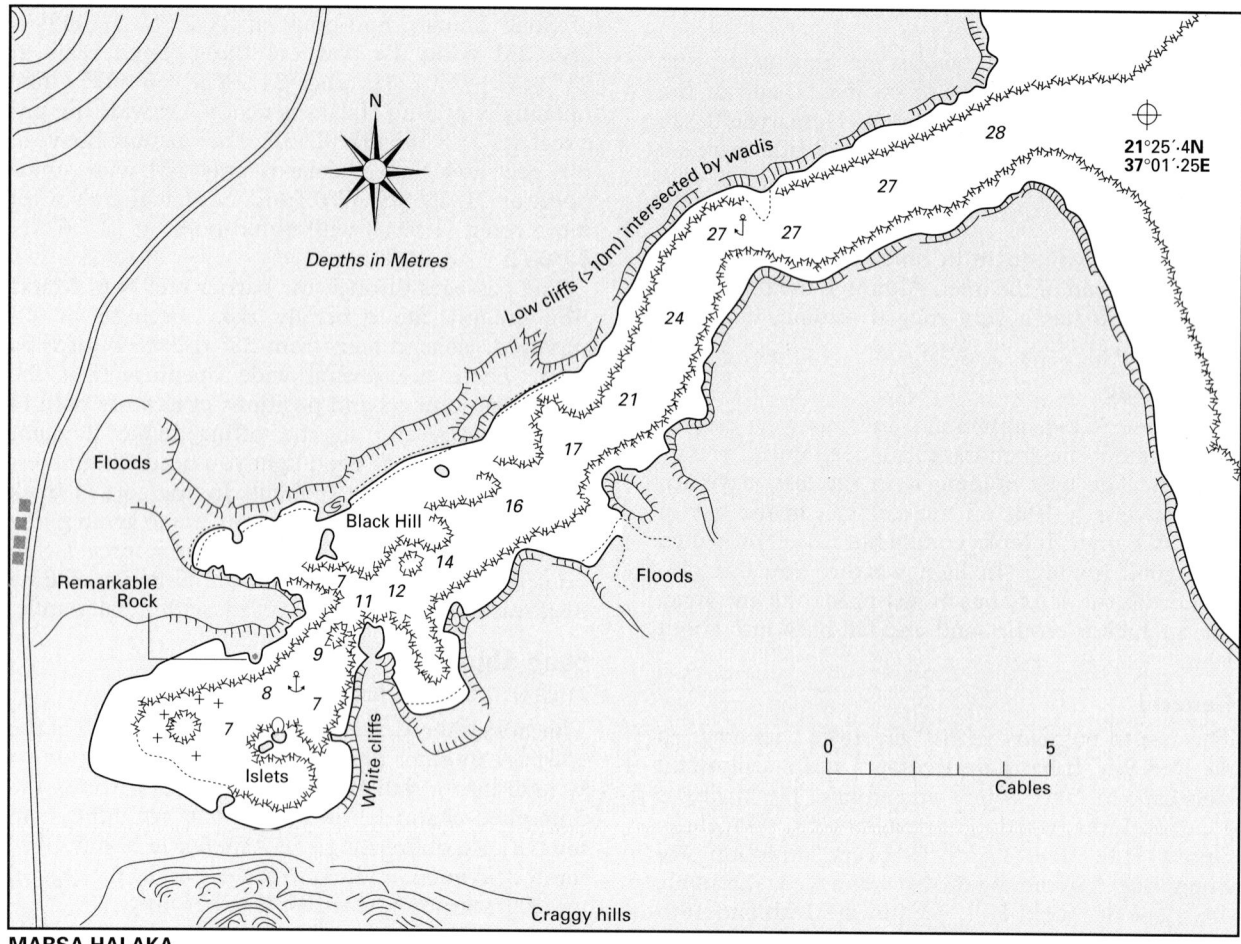

MARSA HALAKA

announced that Marsa Halaka was actually called Khor Delwein the muddle got worse. In around 1950, word came from Sudan that the correct name was Marsa Halaka. The result, as far as we can see, was the invention of a completely mythical *marsa*, Khor Delwein, of which we found no trace except, perhaps, a very open and tricky to enter roadstead. For more, see The Great Marsa Mystery Solved, below.

Marsa Abu Imama

Plan page 169

Approach

The opening in the reef is easy to find and the new beacon on the N side of the entrance is clearly visible beside the stump of the old. Not far from the coast there is a range of hills which is very easy to recognise. Jebel Abu Imama (200m approx) is flat topped and about 2M S of the *marsa*. It looks somewhat like a butte or puy and can be seen for some distance off. A bit further N and inland are three almost conical hills close together.

Dangers

Watch out for an extensive area of shoal ground marked by a stake in the middle of the *marsa*. Two other shallow patches further SW have at least 4m over them.

Anchorage

10–12m, sand and coral, off the sandspit in the inner bay, a remarkably peaceful spot, even in a strong blow. Alternatively, boats anchor close off the sandspit at the entrance to the N bay over a sand bottom, but there's still a lot of coral around and finding really good holding isn't easy. Another possibility is in the lee of the shoal in mid *marsa*. Fishing boats occasionally anchor at the W extreme of the *marsa* near the 6m sounding.

Near the ruined jetty there are two buildings, one a conspicuous white ex-container or Portakabin, which occasionally houses a military patrol. Excellent diving on the reefs and pleasant walking ashore. There is a ruined village to visit, about ½M SW of the huts (conspic).

The 16th-century Portuguese called this *marsa* Rio Farat. In Alexander Dalrymple's first collection of charts for the British East India Company there is a charming rough sketch making the place look like one of the rias around Vigo! Many of these charts were taken from earlier French and Dutch East India Company atlases by van Keulen and Thevenot. Well into the 20th century the charts continued to label Marsa Abu Imama as Marsa (or Khor) Abu Imama Farat.

The Sudan Coast N of Port Sudan

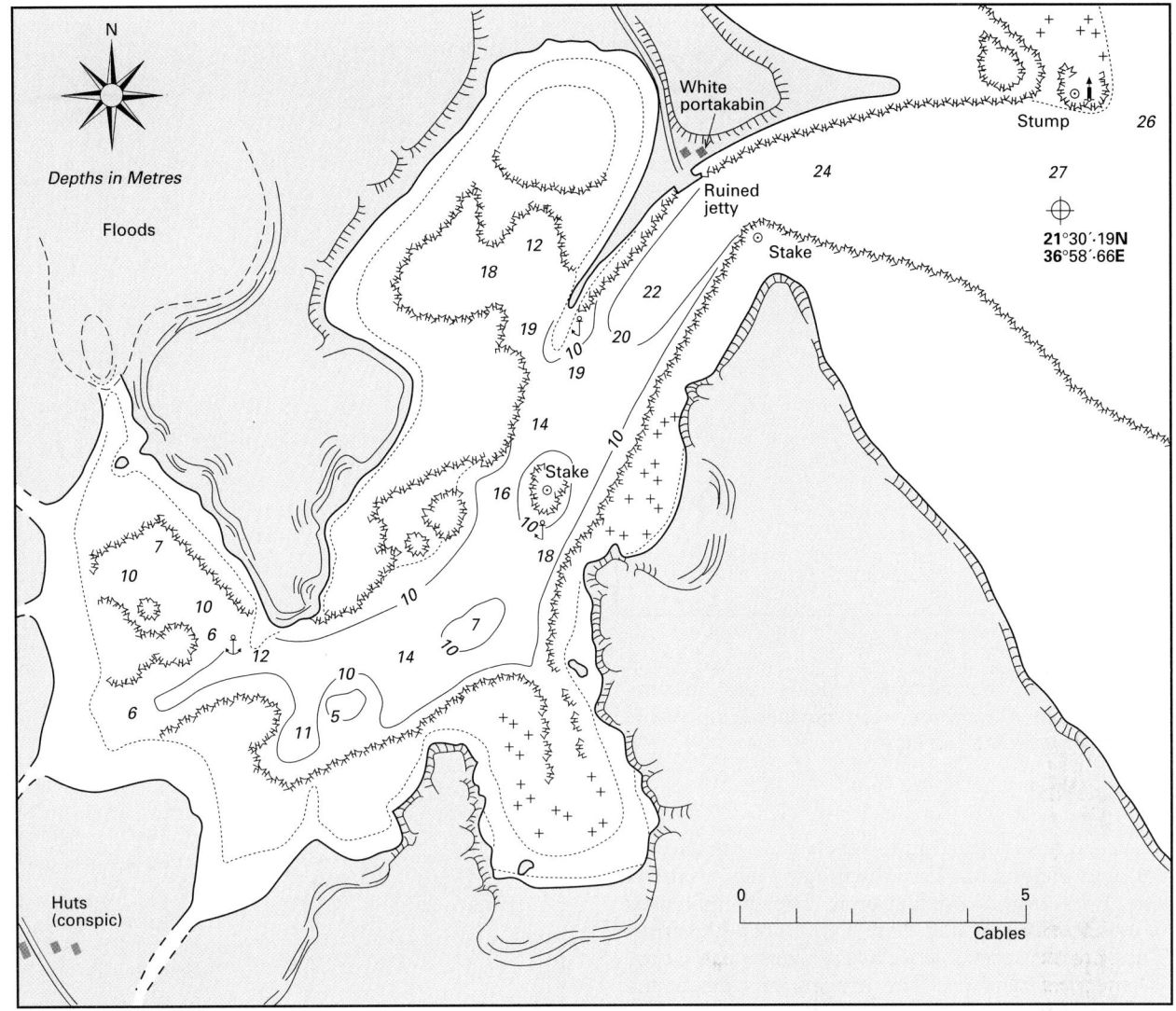

MARSA ABU IMAMA

Inner channel dangers and Khor Delwein
Plan page 170

N of Marsa Abu Imama the fringing reef has an off-lying line of discontinuous reef, which pops up here and there until Marsa Gafatir. The best advice is to stay 1M clear of the land. The most hazardous area is a detached cluster of reefs, in a shallow, E-facing crescent about six cables N/S between 21°32'N 36°57'·75E and 21°32'·65N 36°57'·7E. The next cluster, 1¼M N/S, is closer inshore off the putative Khor Delwein, between 21°33'N and 21°34'·2N, W of longitude 36°57'E. Two more off-liers are at 21°34'·9N 36°56'·65E and 21°35'·4N 36°56'·55E, the second masking the entrance to a pass into the shallow lagoon behind the fringing reef (see Marsa Gafatir below).

As far as we can see Khor Delwein does not exist. If it does it must be a fairly open roadstead, in roughly 21°33'·5N 36°56'·5E, behind the 1¼M long line of awkward off-liers identified above. Our advice is to stay well clear.

Marsa Gafatir
Plan page 170

The entrance trends SW. Beware of off-liers on the S shore and along the fringing reef edge S of the entrance. It then branches into two arms. The N arm runs WNW for 3 cables where a shoal (<3m) crosses, beyond which is a pool at the entrance to a lagoon closed off by a sand bar. The bold may find an anchorage in the pool, the rest of us would anchor E of the shoal in 9–12m, sand, good protection from seas, none from the wind, though there's not much room. The S arm leads W off the pool where the *marsa* divides, through a narrow dogleg S around a conspic boulder, with depths >15m, into an inner pool with depths <12m. Good holding in mud and good shelter from the sea.

1M S of Marsa Gafatir there is another opening in the fringing reef, rather encumbered with two large detached reefs at its seaward entrance, which debouches into the lagoon behind the fringing reef. It looked too shallow for *Fiddler's Green II*'s 2·3m draft, so we didn't probe. Of this and the more N'ly *marsa*, exactly which is Marsa Gafatir we're not sure and there was no one around to ask. Felix Normen

Red Sea Pilot

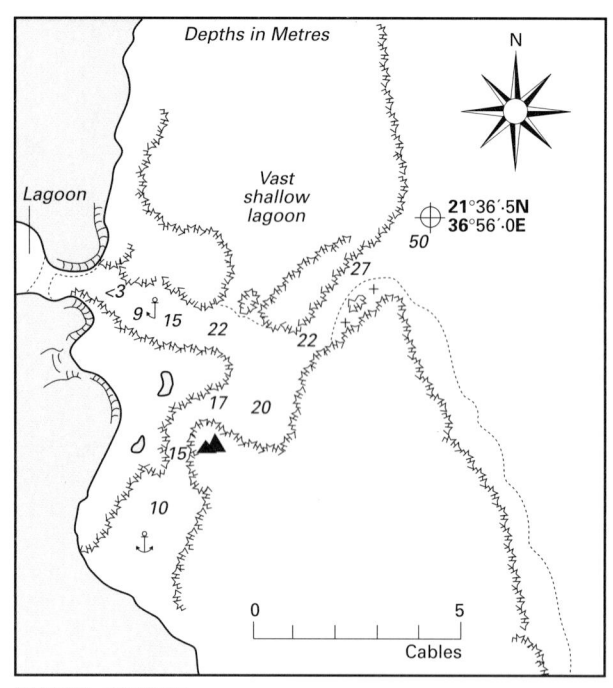

MARSA GAFATIR

gives Gafatir the latitude of the less enticing opening. His description, on the other hand, more closely fits our Marsa Gafatir.

Marsa Wasi

Plan page 171

The entrance is clear and relatively wide but there is a tongue coming out from the fringing reef on the N side. It has a small boulder on it. The channel leads in on 245°–250°M and there are two bay-like arms. The N one is accessible via a very narrow pass at the W end of a sandspit. The fringing reef inside the inlet extends N from the S shore W of the coral-filled bay with the small rocky islet. It is marked by a stake.

Anchorage

12–16m, fine sand and mud, S of the sandspit where marked on the sketch. Very good holding here, with protection from N–NE. In very strong winds a slight swell makes in from seaward, and the reef behind is rather too close for comfort. Nevertheless, this anchorage has served with winds gusting up to 45 knots. The lagoon that forms the N arm of Marsa Wasi is worth a dinghy reconnaissance first. Anchorage is in 6–7m, sand. Currents at mid-tide are likely to be strong in the narrows. In calm weather, the third alternative is to set your anchor on the sand near the entrance on the N side, but watch out for bommies. In S'lies try the S arm.

General

The coast from here to the border is patrolled by the military and is, technically, a military zone. Friendly nomadic fishermen set up temporary camp with their trucks and canoes. Migrating flamingos rest. The coast road, with the occasional vehicle, is a mile or so inland if you go ashore to stretch your legs. In strong winds, the fine dust that forms the top layer

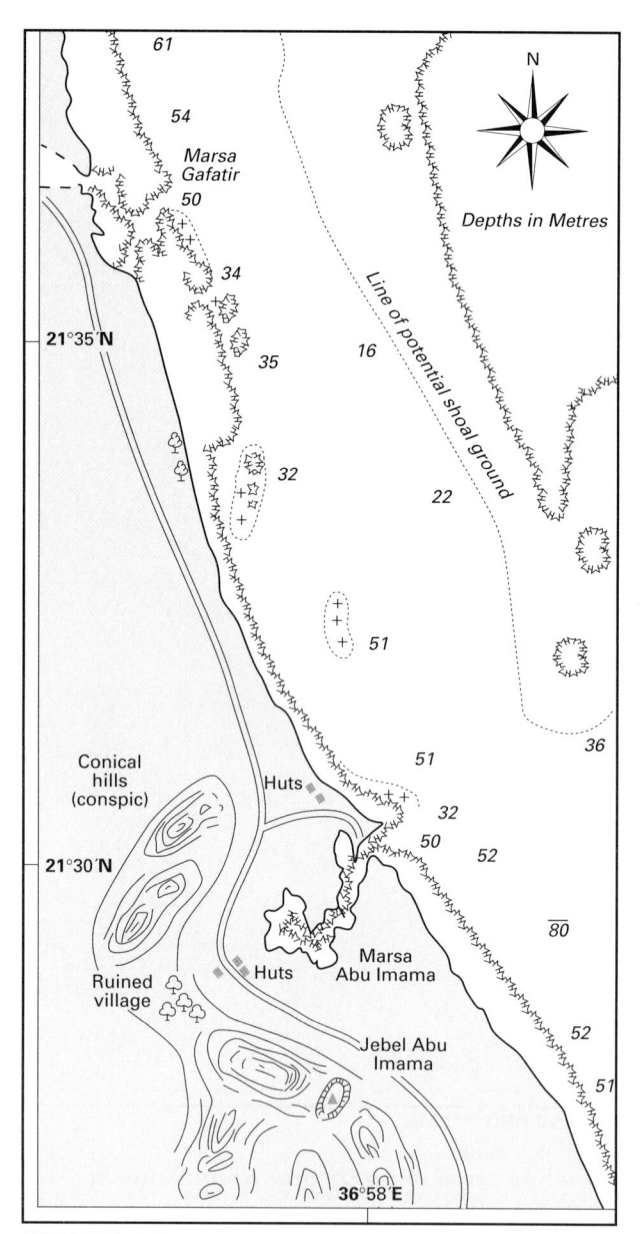

HAZARDS IN THE INNER CHANNEL BETWEEN MARSA ABU IMAMA AND MARSA GAFATIR

of the desert gets everywhere. It happens all along the coast. There's nothing for it but to keep cleaning up, or give in and wait till you're out of the Red Sea!

Khor Abu Fanadir, another of the names looking for a home, would appear to be either a name for the semi landlocked N part of Marsa Wasi, or possibly for the enormous shallow lagoon behind the fringing reef which stretches from here S to Marsa Gafatir.

Marsa Hamsiat

Plan page 171

On the S side at the entrance there is a dangerous tongue of off-liers projecting <1 cable NE from the fringing reef. If you stay in longitude 36°54'·50E until the entrance is well open WSW in the latitude given above, you should be clear. The channel leads inland S of 2 sandspits then into the inner *marsa*, past a tongue of reef to the S and a spit with a rock off it on the N side.

The Sudan Coast N of Port Sudan

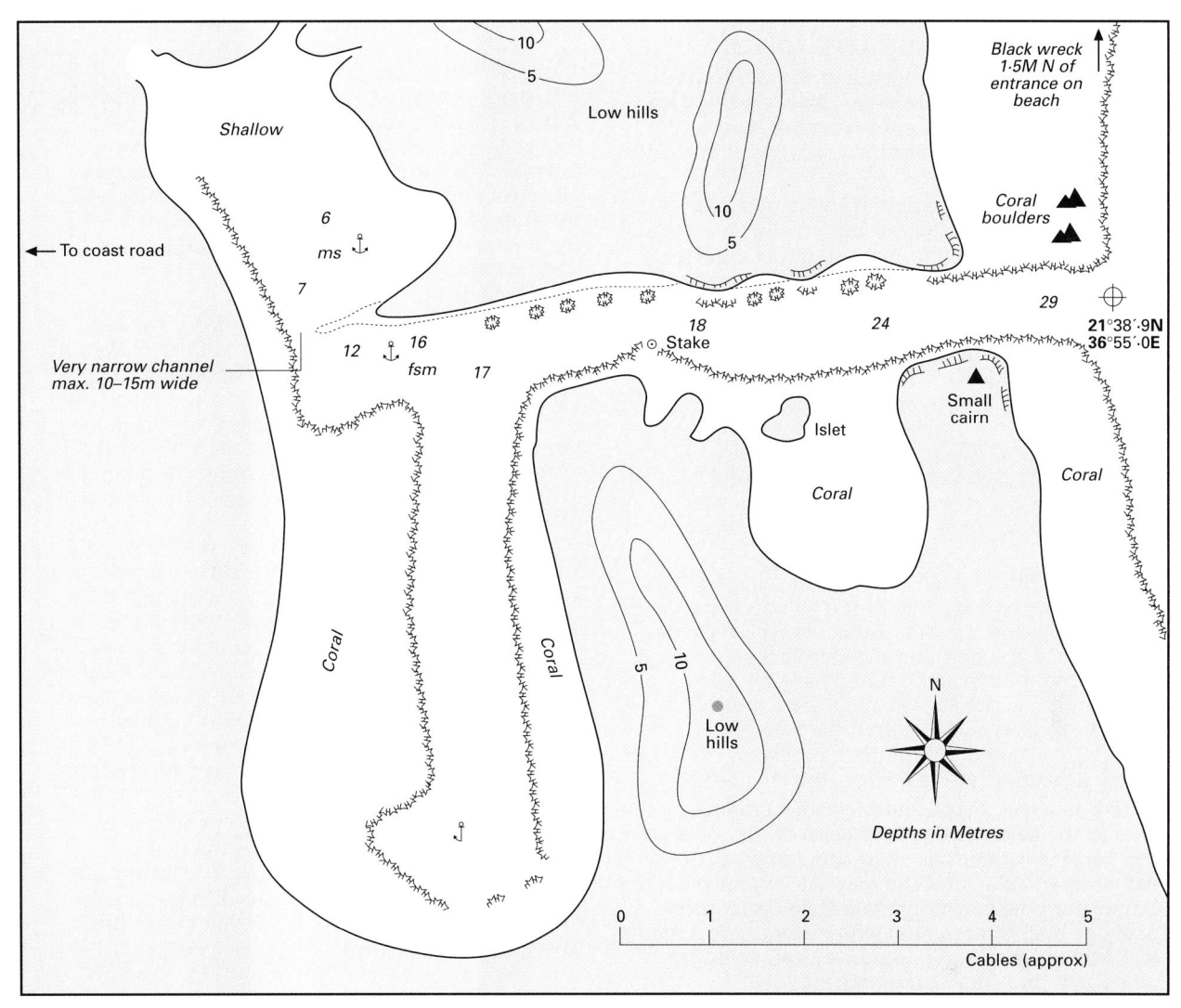

MARSA WASI

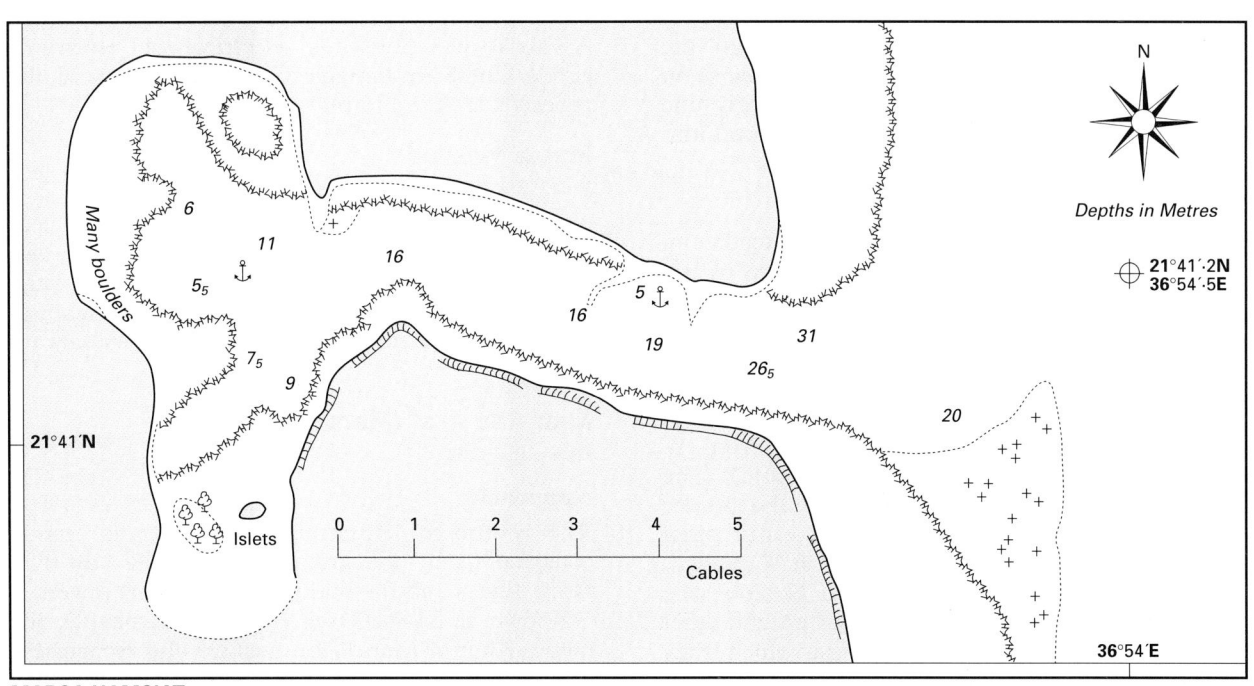

MARSA HAMSIAT

171

Red Sea Pilot

The Great Marsa mystery solved

In the first edition we punned, of the Great Marsa Mystery, 'Surely Sharm Mistake'. Now we think we have solved it. For your entertainment here is the quaint tale which explains the confusion that existed on this stretch of coast.

It starts with the triumph of steam over sail. Steamers could plough non-stop up and down the middle with less need for anchorages and ports of refuge. Hydrographical minds were concentrated instead on marking the safe limits of steamer routes with aids to navigation (lights). There were simple priorities. To chart the Gulf of Suez accurately. To fix the seaward edges of dangers along the E and W coasts precisely. And to establish positions of dangers in open sea. Later the need to chart the N and S Massawa Channels was added because early steamships had proved insufficiently powerful to make headway in the S Red Sea against winter currents and SE winds to seaward of the Dahlak Bank.

In 1870 the British and French hydrographic offices agreed to co-operate. The British volunteered for the Gulf of Suez and Straits of Gubal, in typical cheese-paring fashion suggesting the French took on the rest. Once the Brits had finished their bit, they grudgingly conceded, they might lend the French a hand. Before any work started the Franco-Prussian War (1870–71) and its aftermath paralysed the French military machine. In the end the British did it all.

In consequence, Elwon and Moresby's original surveys give us the best charting ever done of the shores of the Red Sea. Given their ships and equipment they were remarkably good and may actually have been better than the resulting small-scale charts show. Sadly, as is all too common with historical documents, all of the original survey data was lost, eaten by white ants and cockroaches or burned!

Over the following century dribs and drabs of information clarified some details but muddied the waters in others. And nowhere more so than in February 1922. Mr G W Chisholm of the Sudan Government Ports and Railways Department had just produced the first excellent sketch survey of the Shubuk Channel. Now, aboard the Sudan Government Ship *Enterprise*, unassisted by Capt Kirk and Mr Spock, he turned his hand to the northern marsas.

The *Enterprise* did a timed log run from Ras Abu Shagara (then called Sandy Cape) to Khor el Marob corroborated by no astronomical observations. Using the result Mr Chisholm sent to the Admiralty a tracing from the existing BA chart, corrected in terms of number, positions and names of marsas, which disagreed with what was shown, including saying Khor Dhu-I'Lawa was actually Khor Delwen (it's Marsa Halaka!). The result was confusion, as the notes in the 1920s master copy of the BA's *Red Sea and Gulf of Aden Pilot* in the UKHO archives attest. Bizarrely the word came from on high that Mr Chisholm's version was gospel. BA8A (BA138 from 1954 and now BA158) was amended. Then more muddle was added in the early 1950s. A copy of a Sudan Government land survey map revealed Marsa Halaka's true name. Khor Delwen/Delwein was a name looking for a home. The new BA138 tried to cope with everything and, as far as we can see, Khor Delwein was invented.

From the 1970s until now, the only people who knew what was precisely where in the Red Sea were probably the old Soviet Navy. The results of their extensive surveys in the late 20th century are only now becoming available – if you can afford the unbelievable prices!

Anchorage

The inner bay has room for several boats in 11–12m mud, at the head of the bay, or between the sandspits shown on the N shore of the *marsa* in about 5m, sand. If you decide on the latter, in fine sand, there's a risk of dragging off the E one into deep water.

General

Ospreys nest in the mangroves on the mud flats in the S arm and on the coast track camel trains of 100 beasts or so shuffle N. These are apparently part of an intriguing trade triangle run by the Rashaida, a Bedouin tribal grouping which emigrated to Sudan from Saudi Arabia some 400 years ago and still speaks Bedouin Arabic. The Sudanese are no fanciers of camel meat, but the camels from W Sudan are the best in this part of N Africa. The Egyptians love camel meat. However, the border between Sudan and Egypt is disputed and often closed. So the Rashaida buy camels in W Sudan, bring them to the coast, herd them N along the coastal track, turn into the mountains around Jebel Astroiba, the hidden paths amongst which they know well, and smuggle the camels into Egypt where they are sold. They take the money to Saudi Arabia to buy household electrical and electronic goods. These are smuggled into Sudan and sold, the proceeds are used to buy more camels.

Marsa Masdud

21°43'·3N 36°54'·2E See fold-out chart

This is a closed off *marsa* with a typical, clear opening in the reef, which stops at a sand bar. Beyond the sand bar is a large, shallow-looking, closed lagoon. In quiet weather you could anchor in the entrance in 20m mud but any swell will always work in.

Khor Abu Asal (Marsa Oseif)

Plan page 173

Approach

The entrance is straightforward and made conspicuous by the group of fuel tanks, the two radio towers and the buildings around the towers.

Known as Marsa Oseif by the local people, the names Khor or Marsa Abu Asal are also recognised. The officials will want to see your papers but they

The Sudan Coast N of Port Sudan

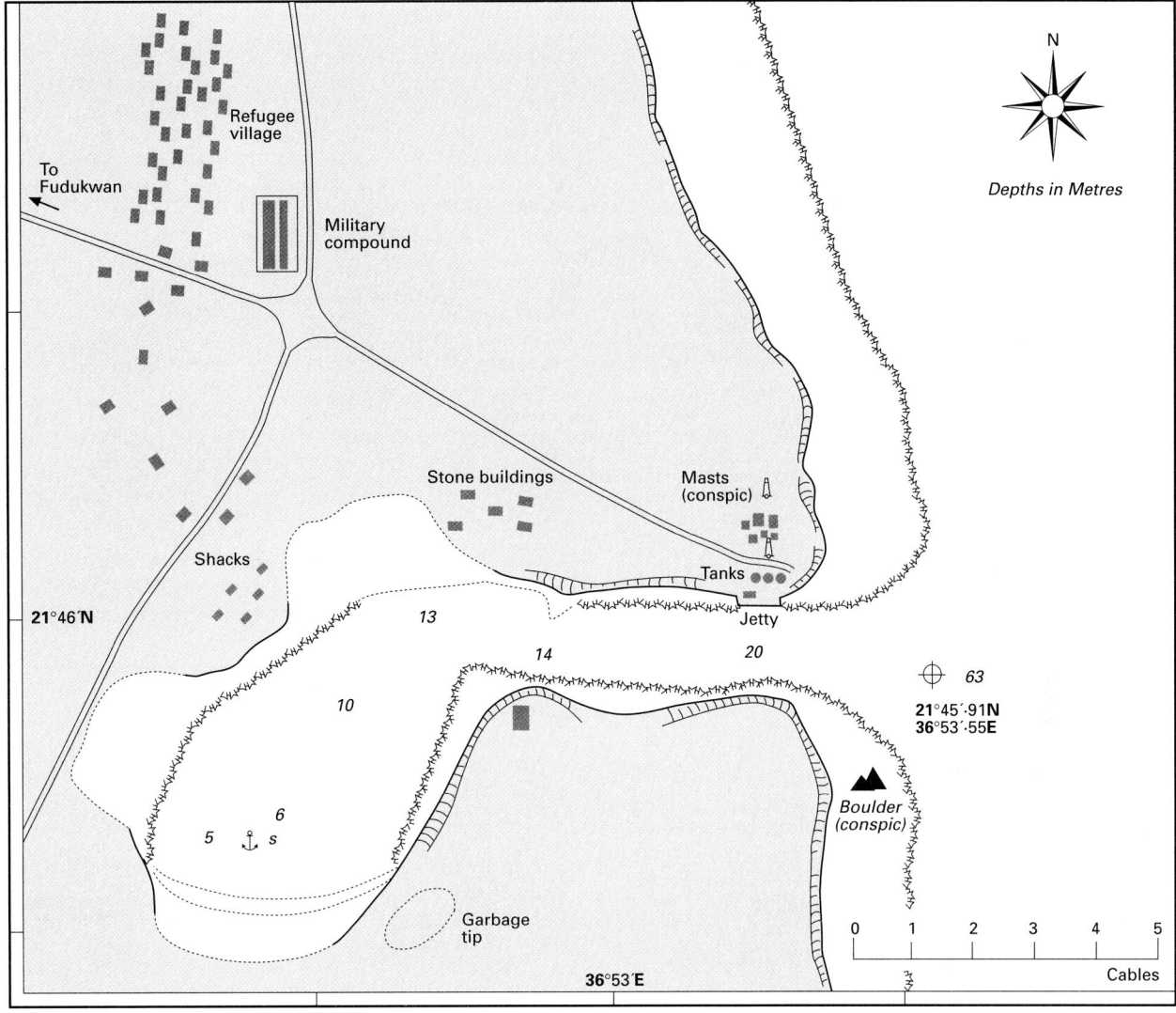

KHOR ABU ASAL OR MARSA OSEIF

are friendly and courteous unless you make life difficult. The three bays of the *marsa* are more like small lobes, the apparent size of which varies with the season. They look larger in winter than in summer. The village N and W of the *marsa* is full of refugees from once Sudanese Marsa Halaib and recently had a population of over 2000.

The anchorage in the bay S of the village is well sheltered with good holding in sand, though the garbage tip doesn't add to its attractions. It may be possible to get transport to the village of Fudukwan in the foothills where more supplies are available sometimes. We were given no indication whether the diesel in the tanks by the jetty could be bought. One could always try.

Marsa Ribda
Plan page 174

Approach
The entrance is wide, clear and trends W to a spit projecting from the N shore. The scant remains of a wreck lie on the reef N of the entrance channel looking like a broken beacon. There is an incongruous set of concrete steps leading nowhere on the S point. At the spit the channel is very narrow, though with 13m, and you must turn sharply. Follow the twisting channel past 2 more sandspits before a large bay, full of coral, opens to the S with an open anchorage to the NW.

Anchorage
The inner *marsa* has good holding in mud, 13m. The far W end of the *marsa* is closed by a sand bar. There is no jetty or village but local officials will probably come from Marsa Oseif (Khor Abu Asal) by boat to see your papers.

Lacking a boat, officials may drive round and signal for you to come ashore. If you don't, quite reasonably, they get agitated. When we anchored here officials visited us in the morning. Then the camp guard changed. Various communication problems ensued. We were resting below and the dinghy was on deck. We noticed some noise ashore. At teatime a very wet visitor in immaculate, soaking long johns appeared over the transom and insisted that one of us accompany him to the village. Eventually we gave in and the mate went for a ride

Red Sea Pilot

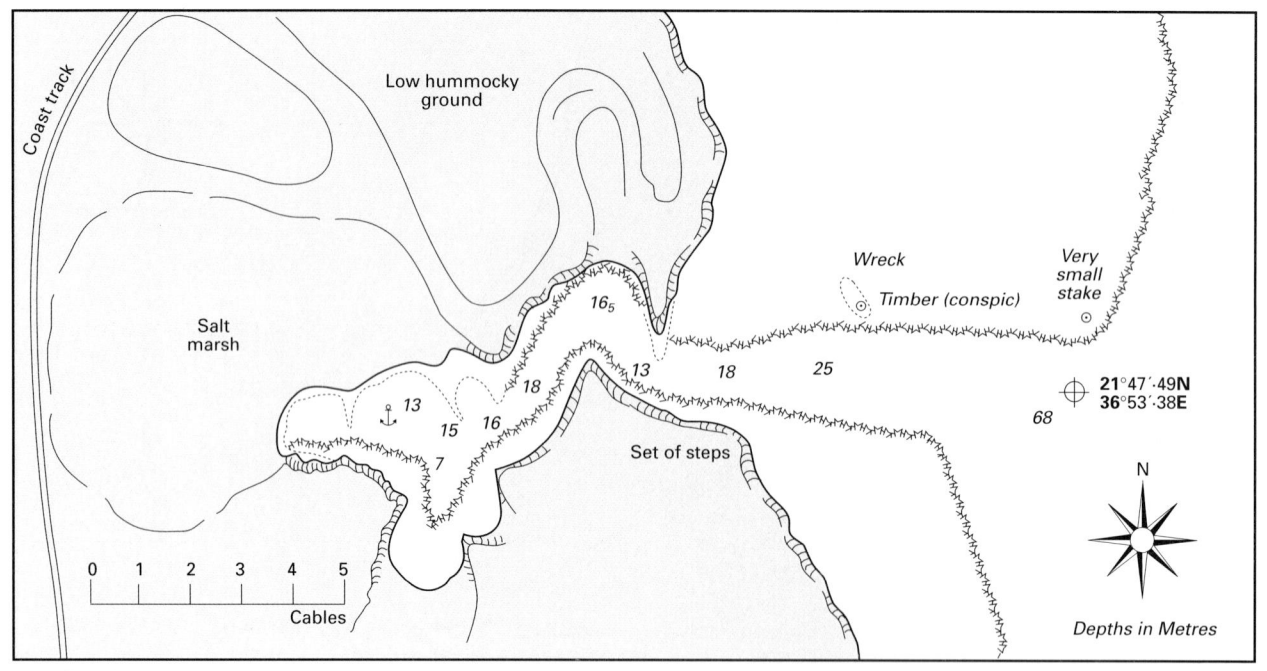

MARSA RIBDA

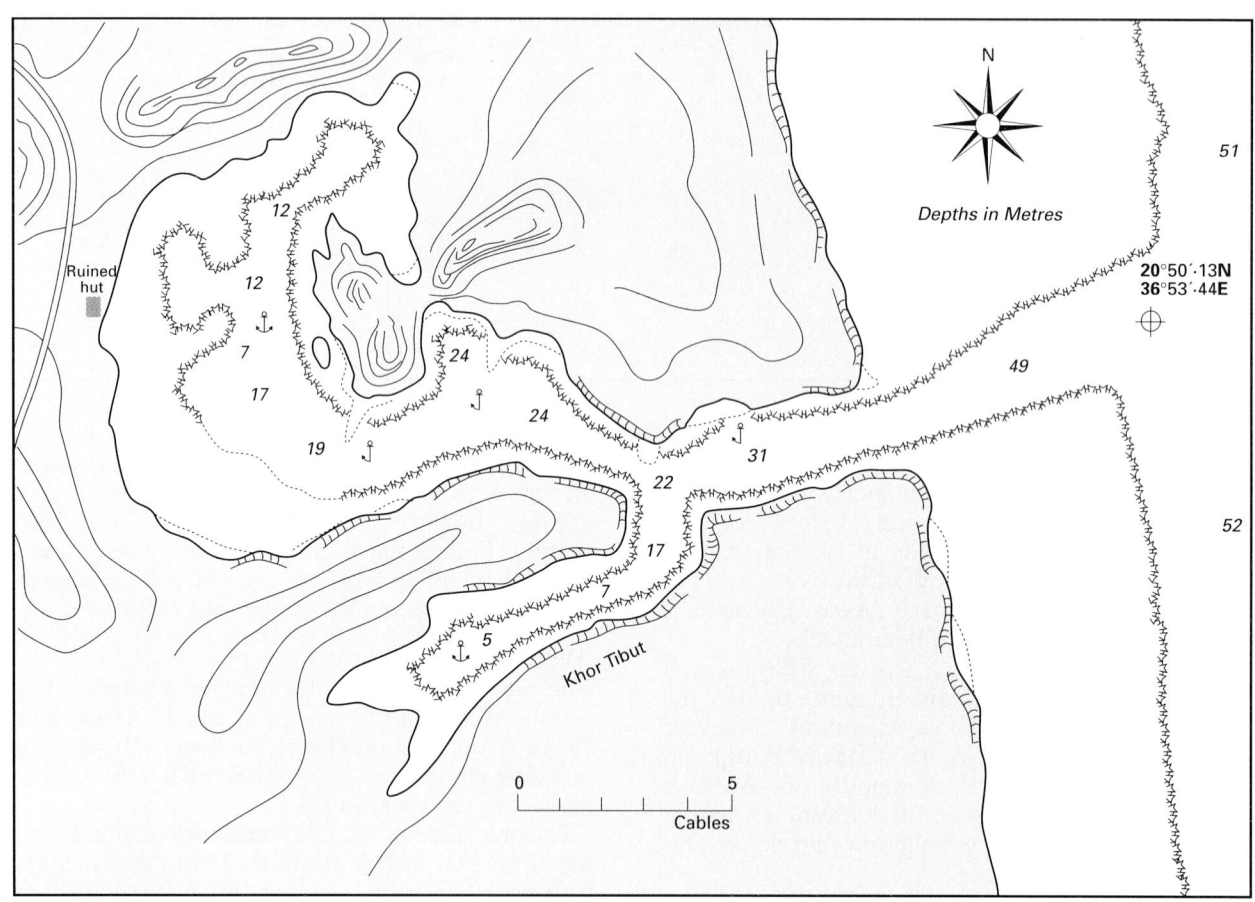

KHOR EL MAROB

through the desert at sunset with a dozen armed guards. Meantime the morning's officer in charge had come home. All was well, except for the embarrassed gentleman in long johns. That was when we realised that a ship's letter in Arabic might be a good idea.

General
This was unaccountably called Marsa Gwilaib for years, a name the locals claim never to have heard of. Initially it was called Eeles Cove after the British naval lieutenant who surveyed it. Mr Chisholm got the name right in 1922 but, eccentrically given the

slavish acceptance of all his other corrections, the BA ignored him and stuck with Marsa Gwilaib.

Khor El Marob
Plan page 174

Approach
The entrance is unbeaconed, but easy enough to find. The channel bears 065°/245°M.

Dangers
Fringing reef extends into the deep channel from the N shore about ½M inside the headlands. Favour the N side of the channel until you are level with the sandy bay where the half-anchor is shown on the sketch, then hold to the middle.

Anchorage
7–12m, sand and coral or sand and mud in the N arm of the inner *marsa*. You may need to hunt around for this depth since in general the *marsa* is deep. Other possibilities are on the sandspit at the entrance to this bay or on the tip of the sandspit between the two 24m depths on the sketch. Fishing boats sometimes moor here. Further E again, one could try Khor Tibut, the S arm, but the channel is narrow and the swinging room is limited. The last alternative is off the sandy beach near the entrance, but unless you put an anchor on the sand and drop back, this is very deep. It is the best if you want an early getaway.

General
Excellent protection, particularly further in. Walking and snorkelling are good but sharks can sometimes find their way in as far as the inner *marsa*. A decade ago dugongs still browsed in the sea grass shallows in the SW corner of the N arm. They may still, but given the economic conditions in Sudan, don't count on it. There used to be a potable well near where the hut (the vestiges of a village) is marked by the coastal track. In 1930 *HMS Dahlia* visited Khor el Marob and found it smaller and 2M further S than charted. Given that they fixed their position by several astronomical sights, that says a lot about the vagaries of astro in Red Sea conditions. Goodness knows where they actually were. As far as we can see there is no *marsa* between Marsa Ribda and Khor el Marob and the nearest anchorage that fits *Dahlia's* description is Marsa Ribda. The thought of even a small ship in that tight corner boggles the mind.

Khor Harbanaikwan
Plan above

The entrance leads NW then doglegs SW at which point it begins to narrow and shoal. The 'V' shaped inlet has sand at the point of the 'V' at the entrance to a wadi with, possibly, a dried out lagoon. There is fair shelter from N'lies in about 5m. The anchorage is tight with not much swinging room and probably easier for smaller (<10m) boats. The dried out lagoon is Khor Harbanaikwan proper.

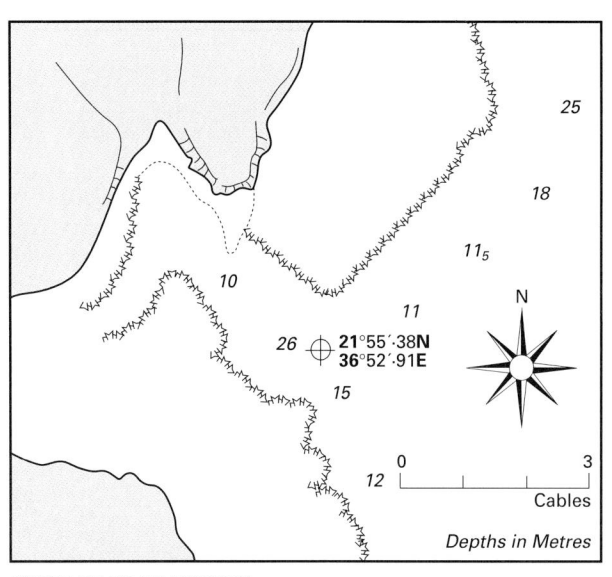

KHOR HARBANAIKWAN

Inner channel dangers between Khor Harbaikwan and Marsa Umbeila
There is a large, detached off-lying reef cluster 3–4 cables N/S centred on 21°56'·85N 36°53'·3E. It lies about a cable clear of the fringing reef about half way between Khor Harbanaikwan and Marsa Umbeila. In the immediate S approaches to Marsa Umbeila both BA138 and US62250 show the fringing reef extending further E along approx 21°58'N than it does.

Marsa Umbeila
Plan page 176

From Ras Hadarba to the N, or from the S, the approach is clear inside the barrier reef. Favour the N shore, through the narrow entrance and into the bay. In good light this is simple enough. The tomb, which looks like a sort of corral on a low bluff N of the entrance, just above a cave-like overhang, is conspicuous from the S.

Anchorage
6–8m, mud. Well protected and very good holding although a swell can make round the point in strong winds, making it rolly. It pays to tuck well in, which means that there is comfortable anchorage for only about three boats.

General
One of the boundaries between Egypt and Sudan lies exactly along 22°N. As you go N (come S) you'll see the Egyptian guardpost on the shore in that latitude. How far whose authority runs either side is disputed, but the area is at present de facto Egyptian, hence the track from the guardpost to the parking area near the tomb, so have your next courtesy flag ready. This is a lovely spot, which feels very secure. There is good snorkelling and a nomad village nearby. You'll sometimes see herds of goats and camels inland and the herder may come to find out whether you have anything he'd like that you want to give away.

Red Sea Pilot

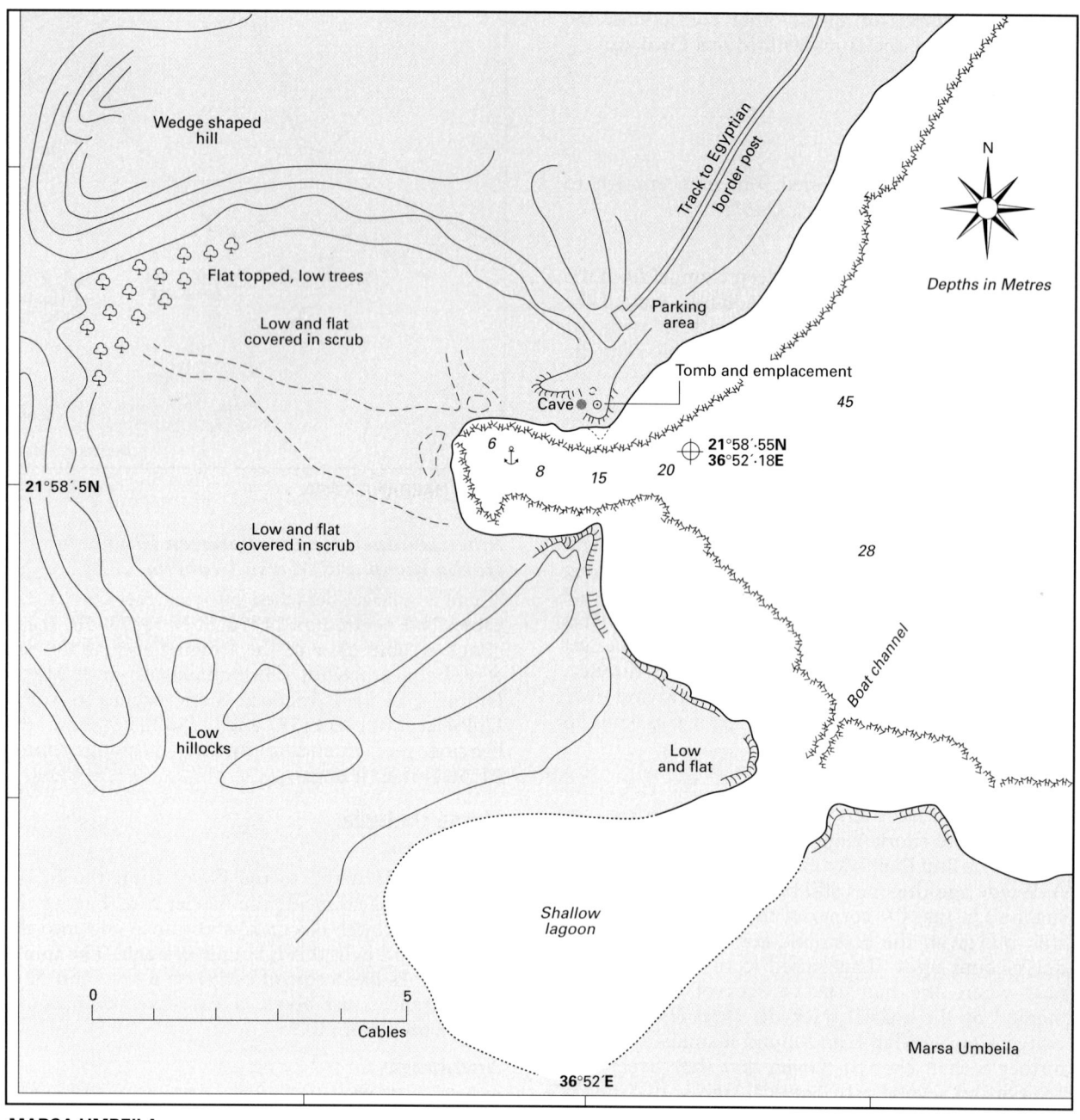

MARSA UMBEILA

Elba Reef
Plan page 177

There is an excellent reef anchorage and dive spot under the lee of the poorly charted, main W part of the Elba Reef complex. From the N, there is a safe entry into the wide passage between the two parts of the Elba Reef complex at 22°00'·8N 37°00'·3E. But be warned that coming from the N in heavy weather you see no sign of either part of Elba Reef until your heartbeat is reaching fibrillation point. From the S, either from seaward or from the Inner Channel, the approach is clear enough provided you take cognisance of the various isolated reefs. There is a beacon, as well as a conspicuous, white (guano covered) wreck 1M SSE of the N tip of the separate E part of Elba Reef. Note the small (about 100m diameter), unmarked reef in approx 21°57'·75N 37°01'·5E. Another unmarked reef has been reported at approx 21°56'·2N 36°59'·2E. A further reported line of reefs, about 1M N–S between 21°57'N and 21°58'N and in 36°55'·2E is of doubtful existence.

The entrance to the anchorage is on the E edge through a pass with 7m at 21°59'·35N 36°59'·38E. Zigzag NNW past a couple of large bommies to a pool SW of a closed lagoon of brilliant turquoise blue water, with 3 large boulders on the reef to the N.

Anchor in the general area of 21°59'·68N 36°59'·42E in 10m sand and coral. Snorkelling and diving are good, though the reef here, as in much of the mid and N Red Sea, is beginning to suffer badly

The Sudan Coast N of Port Sudan

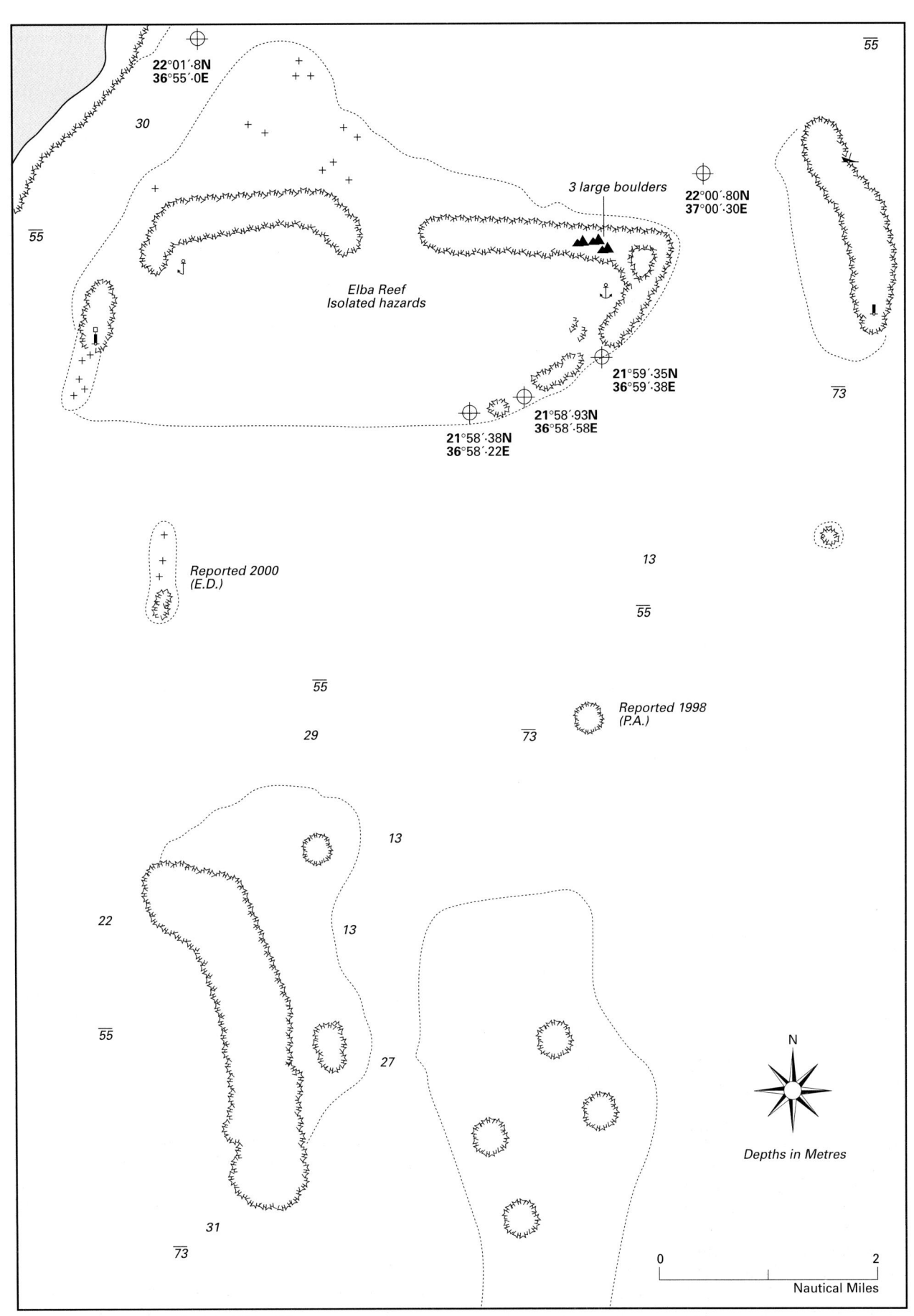

ELBA REEF AND HAZARDS IN THE AREA

Red Sea Pilot

from global warming. The wreck of the *Lavanzo* is somewhere hereabouts, probably on the E, detached reef. Ask a dive outfit in Egypt or Port Sudan for exact co-ordinates. Since the whole wreck is at least 10m down, it is not the visible wreck charted.

The name Elba has nothing to do with the Italian isle, nor the exiled Napoleon. It is a poor romanisation of el Dibia, a name which applies to the whole area W of Ras Hadarba

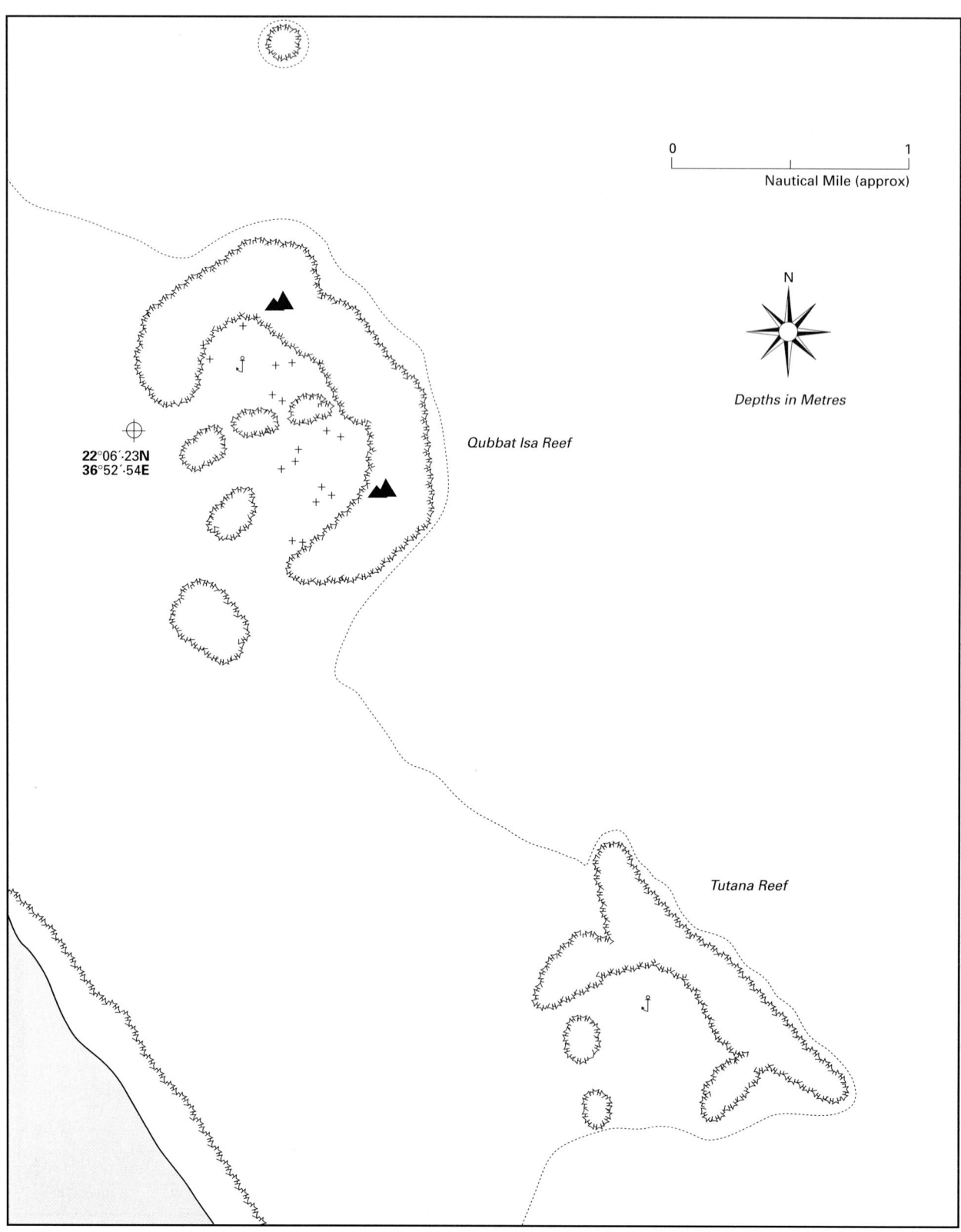

22°06′·23N
36°52′·54E

Qubbat Isa Reef

Tutana Reef

Depths in Metres

QUBBAT ISA AND TUTANA REEFS (SCHEMATIC)

178

The Sudan Coast N of Port Sudan

Qubatt Isa and Tutana Reefs
Plan page 178

These are day anchorages for diving enthusiasts about 5–6½M E and SE from Ras Qubbat Isa (22°10'N 36°46'E). There is a dangerous isolated reef outcrop about 1M N of the conspic rock marked on the N of the reef crescent in the sketch. The sketch is very approximate. There is not much agreement between charted positions and what dive enthusiasts have discovered to be the GPS co-ordinates of the reefs, so choose fair weather and pick your way. Both reefs come with the caution that they are day anchorages.

Don't forget that in any case this is a sensitive area. It is rumoured that the USA has a spy plane base in the Jebel Elba triangle (i.e. the area between the two borders marked on older charts) for monitoring Sudan. We have been assured it is impossible to get a permit to drive into the area, so maybe there's something in the gossip.

Marsa Halaib
Plan below

Sudanese officials sometimes warn yachts to stay offshore between Marsa Umbeila and Ras Baniyas (Ras Banas) on the far side of Foul Bay because the S of Foul Bay is disputed territory. In fact the whole area is firmly in Egyptian hands and has been for several years. Cruisers have been allowed to shelter at Marsa Halaib and Marsa Girid, the last coastal anchorages in this section. Anchorages in Foul Bay, described in the next section, are open to you if you are confident in your ability to navigate in coral with poor charts and rudimentary directions.

Marsa Halaib is a refuge only from stress of weather or if you have other problems. The Egyptian navy has a base here. They will ask you to go alongside and hand over passports and ship's papers. Going ashore is forbidden. However, the navy have been most helpful to yachts with engine trouble. Call them on VHF Ch 16 on approach. Best not to arrive in poor light although the entrance is well beaconed. The watchtower W of Gable Point is a good landmark.

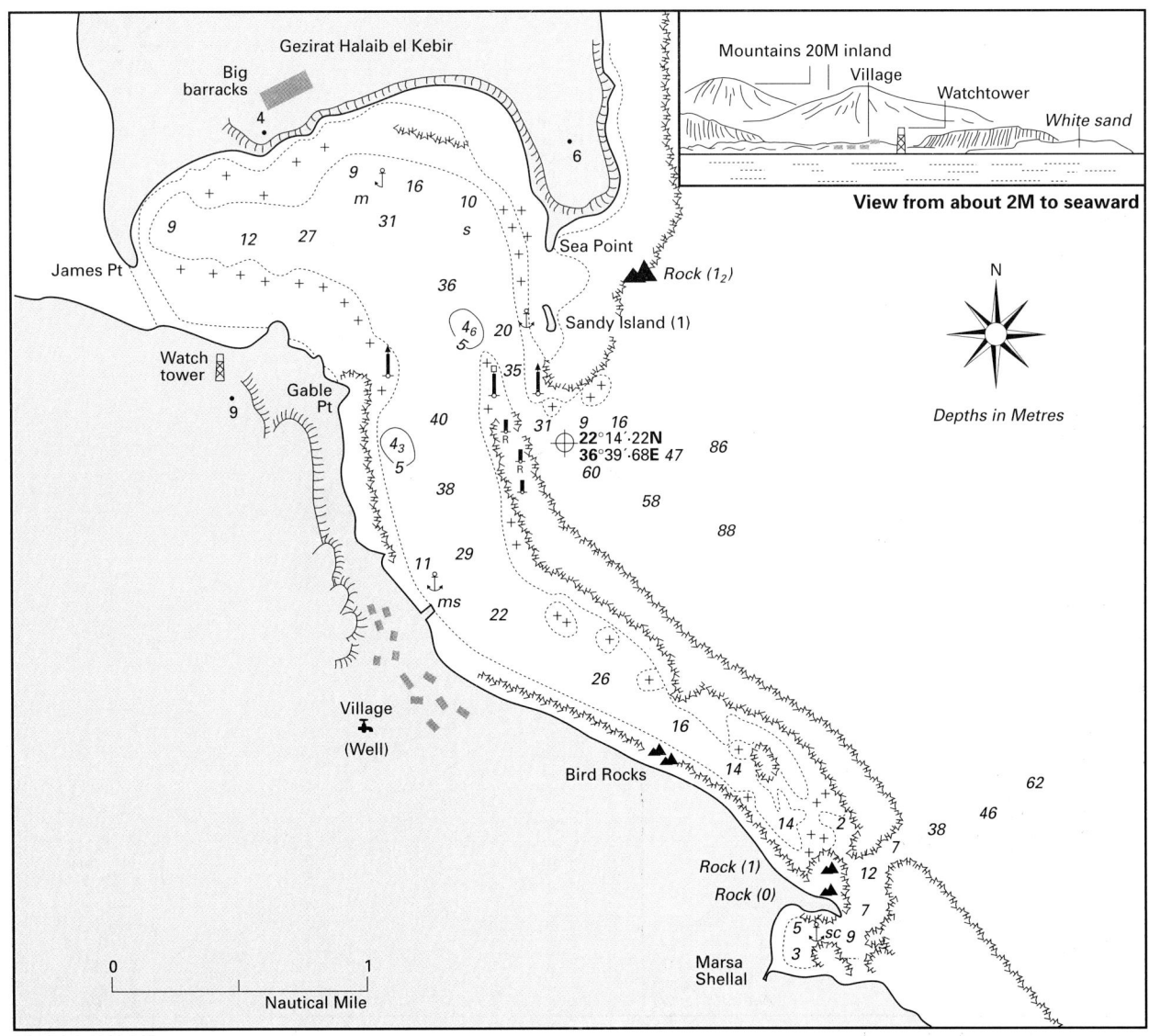

MARSA HALAIB

Red Sea Pilot

Anchorage

As directed, or in 10m, sand, E of Sandy I, N of the entrance channel. An alternative is in 10–15m, mud and sand, off the village. Both have good holding. A half-anchor is marked in the N under the lee of Geziret Halaib el Kebir. This is better protected, but yachts have reported that holding is iffy in a thin layer of sand over coral rock. Marsa Shellal is another possible, but you may excite the military.

Marsa Girid (Sharm Alueda)

Plan below

Approach

The simplest approach to this *marsa* 1½M S of Ras Abu Fatma, is N of the reefs that extend from both Geziret Halaib el Kebira and Geziret el Dibia (see approach plan), although there is a navigable channel S of Gez el Dibia once past the reefs N of Halaib. Once about ½M off, start looking for the narrow pass through the reef, where you will find about 3·5m of water. It is not marked, but the long, low building shown on the sketch serves for orientation. From there eyeball your way past the bommies into the anchorage in the lagoon, N of the jetty. There is a conspic desalination plant with 2 water storage tanks at its E end. Off the jetty are two sunken pontoons.

Anchorage

Anchor in 5–10m, good holding, with shelter from all directions. Diesel and water are available.

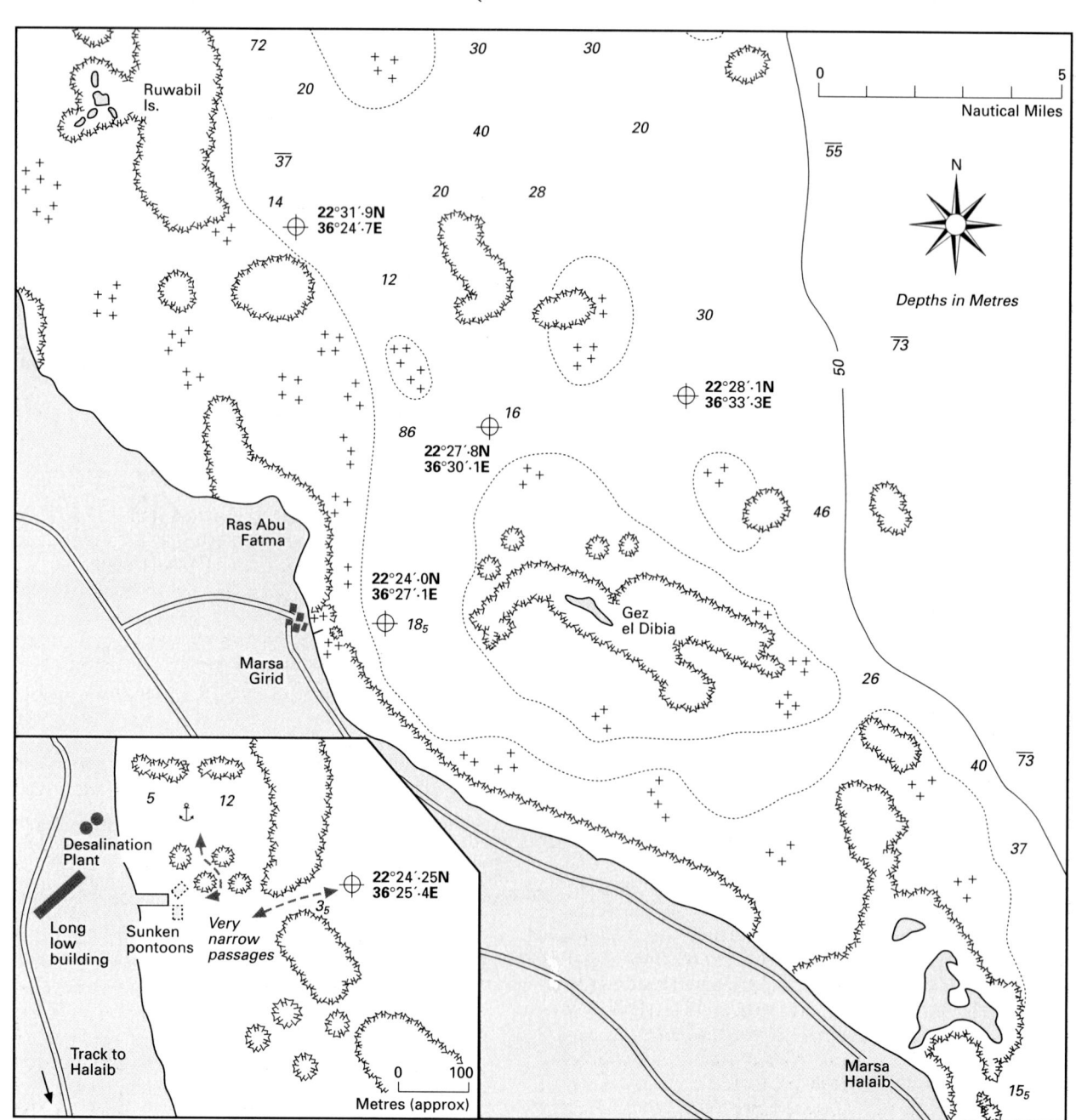

MARSA GIRID AND APPROACHES

General

Not far S, conspicuous because of the wooded country behind it, is all that remains of one of the three medieval Egyptian ports on the route to Suez. This one was called Suakin el Qadim (Suakin-the-Old). Arab geographers called it Aidheb, and it was a major junction point for caravan routes into the Nile Valley. It was superseded by the new Suakin, S of Port Sudan, in the 13th century. The other medieval Egyptian ports were El Quseir and El Tur, further N. On aero maps Aidheb is the name of the settlement, such as it is, at or near the *marsa*.

5. The Egyptian coast: Foul Bay to Hurghada

British Admiralty Charts
Large scale 3043
Medium scale 2375
Small scale 158, 159

US charts
Large scale 62162, 62177, 62188
Small scale 62230

French charts
Large scale 7113
Small scale 7112, 7113

Important notes

1. **Reef anchorages** All of the Egyptian reefs are now, in theory, national parks. According to Law 102 of 1983, it is illegal on the Egyptian Red Sea coast to anchor on reefs (i.e. put anchor or chain on or over any living reef), spearfish whether snorkelling or scuba diving, collect corals or shells (whether the inhabitants are alive or dead), fish with net, line or rod and line, dump rubbish, feed the fish (!) and walk on, disturb, break or damage reefs. This law is little policed because of shortage of funds and almost every offshore island is a rubbish tip of black plastic bags full of wine and spirits bottles and beer cans from dive boats. That's called ecotourism.

    There has been vast, indeed runaway development of hotel and dive resorts along the Egyptian coast. You will see dive boats at anchor on most reefs between Ras Baniyas and Safaga. Resorts of varying degrees of sophistication can be found about every 20M. A positive result has been the laying of moorings to reduce damage to reefs from the out of control dive tourism industry. The project was carried out by the Hurghada Environmental Protection and Conservation Association (HEPCA) funded by the US Government Aid Agency (USAID) and backed by the Egyptian Environmental Affairs Agency (EEAA). Some 570 moorings have been laid on popular but now protected reefs from Ras Baniyas N to the Sinai. Each US made Manta Ray mooring is designed to take high loads (several boats). Some are marked by an orange buoy. Others are merely floating ropes. Sadly, of those we saw, none were tenanted by dive boats which still use their destructive ground tackle. If you see one, please use it, though make a reconnaissance dive to check the condition of ropes, swivels, etc. The buoys are very close up in the lee of reefs. You cannot leave your boat unattended. See also 'Diving' in the Planning guide.

    There are some patrols in the Hurghada and Safaga areas and elsewhere local dive boats, protecting their own turf, can and will make your life difficult if you break the law, including reporting you to the authorities. It follows that if you choose a reef anchorage you should either find one of the HEPCA/EEAA/USAID moorings or find a clear sandy spot in which to anchor. The simple rule is, minimise all damage to coral.

2. **Customs Fee** Egyptian Customs levy a fee on foreign yachts staying over 10 days as follows:
    Minimum: £E100 (US$20–25) for stays of 10 days–1 month.
    £E250 (US$55–60) for stays up to 4 months
    £E500 (US$110–120) for the 5th–8th months
    £E1000 (US$220–225) for the final four months of any year.

    If you intend to stay for the maximum 1 year period you will pay £E1750 (US$380–400) at the outset. Note, if you pay 1 month and stay longer, the first 4 month extension is BACKDATED to your date of first arrival, i.e. you'll pay £E350 (US$75–80) for your first 4 months, instead of US$55–60. Pay in your first port only. Ask for a receipt and obtain a sailing permit for your next port of call.

    It is still possible to avoid the customs fee if you stay on the Egyptian coast only for the maximum 10 days and don't want to call at any marina, eg if you intend to sail from Quseir, Safaga or Hurghada to Suez, stopping only at bay, reef or island anchorages. If you subsequently decide to stop at a marina YOU MUST PAY THE CUSTOMS FEE. You'll be liable for port dues

Warning
Although the most recent editions of charts from all the major hydrographic offices are reconciled to WGS84, this CANNOT compensate for shortcomings in the original 19th-century surveys. Almost NO survey has ever been done on this coast to ANY WGS datum. WGS compatible charts are corrected versions of the old surveys. Three points of navigational importance follow. First, always navigate, even with GPS, with circles of probable error (CEPs) of at least 2 miles, and at night at least 5 miles. Second, navigating by GPS alone in this area is stupid. Do not do it. Use your hand-bearing compass and radar, if you have it, to establish WGS84-to-chart error factors. Third, treat ALL waypoints with appropriate caution, including those given below, and NEVER assume that the waters between, close to or at waypoints are free of hazard.

Red Sea Pilot

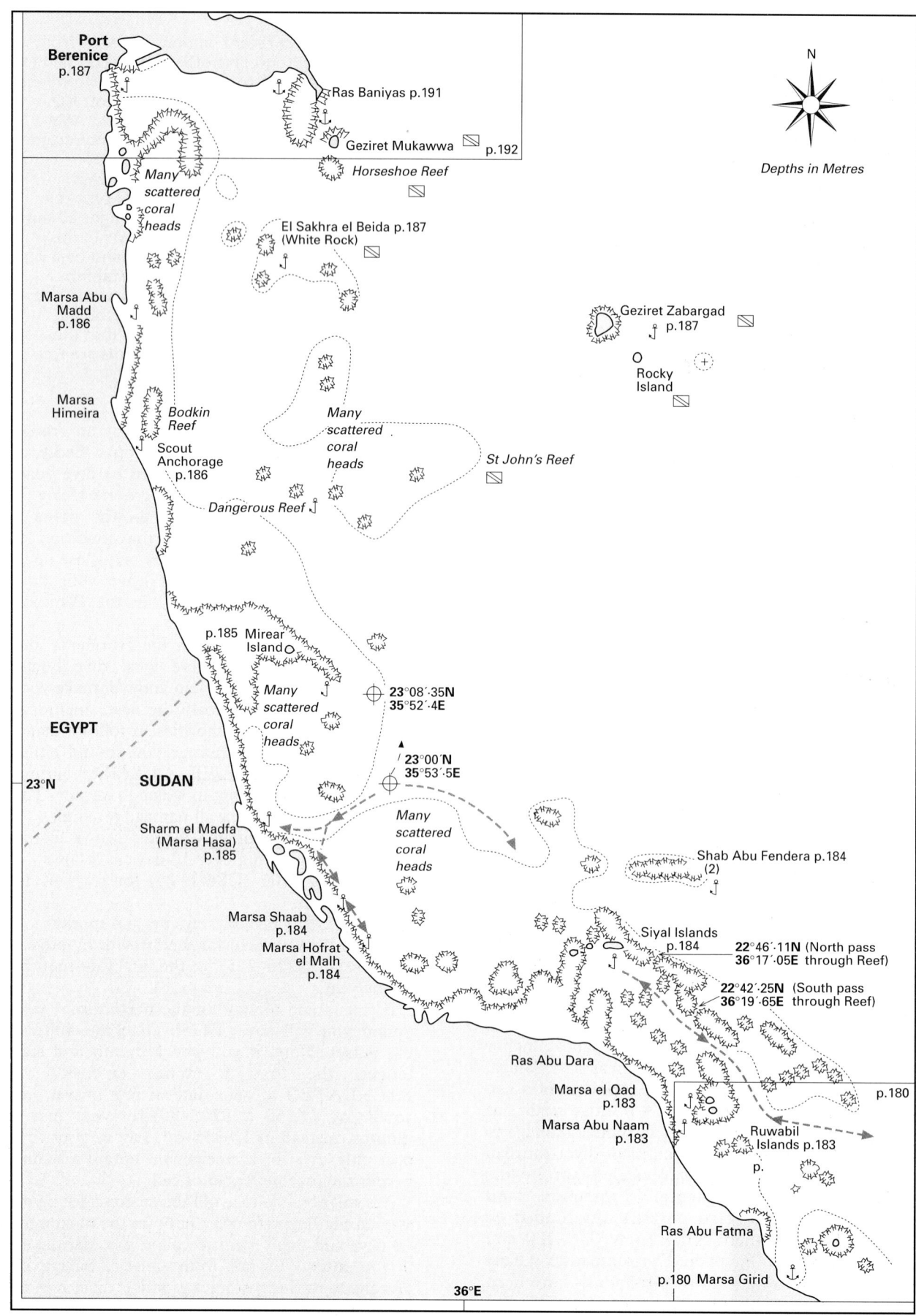

FOUL BAY ANCHORAGES – MARSA GIRID TO RAS BANIYAS
Showing routes into the Siyal Islands etc.

The Egyptian Coast: Foul Bay to Hurghada

and customs fee if you call at any port and stay over 10 days.

3. **Formalities along the coast** Reception for yachts at anchorages between Foul Bay and Safaga varies. You'll get a visit from any nearby military outpost. Do be friendly and civil. Most soldiers are national servicemen sent to the sticks for up to 18 months. A visiting yacht brings someone new to talk to, who may have up-to-date magazines no longer wanted on board. These soldiers are often doing service during, or shortly after graduation from, university degree courses. Imagine yourself in their shoes.

It is OK to cruise up the coast of Egypt, stopping at anchorages en route, before checking in at Safaga. Having a visa will smooth your way. If asked why you have stopped, an excuse of strong winds or engine trouble makes life simpler. Authorities like documentation and will ask for copies of clearance papers from your last port and of visa and entry stamps in passports. They will be pleased if you give them a copy of your crew list or a completed Arabic ship's letter. Note, in theory you are required to check in with the port authority at every port you visit in Egypt. That means paying port dues. This is a good reason for only visiting one port i.e. Safaga, Hurghada or Sham el Sheikh, on your way to or from Suez.

Foul Bay

Foul Bay has been the name of this area since the *Periplus of the Erythraean Sea* and before. The classical Greek is Akathartos Kolpos, though whether, having sailed it, you do or don't experience catharsis is another matter. Navigation requires extra care, since hazards are poorly charted. Stay off the coast if you can, much of it is a restricted area, if little policed.

Foul Bay is becoming a major dive attraction. Good dives are known on Gezirat (Gez) Mukawwa, Gez Zabargad, Rocky I, Horseshoe Reef, St John's Reef and White Rock. For the St John's reef sites you need to ask dive charter operators for co-ordinates. There are no anchorages and, in any case, you should reflect on Note 1 above.

Marsa Abu Naam and Ruwabil Is

Plan above

Approach

Access is from 22°31'·9N 36°24'·7E. Only three islets are charted, though as our plan shows, there are five. If you choose Marsa Abu Naam the two SW-pointing islets give a rough line to it. The 20m line, which penetrates further than charted, runs from deeper water in their direction. Once off the islets, weave through isolated coral outcrops on roughly 240°M for about 2M, in steadily shoaling water, until a group of small sand *cays* appears on the coastal reef. At this point a hill charted as 203m will bear about 195° and you will have found the *marsa*.

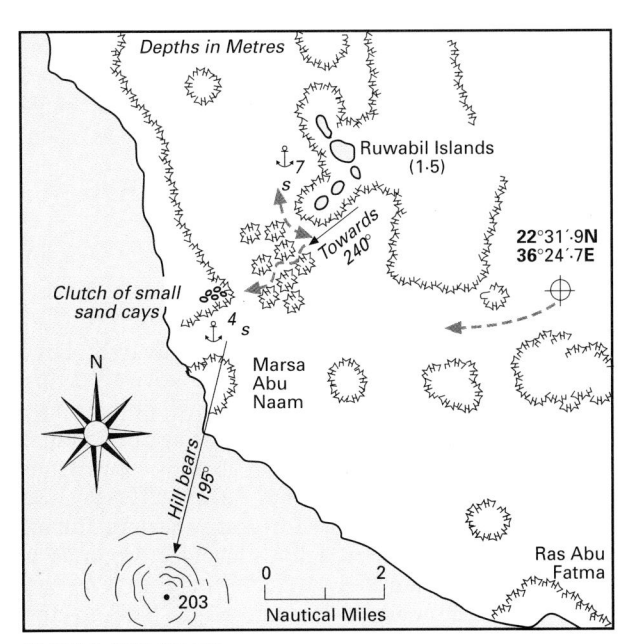

MARSA ABU NAAM

Alternatively, feel round the S of the two SW islets to get into the lee of the reef where shown.

Anchorages

Marsa Abu Naam 4m, sand, well sheltered from sea and wind.
Ruwabil Is 7m, sand, good shelter at 22°33'·97N 36°21'·17E.

Marsa El Qad

Approach.

There is no obvious route here from the previous anchorage, though Lt Moresby's surveyors found one! The best approach is from 22°38'·94N 36°22'·4E, N of the Ruwabil Is. Leave the long reef which stretches NW to the Siyal Is, and on which there may be a sandspit, to starboard. When the S end of the reef bears roughly 080°, head in to the coast on a general line of 255°–260° for 6–7M, with a careful lookout for isolated coral heads (see

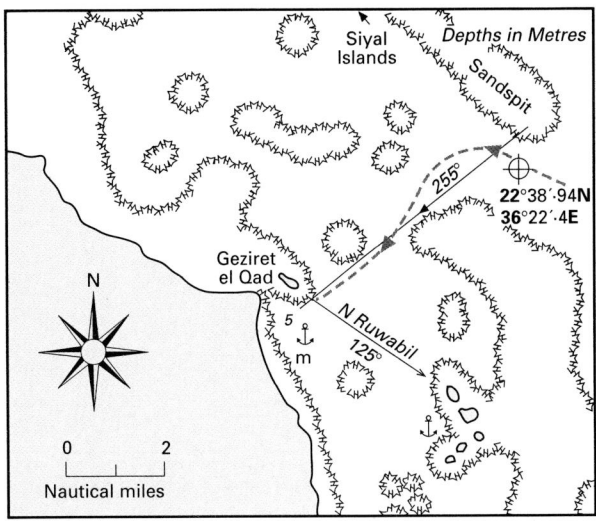

MARSA EL QAD

183

Red Sea Pilot

sketch). The anchorage is about 5–6M SE of Ras Abu Dara, just S of a small, uncharted islet named Gez el Qad, with N Ruwabil I bearing roughly 125°.

Anchorage
5m, mud, between Gez el Qad and the point, less than 3 cables from the low sandy shore.

Siyal Is

Approach
This is tiger country for the adventurous only. Dive charter boats operating in the area advise staying clear unless you know what you're doing. The approach from SE is difficult through scattered coral heads with poor water visibility. There is a pass from seaward at approx 22°42'·25N 36°19'·65E, but the balance of the approach is through waters full of hazards. The best approach, in fair weather only, is shown on the plan NE of 22°46'·11N 36°17'·05E, and then following a clear passage NW to the anchorage over a very uneven bottom. Anchor in 5m sand and coral near isolated bommies S of the crescent-shaped E'most island. There is no safe or easy approach or exit N.

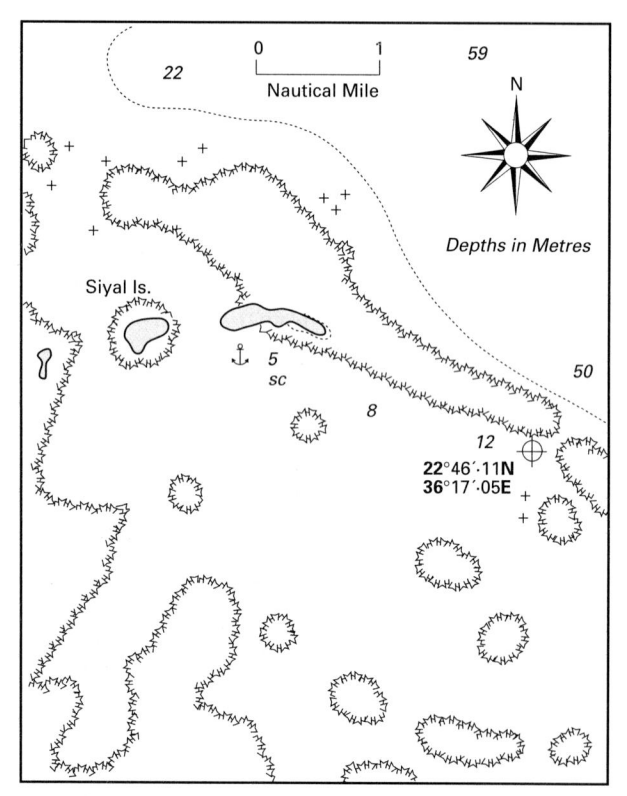

SIYAL ISLANDS

Shab Abu Fendera
Plan page 185

Approach
You need good visibility for Shab Abu Fendera. There is a safe entrance, E of the smaller, 0·8m boulder which lies at the E end of the main reef, at 22°53'·45N 36°18'·95E. The alternative approach is through clear water from the S, but staying EAST of longitude 36°18'·4E. This keeps you E of a large area of dangerous underwater coral pinnacles. Most of these pinnacles are >2m below the surface. The W end of Shab Abu Fendera is submerged.

Anchorage
One anchorage, at 22°53'·3N 36°18'·55E, is at the E end of the main reef S of the boulder (0·8m awash) marked in the sketch. Another is at 22°53'·56N 36°16'·94E in the lee of the central *cay* in 8m, sand and coral. Generally better holding is found E of 36°18'·4E. Good diving at the W end of the reef.

Marsa Hofrat el Malh
Plan below

See the next anchorage for approach details. Once off Marsa Shaab, Marsa Hofrat el Malh is a further 2·3M S along the fringing reef. Anchor about a cable out in approx 4m, mud.

Marsa Shaab

The passage in through the offshore reefs is oriented ENE/WSW and is clear and wide from 23°00'N 35°53'·5E. Once through the deep passage, work your way S parallel to the fringing reef. Marsa Shaab is at the S end of an expanse of relatively unobstructed water. On the shore is the end of a small wadi. Exit is by the same route. Gebel Hamra Dom (207m and 16M inland) is very prominent in good visibility, bearing about 215° from the

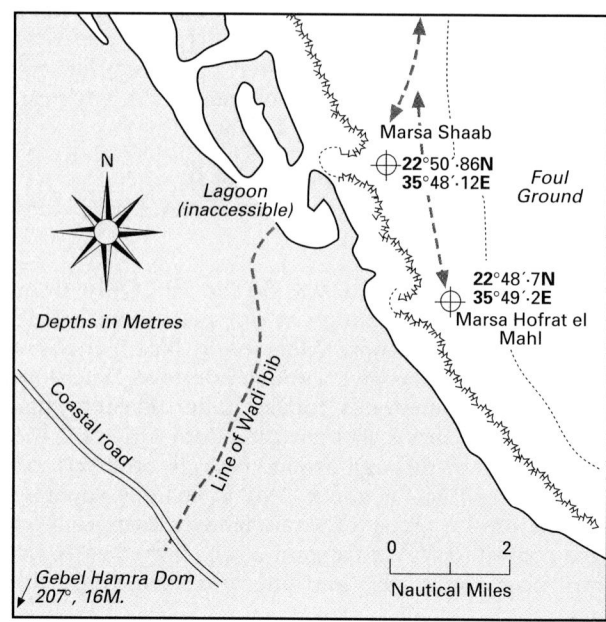

MARSA SHAAB

The Egyptian Coast: Foul Bay to Hurghada

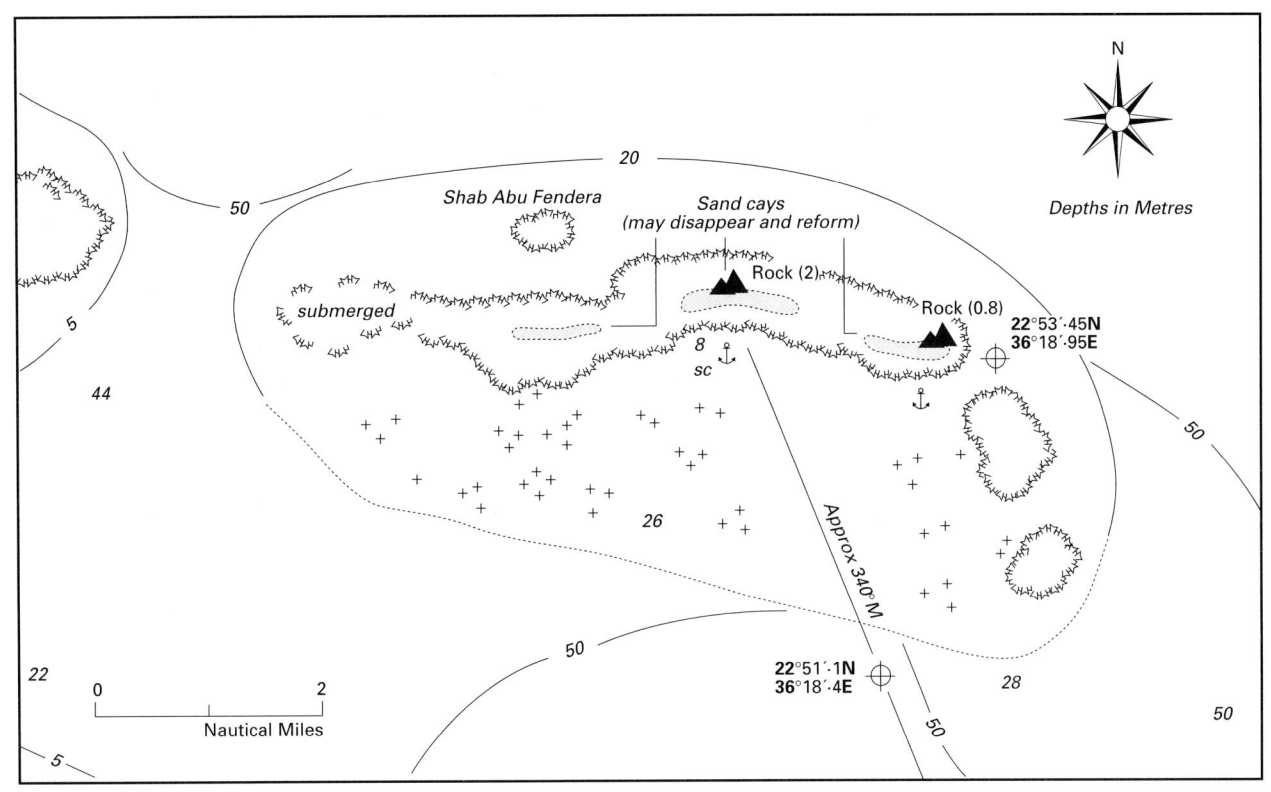

SHAB ABU FENDERA

anchorage in 6m, mud, good holding and well sheltered.

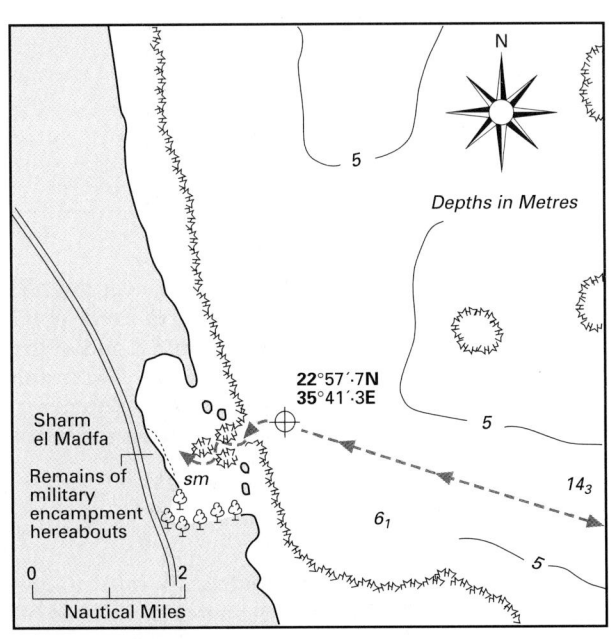

Sharm el Madfa (Marsa Hasa)
Plan below

The pass through the offshore reefs is as described above. The entry to the *marsa* is tricky and needs good light because the water is not very clear and you must weave between coral heads along a narrow, twisting channel. It would make sense to recce ahead in a dinghy, and mark dangers with fenders as buoys. Anchor in 4m, sand and mud, off the beach at the W end of the bay. Vestiges of a Second World War British military camp may still be visible.

Mirear I Reef
Plan page 186

There are 2 beacons on the S tip of a long, NW/SE tending reef. The approach needs good light and MUST be eyeballed. The anchorage is near the beacons NW of 23°08'·38N 35°48'·74E, 4M from low, sandy Mirear I which you can't see from where you drop the hook. The waters between the anchorage and the island are fraught with nasties. This is no area for novices. In late summer 2000 a German boat, fortunately built of steel, was caught out by a wind change at night, banged about on the coral destroying its rudder and prop shaft, and had to limp under sail to Jeddah for repairs, which cost an arm and a leg. Be warned.

Dangerous Reef
Plan above

This isolated reef is clearly marked on the chart and offers a good refuge in the middle of aptly named Foul Bay. It is the first of two, which lie in the middle of the 10M-wide passage between St John's Reef and the foul ground along the coast. The shallow arc of reef is some 1500m end to end. On the reef are two boulders either side of the anchorage, the E'most being the larger. There is a large coral head in the E end of the lagoon in the lee of the reef. S of the lagoon there are three large bommies spaced evenly apart E/W and about a cable

185

Red Sea Pilot

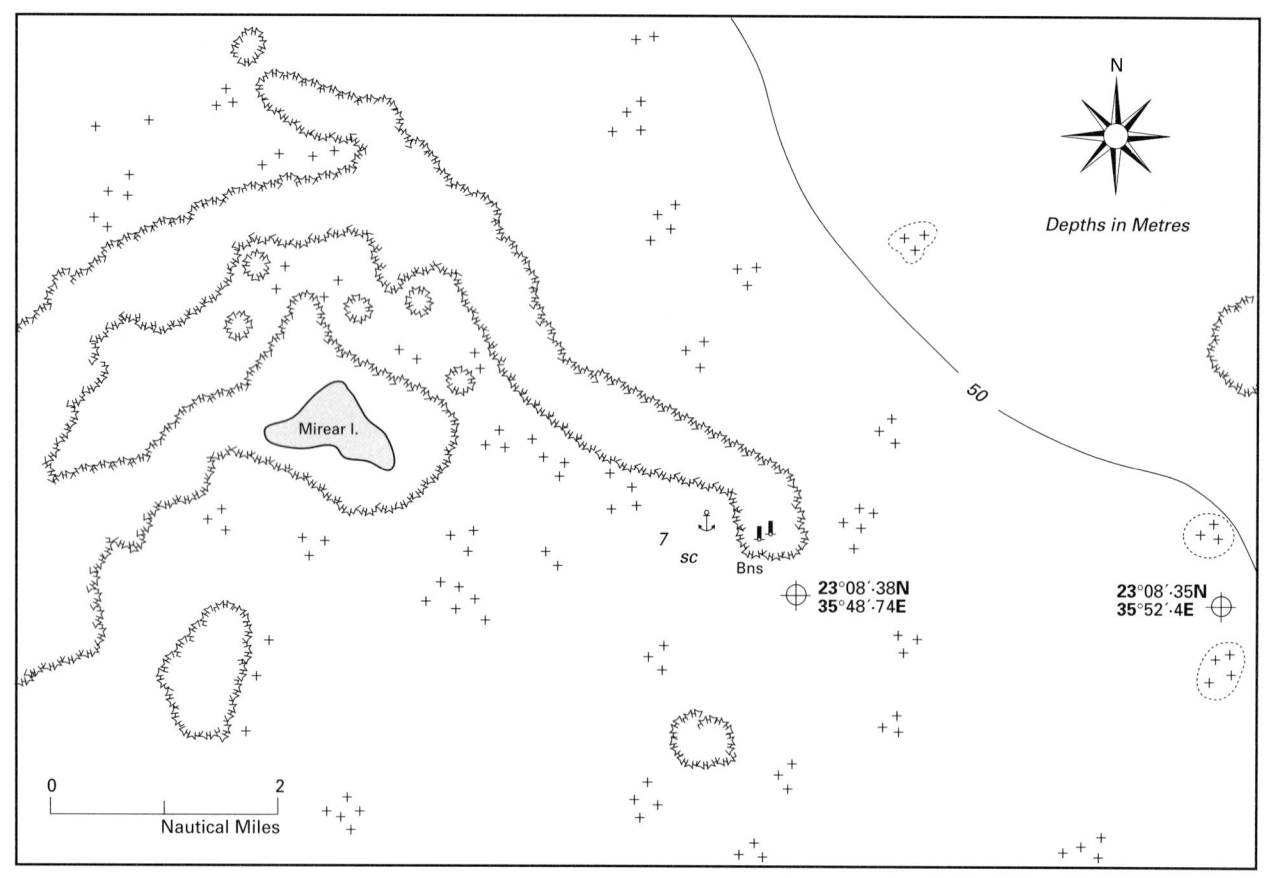

MIREAR ISLAND

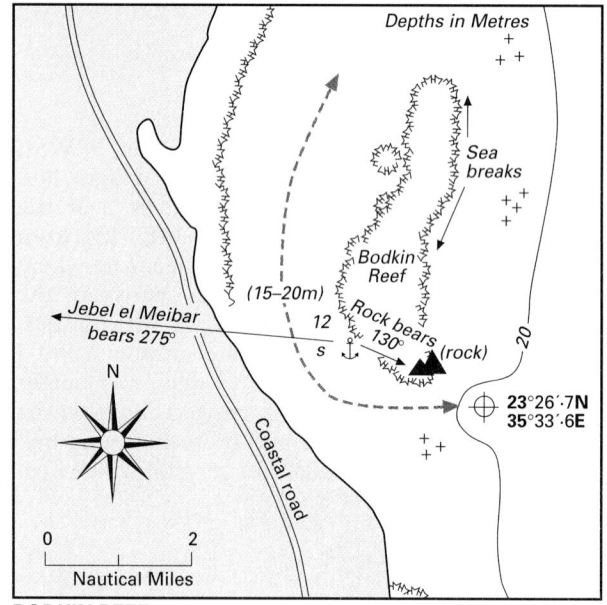

BODKIN REEF

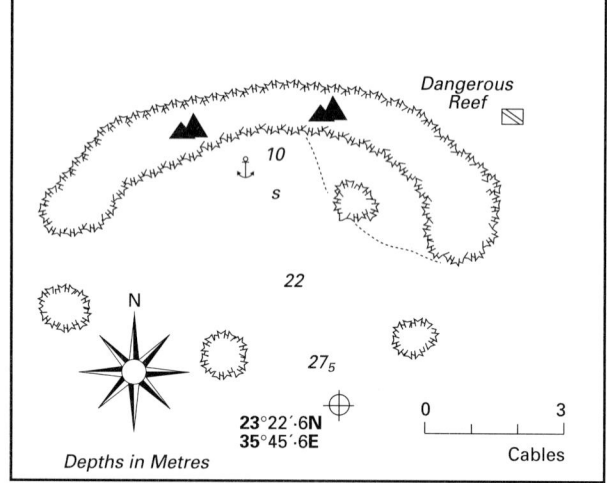

DANGEROUS REEF, FOUL BAY

S of the E/W line joining the horns of the main reef. The anchorage is in 10m sand with some coral.

Scout Anchorage and Bodkin Reef
Plan above

There is a clear passage between the S end of Bodkin Reef and a coral head, which appears isolated, but may be connected to the fringing reef. The pass between Bodkin Reef and the fringing reef is 15–20m deep. The sea usually breaks over the NE edge of the reef. Anchor in approx 12m sand, good holding, with a conspicuous rock on the S end of the reef bearing 130°. The Bodkin, by the by, is the needle-shaped hill inland once called Berenice's Bodkin. There is a village called Klën on the coast road a short distance inland.

Marsa Abu Madd
Plan opposite

From Marsa Himeira and Scout Anchorage, follow the fringing reef N. There is apparently no exit N, and one leaves the way one came in. The *marsa* is

186

The Egyptian Coast: Foul Bay to Hurghada

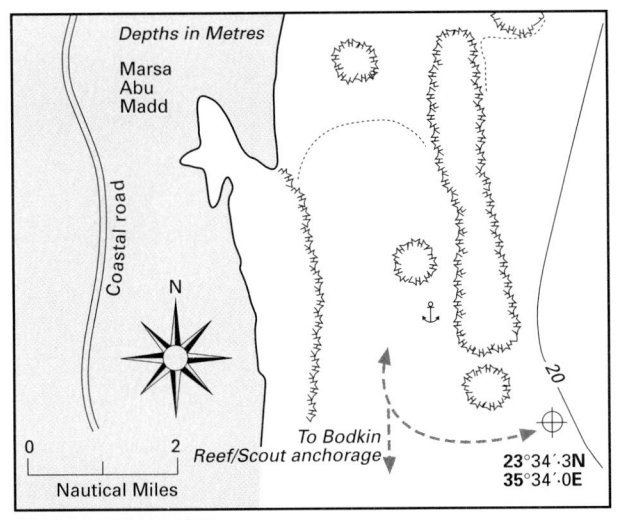

MARSA ABU MADD

filled with coral and there are isolated coral heads S of the main reef, which is about 2M offshore, and another just N of the anchorage, charted as a rock. 3M inland a range of mountains rises abruptly to 1300m. Anchor W of the reef on a mixture of sand and coral with some shell, in about 10m.

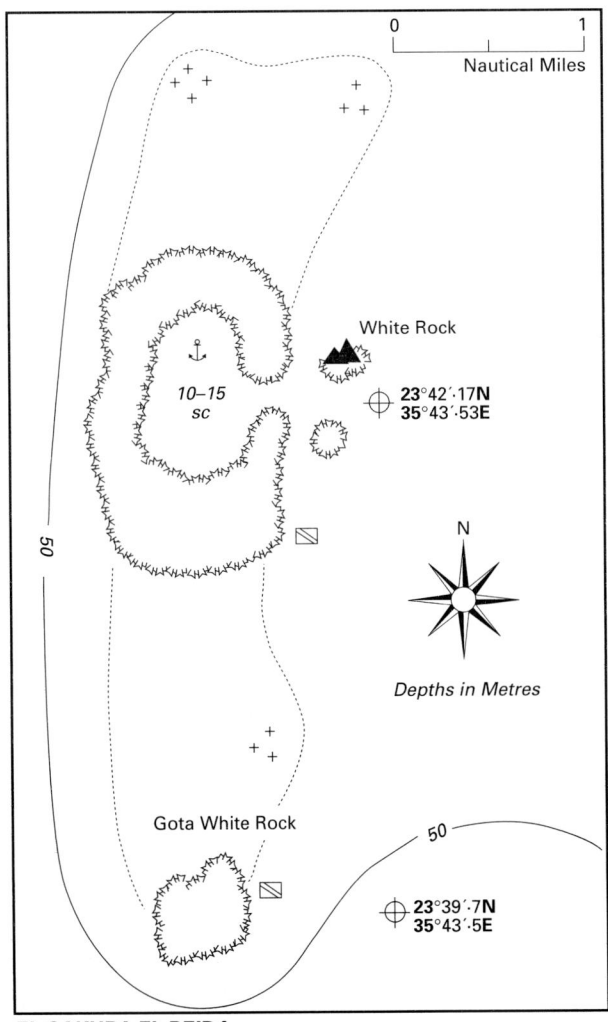

EL SAKHRA EL BEIDA

El Sakhra el Beida (White Rock)

Plan below

The boat-shaped rock makes a good landmark for the entrance and is visible from between 2 and 4M away because it's covered with guano. Coming from the S, pass the reef (Gota White Rock) and keep the white rock bearing about 000°.

The lagoon is formed by the horseshoe-shaped reef. The entrance is on the E side. A waypoint just E of entrance is 23°42'·17N 35°43'·53E, whence steer between two bommies, one with a boulder on it. The pass tends W and has depths <10m. Anchor in 10–15m, sand and coral. Access requires good visibility, steady nerves and a sound eyeball. Good diving, here and on Gota White Rock, the reef a short distance S.

Port Berenice

Admiralty chart 3043, plan page 188

Approach

This is a restricted area but OK in an emergency. Call on VHF Ch 16 first. The safest approach is through Middle Channel, S of Gez Mukawwa. An alternative is through South Channel, S of Horseshoe Reef, keeping clear of the unsurveyed areas. From the N, leave Ras Baniyas to starboard, being careful to give the reef S of the *ras* a good berth. Leading beacons in line, bearing 318°, mark the channel to the port itself. The fairway is about 70m wide, with a least depth of 12m, and ends in a concrete jetty.

Dangers

A submerged rock SW of the fairway.

Anchorage

3–4m, mud and sand, as on the sketch-map. There are off-lying rocks to watch out for along the reef, but there is good protection. About 5½M S of the port another anchorage is in Quaria Kalitat at the mouth of Wadi Kalitat, under Jebel Batuga.

General

The ruins of the ancient town of Berenice, which was of great importance as an entrepôt between Asia and Africa, are in a small bay SW of the present port, overlooked by a low hill. Berenice was the name of Ptolemy II's mother. He founded the port in 275BCE and it had five centuries of prominence. More recently it has been provided by the USA with aid in the form of a sponsored military base. Approach the Egyptian Naval Post only if you need assistance urgently.

Geziret Zabargad (St John's I)

Plan opposite

Approach

Both Zabargad and Rocky I show well on radar. Note the reef 5M E of Rocky I. Otherwise the approaches are clear. The anchorages are poor and only recommended for diving and in settled weather. Eleven moorings have been laid here under

187

Red Sea Pilot

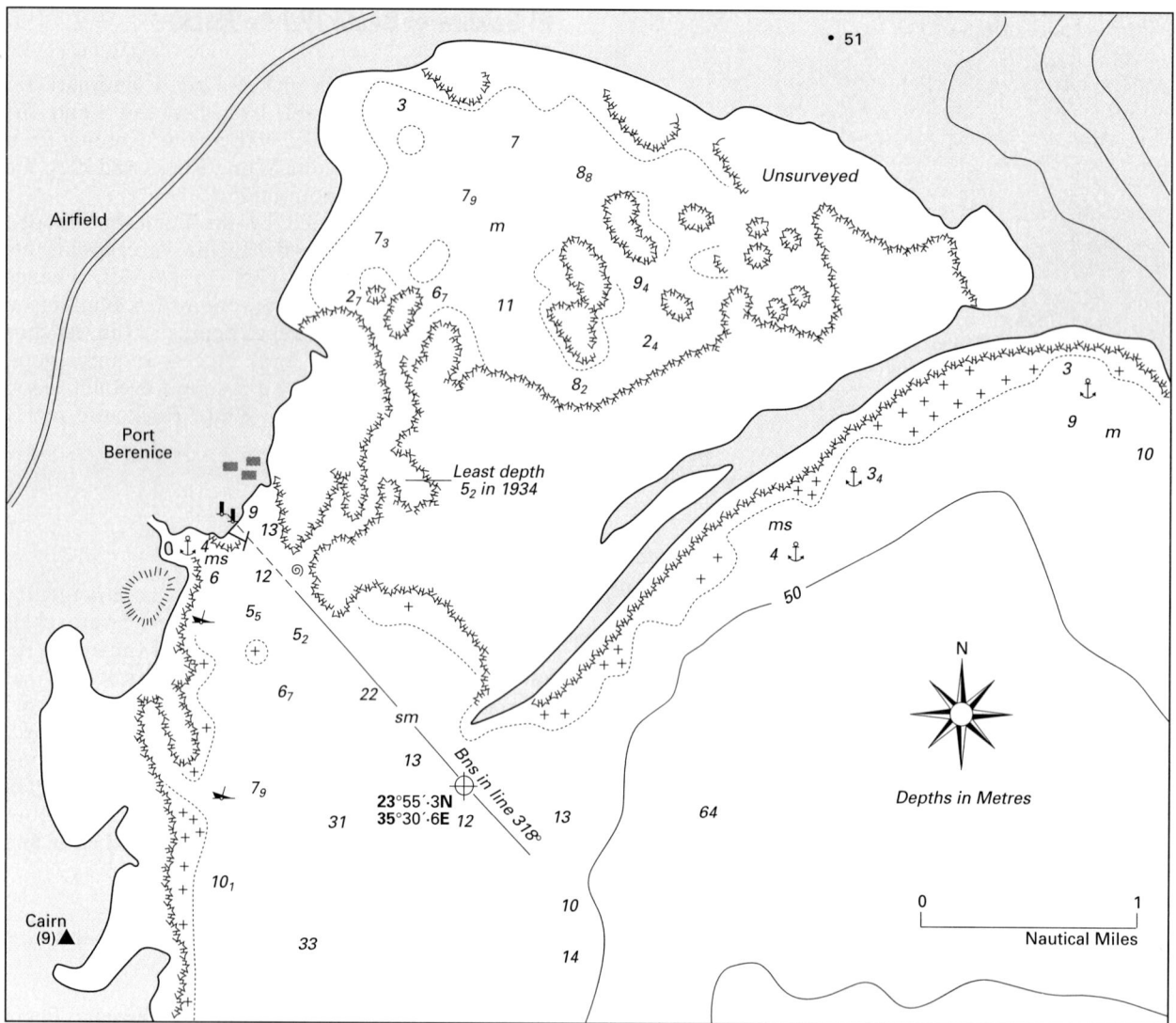

PORT BERENICE

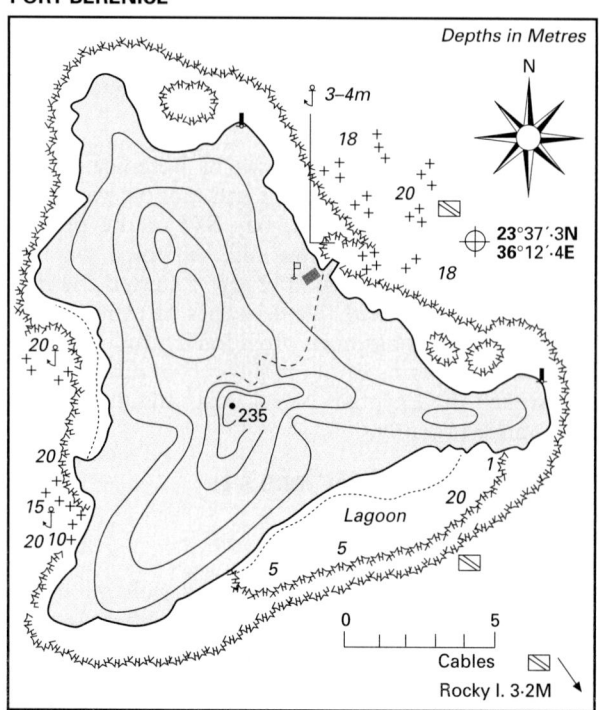

GEZIRET ZABARGAD

the aegis of the US sponsored programme that we mention above. Some may be missing and, in any case, you should inspect carefully before trusting your boat to one. The co-ordinates given to us are 23°36'·08N 36°12'·72E, on the more sheltered S side of the island. There are a further six at Rocky I near 23°33'·76N 36°14'·96E. The islands may sometimes be closed to divers by the authorities.

Dangers

All the anchorages are exposed at most times, the only lee being found during short-lived thermal breezes around dawn and dusk during settled weather. In their absence, fluky winds and currents make any anchorage tenuous.

Anchorages

NE side The anchorage in a narrow cleft in the coral is dangerous in N or S winds. Otherwise anchor in 4m, sand and coral, off the old jetty with the summit bearing 225°.

SW side About 2 cables N of the SW tip of the island not far from a small promontory at the mouth of a wadi. Sand and coral between the bommies in 10–15m. Watch out for chain wraps caused by

The Egyptian Coast: Foul Bay to Hurghada

currents and don't forget that it's strictly illegal to anchor on coral.

NW headland In a small bay under the headland about 50m from the shore in 10m, sand, after threading your way in between the bommies.

General

Good diving and at some times of year turtles nest on the beaches. Underwater passages connect the NE end of the large S lagoon on its S side to the sea. The lagoon itself is a bit stagnant. On the drop-off to seaward of the lagoon is the wreck of the yacht *Hawk*, about 30m down. Elsewhere there is the wreck of the dive boat *Neptune* which sank when it had starting problems and was blown onto the reef. A reminder to us all. There are abandoned gem mines close to the summit of the island, first exploited, reputedly for emeralds, in fact for olivine and peridot, by the Pharaohs. Pliny calls it topaz. The English reopened the mines for a while in the 19th century, but were dissuaded by the difficulties of navigation and anchoring. The island

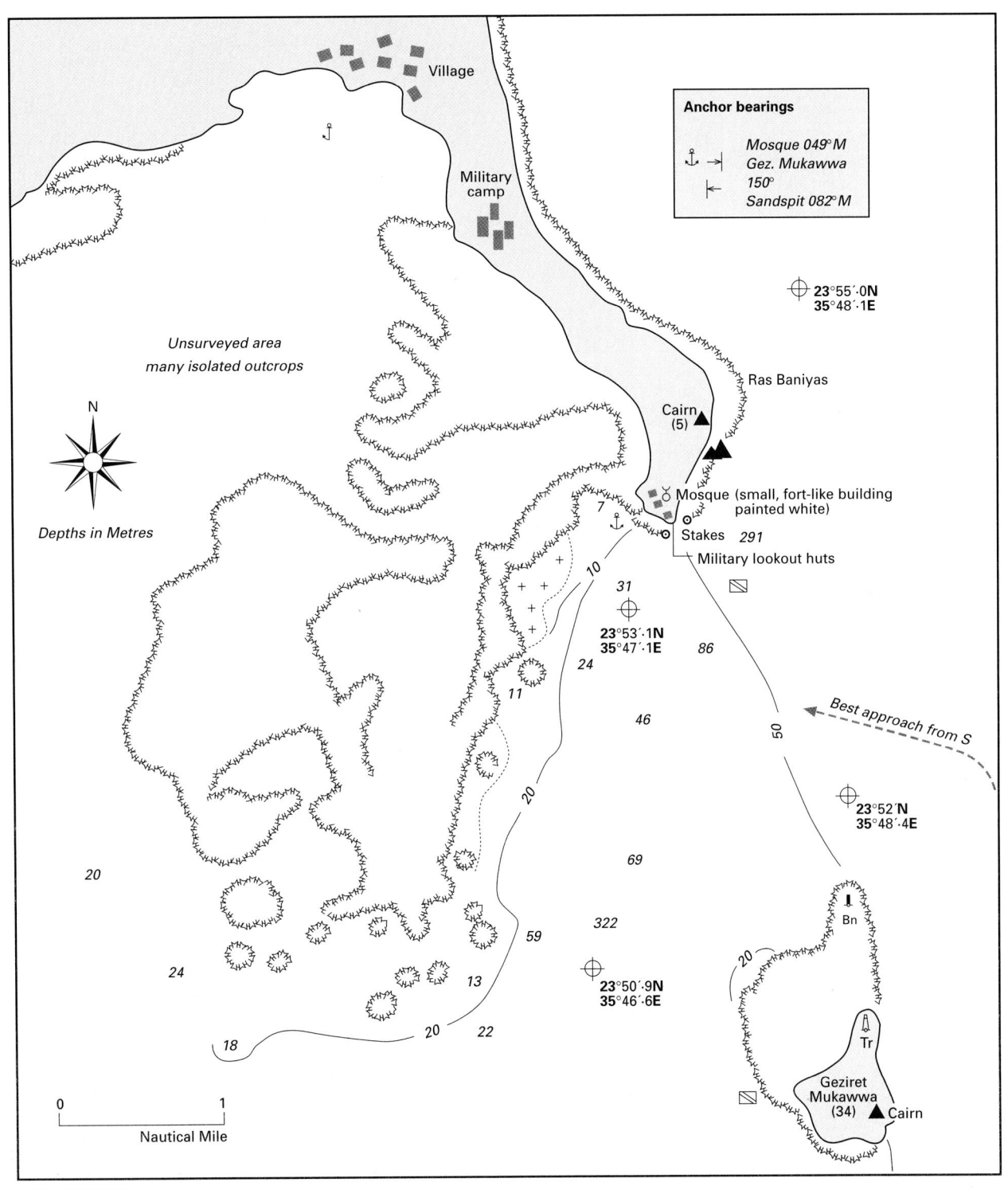

RAS BANIYAS

Red Sea Pilot

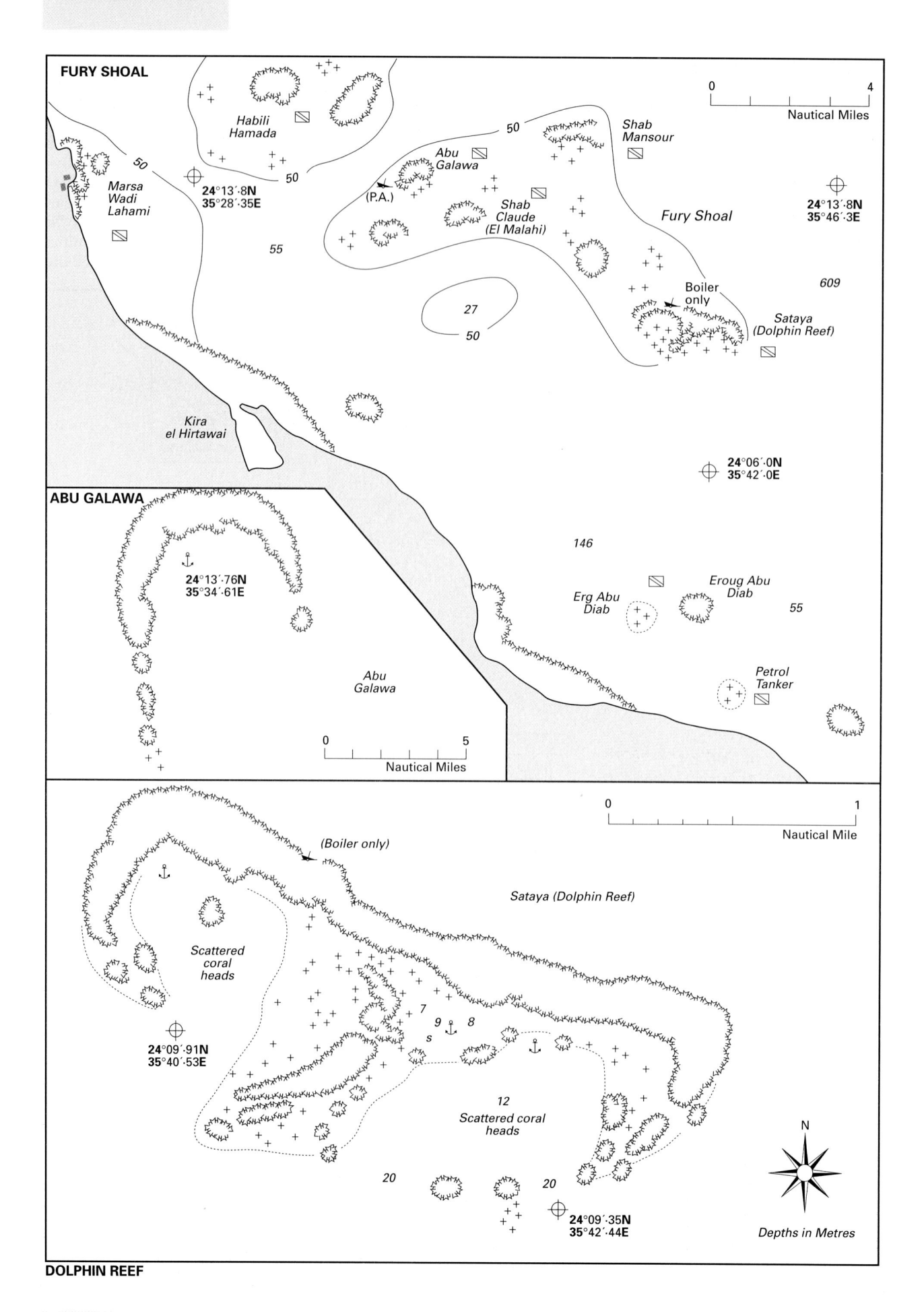

FURY SHOAL

Habili Hamada
Marsa Wadi Lahami
24°13'.8N
35°28'.35E
Abu Galawa (P.A.)
Shab Mansour
Shab Claude (El Malahi)
Fury Shoal
24°13'.8N
35°46'.3E
609
Boiler only
Sataya (Dolphin Reef)

Kira el Hirtawai

24°06'.0N
35°42'.0E

ABU GALAWA

24°13'.76N
35°34'.61E

Abu Galawa

146
Erg Abu Diab
Eroug Abu Diab
55
Petrol Tanker

0 — 4 Nautical Miles
0 — 5 Nautical Miles
0 — 1 Nautical Mile

(Boiler only)
Sataya (Dolphin Reef)
Scattered coral heads
24°09'.91N
35°40'.53E
7
9 8
s
12
Scattered coral heads
20
20
24°09'.35N
35°42'.44E

N

Depths in Metres

DOLPHIN REEF

190

occasionally comes within the convergence zone, when the weather constantly changes. The Arabic *zabargad* denotes the sea mists which often envelop the island during these periods.

Ras Baniyas (Ras Banas)
Plan page 189

Approach
There is a clear and easy approach from the SE, whereas the reefs to the SW call for careful navigation. The anchorage itself is not obvious until you are close.

Conspicuous
The mosque at the military outpost on the spit is prominent and looks like a small, crenellated blockhouse, sometimes with attendant camels.

Anchorage
12–15m, coral, in the lee of the spit. The holding is good in spite of the nature of the bottom. Sometimes it's too good, so buoy your anchor. Snorkel over it shortly before you leave to sort out how to unwrap the inevitable tangles in the coral. Anchor bearings are noted on the sketch.

The military are usually quite friendly but will want to check your papers. If you go ashore ask if the beach is still mined beyond the compound. Excellent diving with water of exceptional clarity. The wind always honks hereabouts but isn't always indicative of how hard it is blowing at sea or further N. Ras Baniyas was once called Cape Nose.

Fury Shoal
Orientation plan page 190

The shoal ground of which the reefs form a part extends further W than charted. For all its fury during strong weather, it is actually named after the schooner *Fury*, which first reported it to the East India Company's hydrographers in the early 1800s. There are several usable anchorages. The two best are in Dolphin Reef, the extreme SE'most reef. Another is in Abu Galawa in the NW. There are more tenuous day anchorages at Shab Claude and Shab Mansour. The whole area, including Habili Hamada, Erg and Eroug Diab and the recently discovered sunken 1940s tanker offers excellent diving and snorkelling. Anchorage details follow.

Dolphin Reef (Sataya)
Plan page 190

A T-shaped reef, orientated NE/SW, separating two lagoons each with a safe anchorage.

Approach
From the S the approach is clear but good light is necessary to pick out isolated reefs across the entrance to the lagoon. The area WNW through NNW of Dolphin Reef is foul. The lagoon in the lee of Dolphin Reef is divided into two parts. Protection in the NW corner of the E part is best. In the last 1M of approach enter the E part of the lagoon in 10–20m between scattered bommies. For final approach to the E lagoon, a GPS position about 2 cables S of the outer bommies is 24°09'·35N 35°42'·44E. Thence head NNW curving N to anchor as below. For the W lagoon the approach position is 24°09'·91N 35°40'·53E, whence head in a curve NNE, N, then NNW to the NW corner.

Conspicuous
The boiler or engine block of a wrecked ship lies towards the E end of Dolphin Reef.

Anchorage
In the E lagoon at 24°09'·7N 35°41'·6E in 9m, sand, good holding with plenty of swinging room. Good protection except from the S. In the W lagoon anchor in the NW corner NW of some isolated bommies.

Diving at the drop-off on E edge of the reef and snorkelling are both outstanding. There are some amphorae at the 'toe' of the leg of the T, so the schooner *Fury* was a bit of a latecomer!

Shab Claude (El Malahi)
24°12'·7N 35°35'·75E See Fury Shoal orientation plan page 190

There is a day anchorage here at the S end of the reef. Better to use the HEPCA/USAID mooring if unoccupied. Dolphin Reef is a better place to wait for quiet conditions if you want to dive Shab Mansour 3M NE. Shab Mansour is tenuous even as a day anchorage.

Abu Galawa
Plan page 190

A horseshoe-shaped reef with a good anchorage in the lagoon. Approach from the S is clear. Anchor under the lee of the reef at 24°13'·76N 35°34'·61E in about 13m. The *Tientsin* was wrecked here in the '50s. The remains are just S of the reef.

Marsa Wadi Lahami
Plan page 194

Approach
The sketch shows the best approach from S and N to avoid the major coral outcrops.

Dangers
The area is poorly charted, and there are shallow patches in the approaches in addition to those drawn in on the sketch. The entrance is narrow and littered with bommies. Hold hard to the spit side of the channel with a lookout in the spreaders and a wary eye on the echo sounder, and turn to port once past the spit.

Conspicuous
There are several good landmarks for this anchorage. Behind the new dive resorts is a range of low black hills, with a cairn on the highest and a bold white stripe at the N end. On the low plateau in the foreground to the N there is a neat notch about one-third of the way from the left end and a clump of trees on a low headland at its right end.

Red Sea Pilot

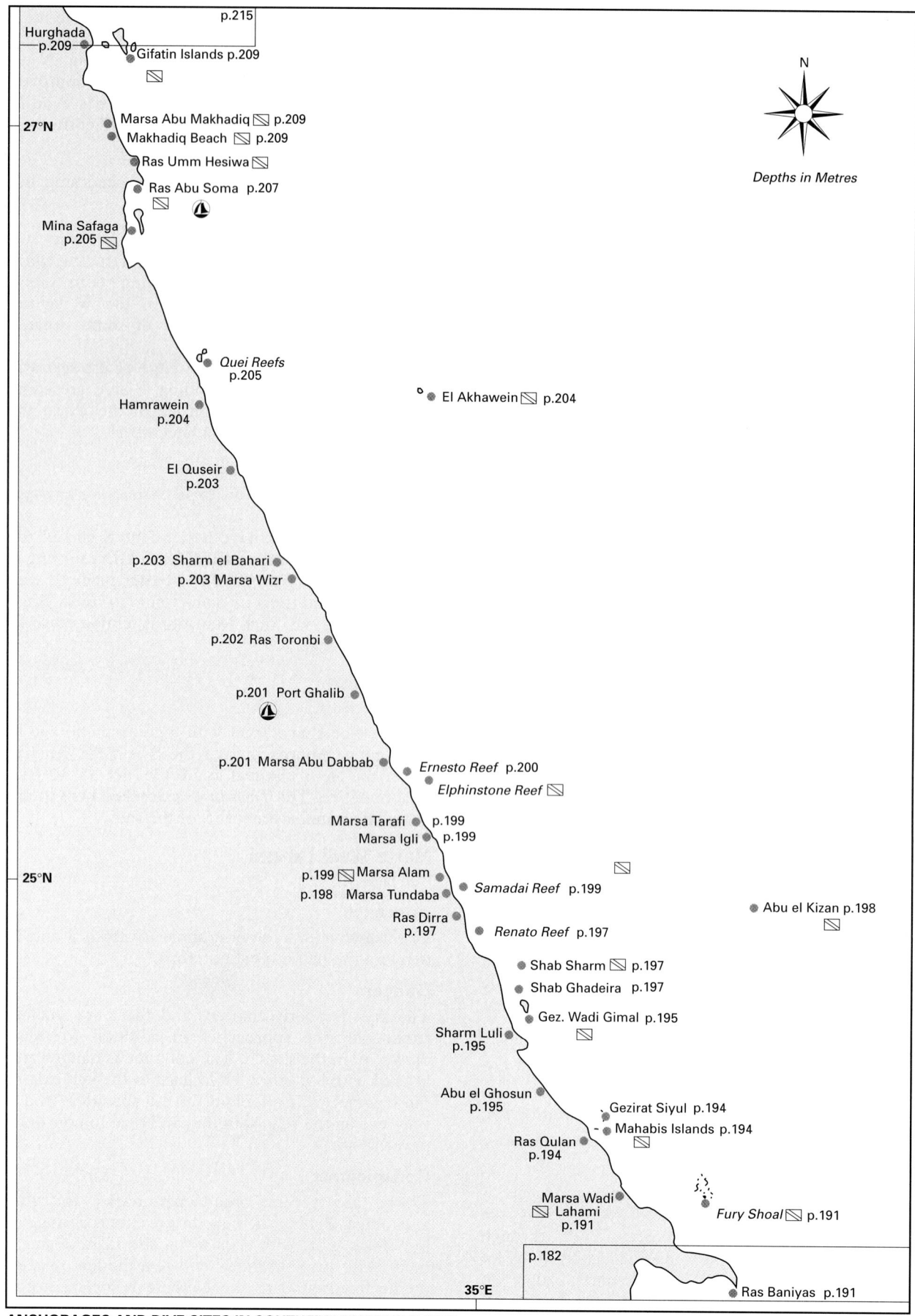

- Hurghada p.209
- Gifatin Islands p.209
- Marsa Abu Makhadiq p.209
- Makhadiq Beach p.209
- Ras Umm Hesiwa
- Ras Abu Soma p.207
- Mina Safaga p.205
- Quei Reefs p.205
- El Akhawein p.204
- Hamrawein p.204
- El Quseir p.203
- Sharm el Bahari p.203
- Marsa Wizr p.203
- Ras Toronbi p.202
- Port Ghalib p.201
- Marsa Abu Dabbab p.201
- Ernesto Reef p.200
- Elphinstone Reef
- Marsa Tarafi p.199
- Marsa Igli p.199
- Marsa Alam p.199
- Marsa Tundaba p.198
- Ras Dirra p.197
- Renato Reef p.197
- Shab Sharm p.197
- Shab Ghadeira p.197
- Gez. Wadi Gimal p.195
- Abu el Kizan p.198
- Sharm Luli p.195
- Abu el Ghosun p.195
- Gezirat Siyul p.194
- Mahabis Islands p.194
- Ras Qulan p.194
- Marsa Wadi Lahami p.191
- Fury Shoal p.191
- Ras Baniyas p.191

Depths in Metres

ANCHORAGES AND DIVE SITES IN SOUTHERN EGYPT KEY MAP

192

The Egyptian Coast: Foul Bay to Hurghada

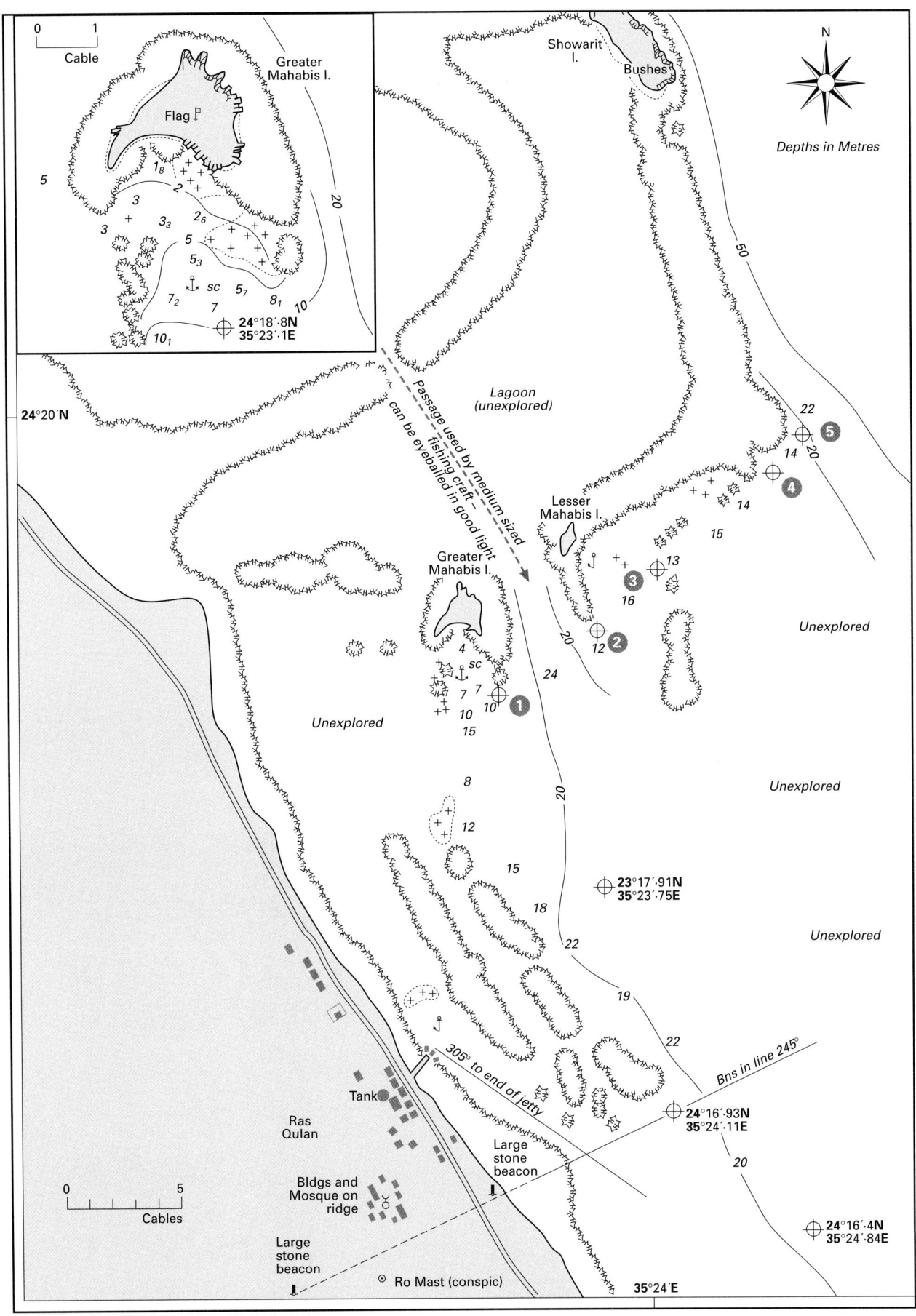

RAS QULAN AND THE MAHABIS ISLANDS

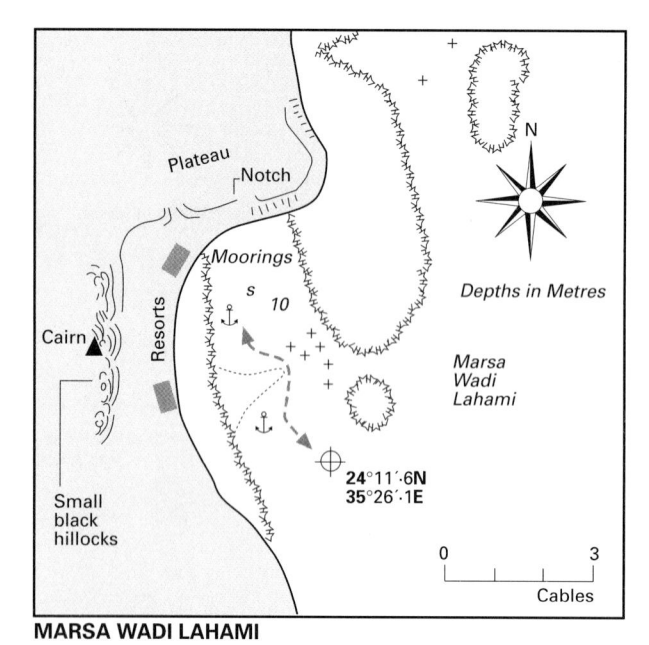

Red Sea Pilot

MARSA WADI LAHAMI

Anchorage

6–10m, sand, N of the spit, or in deeper water S of it in a mixture of sand and coral. Good holding, and well protected in NW'lies.

General

Two dive resorts have been opened here and there are moorings which makes the anchorage rather tight. The recently (1993) wrecked *Hamada* lies on the shelf of the fringing reef just N of Marsa Wadi Lahami. Ask in the resorts for precise directions for diving.

Mahabis Is and Ras Qulan

Plans page 193

Approach

Care is needed to avoid shoal ground off Ras Qulan and there are more isolated coral heads than charted in the approach to the islands from the S and extensive offshore reef to the E through NE. The radio mast and buildings are conspicuous but we don't recommend anchoring off the settlement unless you are in dire need of something. The waypoints on the sketch map pass through safe water.

Anchorage

The anchorage in the lee of Greater Mahabis has room for only two or three boats unless you raft up. It has much to recommend it with flat water even in strong winds and much better shelter from the N than at Gezirat Wadi Gimal further N. An alternative, rather less sheltered, is used by some fishing boats in the lee of Lesser Mahabis where the half anchor is shown.

Medium-sized fishing boats use the passage E of Greater Mahabis into the lagoon to access a break in the reef further N, whence one can pass W of Gez Showarit and Gez Siyul. This is only for the bold, with careful eyeballing and in quiet weather.

From the anchorage at Greater Mahabis safe access from and egress to the NE passes through the following waypoints. This passage, with a least depth of 12m, SHOULD NOT BE ATTEMPTED WITHOUT A CAREFUL LOOKOUT. Waypoint numbers refer to the orientation sketch, page 193.

1. 24°18'·803N 35°23'·254E
2. 24°19'·05N 35°23'·753E
3. 24°19'·346N 35°24'·035E
4. 24°19'·749N 35°24'·576E
5. 24°19'·978N 35°24'·731E

From the last waypoint you're clear seaward or N past Gez Showarit and Gez Siyul. While at anchor here and at other places along the Egyptian coast fishermen may approach wanting something. They will usually offer fish in return which satisfies honour all round.

Gezirat Siyul

There is no safe anchorage in the immediate vicinity of the island. The recommended anchorage is at the N end of the large reef complex N of Gez Siyul accessed best from the passage S of the island. Take the clear, deep (40m) pass between Gez Showarit and Gez Siyul. Steer NW along the W edge of the reef surrounding Gez Siyul and its lagoon to

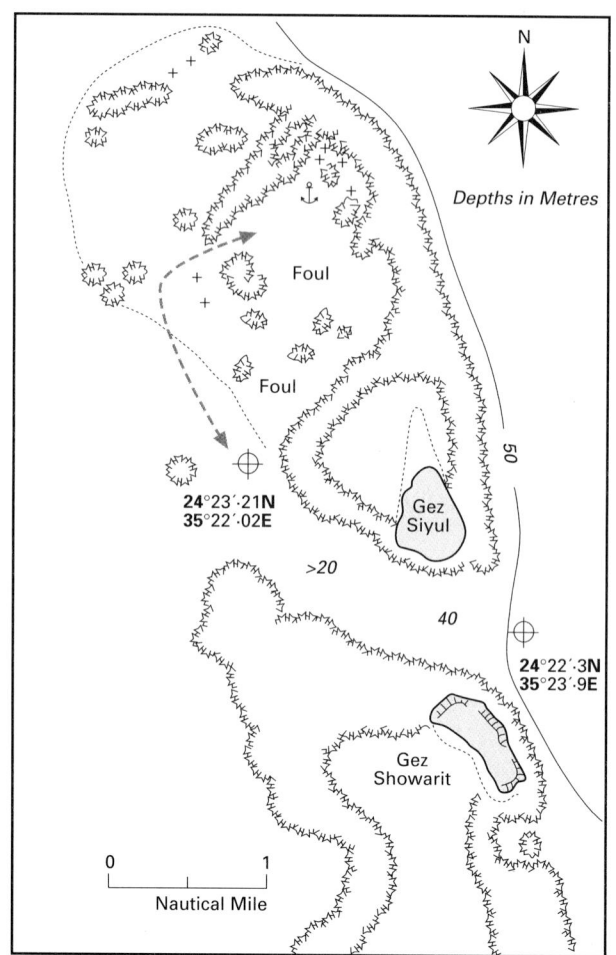

GEZIRAT SIYUL

The Egyptian Coast: Foul Bay to Hurghada

24°23'·21N 35°22'·02E. Thence NNW past a large area of foul ground to starboard to pick your way very gingerly N by E'wards through lots of bommies to anchor in 10–12m in 24°25'·12N 35°22'·27E.

Abu el Ghosun
Plan below

Approach
It is safest to pass N of Small Reef. Aim for 24°30'·72N 35°13'·49E, when you will be 4·5M SE of Ras Honkorab. At that point alter onto 205° until you can see the jetties. The run is approximately 4M. The anchorage is in an open roadstead with only some shelter from the NW. For fair weather, unless you decide to try lying alongside.

Dangers
There is a barely submerged rock about ¾ cable SW of the jetties visible in reasonable light.

Conspicuous
A wind pump near the service station on the road not far inland of the anchorage is clearly visible from some way off.

Anchorage
6m, sand and rock, off the beach, S of the jetties, or stern-to, Mediterranean style.

Facilities
Fuel and water (at low cost) are obtainable on the road.

Ras Honkorab
24°33'N 35°09'·5E
(not illustrated)

There is an anchorage under the lee of Ras Honkorab in approx. 24°33'·64N 35°10'·47E. The shelter is indifferent and you'll generally be better off at Sharm Luli or Gez Wadi Gimal.

Sharm Luli
Plan page 196

Approach
Two sugar-loaf-shaped hills (300–400m), bearing 260° and 270°, mark the entrance in good visibility.

Dangers
There are off-lying, submerged rocks on the N shore of the inlet, the head of which is shallow, backing onto a low sandy shoreline. The channel is 50m wide at its narrowest point.

A shoal has been reported in approx 24°36'N 35°09'E, but its existence is doubtful.

Conspicuous
A 50m radio antenna 2M S of the entrance is easily visible and there is a hut 1M further N. A boulder marks the fringing reef S of the entrance to the inlet, though it is some distance back from the seaward edge.

Anchorage
5–10m, mud, with good holding and shelter.

Formalities
Officials from the army outpost will want to check your passports and draw up a crew list. Yachts have had varying receptions but in general have been welcomed. Unless you need supplies or help choose Gez Wadi Gimal instead.

Facilities
You can get fuel, water and provisions here, although delivery is slow and prices higher than in the towns. Prices have varied, but one yacht paid US$100 for a 40-gallon drum of diesel. The main road to El Quseir runs past the village and army post, but it is a beautiful spot with good diving both in the pass and on the reef outside. A dive-charter operation runs out of Sharm Luli, which explains the availability of supplies.

Gezirat Wadi Gimal
Plan page 196

Approach
From 24°37'·97N 35°10'·83E run in on 315° to 24°39'·16N 35°09'·5E. The passage W of Gez Wadi Gimal past Ras Baghdadi is feasible in good light and clear enough. A mid point in clear water between the reefs W of the N end of Gez Wadi Gimal and Ras Baghdadi is at 24°40'·06N 35°07'·07E. The large resort complex at El Sharm will be on your port bow. There are three US Aid project dive moorings marked by bright orange buoys. One in 24°38'·6N 35°10'·53E, the next at 24°39'·22N 35°07'·98E, and the last at 24°41'·3N 35°08'·4E. We didn't use them but there's no reason why you shouldn't unless you think your boat is too

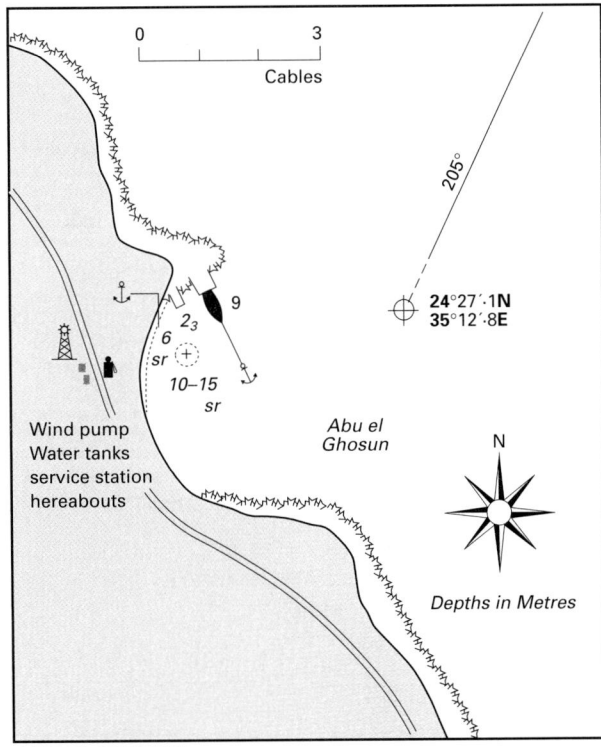

ABU EL GHOSUN

Red Sea Pilot

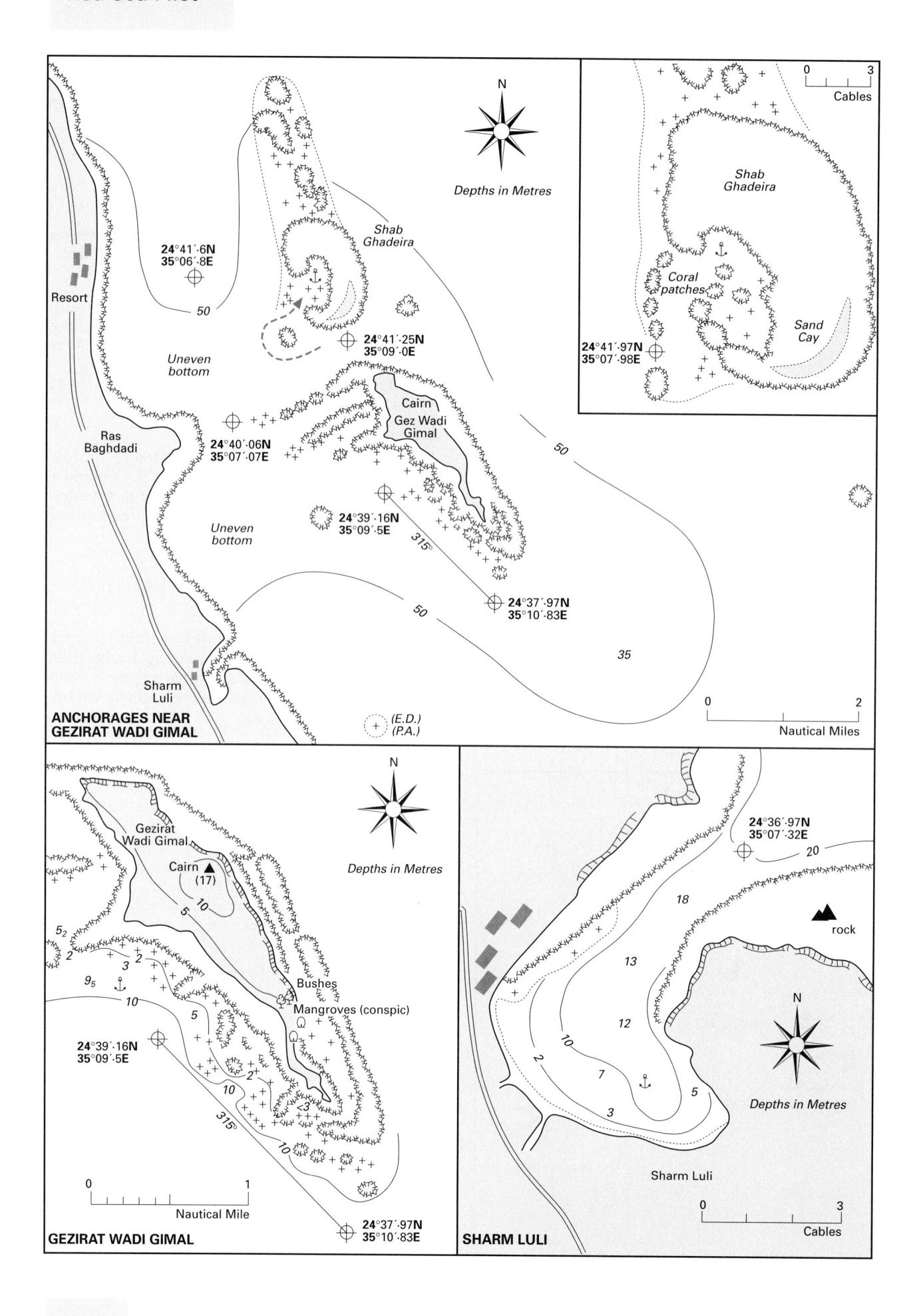

ANCHORAGES NEAR GEZIRAT WADI GIMAL

Depths in Metres

- Resort
- 24°41′·6N 35°06′·8E
- Shab Ghadeira
- 24°41′·25N 35°09′·0E
- Uneven bottom
- Ras Baghdadi
- 24°40′·06N 35°07′·07E
- Cairn Gez Wadi Gimal
- Uneven bottom
- 24°39′·16N 35°09′·5E
- 315°
- 24°37′·97N 35°10′·83E
- Sharm Luli
- (E.D.) (P.A.)

Inset: Shab Ghadeira — Coral patches, Sand Cay
- 24°41′·97N 35°07′·98E
- 0 — 3 Cables

0 — 2 Nautical Miles

GEZIRAT WADI GIMAL

Gezirat Wadi Gimal
- Cairn ▲ (17)
- Bushes
- Mangroves (conspic)
- 24°39′·16N 35°09′·5E
- 315°
- 24°37′·97N 35°10′·83E

Depths in Metres

0 — 1 Nautical Mile

SHARM LULI

- 24°36′·97N 35°07′·32E
- rock
- Sharm Luli

Depths in Metres

0 — 3 Cables

The Egyptian Coast: Foul Bay to Hurghada

heavy. However, be aware that these are dive boat buoys and very close up in the lee of reefs. You cannot leave your boat unattended.

Danger
The reef fringing the island extends well beyond the S tip. An uncharted outcrop has been reported in 24°36'·65N 35°06'·95E and there are bommies within 0·5M of the SE part of the reef fringing the island.

Anchorage
There are many possible anchorages WSW of the island in sand among coral outcrops, good holding. In strong N winds it can be pretty bouncy, even at night, so tuck as far in and N as possible.

Shab Ghadeira
Plan page 196

The S end of Shab Ghadeira has a large, conspicuous sand *cay*. N of it is a lagoon with a safe anchorage accessible from the W. Note the N/S line of off-liers W of the sand *cay* through which you head to enter the lagoon. Pass N of Gez Wadi Gimal and S of Shab Ghadeira via 24°41'·25N 35°09'·00E, thence W then N towards the entrance that runs W to E through the off-liers in 24°41'·97N 35°07'·98E, eyeballing as you go. Once inside the line of off-liers turn N between them and a hook of reef extending NW from the sand *cay* to enter the lagoon. There are many bommies inside so feel your way to the NE corner to anchor in 10–12m, sand and coral.

Shab Sharm
24°47'·3N 35°10'·7E
(not illustrated)

This kidney-shaped reef offers good diving and although anchoring is difficult there is a mooring which you might be able to use, position 24°47'·26N 35°10'·89E.

Renato Reef
24°50'·2N 35°02'E, plan opposite

The reef is named after the divers' pilot for the Red Sea, Renato Marchesan. He has been working his 1920's *navicello* Apuano, the *Ernesto Leoni*, along the coast N of Port Sudan for years and knows these waters like the back of his hand. The reef is 2M SE of Ras Dirra, between Ras Baghdadi and Marsa Tundaba, W of the passage inside the offshore reefs and S of an area charted as full of reefs. It is a flat E/W bar with a slight S-tending hook at its E end. S of the W tip, which may be marked by a stake, there is an off-lying bommie. Anchor E of the bommie under the main reef. Fair shelter in N winds, though iffy in strong weather. Some slight swell may work round, but it is no more than a nuisance.

The reef is inshore of the reef complex of Radir North, Radir East and Radir el Bahr all of which, especially Radir East, are rated excellent dives.

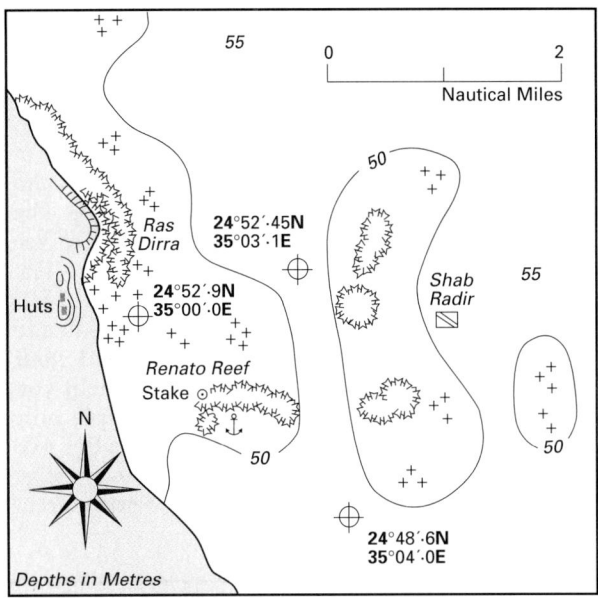

RAS DIRRA, RENATO REEF AND SHAB RADIR

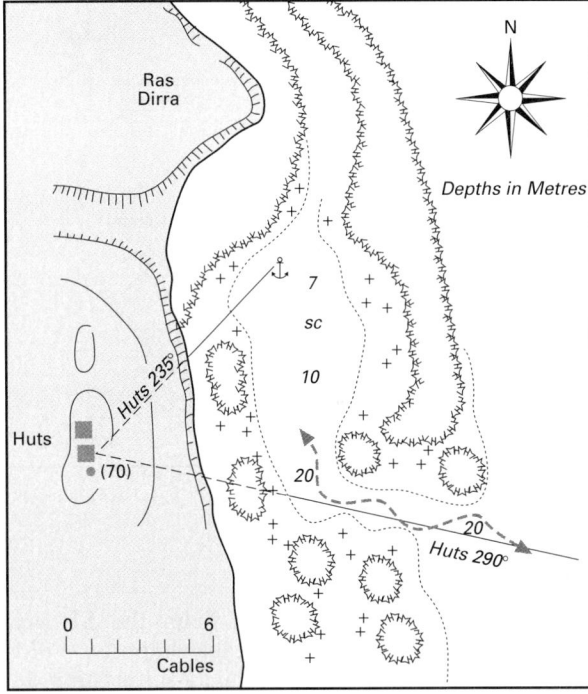

RAS DIRRA

Ras Dirra (Bir Ghadir)

The low headland is not easy to see from any distance, and the approach, through offshore reefs for some 5M to the E, should only be made in good light towards 24°52'·9N 35°00'E. At that point the huts on the hill (70m) behind the inlet bear approximately 290° and you weave your way in to the anchorage, where the huts bear approximately 235°. Isolated coral heads in the approaches. A lookout in the spreaders and a morning arrival are highly recommended. Leaving before dawn would be tricky even with careful GPS waypointing and you could get stuck in bad weather. 7–10m in sand and coral, with reasonably good holding.

197

Red Sea Pilot

Abu El Kizan (Daedalus Reef)

This was once called Centurion Reef, then Daedalus before reverting to the Arabic name. It is steep-to and the approaches all round are clear. There is no real anchorage. Pick up the mooring at 24°55'·925N 35°52'·276E. In steady N'lies or calm weather you can also tie to the end of the jetty. The five lighthouse-keepers will be very pleased to see you, especially if you take some goodies. The French-built light dates from the late 19th century. Let the keepers know in advance if you would like to visit. They must report your presence and your agreement to some gentle subterfuge can help you all enjoy your stay. Otherwise they see people only when they are reprovisioned from Suez every two months or when they are relieved every six months. This is one of Egypt's National Marine Parks. Great diving, especially along the N wall.

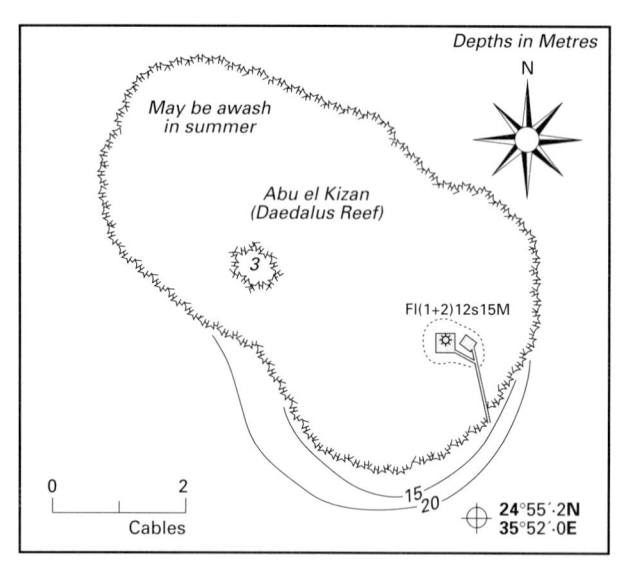

DAEDALUS REEF

Marsa Tundaba

Co-ordinates for the approach from the SE are 24°57'·2N 34°57'E. The anchorage is approx 0·8M NW of here in 8m, sand, opposite a fishing camp on the beach. There are a couple of white buildings to the S and a conspic army lookout and radio mast on the high ground to the N above Ras Samadai. There are isolated bommies off the fringing reef so approach in good light and pick your spot carefully for good holding. It is surprisingly sheltered from swell.

Marsa Nekari, 2·5M SE of Marsa Tundaba in approx 24°55'·6N 34°57'·7E, is a dive resort. Their boats are moored on permanent moorings behind a small *ras*. It may be possible to find shelter here in quietish weather. There are HEPCA/USAID moorings on the reefs at 24°55'·34N 34°57'·44E and 24°56'·36N 34°59'·01E. The second is on Shab Nekari, which lies offshore in the SE approaches to Marsa Tundaba.

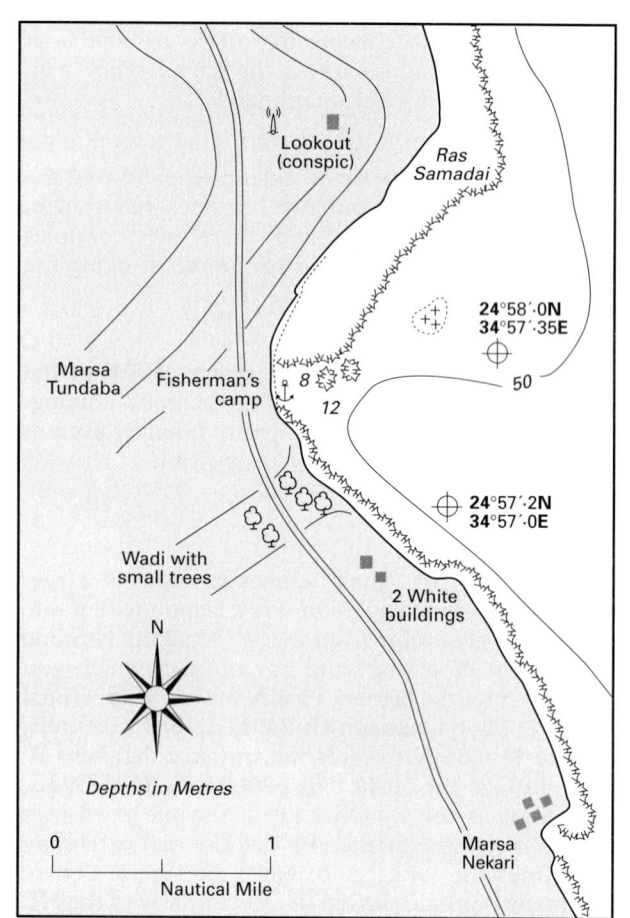

MARSA TUNDABA

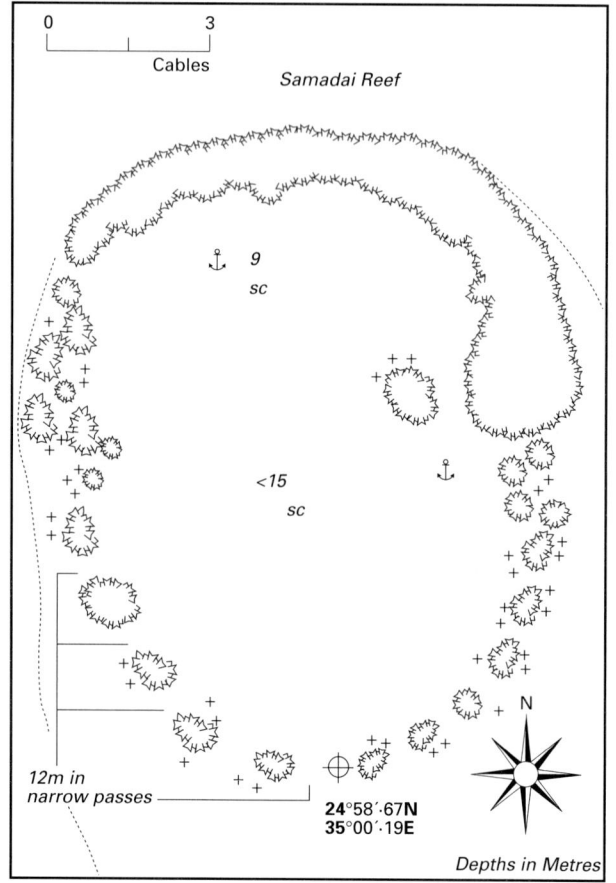

SAMADAI REEF

198

Samadai Reef
Plan page 198

The horseshoe-shaped reef and a chain of coral heads enclose a large lagoon about 1M N/S. A GPS position just S of the narrow pass between bommies at the S end of the lagoon is 24°58'·67N 35°00'·19E. Keep a good lookout from aloft. Depths are generally 10–15m inside the lagoon, sand with coral patches. Fair protection in good to moderate weather, though Renato Marchesan of *Ernesto Leoni*, with more than a decade of Red Sea experience, says shelter is adequate even in strong N'lies and you can be comfortable here overnight, though he commends the more SE'ly anchorage at high tide especially in strong winds. Diving is reported to be excellent on barrier reefs in NW and SE corners and 10 HEPCA/USAID moorings have been laid towards the N end of the lagoon 24°59'·04N 35°00'·19E. Dolphins often visit the lagoon.

Marsa Alam
Plan page 200

Approach
The entrance to the *marsa* looks more like a bay than an inlet until you are on close approach.

Dangers
There are two detached bommies off the fringing reef in the approach from the ENE, N and NE of the GPS position on the sketch. An approach approx 270° W from 25°04'·65N 34°54'·5E to 25°04'·60N 34°53'·93E will bring you S of them to a position 1 cable SE of the entrance. It is best to time your entrance for just after low tide with the sun well up and S of SW. Beware 2–3 knot tidal streams with eddies in the channel at springs and watch out for leeway. The channel is very narrow and has a least depth of 2·6m if you hold to starboard. The S'most of the two stakes shown on the reef E of the entrance is short and not easily visible. A patch with less than 2m blocks the natural turn W at the first gap. Don't turn W until in approx 24°04'·85N 34°53'·74E, in 4·5m. The point should be marked by a small stake on your starboard side.

For a rough back bearing/exit line, keep the right edge of the building with a tower in front of it in line with the left edge of a flat toppped hill S of the entrance on approx 165°.

Conspicuous
The village stands out as do the radio masts, water tower and bizarre, domed structure, perhaps a kiln, near the harbourmaster's office. At night the radio masts have flashing red lights.

Anchorage
Holding is good in 4–5m sand and weed, but, as with similar anchorages in the Med, you need to be sure your anchor is well in. The *marsa* is extensive and there is room for many boats. There are two or three boats on permanent moorings. If you arrive in the dark heave-to in the offing or anchor overnight just S of the entrance and wait for morning.

Formalities
We were surprised by the level of bureaucracy. If you go ashore you will be probably be met by an 'agent' who will take you to the Harbourmaster who will want to hold passports, may restrict movement and will probably ask why you have stopped. Excuses such as engine trouble or bad weather seem to suffice even if you have no Egyptian visa. A little English is spoken by the 'agents' but a ship's letter in Arabic helps.

Facilities
We were told that the jetty is ostensibly 'closed to foreigners'. It is extremely shallow at low tide, even for a small RIB, and covered with sharp oysters as well as bits of sharp, rusty iron. The village has a clinic with a doctor on duty, a fuel station, basic groceries and a very rustic PO. If you don't get that far you can buy fresh bread at the small store near the jetty. An airport is due to open not far away, near Port Ghalib (see below). Outside the *marsa* there is a mooring buoy for dive boats on a reef, Shab Marsa Alam and Erg Marsa Alam at 25°04'·17N 34°56'·2E. The diving is recommended.

Marsa Igli (Marsa Zebara)
25°10'·5N 34°50'·5E (not illustrated)

This is a small *marsa* with a pass through the reef about 1½ cables wide. A small twin-peaked yellow hillock marks it from seaward. Anchor in 7–10m and think about a stern anchor to control swing, since there is not much room. There is a small military base on the N shore of the inlet, but there have been no problems for boats in the past.

Marsa Tarafi
Plan age 201

Approach
There is a clear and easy approach to the entrance waypoint at 25°12'·42N 34°48'·80E. The bay itself is small and access to its W end is often obstructed by a water barge moored off the jetty with a stern line to the reef on the opposite shore. A second barge sometimes encumbers manoeuvring even further.

Conspicuous
Three tall silver silos on the N shore are visible from some distance. Four low tanks are nearby.

Anchorage
12–15m in the outer part of the inlet, N of the beacon. There is good protection from the N. It is best to anchor fore and aft so there'll be room for three to four boats, or more if they raft up. Water barge movements may mean you have to shift anchor.

Red Sea Pilot

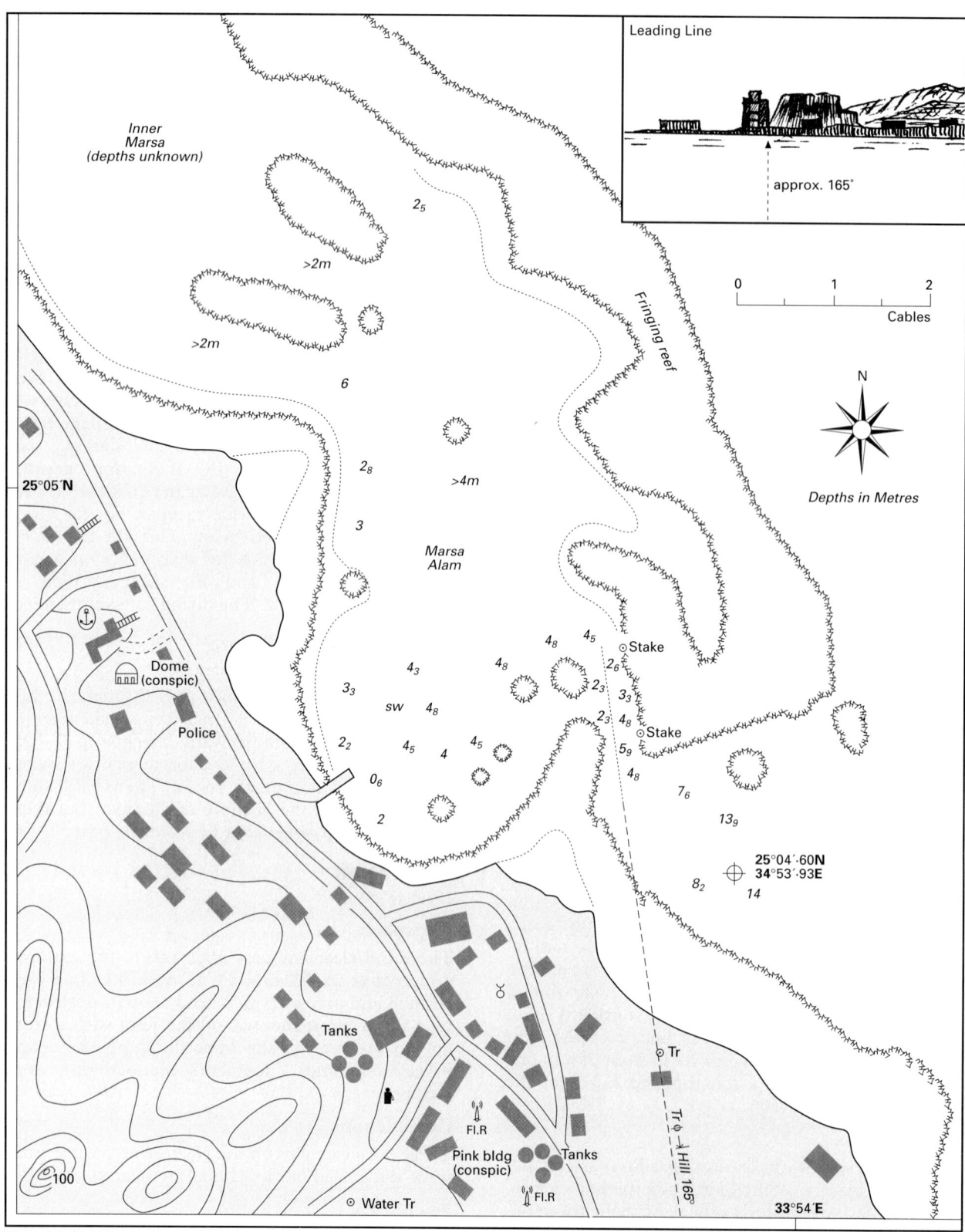

MARSA ALAM

Facilities
Water, available through the army base, compensates for the obstacles to anchoring. If the military seem less than helpful try smiling and produce your Arabic ship's letter.

Ernesto Reef
Plan page 201

This reef, just S of Marsa Abu Dabbab, offers good shelter in N'lies. On the chart it is the SW'most of the three detached reefs marked on the inshore shelf W of Elphinstone Reef. The largest, outer reef is a

200

The Egyptian Coast: Foul Bay to Hurghada

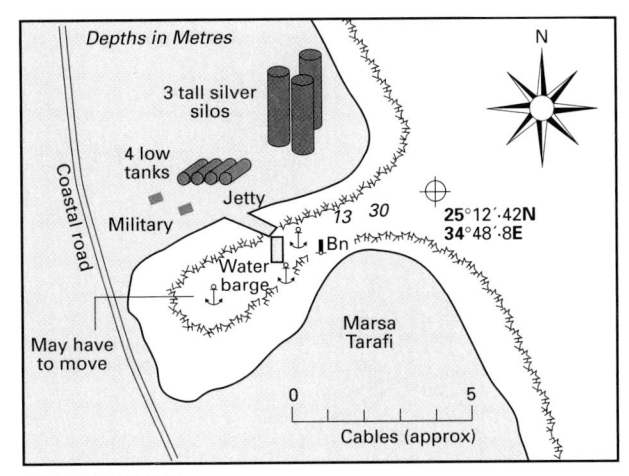

MARSA TARAFI

good dive site. Ernesto Reef extends broadly NW/SE in the shape of the left half of an arrowhead on a shaft broken off short. There is a detached bommie near where the half head meets the shaft. At the tip of the arrowhead is a dense cluster of bommies. Access and egress is from the S. The anchorage is at 25°18'·83N 34°46'·10E in sand with some coral. Preferable are half a dozen US aid project moorings here, around 25°20'N and between 34°46'·2E and 34°47'·6E. Elphinstone Reef to the E also has good diving and five more HEPCA/USAID moorings.

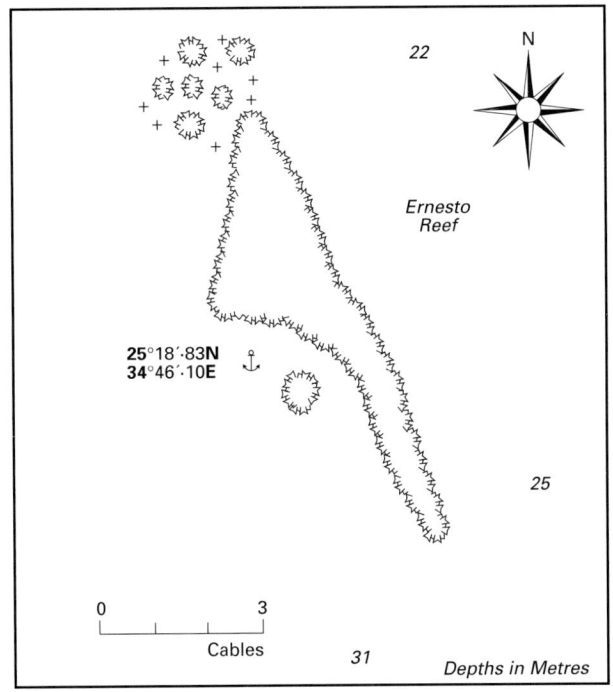

ERNESTO REEF

Marsa Abu Dabbab

From the N the approach is clear, but Elphinstone Reef makes access from the S and E a matter of picking a course through the reefs, which are not well charted but do have wide navigable channels.

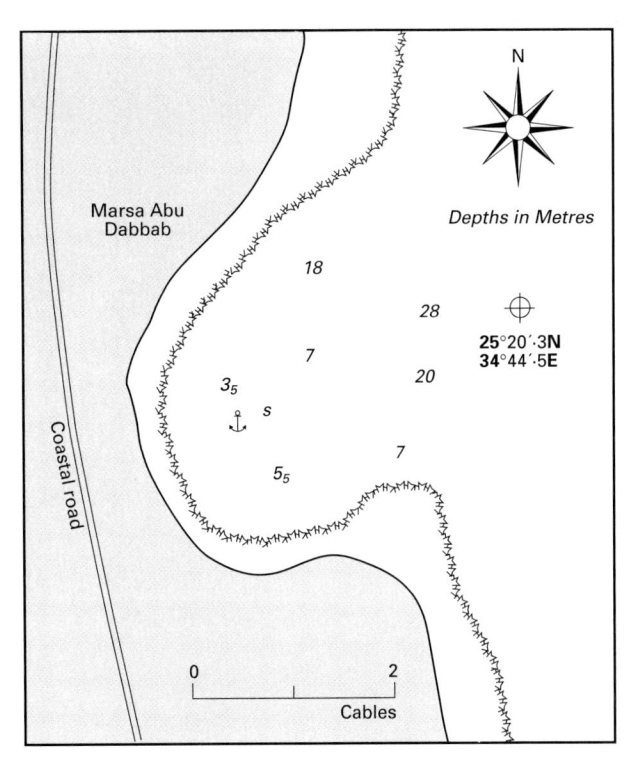

MARSA ABU DABBAB

The *marsa* is a shallow indent, and only protected from the W. It can be difficult to identify. Anchor in about 6m, on a band of sand which runs parallel to the beach, as close to the reef as possible with a stern anchor in deeper water to minimise roll if a swell makes in. There is good fishing here but many jellyfish at certain times of year. The military ashore are friendly.

Port Ghalib (Marsa Mubarak)

Plan page 202

A large marina is being built here along the lines of France's Port Grimaud. The old shallow lagoon has been dredged to form the marina's outer basin. Further work on the 2500 hectare site with its 20km of beachfront will create a 1000-berth marina village with hotels, golf course, etc. It will be a new port of entry. The international airport 4 kms to the N which serves the Marsa Alam area, offers ready access. As of April 2002 the outer basin was ready but moorings are not yet in place so berthing is alongside. There is room for 300 boats (max 50m loa) where dredged depth is 5m. Fuel, water and electricity are available. For now other supplies have to be brought from Marsa Alam. There is a fairway buoy marking the entrance channel which is 400m long and 60m wide between channel marker buoys. VHF radio call sign is Port Ghalib, working channel 10, listening channel 16. The daily rate for boats 6–14m LOA is US$10.50, weekly, US$77.
Contact: ☎/Fax +02(0)65700240
email portghalibmarina@marsa-alam-airport.com
A HEPCA/USAID mooring is just S of the new development at 25°31'·95N 34°38'·56E.

Red Sea Pilot

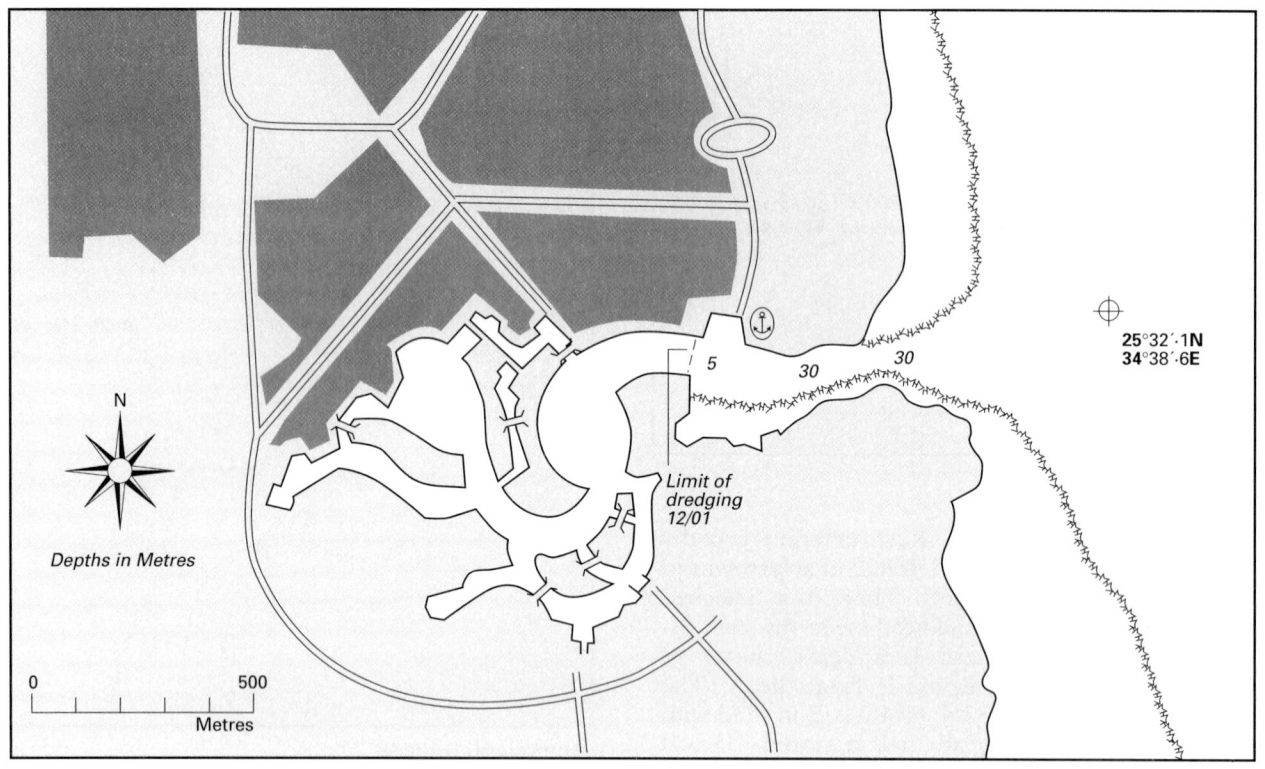

PORT GHALIB

Ras Toronbi

Plan page 202

Note Our GPS co-ordinates differ from those in the Admiralty Pilot and on older charts by 1–2M. The difference is diminished on the new BA159, but is still <0·5M, the larger discrepancy being longitude.

Approach

The anchorage is not easy to identify from any distance. A group of buildings on a low cliff 1–2M S of Ras Toronbi can cause confusion. There are two channels through the reef towards the army post. The fringing reef around the headland extends further E than the two off-lying reefs that give the anchorage protection from the E and S. The connection between the two anchorages is reported as having a maximum depth of 3m. Go carefully if go you must.

Conspicuous

On closer approach the tall radio aerial makes for positive identification. Dive charter boats tend not to use the anchorage but to hang off in the lee of the reef off the *ras*, just outside the N entrance, where they lie in quiet water, out of the swell. There is a HEPCA/USAID mooring which you can use if it is free at 25°39'·69N 34°35'·24E.

Anchorage

7–10m, sand and weed, N of the jetty. Depths in the S anchorage are shallower and the bottom is sand. The reefs are easy to see in good light. It can be rolly, especially in light weather. Stronger winds seem to blow the swell straight past!

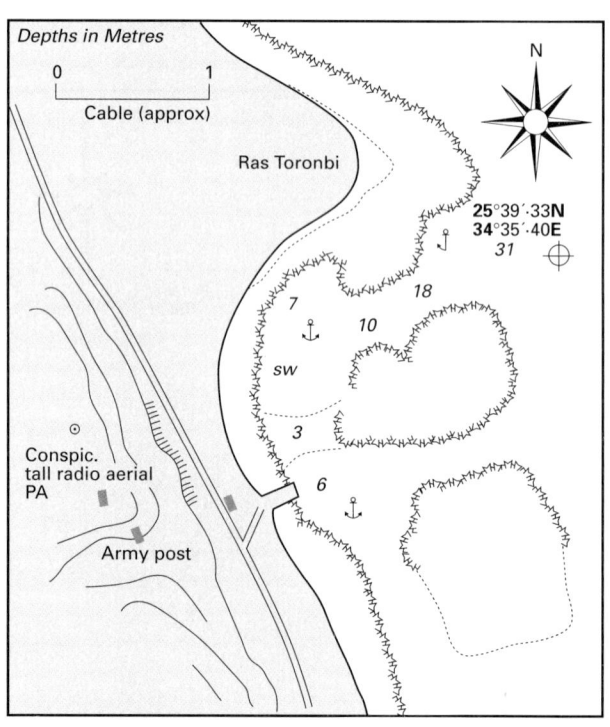

RAS TORONBI

Formalities

Soldiers, who may not be in uniform, will visit to check papers. They may need ferrying out to the boat. The soldiers at the post have usually been helpful and friendly.

The Egyptian Coast: Foul Bay to Hurghada

Marsa Wizr
25°47'N 34°29'·5E (not illustrated)

A good anchorage is reported here in a small bay with easy access and egress, protected from all but E'lies. Anchor fore and aft in 6m on sand, between the reefs. This should hold your bow against the swell, common in even moderate N'lies. A HEPCA/USAID mooring is the alternative.

Sharm El Bahari
25°52'N 34°26'E (not illustrated)

There is a wide entrance through the opening in the fringing reef, which breaks in any swell. There is a new, conspicuous hotel complex and dive resort with a jetty. Anchoring is difficult because the resort has laid moorings. If a mooring is free there seem to be no objections to yachts using it, but you'll need to lay a stern anchor to reduce swing.

El Quseir
Admiralty chart 3043, plan page 203

Approach
The jetty is crowded with dive boats and you may not find anywhere to get alongside or even stern-to. Care is needed when manoeuvring to avoid the dangers off the pier. The entire area can be affected by a large swell in certain conditions. There are now only two mooring buoys in the bay. There may be charges for using the jetty.

Conspicuous
There is a conspicuous radome 1M S of Quseir Bay and a new hotel complex to the N.

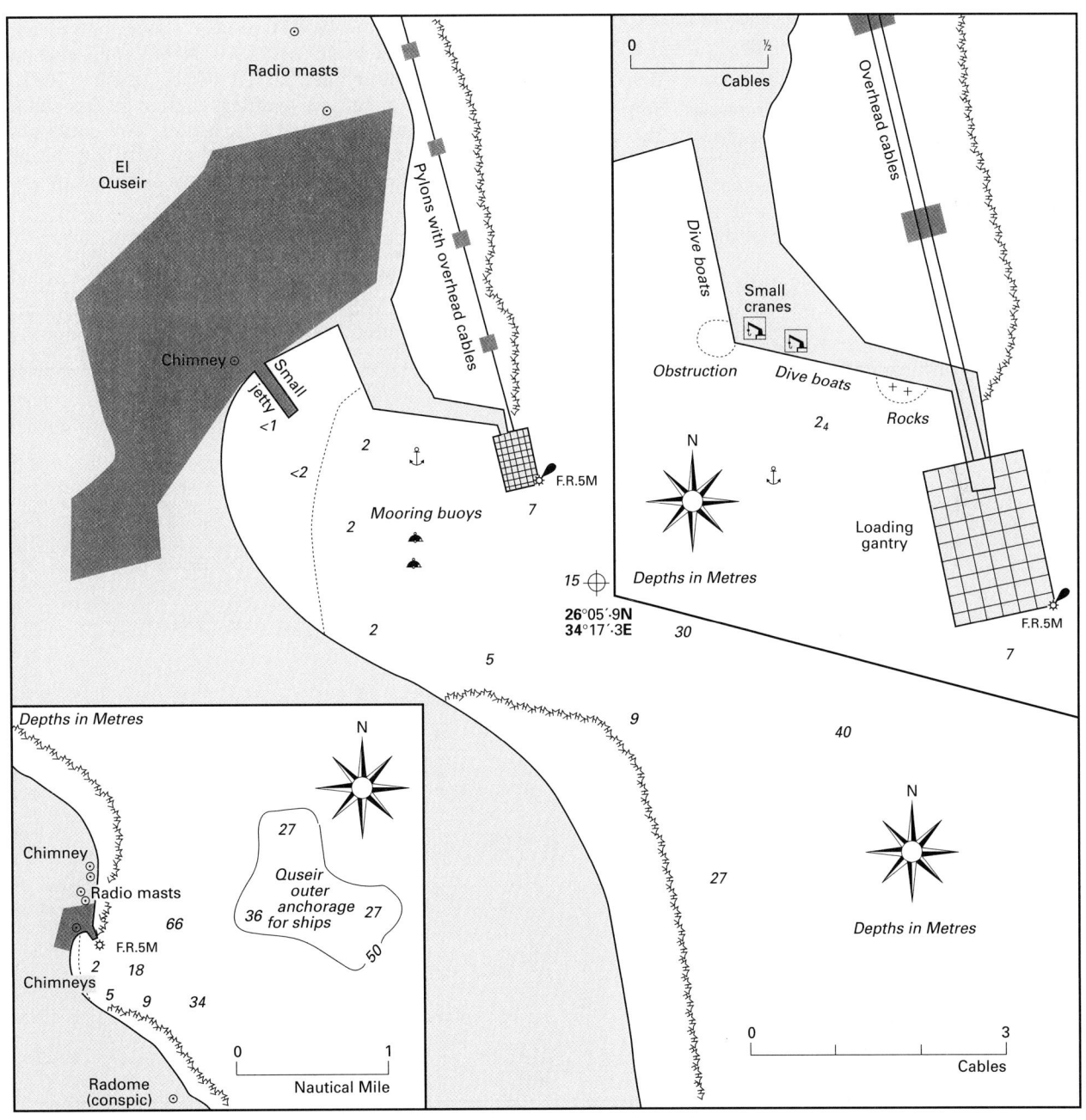

EL QUSEIR

Red Sea Pilot

Formalities
This is now a port of entry but clearance is easier at Safaga where some of your documentation will have to be processed anyway. If you use an agent, fees are approximately US$40 (basic charge) plus service fees, eg US$25 for taking passports to the immigration authorities in Safaga.

Facilities
Fuel is usually the primary reason for calling at Quseir and is usually available even if you cannot check in, though you may have to clear with health and customs authorities. You can order it through a ship chandler or an agent but the former is usually cheaper and it can be delivered within a couple of hours by cart in 40-gallon drums. Water is available but expensive at about US$20 per ton, minimum charge. Stores are easily available in the town as are other services such as telecoms and laundries. There are buses to Hurghada and Safaga.

General
The ancient port of Leukos Hormos was built near here by Ptolemy II. Quseir was one of the most important departure points for pilgrims to Mecca until the 10th century, and in medieval times it was an entrepôt for the Nile Valley and the Red Sea. Its decline began with the opening of the Suez Canal.

Hamrawein
26°15'·5N 34°13'E (not illustrated)

There is an anchorage in the lee of the large phosphate loading terminal. However, if there is a ship loading, don't even think of it. Thick, choking clouds of red phosphate dust swirl downwind coating everything in their way. Only for the desperate who, in any case, may be moved on by security paranoia.

El Akhawein (The Brothers)
Plan below

These two islets lie along the central route in the Red Sea and are normally used only as a landmark though there is good diving. The Admiralty Pilot tells us that 'owing to abnormal refraction El Akhawein have been seen from a distance of over 100 miles'. For a fair-weather day-only stop use one of the HEPCA/USAID project moorings on the Large Brother at 26°18'·733N 34°50'·817E and on Small Brother at 26°18'·043N 34°51'·768E. Otherwise, in settled weather, anchor in 7m, sand, ½ cable from the N tip of the S islet. You could put a long line to either bollard or jetty on Large Brother. The authorities sometimes close the area to divers.

Quei Reefs
26°23'N 34°11'E (not illustrated)

The channel inshore of Quei Reefs is deep and clear. There are three reefs with deep water separating

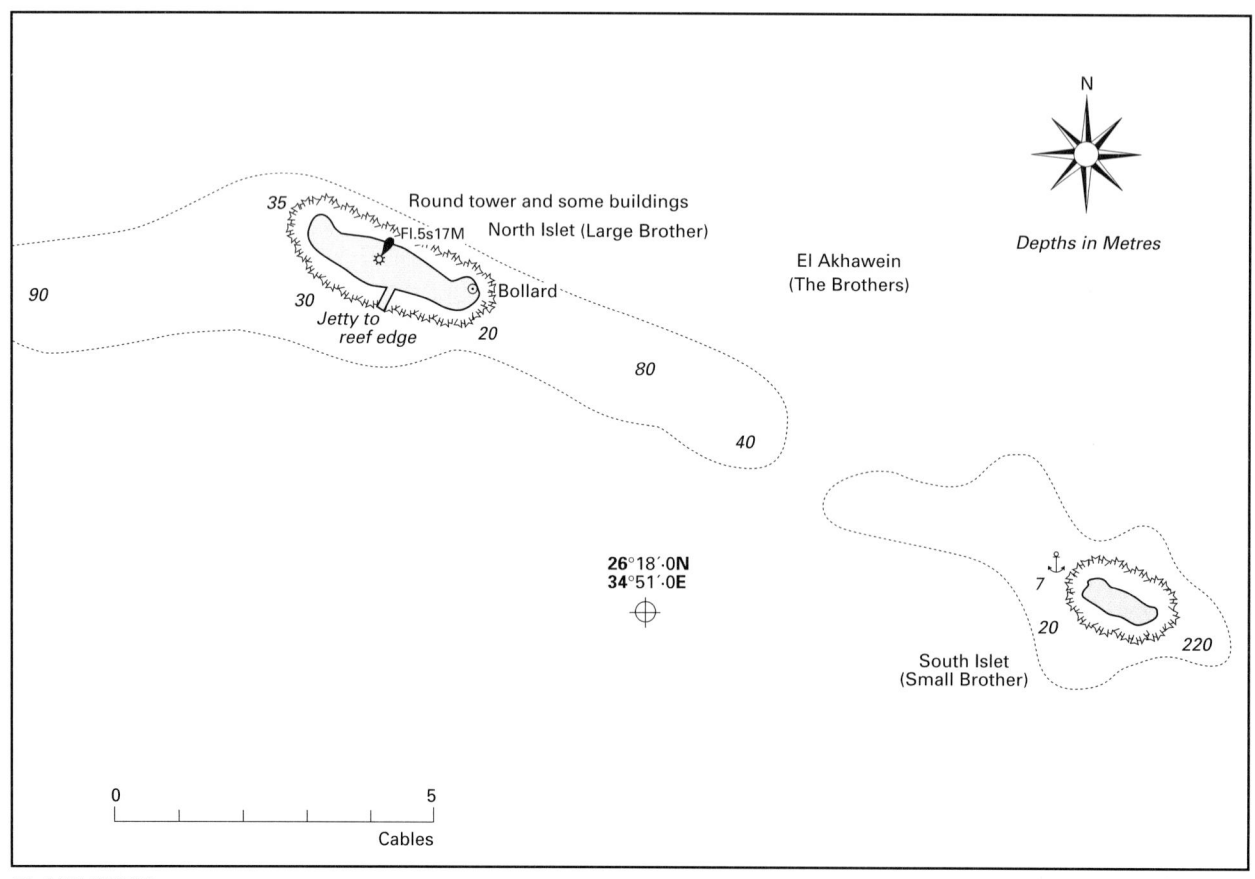

EL AKHAWEIN

The Egyptian Coast: Foul Bay to Hurghada

them and islets that are exposed at low tide, NE of the village at Bir Quei. The area is not well charted and the bottom is mostly a mixture of coral, sand and rock, so holding is questionable. The anchorage in 8m, behind the middle islet, gives shelter from the sea and is very tight so you can only use a short scope. Suitable in calm weather only. Of interest primarily to divers as a temporary stop, though the area isn't rated by the local dive outfits.

Mina Safaga
Admiralty chart 3043, plans below and page 206.

Approach
From the N enter at Ras Abu Soma (see next entry.)
The S and E approaches are littered with reefs, some of which have new navigational marks. Unless you want a break from sea and wind, there is no need to go via the harbour. You can go N of Gez Safaga instead. See sketches for waypoints. If you do go through the harbour, in the narrow passage which leads N to the anchorage, keep 20–30m off the sand spit near the naval base and follow the coastal fringing reef N. One report recommends steering approx 315°M along the inshore reef until near a large dive boat jetty E of a radio aerial. Least depth approx 3m at LW. A minaret and an aerial are inland NW of the resorts.

Anchorage
For both checking in and staying a few days the only reasonable anchorage near the town is at 26°47'·5N

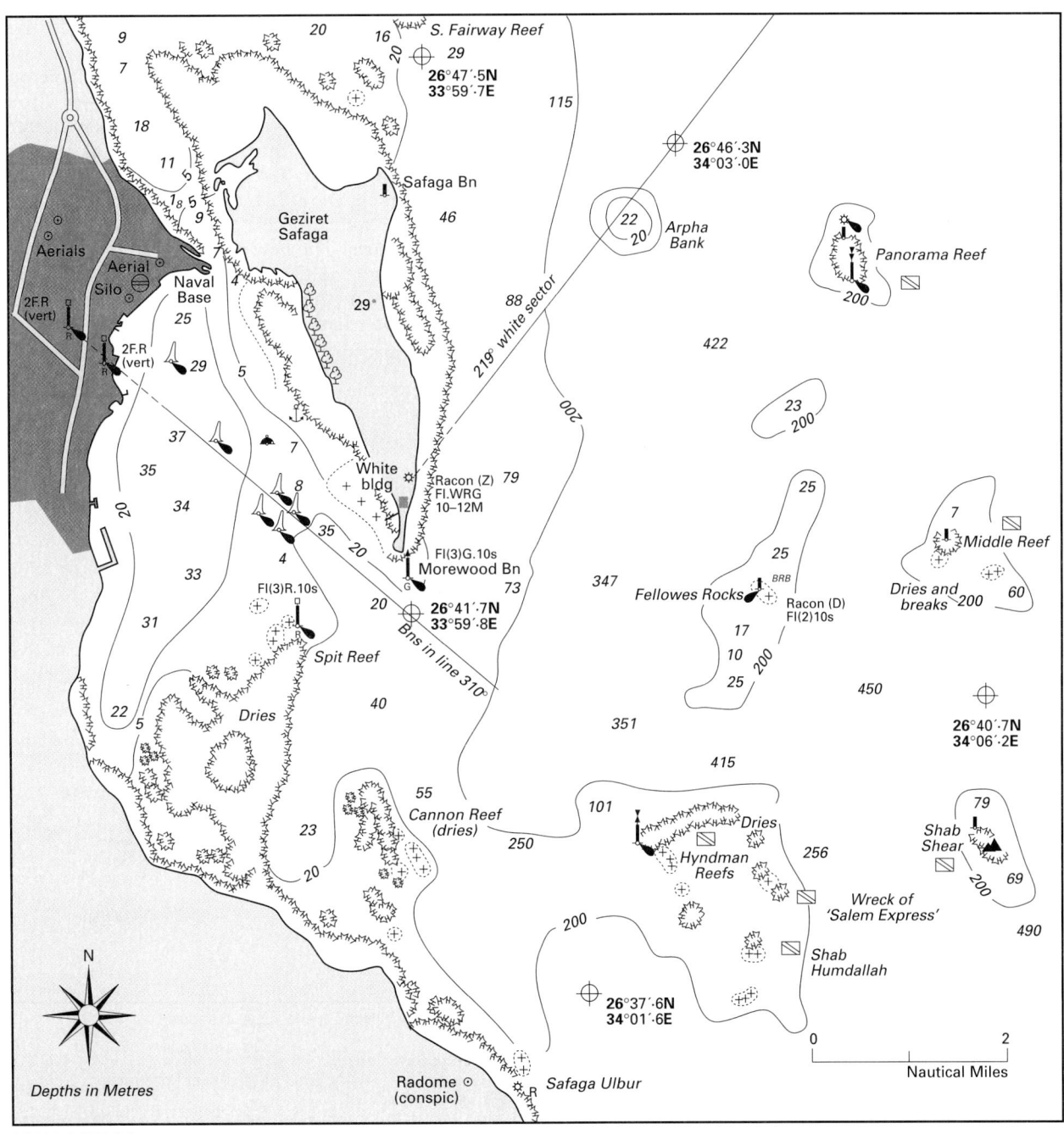

MINA SAFAGA

Red Sea Pilot

33°56'·5E off the hotel strip, good holding in 7–8m, sand patches in coral. It can be rolly and the anchorage area gets pushed out towards the 10m line each year by increasing numbers of dive boat moorings. Beware of floating lines attached to mooring buoys near the dive boat jetties. An alternative overnight anchorage off Gez Safaga at 26°43'·58N 33°58'·40E, NW of Morewood Beacon, has been used, in 8m, coral and sand.

Formalities

This has been the most popular first check-in port in Egypt in the past. However, if you intend to call at any marina you will be liable for the Customs fee which at present cannot be processed here (see details of the fees at the beginning of this chapter and the rest of this section below).

At the anchorage off the hotel strip, the 'beach police' or coastguard will come out and ask you to go to their base at the Orca Village Dive Resort in the NW corner of the bay. Tie up on the W side of the end of the solid jetty, not the floating pontoon, but please ask if it is OK. Once you've seen the coastguard, if you want to DIY you will have to go into town. Walk past the resort, up the dirt road to the main road and catch a minibus or taxi. A minibus fare was 50 piastres in late 2000. For the Port Offices and Immigration get out at the big crossroads after Bank of Egypt. Walk down past the checkpoints. The Immigration Office is inside the second, and less elaborate gates into the port in the small yellow, two-storied building about 100m on the right. You will have to show identity, may be kept waiting in the guard post at the gate, and will be given an escort.

Port dues and health clearance should be about US$40. Visas and the customs fee are extra. Before you clear in decide how long you intend staying in Egyptian waters.

a. *if you are making a fast run and won't call at any marina*: Ask for a 10 day 'Sailing Permission' marked 'Destination Suez'. This allows you to anchor in bays, reefs and islands BUT NOT TO CALL AT ANY MARINA. In theory such permission means you must get to Suez in 10 days, but at the Suez end this is looked upon flexibly as far as we know. Once you've cleared in and have your visas, entry stamp, receipts and sailing permit marked 'Destination Suez', you don't have to call at any other port in Egypt before Suez and it's wise not to. If you do you'll be liable for port dues again.

b. *if you intend to take your time (for example to travel inland) and/or call at a marina*: You must pay the Customs fee (scale on page 204). So consider NOT checking in at Safaga since you cannot pay the customs fee here. Customswise Safaga is in Hurghada's bailiwick. So it's easier to clear in at Hurghada to begin with (see also Hurghada and El Gouna below.)

When you leave Safaga the coastguard will want to see your papers again. If headed S the authorities prefer you to get an exit stamp in your passport even if you intend to stop at anchorages before the border with Sudan. The coastguard will want a photocopy of the exit stamps in your passports before you leave. Checking out of Egypt was free in 2000.

Fuel and water Agents are not needed or obligatory though they will offer help with paperwork, fuel and water. Work out what you want to do and agree on a price in advance. We paid E£30 (US$9) for a half-day taxi service which included being taken to immigration to check out, the bank, the fuel station, the PO, the market etc. Diesel and petrol are available by jerry and LPG refills are possible but slow.

Laundry and garbage Agents will deal with laundry for you. Alternatively there are laundries in the town. There is a large, unsightly garbage tip on the dirt road that leads from the beach to the main road.

Money and provisions Cash advances are available on *Visa* and *MasterCard* at the Bank of Egypt. It is open in the morning every day though for shorter hours on Friday and Saturday. Take your passport with you. Groceries are available from supermarkets about half way to the port area and from the *souk* in the old town. Safaga is divided into two parts. The new town is NW and N of the port (though 'new' is here a term of art) and the old town is at the S end of the port. The latter tends to be what locals mean by Safaga.

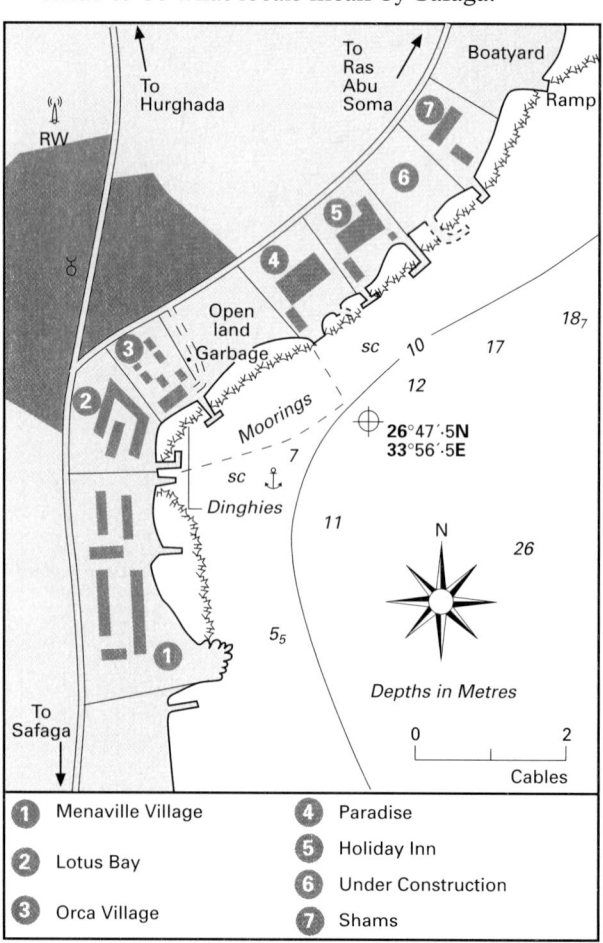

SAFAGA RESORT ANCHORAGE

1. Menaville Village
2. Lotus Bay
3. Orca Village
4. Paradise
5. Holiday Inn
6. Under Construction
7. Shams

The Egyptian Coast: Foul Bay to Hurghada

Communications There are internet cafés near the anchorage and in the town. There are card phones but you can also make international calls from shops.

General

Safaga stands on the ruins of the old port of Philotera, built by Ptolemy II in the 3rd century BCE. The mountain range inland running N is truly spectacular. It is possible to leave your boat here for a sightseeing trip to Luxor. Safaga is a brilliant place for windsurfing. The world championships were held here a few years back. The bay E of the hotel strip is full of windsurfers blasting to and fro all day once the sea breeze has packed in behind the gradient wind.

Diving is highly rated but, as with the Hurghada area, hopelessly overcrowded. Known sites are marked on the approaches plans for this and the next entry. For some reason dive books tend to refer to Hyndman Reefs as Shab Shear and Shab Shear as Abu Kazan.

Ras Abu Soma

Approach

The area is now very conspicuous because of a new resort complex, with at least three hotels and a golf course! The greens of the latter and a large, lit water tower further N are hard to miss. There is a long walkway to the SE edge of the reef around the *ras*. Dive boats on the reefs serve as good marks but there are unmarked shallows in the approach. Enter only in good visibility. An approach position is 26°50'·16N 33°59'·29E. There is a black and white beacon, Fl(4)20s16M, Racon (C) in 26°51'·32N 34°00'·18E, on the N of the *ras*.

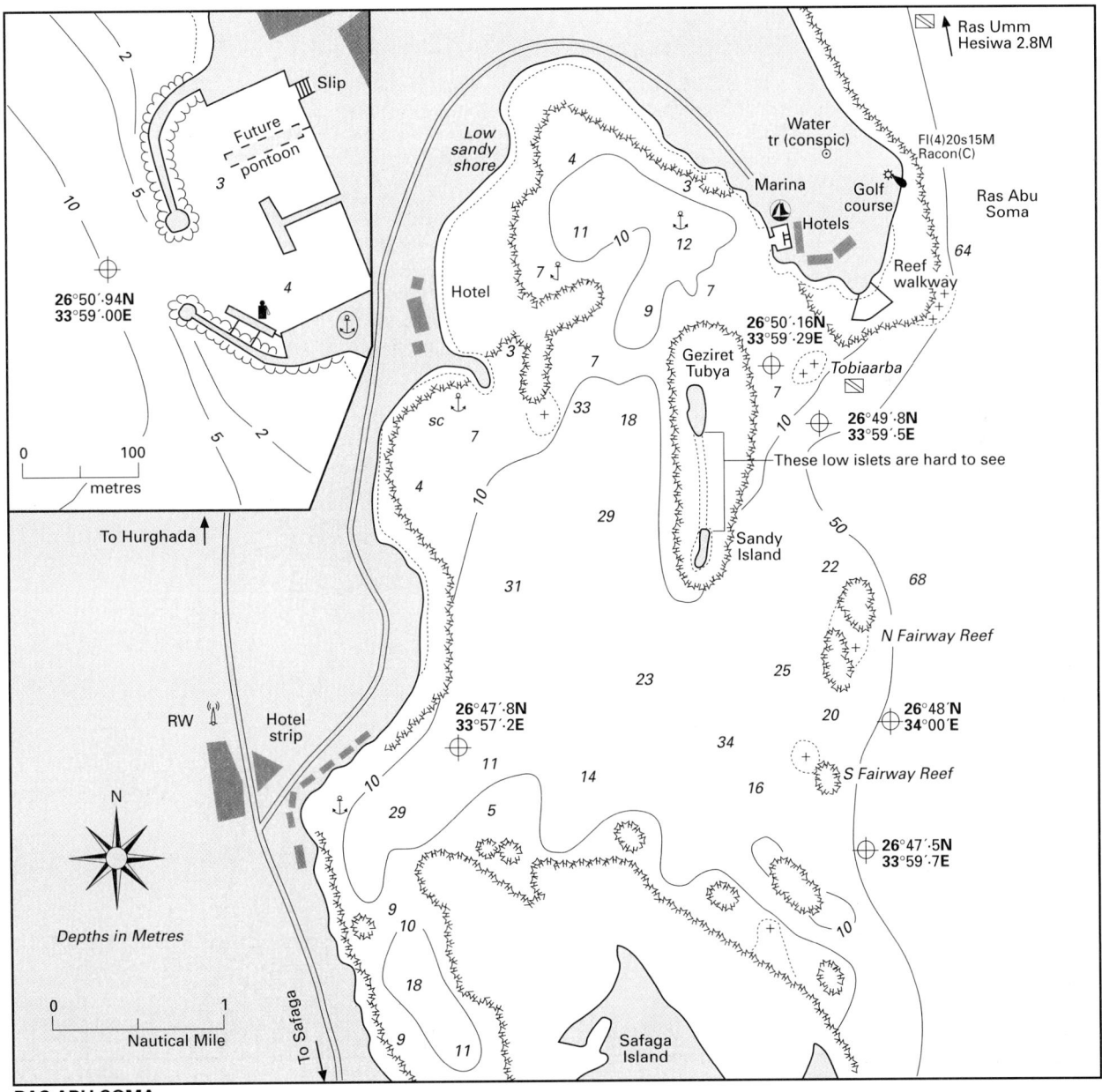

RAS ABU SOMA

Red Sea Pilot

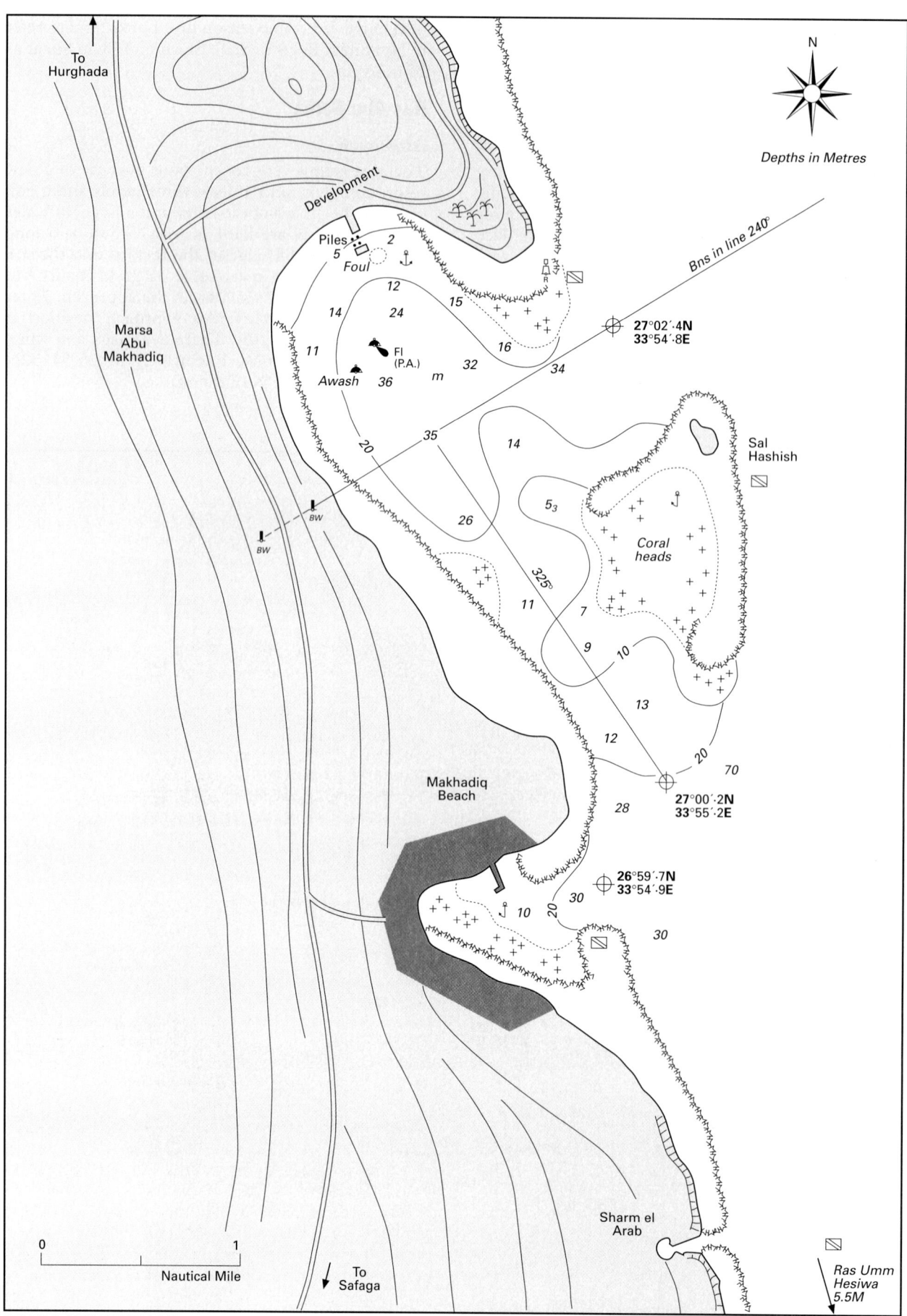

MARSA ABU MAKHADIQ

The Egyptian Coast: Foul Bay to Hurghada

Anchorage

The anchorage outside the marina, in 8–10m sand, good holding, is quieter than the one off the Safaga hotel strip, and although the winds are often strong, a lot more comfortable. If you anchor here transport into the town will not be as easy but things will improve as the hotel and marina development is completed.

Facilities

There is a small new marina here, incomplete in 2002. The entrance is at 26°50'·9N 33°59'·00E, depths inside around 4m shoaling abruptly to 1·25m close to the walls. It has 70 stern-to berths for boats up to 45m. There are a few visitors' berths. Call on VHF Ch 12 on approach. Berthing is possible at the dock opposite the entrance but it can be very dusty. Facilities are scanty as yet and there's a basic charge of US$10 per day. When complete, berthing will probably be more expensive. Water is available. Fuel, showers, provisions and laundry will be available soon but everything is more expensive than in Safaga. *Email* efc@internetegypt.com

There is good diving on the nearby reefs with HEPCA/USAID moorings off Gez Tubya, S of the anchorage and off the Fairway Reefs, N of Gez Safaga but currents can be very strong.

Panorama Reef and Middle Reef, in the E approaches to Safaga and Ras Abu Soma also have moorings at 26°45'·08N 34°04'·68E and 26°42'·31N 34°05'·92E. And, finally, there are moorings and a dive site at Ras Umm Hesiwa and Shab Saiman (26°54'·2N 33°59'·9E).

Makhadiq Beach and Sal Hashish
Plan page 208

The next anchorage N'wards is what the locals call Makhadiq Beach. It is surrounded with new hotels and has a long new jetty projecting into the bay from the N shore, which given the dive boats, the windsurfers, Hobies and the reef on the S shore doesn't leave much room. It is best avoided. Sailing N from here to Marsa Abu Makhadiq beware the fingers of reef across the channel from the patchy reef extending SSW from Sal Hashish It. Isolated bommies also stretch a good way across the E side of this channel. The line of the channel from 27°00'·2N 33°55'·2E is 325°. The deeper water is on the inshore side, though beware of the isolated bommie on the W side of the N entrance to the channel. Note there is a good anchorage in the lagoon S of Sal Hashish It. Access it with a careful lookout through the bommies of the patchy ground extending SSW of the islet or through clearer water from the S. The diving on the E drop off of Sal Hashish is said to be excellent. There is no anchorage in the small *marsa* called Sharm el Arab (26°58'N 33°55'E).

Marsa Abu Makhadiq
Plan page 208

Approach

For approach from the S see above. There are two mooring buoys in the middle of the bay in the approximate positions shown. The N buoy is lit, the other buoy is awash and hard to see. There is a lattice work beacon on the E side of reef on the NE entrance to the bay. From the N there are also leading marks either side of the road on the SW side of *marsa* in line bearing approx 240°. They are not easy to see either. There is an old jetty at the head of the bay but the end, with a derrick, is separated from the jetty. There are four bare piles in a square between them. The whole N shore is under development for hotel complexes, the coastal bane from Safaga to the Gulf of Suez.

Anchorage

Best shelter, out of any swell, is E of the pier. One position is 27°02'·74N 33°53'·72E in about 10m sand, good holding and well protected in strong N winds. Note that the bottom is fouled by an old anchor and chain for 30m around 27°02'·71N 33°53'·53E, just SE of the pier end.

Gifatin (Giftun) Is
27°12'N 33°57'E

Note Anchoring is now prohibited in the Gifatins. Gifatin el-Saghir, Gifatin Kebir, Abu Rimathi, Shab Abu Rimathi, Umm Agawish el-Kebir, Abu Minqar and all other reefs and waters in this area are a marine reserve. Hefty fines and jail sentences for offenders. Free moorings have been laid in the area and though they tend to be heavily used by dive boats during the day they can often be vacant at night. For a list of moorings see the Diving section in the Planning Guide. There is a restaurant with a reportedly pleasant ambience on the S tip of Gifatin el Saghir.

Hurghada
Admiralty chart 3043, plan p10

Approach

In general your final approach is best made in daylight.

Use the S pass either side of Shab Abu Rimathi or take the inshore passage between the mainland coast and Umm Agawish el-Kebir between the waypoints and on the courses shown on the sketch. Note that none of the charted beacons on Umm Agawish el-Kebir and the small reef in the S entrance to the channel existed in October 2000. The beacon charted on the NE tip of the reef E of the hotel strip is a yellow buoy.

The leading marks for approach from the N are very clear and the Minqar Channel is straightforward. The reef off the W side of Abu Minqar is beaconed. See also under Dangers, below, and under Strait of Gubal anchorages in the next section.

209

Red Sea Pilot

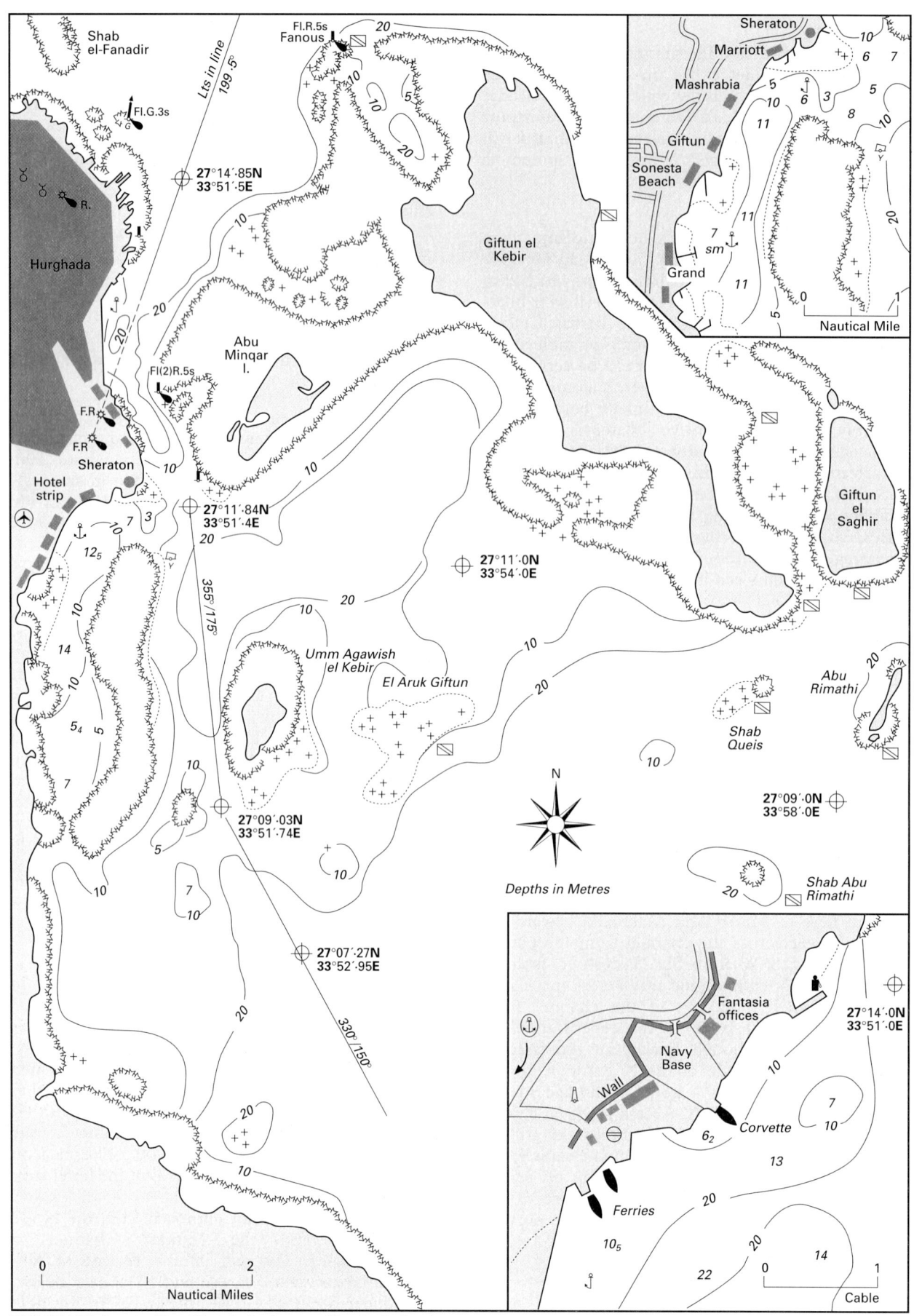

HURGHADA (Insets: Hotel Beach Anchorage, Hurghada Port)

210

The Egyptian Coast: Foul Bay to Hurghada

Dangers

From the S the unmarked reefs of Shab Abu Rimathi, Shab Queis and El Aruk Giftun are the major pitfalls but by day they are marked by dive boats. From the N unmarked Carless Reef and Umm Qamar are the only real hazards, all the rest are visible in daylight. There is shoal water off the point on which the Sheraton stands, and further S another shoal extends from the shore N of the Sonesta Hotel. In calm weather in the spring there can be early morning radiation fog in the approaches to Hurghada. Never a dull moment.

Conspicuous

There is a conspicuous new water tower NW of the town in 27°17'·7N 33°44'·6E. In the port there is a new ferry pier and a jetty under construction SW of the Port Control building. The Sharm el Sheikh jetcat berths N of this new jetty. Extensive reclamation is also underway along the fringing reef between Merlin Pt and the port. There are many new piers along the whole shoreline N and NW of the port which are used by hotel dive boats, the numbers of which in Hurghada are now well over the 1200 mark!

Formalities

You are obliged to check in at every Egyptian port of entry that you visit and you have to pay port fees (see below). Some yachts try to avoid the procedures and the associated fees, but this is illegal. Yachts are required to go alongside the jetty at the port authority (see inset of Hurghada port area) to clear in and to clear out. Police, ferries and other craft are anchored stern-to nearby, and care is needed on approach to avoid their anchor lines. It is easier to anchor off and dinghy in.

Clearing can be done with or without an agent. Having one makes life a lot easier, but it obviously costs more. The only agent at present is Fantasia. Their office is in the New Harbour. They sometimes listen on Ch 16. ☎ +20 65 443675, email fantasia60@hotmail.com. If you are thinking of this we suggest you contact Abu Tig Marina (see next chapter for contact details) BEFORE you get to Hurghada or El Gouna. This should enable you to minimise expense.

If you do your own clearance, procedures are straightforward but time-consuming because the offices are scattered and you'll have to find your way around. Officials are on duty from approximately 0900–1400. They are available at other times, but will charge extra. Overtime is also payable if you want to check in on a Friday or a public holiday.

Fantasia's agent's fee structure has been broadly agreed with Abu Tig Marina as follows but limited negotiation may be possible:

Port dues, health clearance Official rate is calculated on loa and tonnage. Abu Tig negotiated rate = £E300 (US$65). But expect to be asked for <US$120.

Agents fee US$40. You may be asked up to US$80.

Customs fee and duty stamps £E110 (US$25). (See also page 181.)

Visas (if you don't already have them): US$15 each

If you use Fantasia and are going to El Gouna you can normally leave passports and ships' papers with them and go on directly to the marina. Your papers will be delivered to El Gouna later. This avoids more than a brief stop in Hurghada to wait till you have visas. Pay when you leave the marina. For details of marina berthing rates and the current offer of free berthing for those who stay between 10 and 30 days, see the next chapter.

Health officials will usually come to the boat and may ask to see vaccination certificates, but these are not obligatory unless you have come from an area where infectious diseases, such as yellow fever, are endemic. Multiple crew lists are required. You must have a visa if you want to go ashore. If you arrive without one, immigration officials will issue one.

Before moving from the clearance jetty, ask port control for an authorisation note to anchor off any of the hotels or use any of the docks. To get the authorisation note, and to get port clearance, you must write a short letter saying who you are and what you want. This can be done on the spot on almost any piece of paper. Much face is given all round if the ship's rubber stamp is wielded fiercely. The Greater Gods of Bureaucracy were invented in either Ancient Egypt or Ancient China. They still preside in both countries.

Anchorage

The best anchorage is about 1M S of Merlin Point, near the Sonesta and Grand Hotels. Good holding in 7–10m, sand and mud, with reasonable protection from the very strong winds that often blow. The bottom shoals rapidly so anchor well out. This is a nuisance in bad weather when getting ashore is a wet business. See Facilities for hotel policy on yachts anchored off. The only other possibility with easy access to the town is where the half-anchor is shown SE of the Sheraton, but holding is poor and there is less shelter.

Facilities

Money Local currency is available from ATMs.

Fuel An agent will obtain fuel for you, or you can organise it yourself. There is a fuel dock at the New Harbour, and one can go alongside here, but the bottom is 'interesting' and the dock often crowded. In addition to agents' fees, the cost of fuel consists of duty to customs and the pump price at a fuel dock. If you obtain fuel without an agent's services, you must:

1. Go to a bank and change funds sufficient to cover the cost of the fuel required.
2. Take the exchange receipt from the bank to customs at the port.
3. Tell them how much fuel you need. They will calculate the price, which must match or be less than the amount of the funds changed at the bank. You then pay them the appropriate duty and they issue a receipt.

4. Take the receipt issued by customs to a fuel dock, where the stated amount of fuel will be delivered to, and only to, the boat named on the receipt for customs. This means that you cannot pay for fuel for another boat, except by filling jerry cans for them.

5. Finally, pay the pump price for your fuel.

In most respects this procedure is similar to that used in Europe for obtaining duty-free fuel. The difference is that in Egypt 'duty free' fuel is duty paid at some twice the price of the stuff from the pump! The good news is that diesel is still relatively cheap. Another possibility is to take your jerry cans to a roadside service station. This is cheaper, but illegal. You may find a local person quite willing to perform this service for you, for a consideration of course.

Petrol is easily available in small amounts from service stations. LPG is very difficult to come by if you have US or Australian/NZ bottle fitments. A pigtail is no help; the LPG depot scorns them.

Water The potability of mains water is dubious. If necessary you could boil or filter it for drinking. Alternatively buy bottled water. For tap water ask at one of the hotels or try using an agent. The best reception for yachts wanting water recently was from the Sonesta Hotel. However, the hotel has begun charging boats which anchor off its waterfront. In exchange for a payment of E£10 per day, boat crews were welcome to go ashore via the hotel jetty, go alongside to take on water, and use the pool and the showers in the hotel grounds. For those who have spent some weeks getting to Hurghada, this can be an extremely attractive proposition!

Stores It is easy to reprovision. There are shops, restaurants and a bank not far S of the port, or you can take a minibus into the town, where there are bigger groceries and a market.

Communications There are cardphones and hotels offer phone and fax services. The GSM mobile phone network works well. Alternatively, there is a 24-hour phone office in the town. The hotels will hold mail and usually offer a gracious reception, which can be reciprocated by making arrangements in advance if one wants to use their facilities. There are cyber-cafés in the town for email. The PO is generally crowded and busy. Mail sent from there to Europe should take about a week but mail has been known to go missing. Poste Restante is not recommended by local residents.

Medical Hospital at Hurghada with recompression facilities.

Transport Minibuses ply between the strip of hotels and the town. Their rates are fixed. Metered taxis are easy to find, and you can negotiate an hourly rate if you have a number of errands to run. Fares are expensive in Hurghada because of the tourist trade. There are planes and buses to Cairo and buses to El Gouna.

Laundry and garbage There are laundries in the town, and occasionally a service is offered by an entrepreneur who chugs through the anchorage off the hotels. The hotels themselves will also take laundry, but the price is, naturally, higher.

Once you have cleared in you can take garbage to a receptacle on the street. The New Harbour has garbage bins. Ask security guards at the hotels where to leave garbage for disposal if you are going ashore there.

General

Hurghada is in fact a compound of two villages. Dahar is 'downtown' where the market, banks, GPO and most shops are. Sekala is the 'new' town that has sprung up around the port and on the hill to the S. Thirty years ago there was virtually nothing on the waterfront at all bar a few fishermen's shacks.

There is a marine-biology station about 3M N of the town. It has a large collection of fish and is worth a visit. Buses run out there. The ancient town on the site of which Hurghada is built may have been Myos Hormos, founded by Ptolemy II, however, scholars are equally inclined to think the place may have been at El Gouna in Abu Shar Bay between present day Hurghada and El Gouna.

El Gouna to Suez

Admiralty charts
Large scale 333, 2090, 2098, 2373, 2374, 3043, 3214, 3492
Medium scale 2373, 2374, 2375
Small scale 159, 5501

US charts
Large scale 62188, 62191, 62194, 62195, 62193
Medium scale 62191, 62195, 62188

French charts
Large scale 6878, 6908, 7013
Medium scale 6878, 6908
Small scale 7113

El Gouna
Admiralty chart 3043, Approach plan p214 plan p215

There are two facilities. Abu Tig Marina is a full service, international standard marina. Abydos Marina, is a low-cost, bottom budget operation.

Approach
From Hurghada
WPT 11 27°16'·90N 33°52'·00E
in open water 3M NNE of Hurghada harbour jetty.

WPT 10 27°21'N 33°48'·20E
0·9M W of W edge of Shab Abu Nigara.

WPT 9 27°25'·6N 33°43'·85E
0·6M SW of a beacon on an isolated reef, you'll be in around 10m at this point. It's just a tongue of shoal ground which may give you as little as 5·5m as you head for WPT 7, but don't fret.

WPT 7 27°24'·60N 33°41'·4E
At this point you are off Abu Tig Marina. The entrance to the buoyed channel, least depth 3·6m, is marked by lit beacons on the reef each side of the entrance, starboard, Fl.G.5s.4M (in good weather), at approx 27°24'·36N 33°40'·54E. There are 2 more pairs of buoys R/G (F.R and F.G) in the dredged channel and beacons (Fl.R.5s1·5M, Fl.G.5s1·5M) at the marina entrance.

To keep going to Abydos note the eccentric buoyage:

WPT 8 27°23'·25N 33°41'·69E buoy Fl.G. You have a choice. Either leave this one to port and the beacon to starboard, least depth >4m, or leave it well to starboard to pass E of the reef it marks.

Thence about 0·75M S to Marina Buoy Fl.R in approx 27°22'·8N 33°41'·71E. LEAVE TO STBD then turn W and the two piers of Abydos are 4 cables ahead.

Via the Tawila Channel
WPT 1 27°48'·85N 33°43'·10E
in open water N of the N entrance to the Tawila Channel 1·4M NE of Ashrafi I Lt.

WPT 2 27°40'70N 33°45'·70E
between Gubal Saqhira and South Qeisum, note that charted beacons on the reef N of Bluff I and SE of S Qeisum were missing in late 2000. There are many oil production installations on the N side of the SW end of the Tawila Channel.

WPT 3 27°35'·90N 33°42'·85E
N entrance to Bahriya Tawila, the narrows between Tawila I and the reef extending SSE of Umm el-Heimat Saghira marked by beacons and buoys. In approx 27°36'·02N 33°43'·2E there is a green starboard-hand buoy. In approx 27°36'·15N 33°42'·4E and 27°36'·2N 33°43'·0E there are red port-hand buoys. These mark the deep water channel on a rough line 060°/240°, with the highest point on Gubal I (121m) as leading or back mark.

WPT 4 27°35'·20N 33°42'·40E
S entrance to Dahriya Tawila. The exit to the narrows is marked by eccentric buoyage. In approx. 27°35'·25N 33°42'·6E there is a horizontally marked RW buoy. Given its location it is presumably supposed to be an E quadrant buoy on the old uniform cardinal system, though the colours make it a S quadrant one! Whatever, leave it to starboard going N, to port coming S.

WPT 5 27°30'·00N 33°41'·87E
off NE tip of reef WSW of Shab Tawila.

WPT 6 27°26'·20N 33°43'·20E
ESE of SE tip of Shab Esh, there is an unlit, orange buoy at 27°26'·42N 33°42'·44E.

WPT 7 27°24'·60N 33°41'·40E
and as above for Abu Tig Marina and for Abydos.

Abu Tig Marina
The arrivals pier is at the fuel dock on the port side of the entrance. Call on VHF Ch 16 or 73 on approach.

Formalities
Officially, visiting vessels in Egypt should check in with the regional port authority for each port or sub-port they visit. El Gouna is in Hurghada's bailiwick. You have the following alternatives for clearance:
1. *From S.* EVEN IF YOU HAVE ALREADY CHECKED IN AT SAFAGA check in at Hurghada as well (see above). Let Abu Tig Marina know your eta Hurghada beforehand. You can leave your papers for processing with the Hurghada agent. They will be sent on to you at El Gouna. Pay when you leave the marina. Give the marina office 24 hours' warning. They will need

---

Warning again, experience says even when it's blowing, winds on the E side of the Gulf of Suez, especially N of El Tur, are not as strong on the W side and seas are not as bad.

In this section we continue listing anchorages in order of latitude, S to N, apart from in the Strait of Gubal. There is no ideal arrangement.

Red Sea Pilot

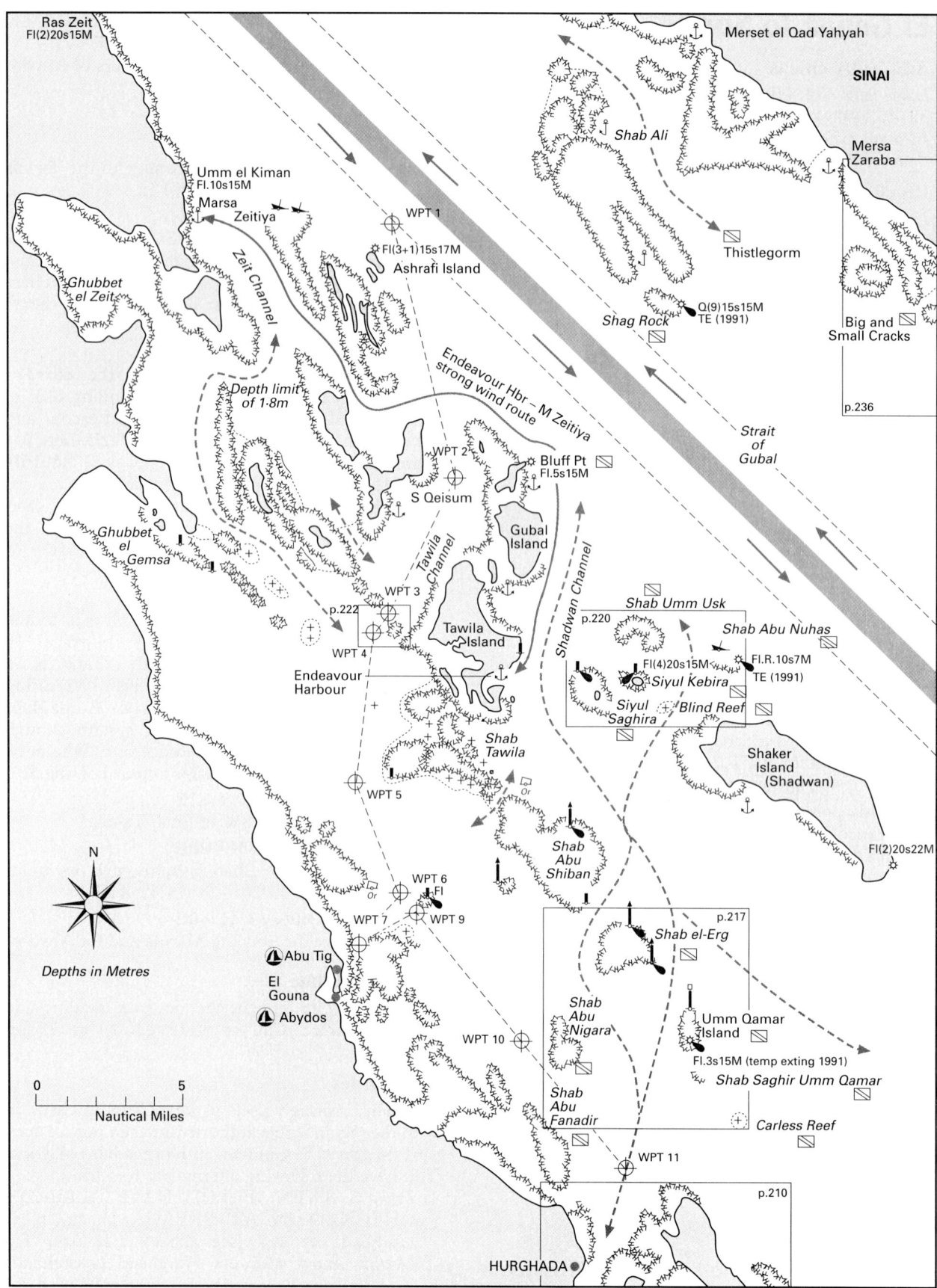

ANCHORAGES, DIVE SITES AND ROUTES IN THE STRAIT OF GUBAL: KEY MAP

214

The Egyptian Coast: El Gouna to Suez

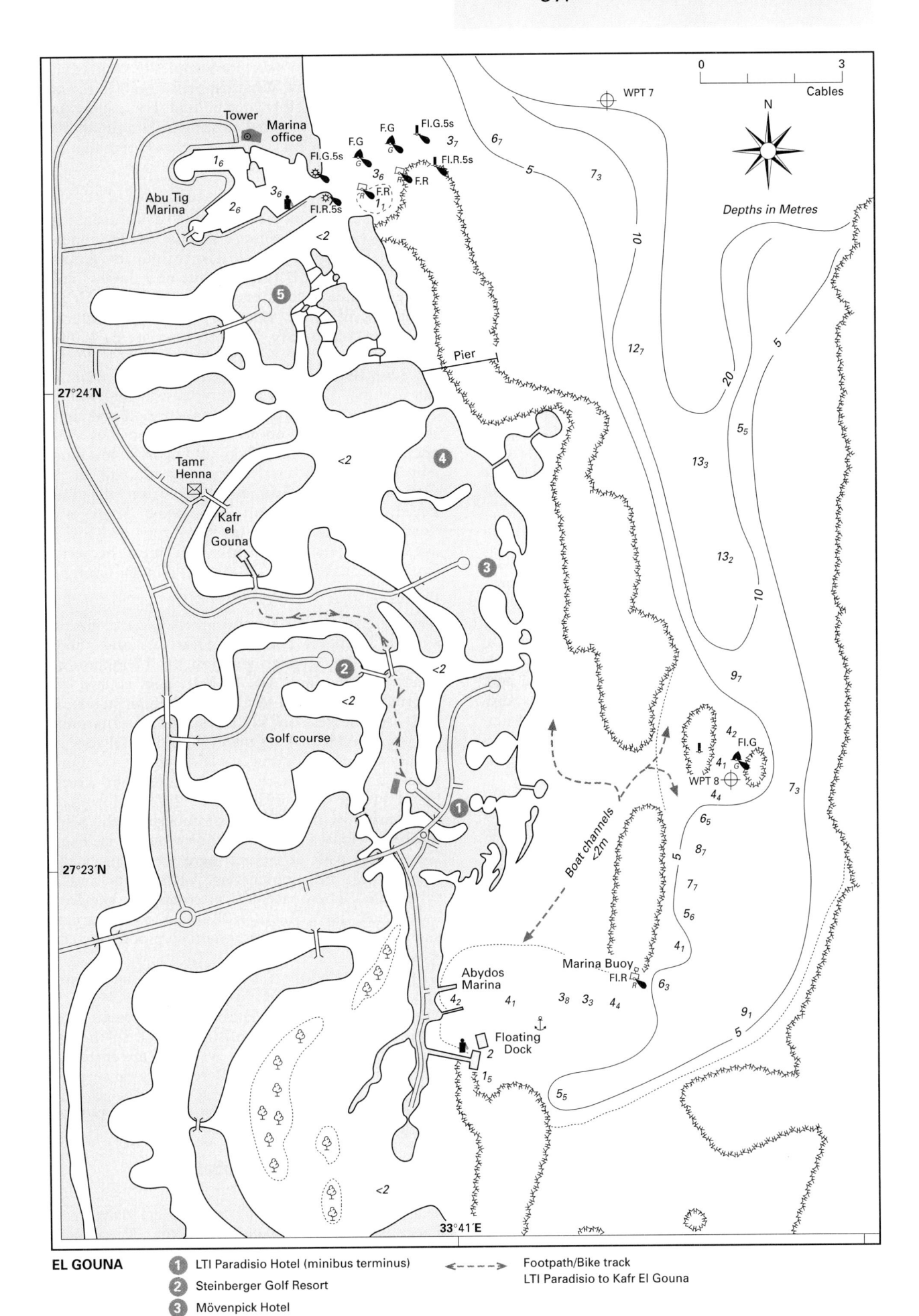

EL GOUNA

1. LTI Paradisio Hotel (minibus terminus)
2. Steinberger Golf Resort
3. Mövenpick Hotel
4. Hill Hotel
5. Sheraton Hotel

- - - → Footpath/Bike track
LTI Paradisio to Kafr El Gouna

passports and ship's papers and will issue a 10-day dailing permit for your passage to Suez. Pay all fees on their invoice once you have your 10-day permit. With this clearance you have 24 hours within which to leave. Make sure you have a good forecast. The marina office will help with that too.

Note: Make arrangements for your Suez transfer from here if not before. Yachts coming from El Gouna may get a good discount from Prince of the Red Sea (see below).

2. *From Suez.* Get a sailing permit stamped 'Destination Hurghada'. Go direct to El Gouna. An agent will be organised to come from Hurghada to process your papers.

Official fees
These will vary depending on your stay. Using Fantasia, the only agent at present either in Hurghada or El Gouna, agent's fee US$40, port dues and health clearance £E300 (US$65), one month customs fee and duty stamps £E110 (US$25 – see page 181 for fees for longer stays). Visas (if you don't already have them) US$15 each.

Facilities
There are 126 berths for boats up to 40m and 12 visitors berths dredged to approx 2·6–3·6m. Mooring is Mediterranean style with service posts for water, electricity, phone and cable TV. There are dedicated showers, toilets, cafés, bakeries and a laundry for visiting boats. Fuel is available and there will be a chandlery, and mechanical, electrical and electronics workshops. The helpful and knowledgeable marina manager and harbour master is Philip Jones.

Rates
El Gouna's current rate for visiting yachts <18m LOA is US$15 a day including power and water with limits of 1·5 cubic metres of water and 15kWh per week. If you are <18m LOA, and stay over 10 days but less than 30 days berthing is free, water US$4·50 per cubic metre, electricity US13·5c per kWh. The monthly rate thereon is US$15 per metre LOA. These rates are intended to encourage longer term berthing. There are shuttle buses to downtown El Gouna.

Contact
☎ +20 65 580 073. Mobile +20 12 223 0090
Fax +20 65 580 040
www.elgouna.com/www.abutig-marina.com
Email info@abutig-marina.com

Abydos Marina

Approach
Note the eccentric buoyage in the access route above. Once you are near the piers hover until someone waves you to a berth.

Formalities
Until El Gouna's port status is established, a coastguard will probably come and ask for your papers. It is best to have photocopies of your sailing permit to give away. What happens next is in the lap of the gods or possibly your billfold. For a short stay you may be able to avoid clearing in Hurghada. For a long stay, you may have to go to Hurghada, for details see Abu Tig above.

Facilities
There are 2 floating piers. Visiting yachts usually lie on the N side of the S'most. Anchor Mediterranean style. It is near impossible not to foul the ground tackle of boats in berths opposite and there is only space for three or four visitors. Daily rates are E£50, regardless of LOA, electricity included. Water is E£10 per cubic metre. Monthly rates are E£450 for up to 7m, E£800 for 8–14m and E£1000 for 14–20m. Services are basic and ablutions facilities not inviting. The staff are very friendly and will do their best to help. Alternatively anchor off and come to an arrangement about using the facilities. The anchorage is sheltered, good holding in 3–4m, mud. Abydos was the first marina operational in El Gouna. The El Gouna shuttle from the LTI Paradiso Hotel is about 1km away. There is a floating dock for vessels up to 60 tons and fuel is available at reasonable prices. It must be jerry-jugged because depths off the fuel pier are limited.

General
Modern El Gouna is an amazing mega development by the Egyptian company Orascom and much patronised by Egyptian glitterati. Its 17 million m^2 boast luxury villas, an 18-hole golf course, an international hospital to US and German standards with recompression facilities, an international airport, a water-making plant, an industrial zone, a winery, a brewery, a power station, an international school and almost a cordon sanitaire from scruffy Egypt proper with which you'll have a love-hate relationship by now. You can rubberneck the place from the canals interlacing it. Downtown at Tamr Henna and Kafr el Gouna there are shops, a PO (unreliable), film processing, restaurants and a cyber-café. There are six international standard hotels including a Sheraton-Miramar, a Mövenpick, and the LTI Paradisio. After hard commons you can gross out.

In his translation of the *Periplus* Lionel Casson says that Abu Shar Bay, just south of El Gouna, is Muos Hormos (Mussel Harbour) but it looks a bit improbable given all the offshore reefs. In the 1st century AD Muos Hormos was still an entrepôt, though much upstaged by Berenice because even then no one really liked flogging to windward, and by-passing Berenice for Muos Hormos added weeks to the return voyage.

Strait of Gubal (Madiq Gubal)
Key map page 214

The Strait is about 6½M wide at its narrowest. There is a traffic separation scheme. Keep clear when sailing and stay outside the shipping lanes when motoring. Tidal streams run 1½–2kt. The N-

The Egyptian Coast: El Gouna to Suez

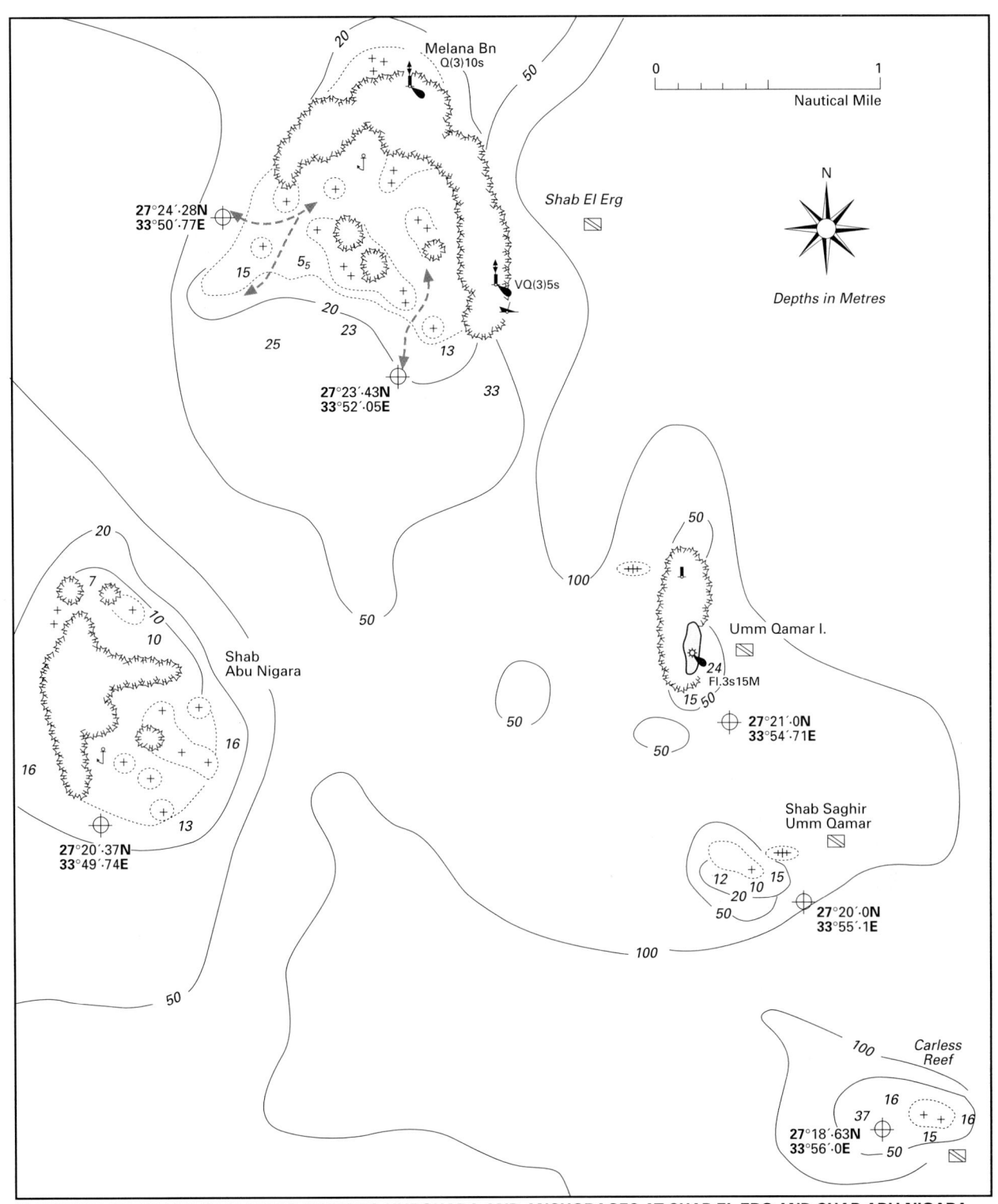

MAIN DIVE SITE REEFS IN N APPROACHES TO HURGHADA AND ANCHORAGES AT SHAB EL ERG AND SHAB ABU NIGARA

going current generally sets longer than the S-going. Tides near the reefs may set towards them. Most oilfields and rigs in the area are charted but frequent changes date charts quickly and there are uncharted and unlit hazards so sail in daylight. This advice applies also to the Gulf of Suez in general.

In strong N'lies keep out of open water, particularly during a windward tide. The plan shows sheltered passages by which you can make progress to windward when otherwise you'd be weather-bound. In shallower water closer to the reefs' edges you get even more protection, though in fresh winds cloudy water from stirred-up sand makes reef edges hard to see. Dust haze in the air also reduces visibility and makes close-waters navigation a little tense.

Adrian Hayter, single-handing in *Sheila*, entered the Strait of Gubal from the N:

217

Red Sea Pilot

'Dawn broke on the wildest scene I've ever known
. . .On either side were the great barren mountains, giving a sense of timelessness as if they had outlived their own souls, and in between lay *Sheila* tossed by a furiously angry sea. . .We were fleeing headlong towards a narrow exit. . .'

Diving N of Hurghada

Diving and snorkelling trips at the S end of the Gulf of Suez are now big business and dive boats are common. Where there are anchorages they are dealt with separately.

In calm weather Carless Reef, 27°18'·8N 33°56'·45E, is highly recommended for its two coral towers, soft corals and numerous moray eels but beware the fire coral. Locals carelessly call it Careless Reef, perhaps not knowing that its real name commemorates one of the Red Sea's pioneer hydrographers, Lt TG Carless IN. In his mid-20s during the survey, his *Memoir on the Gulf of Aqaba, from notes during the survey by Moresby in 1833*, published in the Proceedings of the Bombay Geographical Society in 1836, makes fascinating reading even today. He went on to further surveying laurels in the Gulf of Aden and on the Somali coast and died of smallpox, aged 42, whilst Commodore of the Indian Navy's Persian Gulf Squadron.

Between Shab Ali and the mainland is the Second World War wreck of the *Thistlegorm* 27°49'·0N 33°55'·28E, sunk in 1941 by long range bombers from German-occupied Crete. The late Jacques Cousteau discovered it in 1956. Her deck cargo of railway wagons and holds full of trucks, jeeps, motorcycles and tyres is all there. The wreck of the *Sarah H* lies off the W side of Shag Rock in approx 27°46'·7N 33°52'·5E. Shab Mahmud, a long, submerged barrier reef at approx 27°45'N 34°05'E, offers a number of good dives especially round Beacon Rock, 27°42'·3N 34°07'·6E, including the 19th-century wreck of the *Dunraven* in approx 27°42'·19N 34°07'·35E. Note, you are not allowed to anchor anywhere within the Ras Muhammad National Park unless you have first obtained permission from customs at Sharm el Sheikh. In practice you will be required to use the buoys now laid to protect the reef environment. It is probably cheaper and easier, if dive Ras Muhammad you must, to leave your boat in El Gouna and go on a dive safari.

Outside the park area, for the intrepid in small yachts with reliable engines, there is an inshore passage inside Shab Mahmud to a narrow pass in approx 27°46'·54N 34°03'·37E with a least depth of 2–3m. THIS HAS TO BE CAREFULLY EYEBALLED IN QUIET WEATHER, ideally after a recce in the dinghy. The wall to seaward and the pass itself, known as the Big Crack, offers a good drift dive, as does the Small Crack not far away. All are outside the park boundaries.

Shab Abu Nigara

Plan page 217.

Keep a good lookout for isolated reefs on either side of the entrance and for bommies once you are inside the anchorage. You should have 7–10m, sand, with good holding and protection from the reef. Use a HEPCA/USAID buoy if there's one free. If there's not and there's no clear sand to anchor go elsewhere.

Shab El-Erg

Plan page 217

The reef, covered at high water, is steep-to on its N and E sides. Enter the lagoon close S of the tip of the W arm of the main reef or between two isolated coral outcrops W of the S tip of the E arm. Once inside, eyeball your way N until close under the main reef SSW of Melana beacon, E cardinal. The anchorage is in 7–10m, sand and coral. Look for a clear sand patch or use a buoy, there are at least 6 HEPCA/USAID buoys. An alternative is formed by a break inside the reef's W arm, approximately 1M SW by S from Melana Bn. Good visibility or a waypointed exit track is needed for anchoring and leaving from further in. There is another E cardinal mark on the SE side. Excellent snorkelling. The dive highlights are Dolphin House or Circus on the SW side, and Manta Point and Poseidon's Garden outside on the E side of the main reef.

Shaker (Shadwan) Island

Plan below

The light at the S end of the island is weak and not easily distinguishable against the background hills. The leading marks for the approach are difficult to see. The chart shows a leading line passing between two buoys. The buoys no longer exist nor does the leading mark. An alternative set has been erected leading 355° as shown on the sketch, but don't rely on them. The charted bommie in the middle of the bay's entrance seems to have disappeared or is very

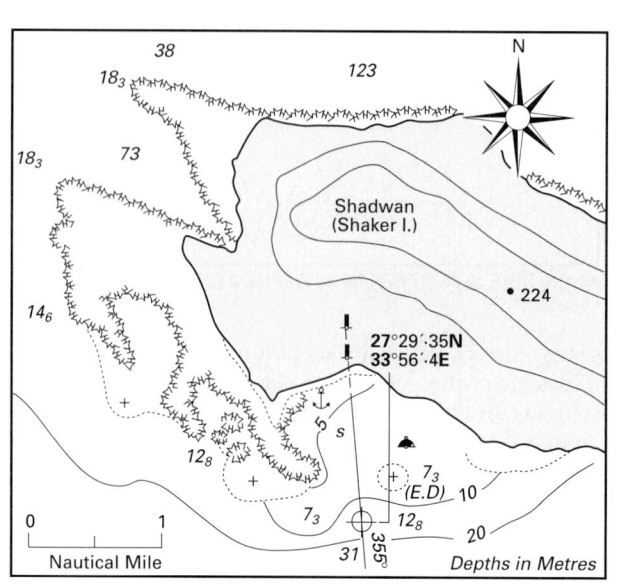

SHADWAN

The Egyptian Coast: El Gouna to Suez

difficult to see. There is a mooring buoy at 27°29'·87N 33°56'·70E. A magnetic anomaly has been reported E of the island.

Danger
The island is a prohibited area and is still mined. A soldier was recently killed or wounded by one as he came to warn some yachties, who'd been ashore walking. They left, apparently without noticing his fate, or if they did, they didn't care. Don't go ashore.

Anchorage
5–8m, sand, off the beach where marked on the sketch. There is good protection.

General
The military is not always tolerant of yachts. Avoid the SE corner near the light, where they are based. Check with dive-charter boats operating in the area if you want to visit. Lady Brassey tells us that Shaker Island was named the Isle of Seals by the ancients because the gulf apparently once abounded with them.

Endeavour Harbour and Tawila island

Approach
In the Shadwan Channel depths may be less than charted. Enter the anchorage with the beacon on Siyul Kebira bearing 090° (see sketch inset). The beacon on the hill is defunct, apparently there's a bird's nest in its works. The new light (Fl(2)G.5s) is on the E tip. The remains of a landing-craft dock is on the shore between the two.

Dangers
There is a shoal patch off the SE side of the small islet ½M S of the entrance and another just to the N of the entrance itself.

Anchorage
8–12m, sand, SW of the disused jetty for excellent protection in all winds. A second good anchorage, less sheltered but with access in poor light, is S of the ruined landing craft dock.

General
There is good walking here, which is as well since you can get stuck for days waiting for a weather window. It is stark and deserted with a few ruined buildings and the occasional fisherman. Sadly, the windward shores are a garbage tip, with inlets

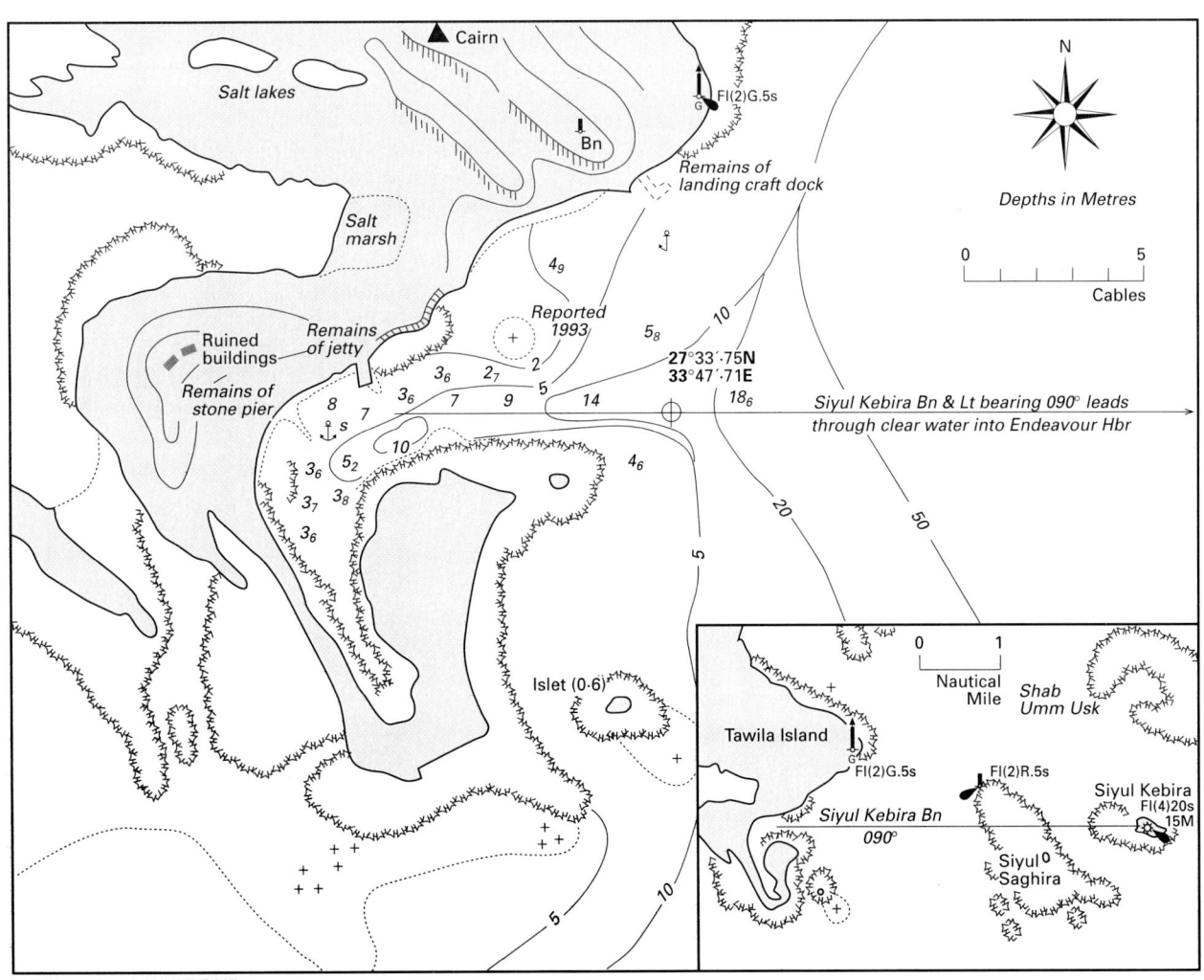

ENDEAVOUR HARBOUR

Red Sea Pilot

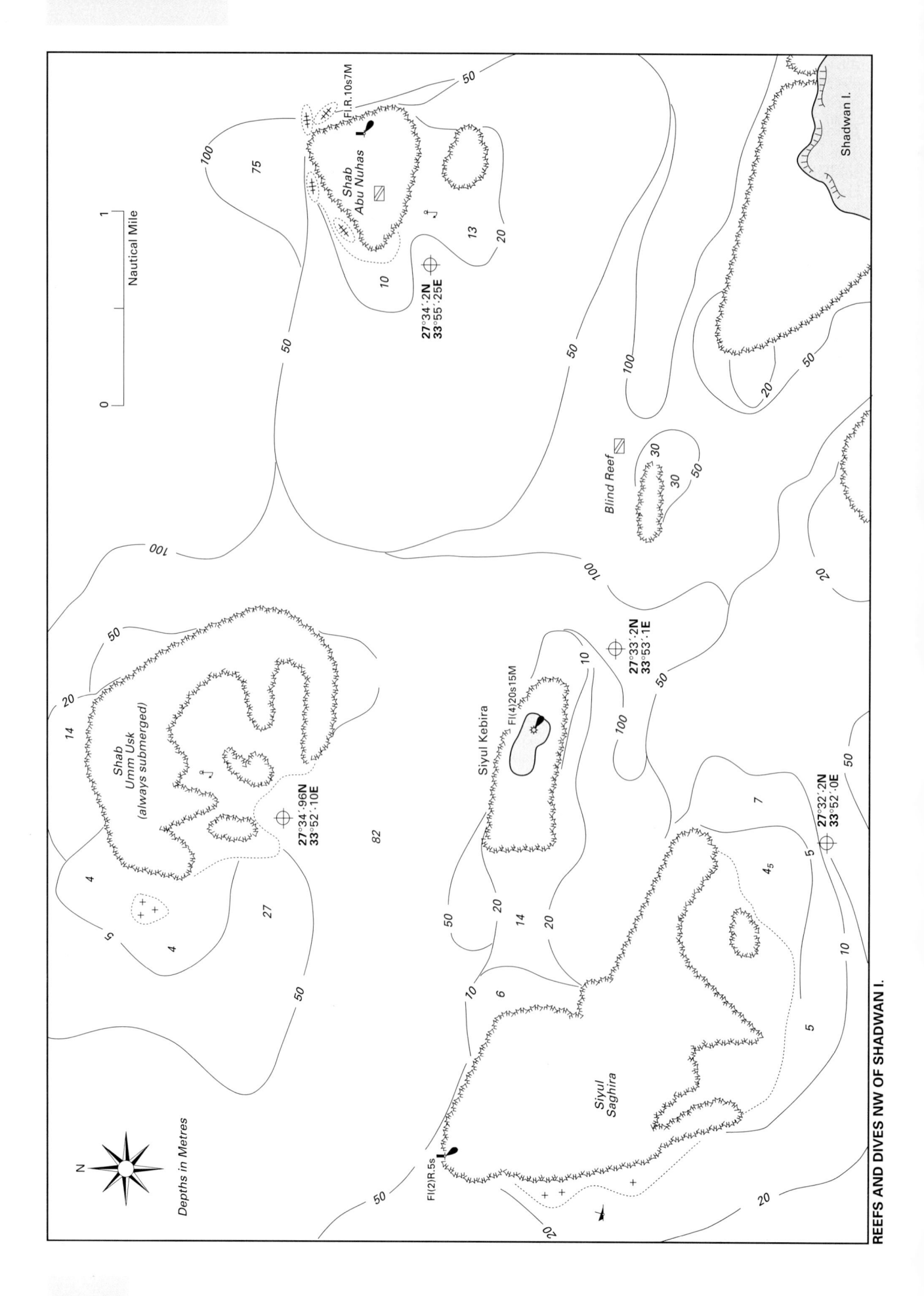

REEFS AND DIVES NW OF SHADWAN I.

The Egyptian Coast: El Gouna to Suez

littered with junk, knee-deep in places, and a tide line thick with gooey tar.

Shab Umm Usk
Plan page 220

Approach
The reef is more submerged than most, but visible in daylight. The entrance has 4–9m of water but must be eyeballed since a coral head partially blocks it. A waypoint just off the entrance is 27°34'·96N 33°52'·10E and you're into the lagoon at approx 27°35'·14N 33°52'·26E.

Dangers
Blind Reef, in the channel between Siyul Kebira and Shaker Island, is difficult to see. Off-liers extend at least ½M S of Shab Abu Nuhas, if you're making in from the E. There are strong currents with overfalls on the N side of Shab Abu Nuhas.

Anchorage
8–10m, sand or sand and coral. Dive boats anchor here overnight. A friendly family of dolphins will sometimes let you swim with them, an experience worth many an unbeaconed reef.

Dives nearby
Siyul Kebira means Big Siyul in Arabic, perhaps because it has the island. There are some HEPCA/USAID moorings on the SE corner. Drift diving along the reef walls with good snorkelling on the reef. The dive at Blind Reef nearby is around the E end. Siyul Saghira, Small Siyul in Arabic, the bigger reef has four HEPCA moorings on the sheltered S side. Divers drift dive the wall on the N side of the ESE-pointing finger at the SW end.

Shab Abu Nuhas
Plan page 220

There is only a fair weather day anchorage unless overnighting in vestigial shelter with dodgy holding leaves you unflurried. 2 HEPCA/USAID dive buoys are a better choice. Ask at El Gouna for precise information. The reef is famous for wrecks, all up there are seven, but currents are strong (see above). The diveable ones are accessible only in mild weather. The reef is almost always beneath the surface, which accounts for its ship-munching performance. On the N flank of Shab Abu Nuhas, 27°34'·9N 33°55'·65E, there is the late-19th-century *Carnatic* and the 20th-century *Giannis D*, *Kimon M* and *Chrisoula K*, though the identity of the last two is a matter of some dispute.

Gubal I South
Plan opposite

This is not the world's most appealing anchorage, but it does at a pinch. You need a lookout aloft. Enter between the reefs in the centre of the large bay and the sandspit under the beacon on the S tip of Gubal I. Keep heading NW and anchor in 10m, coral and sand, just before the shoal water. The reef hereabouts is all dead and covered with algae, though there are plenty of small fish.

South Qeisum
Plan opposite

Approach
From El Gouna and the S enter via the buoyed channel (see sketch of SW exit to Tawila Channel, p000). From the N find the entrance to the Tawila Channel at 27°40'·70N 33°45'·70E. All the charted beacons were missing in late 2000. Towards the W end of the Tawila Channel there are many oil production installations.

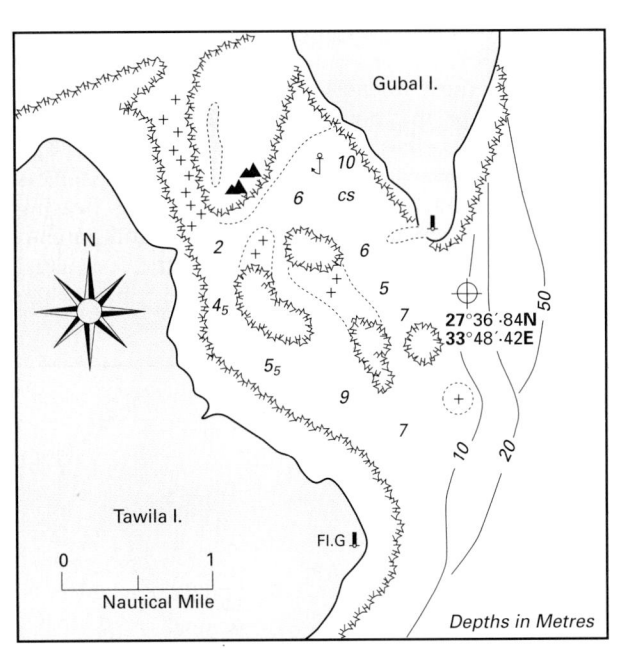

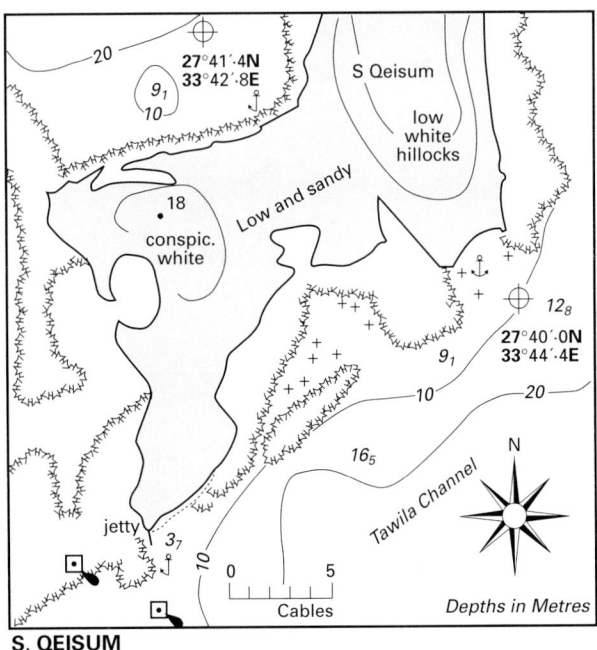

GUBAL I. SOUTH ANCHORAGE

S. QEISUM

Red Sea Pilot

Anchorage
There are three anchorages marked on the sketch. Off the jetty the bottom looks like coral and is deeper than charted. We didn't fancy it. The anchorage towards the SE tip of South Qeisum is better. Approach with the low whitish hills bearing roughly N. Anchor in 6–7m with good sand patches between coral. It's a pretty spot. In the less usual event of S winds, the anchorage in the N bay of the island is well sheltered.

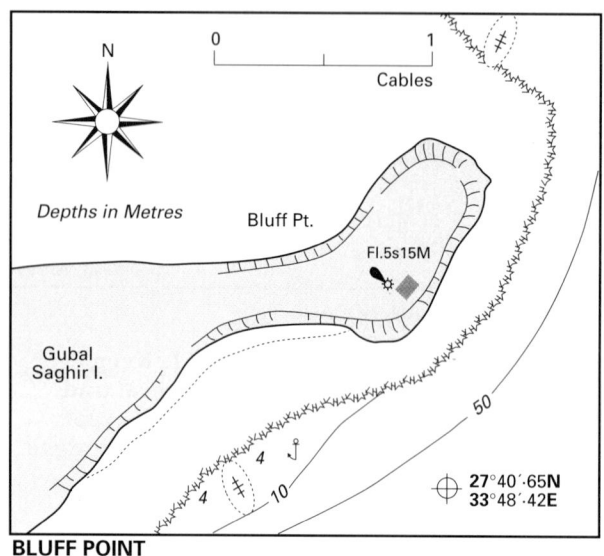

BLUFF POINT

Bluff Point

Approach
If coming from the S in strong headwinds, try to sail with the flood tide which sets N'wards. From Endeavour, follow the coast close inshore to get flatter water. The same applies to reefs. The closer you are the flatter the water. You can keep the E coast of Gubal I very close aboard.

Anchorage
11–16m, with good holding and protection from N. It is possible to enter the anchorage even at night but you need to tuck in close to find shallow enough water, meanwhile watch the sounder closely.

General
Dive boats often anchor here, where there is a good wall dive and there are two HEPCA/USAID moorings. There is dinghy access to the beach through a gap in the reef. A causeway, or some such, is marked on the chart between Gubal Saghira and Gubal I. We haven't explored it. The wreck N of the point attracts divers.

Marsa Zeitiya

Approach
If there are strong winds and you are sailing up the sheltered passage N of South Qeisum I to the Zeit

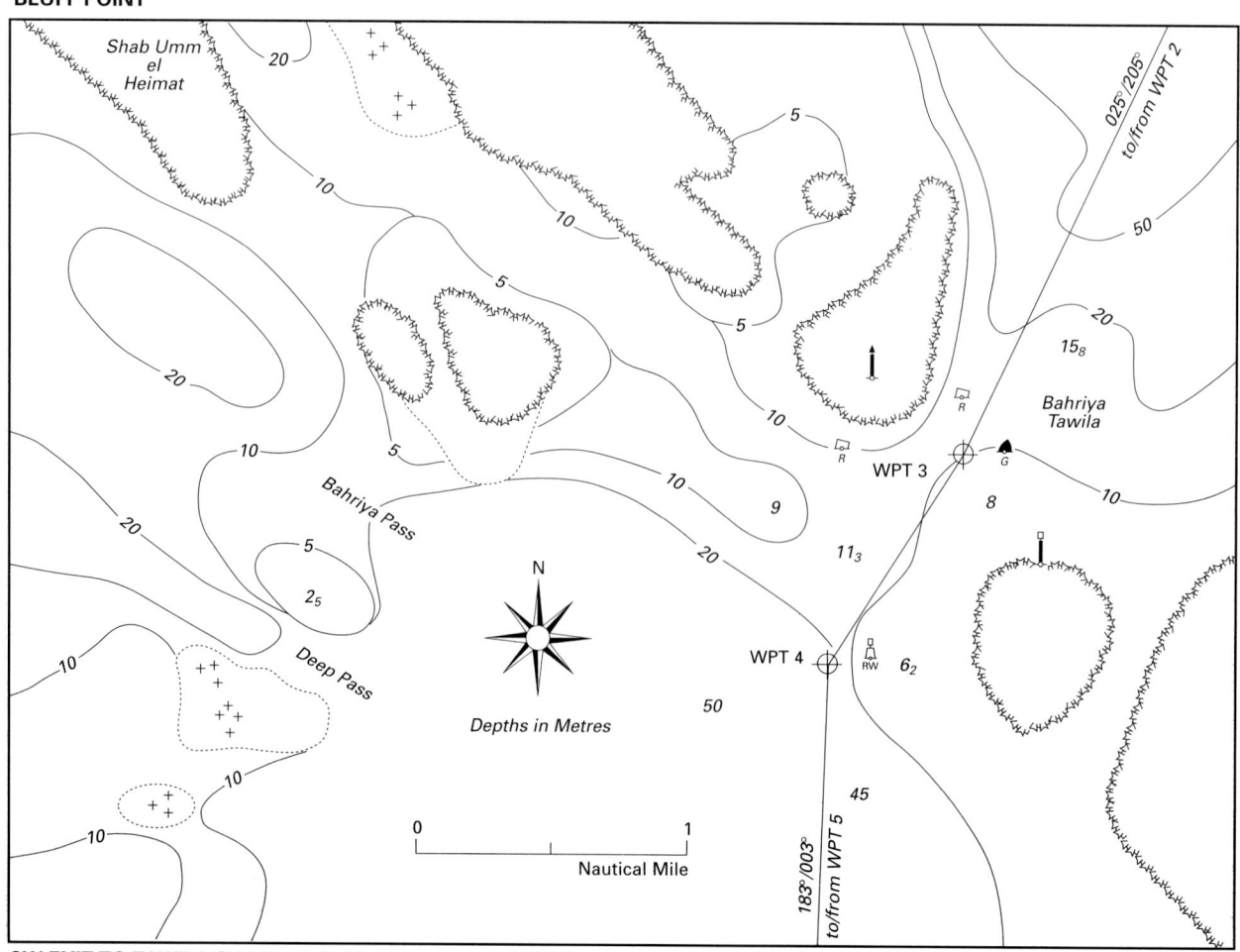

SW EXIT TO TAWILA CHANNEL (El Gouna approach WPTS)

The Egyptian Coast: El Gouna to Suez

Channel, hold a course to keep you in about 10m of flat water close to the W shore of Shab Ashrafi until you can make Zeitiya in an easy fetch. If the flood tide is running and you have a GPS, juggle between comfort and VMG. In rougher water boat speed drops but the tide runs faster because it's deeper. Alternatively, use the Ashrafi Channel but this takes you through unprotected waters at its N end. You may see orange buoys by the reefs in this area, probably marking dive sites. If you are approaching from the N in strong winds you will also find that keeping close to the coast makes life easier.

Conspicuous

Several wrecks on Shab Ashrafi (the reef on the opposite side of the Zeit Channel) make a very visible reminder of the dangers. When approaching the anchorage the military presence is obvious by the gun emplacements on the hill to the S. There are many buildings and oil tanks. The position of the light (Fl.10s24m15M) is uncertain. Official publications have it on Umm el Kiman where it was when we were last here. Some reports place it on the reef S of the anchorage where the sketch has a symbol in brackets. A radio mast on the ridge to the SW is conspic.

Anchorage

7–8m, sand, in the SW corner of the bay clear of the area used by work boats and tugs near the jetty. Keep well clear S of the two big SPLM buoys for tankers with their floating hoses and long pick-up ropes.

General

This is a busy but well protected anchorage, peaceful in N'lies but brightly lit at night. Wind strength in the anchorage is generally stronger and gustier than conditions offshore.

Mersa Zaraba

Plan page 224

Gebel Mezraiya (477m) bearing 023° leads through the entrance between two isolated reefs, with a least depth of 11m in the pass. You need to turn N shortly after the entrance. There are submerged rocks E of the reef on the N side of the entrance, and several coral heads to watch out for once inside the

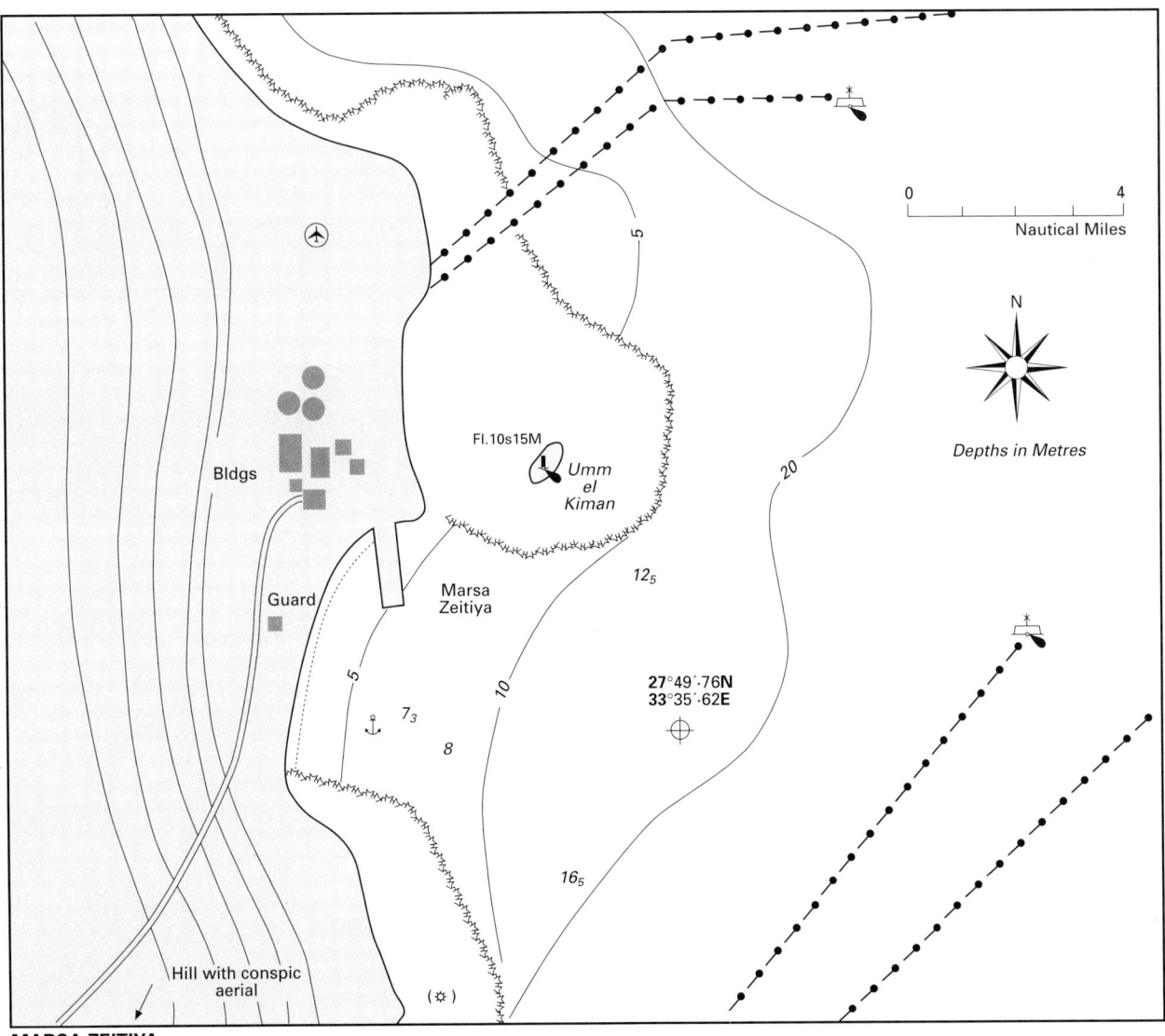

MARSA ZEITIYA

223

Red Sea Pilot

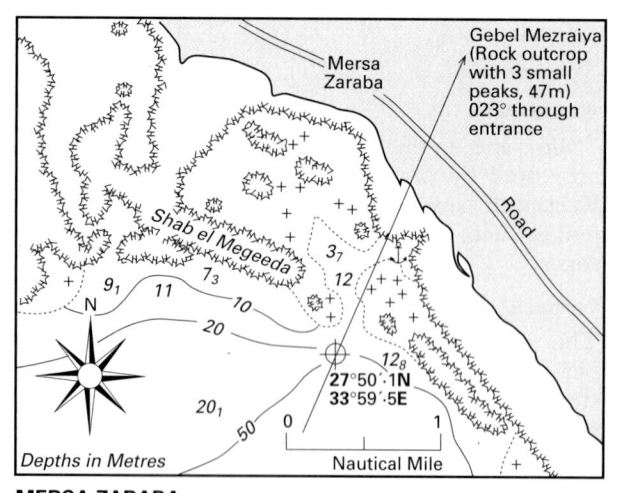

mersa. Anchor in 4–7m, sand and coral; best to buoy your anchor.

Shab Ali

Approach

In good light the plethora of bommies does not pose too great a problem. Stakes mark one of the coral heads in the final approach to the N anchorage in the lee of a hook of reef marked by a beacon. A lit green mark, Fl(3)15s, off the NE side of the reef makes a useful datum. The oil rigs are visible from some distance. So are Shag Rock and its beacon and the wrecked tug and barge on the W and N of the reef respectively.

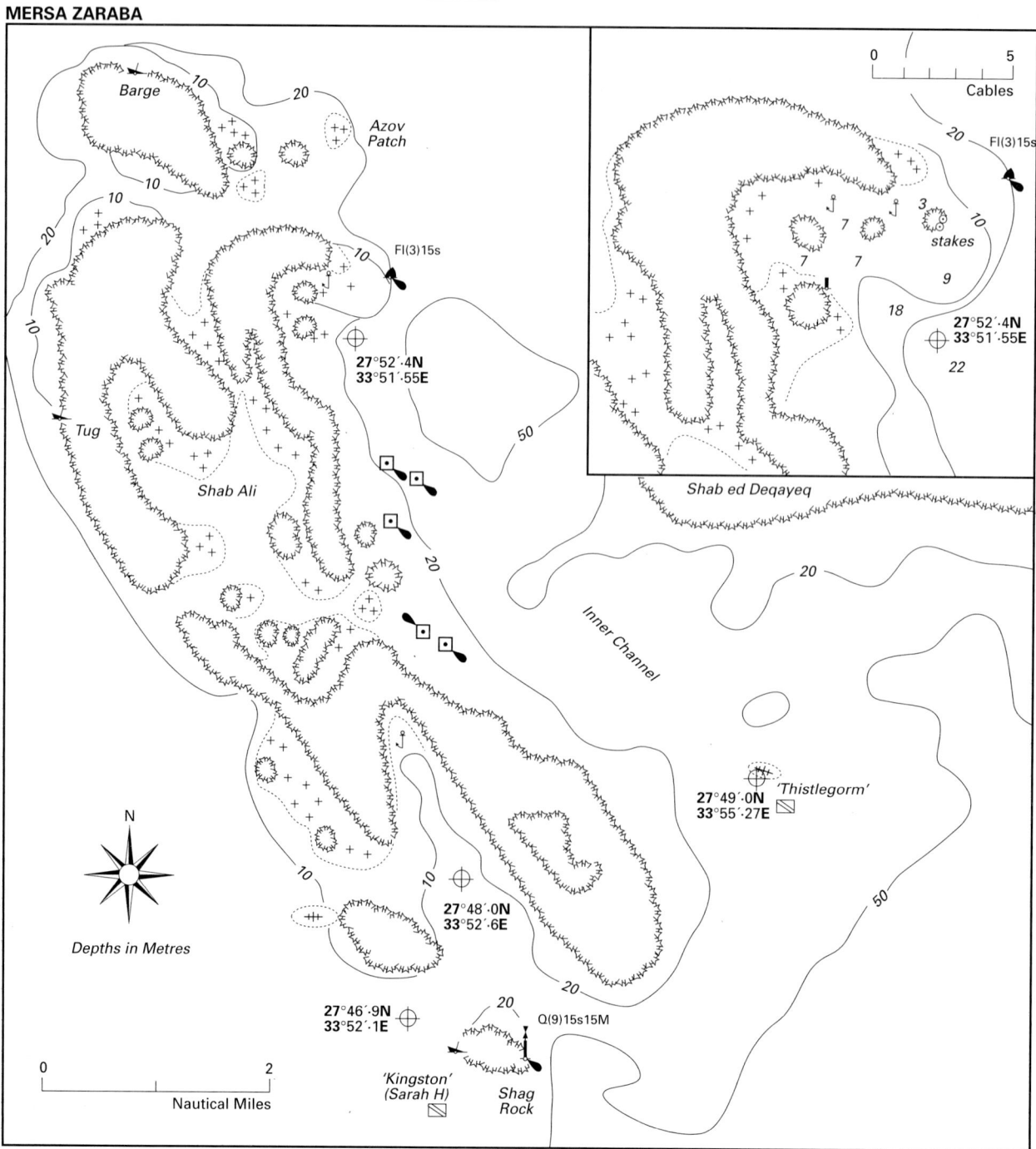

MERSA ZARABA

SHAB ALI: ANCHORAGES AND DIVE SITES

224

The Egyptian Coast: El Gouna to Suez

Anchorage

7m, sand and coral, in the N anchorage at approx 27°53'·0N 33°51'·4E in the NE corner of the reef complex (see inset on sketch). Approach from the green buoy on 260° or feel your way NW to NNW from the waypoint. The anchorage lies within a prohibited area, according to some charts, but no yachts have been moved on to our knowledge. In the large inlet on the S side of Shab Ali, another anchorage at 27°49'·27N 33°51'·97E has reasonable holding in sand and coral, is a lot easier of access and a better spot for waiting for good conditions for diving on the *Thistlegorm* or *Sarah H*. The latter is actually called the *Kingston*, but known as the *Sarah H* after the captain's wife!

Merset El-Qad Yahya

Approach

Use the channel N of the conspic wrecked tug high and dry on the reef blocking the S part of the *marsa* entrance. Run in on 083° on Gebel Mezraiya, the 3-peaked hill about 6M inland. If you use the S entrance, hold towards the E side of Shab Rayis. Depths in the N part of the *marsa* are consistently deep right up to within 20m of the shore. The inner E tip of the reef N of the entrance is marked by a small stake. The jetty is used by fishing boats. There are a few huts and a police post, but they ignored us. There is another small pier and a hut on Ras Kenisa. Access is via a passage through the reef. The W tending creek leading to it from the first jetty is initially deep and then very foul.

Anchorage

13m, sand and weed, about 30m from the N shore, 200m E of the small jetty. The W-tending creek is no place to anchor. It is fairly narrow, 12–13m at the E end, and full of coral where it shoals to 5m about two thirds the way along. The road to Sharm el Sheikh, a mile or so N, is very busy.

This section continues up the Gulf of Suez. If you are lucky enough to have calm weather, go hell for leather for Suez remembering to keep a careful lookout given the constantly changing oil industry installations. Watch for steel structures cut off near water level. They are almost invisible and quite lethal. By motoring just outside the shipping lanes you should avoid them. If the wind won't play most yachts day-sail given the oilfields and associated hazards. In daylight you will also have the bonus of

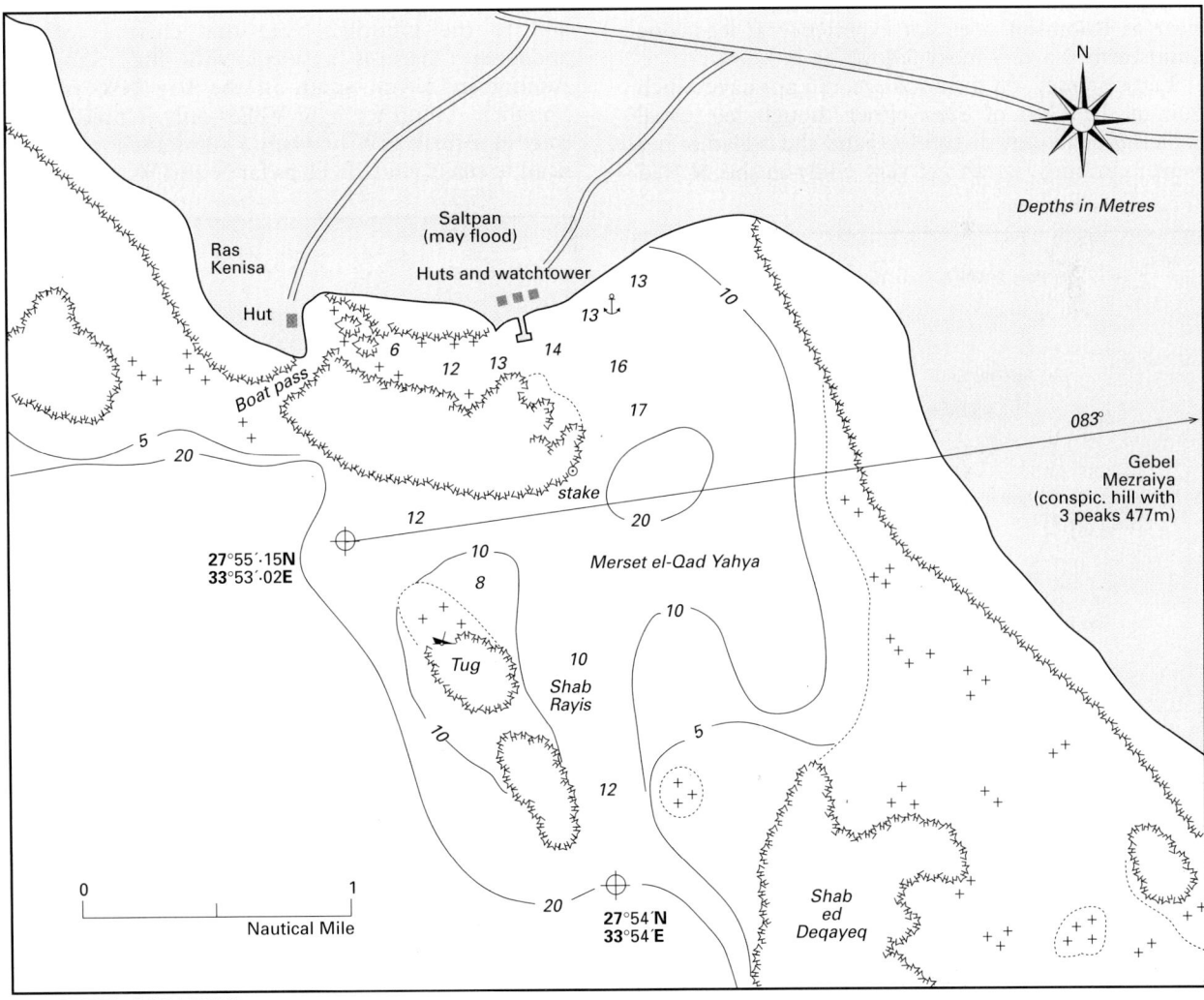

MERSET EL-QAD YAHYA

Red Sea Pilot

magnificent mountain scenery, especially on the Sinai Peninsula.

E. M. Forster, in *Pharos and Pharillon*, reminds us:

'The beauty of the Gulf of Suez – and surely it is most beautiful – has never received full appreciation from the traveller. He is in too much of a hurry to arrive or to depart, his eyes are too ardently bent on England or India for him to enjoy that exquisite corridor of tinted mountains and radiant water.'

W. A. Robinson's account of sailing here in *Svaap* goes some way towards explaining why yachts tend to keep moving while the going is good:

'Like a battering ram pounding at the gates of a medieval fortress we hammered away against our indefatigable enemy, the NW wind...There, only 25 miles from Tor, safety, and fresh supplies, we nearly encountered disaster...We had entered the Strait of Jubal, Asia and the Sinai Mountains towered abruptly...At 4pm, there was a sudden shift of the wind and a strange yellowish cloudbank bore rapidly down upon us...It was a sandstorm, the first of our experience, and I hope the last. For two hours we were buried in a yellow murk, while wind of hurricane violence tore at us...'

Fortunately, by then *Svaap* had taken shelter. Robinson tells us, nevertheless, that it was

'...one of the most distressing experiences I have ever had, for one could hardly breathe...'.

Anyone who cruises in the Red Sea as late in the year as Robinson does not experience at least one sandstorm is a very lucky fellow.

Lady Brassey, on the *Sunbeam*, did not have much fun in the Gulf of Suez either though her usual stoicism belies any discomfort, and she is glad of her warm clothing (it can get very chilly in the N Red Sea in the spring). She also remarks on the novelty of it all:

'...I suppose that for years no sailing ships have been in the Gulf of Suez. The wind blows so steadily for months together, that for six months in the year you cannot get into the Red Sea, and for the other six months you cannot get out of it.'

Are we a hardier breed or is it just another form of masochism? Perhaps it's just having more weatherly craft with powerful donkeys. You can keep out of the worst seas by following the technique, mentioned above, of hugging the coast, keeping no more than 50m off, with an eagle eye on your echo sounder. You will see local boats doing this, and the difference is remarkable.

Note An uncharted, unmarked rock or obstruction at 28°04'·2N 33°40'·14E, some 2·39M, 262° from the small, hooked spit at Ras el-Sebil has been reported. In general the area to seaward of the SW side of Shab Gana (28°00'N 33°43'·5) should be avoided at night.

Ras Shukheir
Admiralty chart 333

The approach is no problem by day but at night the plethora of lights can be confusing. There will usually be supply boats at anchor in the bay just to add to the muddle. Note that close E of the anchorage there is a wreck with little showing. Anchor in 2–4m, sand, in the bay NW of Ras Shukheir. Good for light W'lies only. Unpleasantly rolly in a fresh NW'ly when a stern anchor to hold head to sea is vital. Tuck as far N and W as possible.

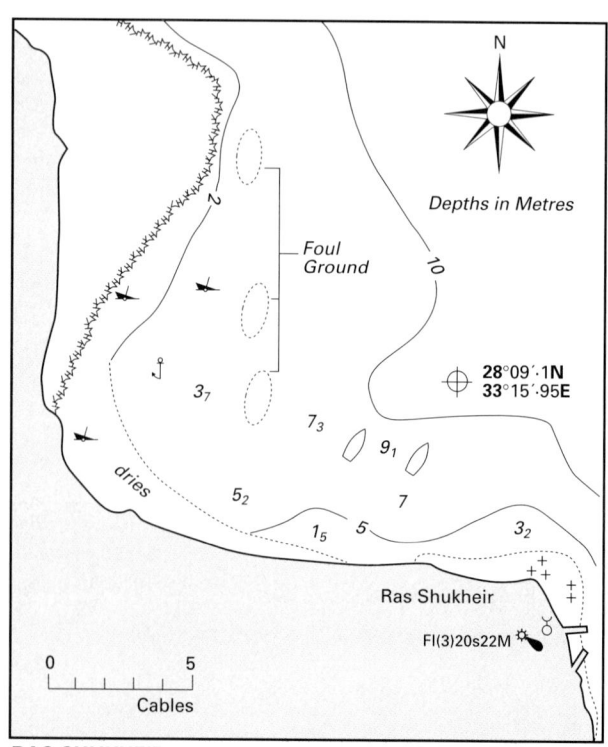

ANCHORAGES IN THE GULF OF SUEZ: KEY MAP

RAS SHUKHEIR

The Egyptian Coast: El Gouna to Suez

Sheikh Riyah Harbour
Plan below

Approach
Shab Riyah, with a minimum depth of 4m, partially closes off the entrance. It is marked by a beacon.

Danger
There is a pile in the middle of the harbour on a line 056°M from the beacon on Shab Riyah.

Anchorage
7–8m, hard sand, with good holding in strong winds, although some swell can work in. A good, well-protected anchorage but you probably won't be allowed to land.

General
Service boats from the oilfields use the bay from time to time, so it may not always be quiet here. There is a small jetty used by fishermen in the NW corner which has a police guardpost and an ugly, part-completed hotel development on the E shore. There's a village at the foot of a low scarp N of the N shore.

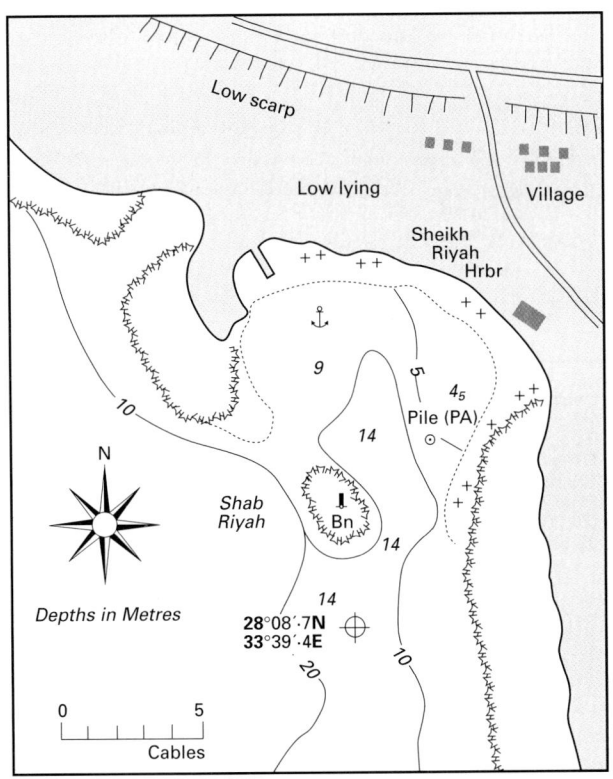

SHEIKH RIYAH HARBOUR

El Tur (Tor Harbour)
Plan page 228

Approach
Erg Riyah has a wreck on its NW side (not where charted) and two lattice work beacons, side by side. From the N the leading beacons, 095°, are not lit but are easily visible by day. Both are lattice towers with identical inverted truncated triangle leading marks. Just to confuse you there is an additional pole beacon, triangle point up, in front of the front leading beacon. Grafton Reef has two beacons, one old and one new with a black cone topmark. Neither was lit in 2000. Give them a wide berth, the reef extends 50m beyond them. From the S it's safe to sail on 005° from the waypoint given. There is an abandoned, unlit rig service boat in 28°13'·76N 33°36'·93E, at the entrance to the harbour proper. At night the lights of El Tur are very conspicuous and extend all around the bay to Gebeil about 1½M southwards. There is a production platform, MO(U), 1·8M WSW of Erg Riyah.

Dangers
The edges of Erg Riyah seldom break and do not show up well even in good visibility. The wreck on the reef masks the beacon when you are coming from the S, and sometimes appears to be a ship at anchor.

Anchorage
3–4m, mud and sand, just S of the fishing jetty at the E end of which there is a wreck. The holding is good and there is excellent protection except from the SE.

General
The port is a secure area. There is a large military airfield behind the leading beacons. You can also see the wall separating the port from the town if you look hard. The message is, don't go ashore though as an overnight anchorage it's excellent. We arrived after dark, surveyed around in the morning to check depths, and were left completely alone. Others with valid visas decided they had a right to land. They were arrested. If you really need something take your dinghy to the primitive pier E of the boatyards. Present yourself at the guard post and explain. You will probably be denied permission to land. If allowed ashore, you will probably have an escort. Permission to go ashore on the Sinai coast is difficult even with a valid visa, best not to bother.

With friends in the right places and ample warning, some have managed to use the cradles on the slip railway behind the building on the navy/fisheries pier. Negotiations take time. If you get the green light, supervise. The workers are not accustomed to fin keels.

Ras Gharib
Admiralty chart 2374, plan page 229

Approach
Stay outside the 5m line until you're in the near offing. The tank farms, mooring buoys, leading lights (not shown on the sketch, they're for the tanker SPM) and buildings are obvious. Ras Gharib light (39m) is close NE of the old light tower with a mosque to its NW.

Anchorage
4m, sand and mud, with good holding and fair shelter from the N and NW. The space between the

227

Red Sea Pilot

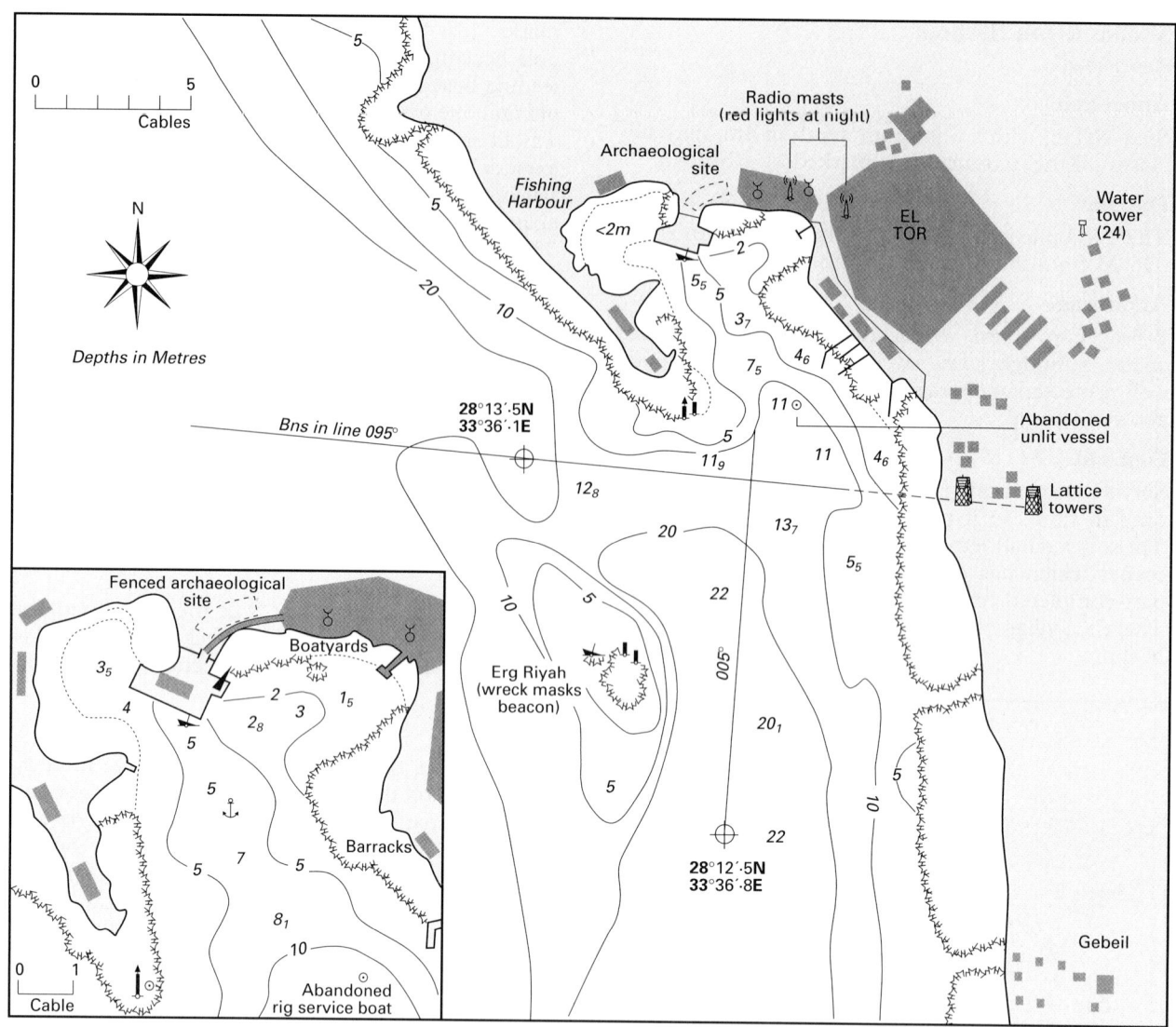

EL TOR

jetties is now filled with five moorings and local fishing boats. It can be very rolly.

Facilities

Water is available in the town, but must be jerry jugged. For fuel, go to the Total wharf, a mile or so S of the anchorage, as shown on the inset.

False Ras Gharib

Admiralty chart 2374

The anchorage can be hard to pick out in poor light unless you are coast-hugging, in which case watch out for isolated coral outcrops as you close the coast. The farm of nodding oil pumps (see sketch) are immediately W of the anchorage. Note that False Ras Gharib light (10m) is 2¾M NW of the point. The anchorage, in 5m, sand with good holding is good enough in light W'lies but rolly and uncomfortable in N to NW'lies and untenable in strong NE'lies.

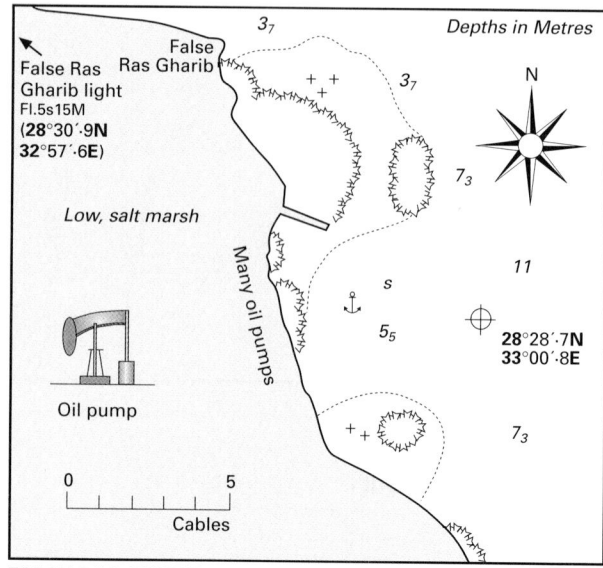

FALSE RAS GHARIB

Shab Alleda and Ras Gihan

Plan page 230

Shab Alleda is only a last-ditch stop, in 2m or less, suitable for shoal-draught boats and only really safe

The Egyptian Coast: El Gouna to Suez

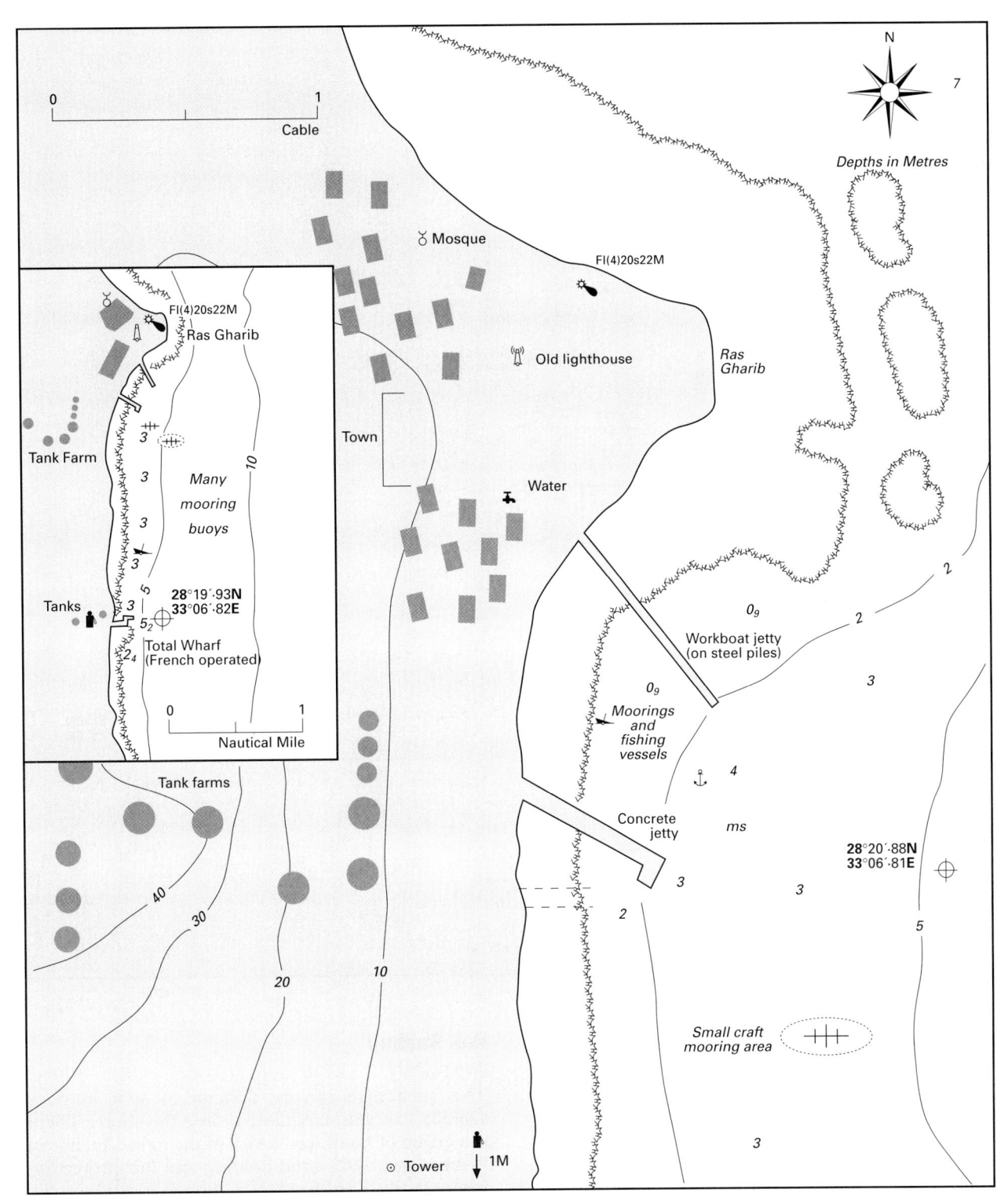

RAS GHARIB

for those which draw about 1m. Anchor in sand with good holding and protection from the sea under the lee of the reef where shown on the inset to the plan. The anchorage at Ras Gihan, a few miles NW, is hard to pick. The most likely spot is near 28°29'·25N 33°18'·65E, NW of a couple of rusty old storage tanks and a small row of battered, abandoned huts called Abu Durba. At the S end of a stunning ridge of jagged, red mountains which glow in the sunset, there are three white patches. Close S of them are two battered wooden leading marks. These seem to lead into an anchorage, 3–4m, in the tenuous lee of a small headland just N of a charted shoal patch (1·5m). The alternative, near Ras Gihan itself, is in approx 28°30'·3N 33°17'·3E in the lee of a projecting tongue of reef a mile or so further NW. It looks decidedly iffy even in quiet weather and we chickened out in a flat calm! Maybe it looks more obviously welcoming, and easier to spot, in a blow.

229

Red Sea Pilot

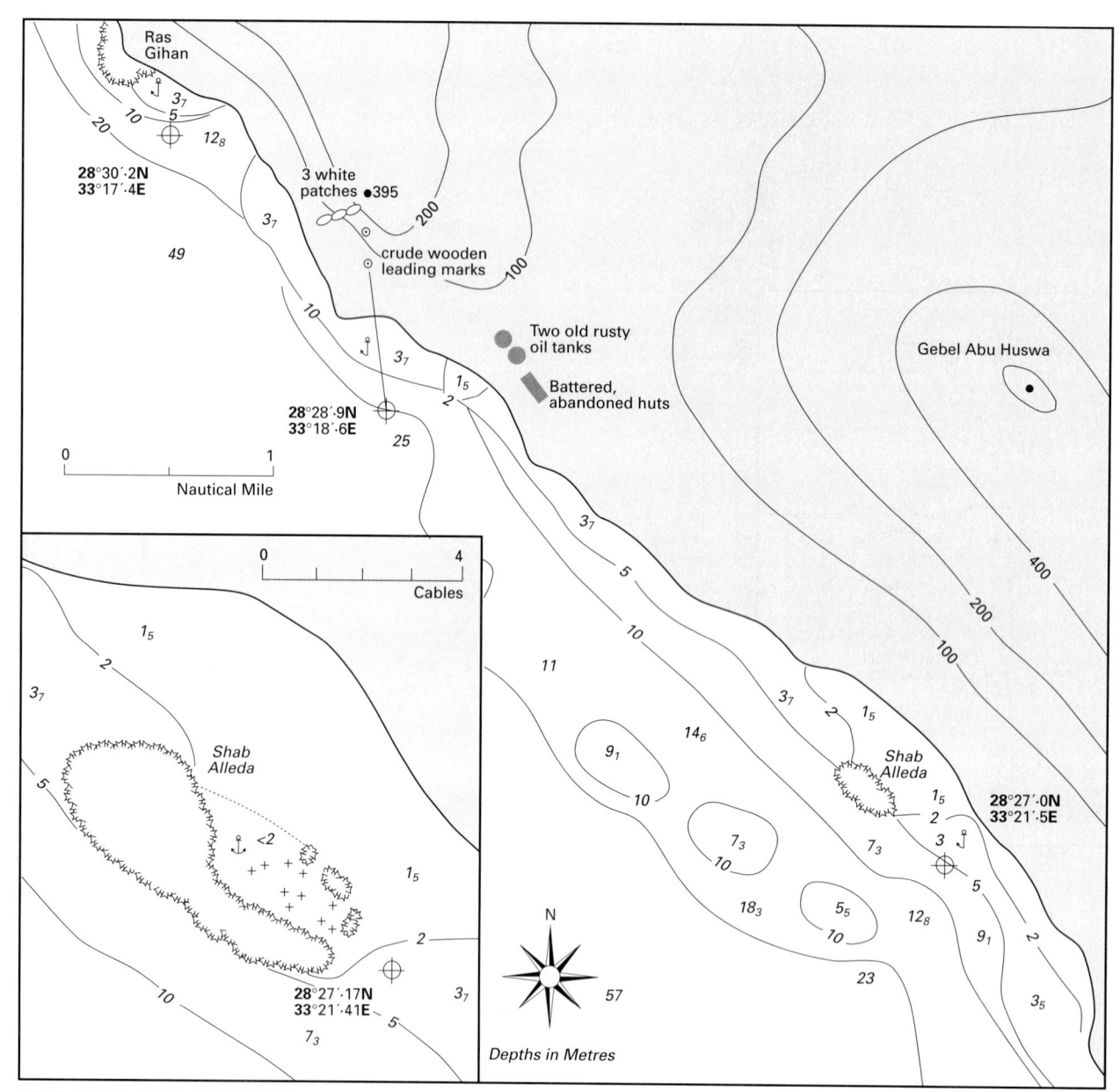

SHAB ALLEDA AND RAS GIHAN

Ras Sheratib (Shab El Hasa)

Plan page 232

You can cross the inner part of Belayim oilfield without restriction but look out for hazards. The anchorage in 4–5m, sand or sand and coral, E of the reef, is well protected and is used by rig support craft. An oil production structure, at approx 28°36'·82N 33°11'·7E, marks the N exit/entrance to the pass between Shab el Hasa and the shoal ground to the E. There are strong N and S-going currents in the area. A stern anchor helps counteract roll if there's any swell. Set it well because of the tidal currents. A possibility for S'lies, about 2M NE is marked by a half-anchor. Approach with care from the NW. Suitable only for shallow-draught boats.

Ras Ruahmi

Plan page 231

The anchorage can be difficult to spot in poor visibility, so you may need to hug the shore. There is a group of buildings ½M S of the light. Anchor in 2–4m, sand, with good holding and fair protection where marked on the sketch, though it can be rolly. Good in light W'lies but very unpleasant in N to NW'lies. Shallow-draught boats can tuck further in for better shelter.

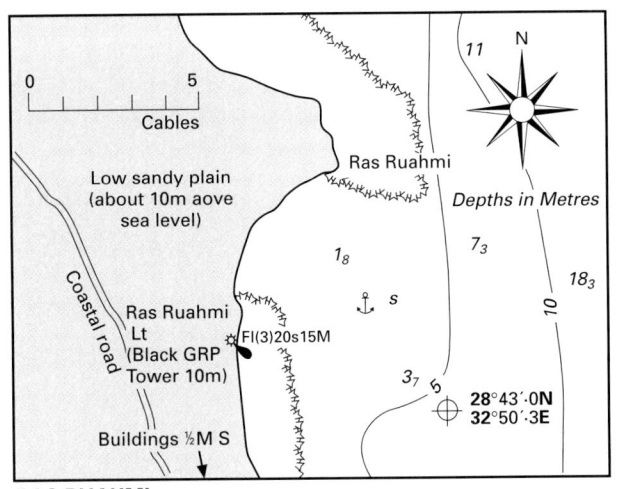

The Egyptian Coast: El Gouna to Suez

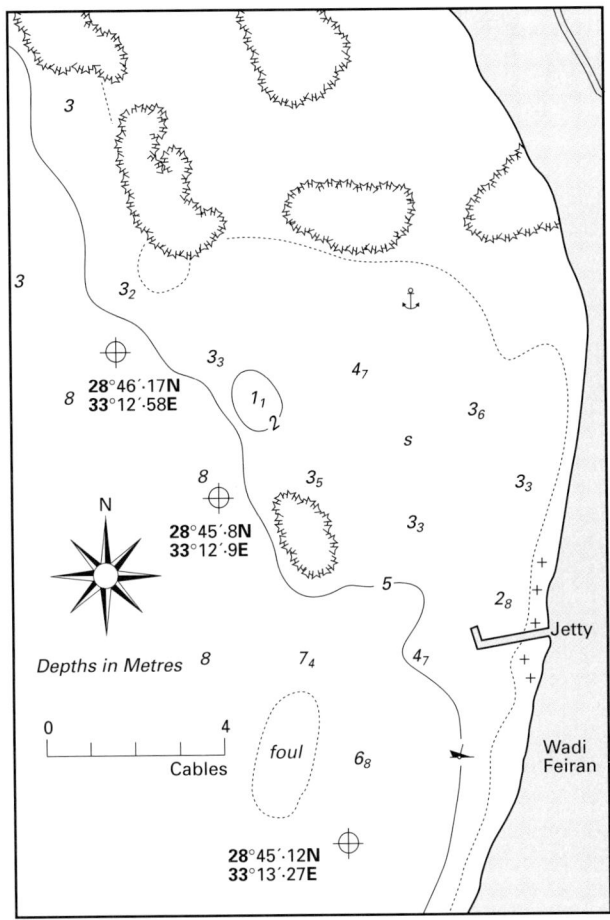

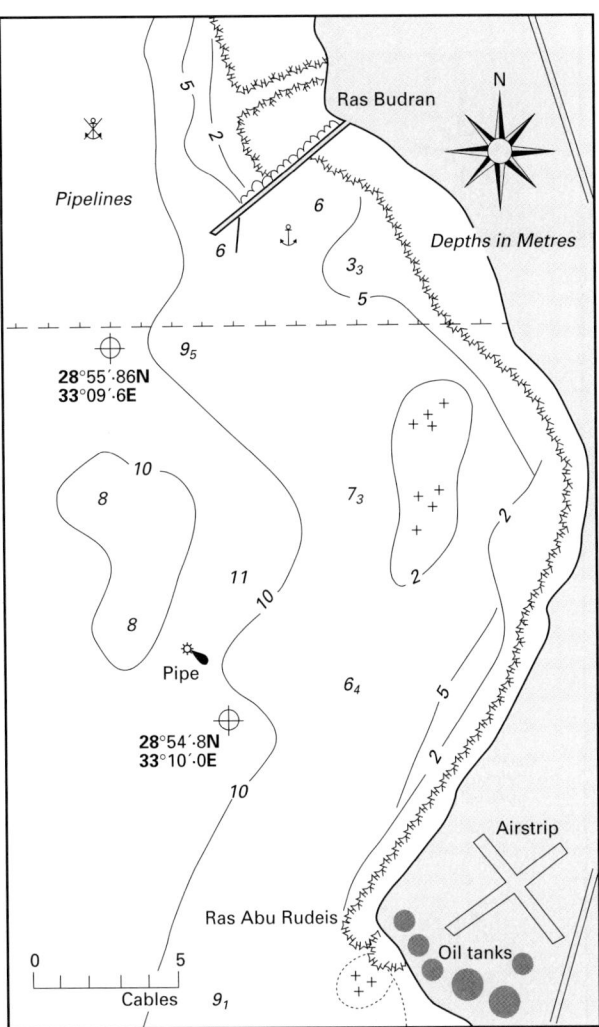

Wadi Feiran
Admiralty chart 2373

This is for fair weather only unless you are very shoal draft. It is untenable in even a moderate NW'ly unless you can tuck right up and into the NE corner and it's rolly even then.

Ras Budran and Ras Abu Rudeis
Admiralty chart 2373

The tanks and buildings of the village S of Ras Abu Rudeis are obvious. Note the pipe, in theory lit (Fl.R), in the immediate approaches at 28°55'N 33°09'·94E. The main dangers from the S are the unmarked, unlit pipes sticking out of the sea S of Ras Abu Rudeis in approx 28°52'·2N 33°09'E and 28°51'·44N 33°09'E. If you stay in over 20m you should be clear.

Anchorage
The best spot for fresh weather is as close under the breakwaters and jetty as visible or audible agitation ashore seems to allow. Fair shelter but bouncy.

231

Red Sea Pilot

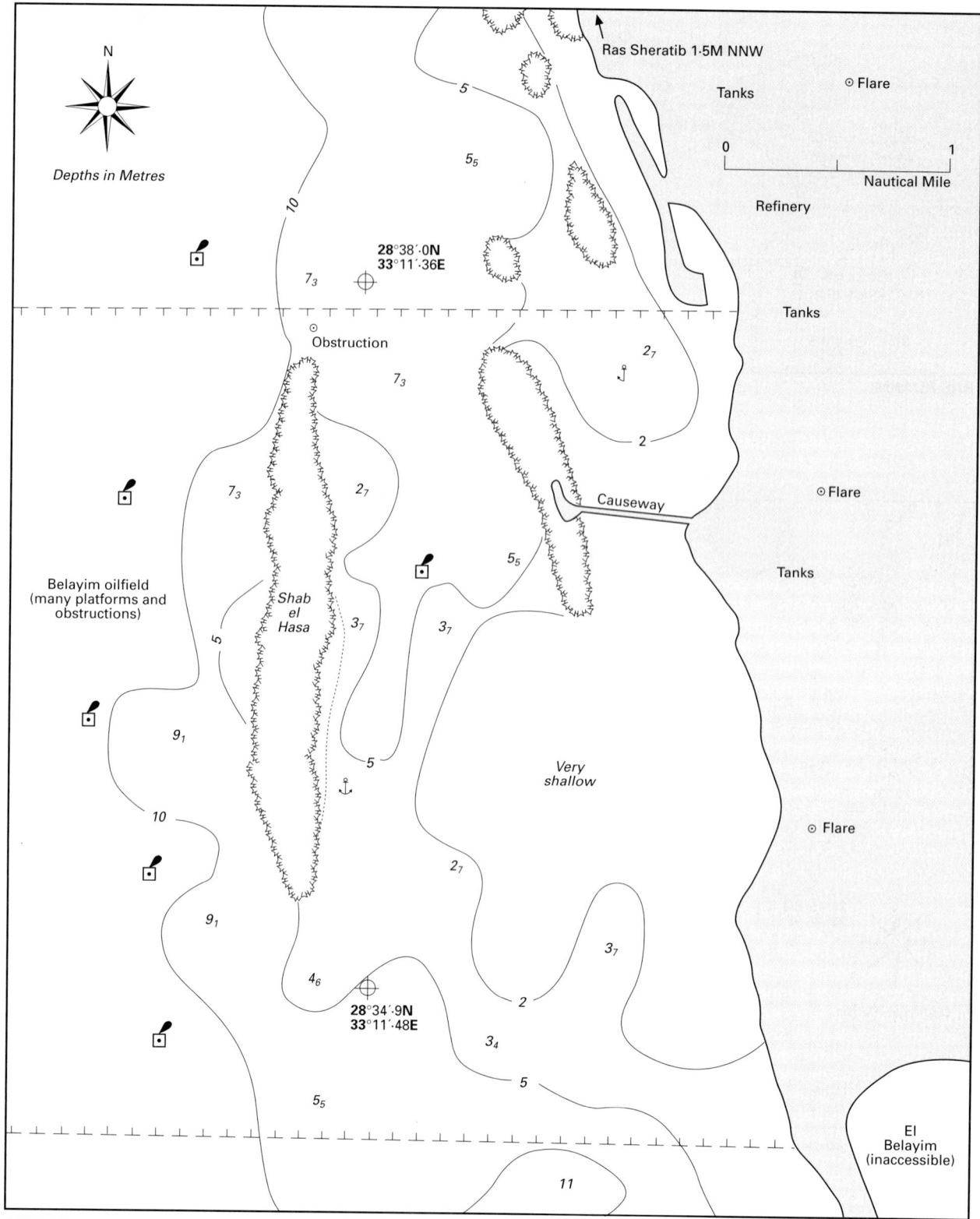

SHERATIB SHOALS AND SHAB EL HASA

Marsa Thelemet

Plan page 233

Approach

There is a small, unlit buoy at 29°02'·37N 32°38'·26E, S of the leading line on 302°. It marks the fairway entry. The unlit leading beacons are doubled, the old ones in front of the new, but even so hard to see. Once in 29°02'·52N 32°38'·12E turn onto the secondary leading line to steer 005° to the head of the *marsa*. The main jetty is weakly lit on its outer ends (Fl(3)R.10s on the S tip, F.G on the N) and has two modest floodlights illuminating the

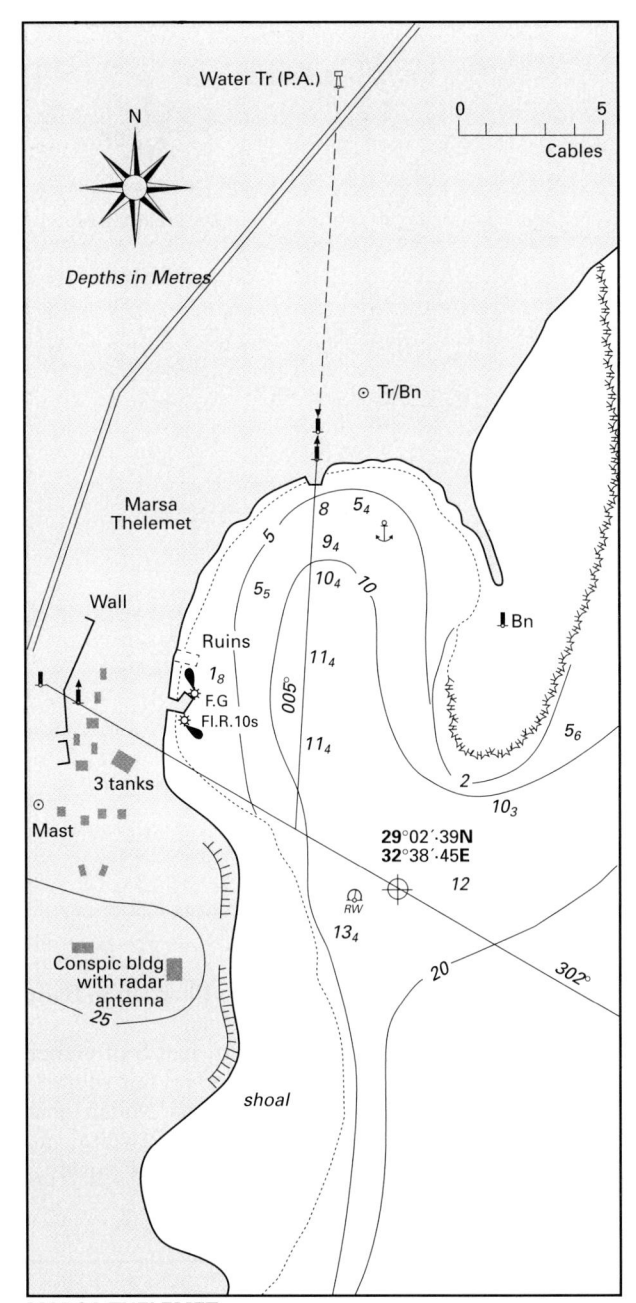

The Egyptian Coast: El Gouna to Suez

MARSA THELEMET

Danger
The spit, which mostly covers at HW is marked only by the unlit beacon 0·5M N of the S tip of the reef. On approach watch out for trawlers and fishing boats with long lines and nets that are difficult to see, especially at night.

Anchorage
Anchor E of the ramp jetty in 5m–6m sand, with good holding in approx 29°03'·67N 32°38'·23E. Some yachts have reported difficulty getting their anchors to dig in, so holding may not be good everywhere. Alternatively, there may be a soft top layer through which you need to let your anchor sink. Our 16kg Bruce, on a 7:1 scope 10mm chain, held well in 45kts. There is a shifting patch of discoloured water in the middle of the *marsa*, though there are consistent depths of 11m.

Formalities
The officer in charge of the garrison may come out and ask for passports and ship's papers. If you're polite and co-operative, he is friendly and helpful. A rig service boat occasionally uses the jetty. Fishing boats shelter here in bad weather and camels sometimes grace the shore.

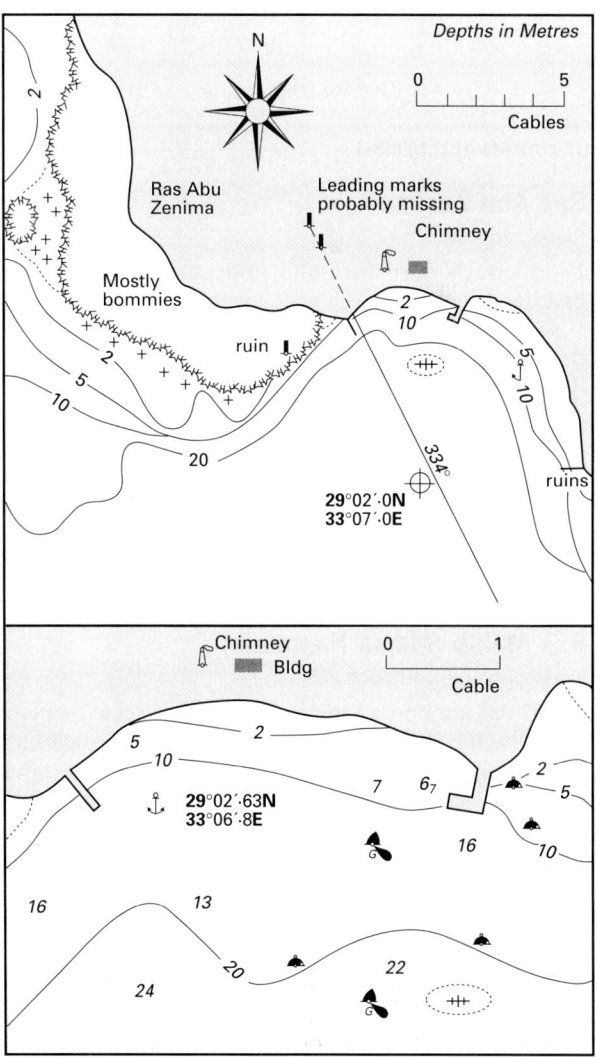

outer part. The second leading marks, unlit, are lattice towers (front: arrow head point up, rear: arrow head point down) leading to a ramp jetty. These line up with a water tower some distance inland. Immediately S on the coast in approx 29°N, there is a large resort, brightly lit at night. When approaching from Ras Zafarana in the NE, the ramp jetty and second leading marks are much more obvious than the main jetty and leading marks. There is a conspicuous building with a radar antenna on its flat roof on the ridge ending in low bluffs about 0·8M S of the main jetty. There is a manned garrison in the scruffy buildings surrounding the main leading marks, fairly brightly lit at night.

RAS ABU ZENIMA

233

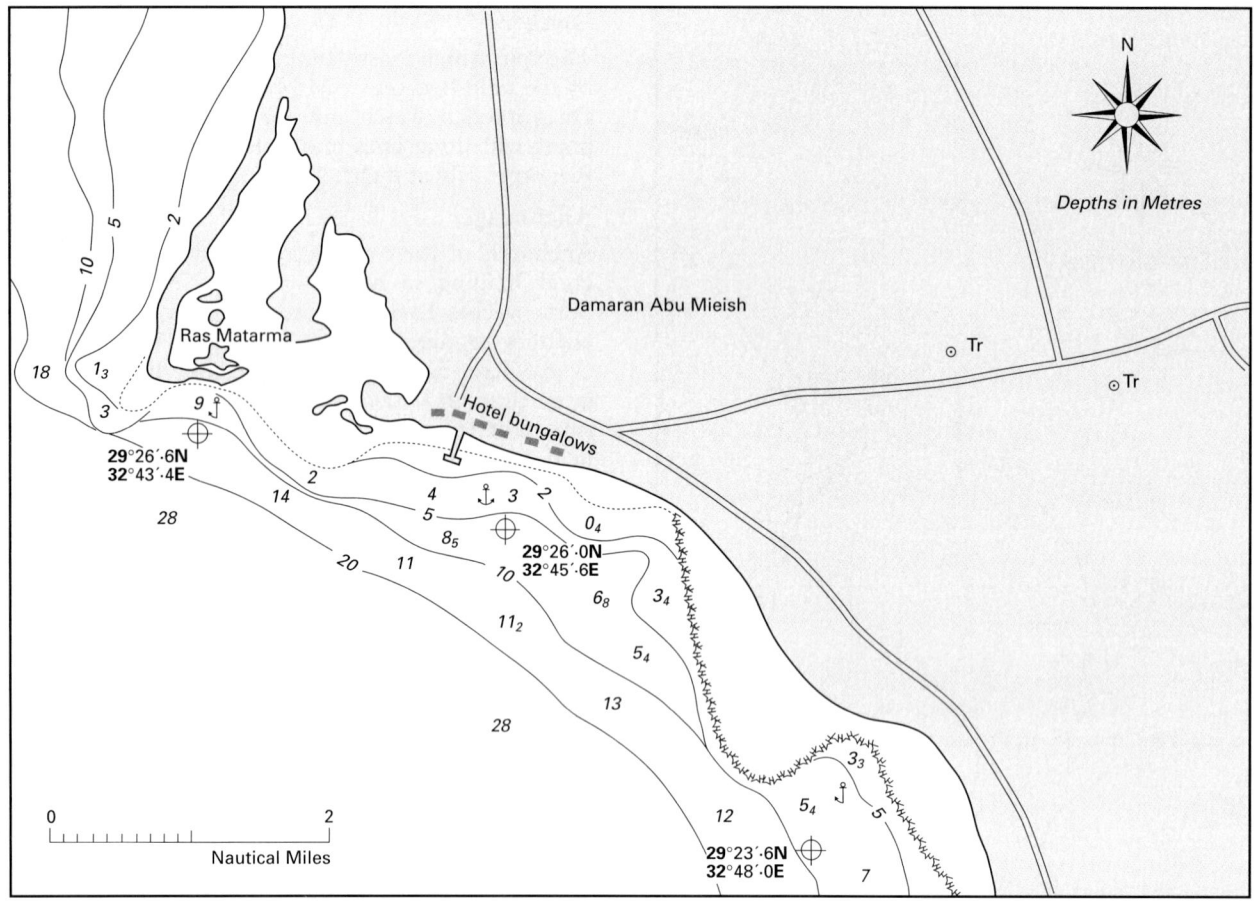

Red Sea Pilot

DAMARAN ABU MIEISH

Ras Abu Zenima
Plans page 233

From the N give the sand and coral bank off the headland at least half a mile clearance.

Anchor in 8–16m, sand, between the broken down old loading jetty in the NW of the bay and the 'J' shaped jetty, sometimes used by rig supply craft, at the head of the bay. There's another ruined jetty at the S of the bay. Tuck well in because the 20m line comes within 200m of the shore. Getting the anchor to hold can be tricky. As at Ras Malab you must pay out scope quickly after getting as close to shore as you dare. Well sheltered in N'lies, with good holding and no swell.

Ras Malab (Mersa Hammam)
Plans opposite and page 235.

From the anchorage the large conspic 'rock', which is about 30m high, bears approx 080°. The rock has a pale stripe on its upper surface. There is a new jetty and leading marks near a building with high fencing and floodlights, only lit occasionally. Berthing alongside the jetty is not possible. The anchorage has room for 6–8 boats in about 4–5m sand where the bottom shelves rapidly. Pay out scope quickly when anchoring in strong winds or by the time the anchor has set, you'll have dragged into deeper water.

The anchorage is probably called Mersa Hammam because of the hot springs near Gebel Hammam Faraoun, the Hill of Pharoah's Bath giving, in translation, Bath Bay.

There is also a tuck behind a spit, just S of Gebel Thal in 29°08'N 33°01'·5E which offers fair shelter. Another possible refuge, between Ras Malab and Ras Matarma, is under Ras Abu Suweira, at 29°18'·4N 32°50'·05E just E of the shoal pushing SSW from the *ras*.

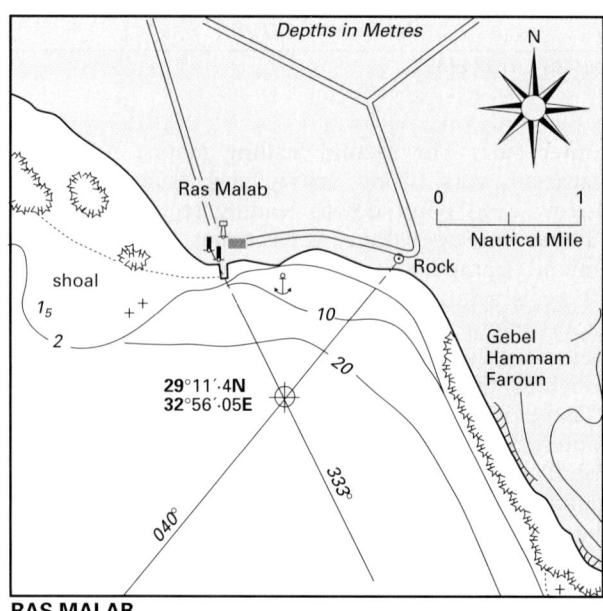

RAS MALAB

234

The Egyptian Coast: El Gouna to Suez

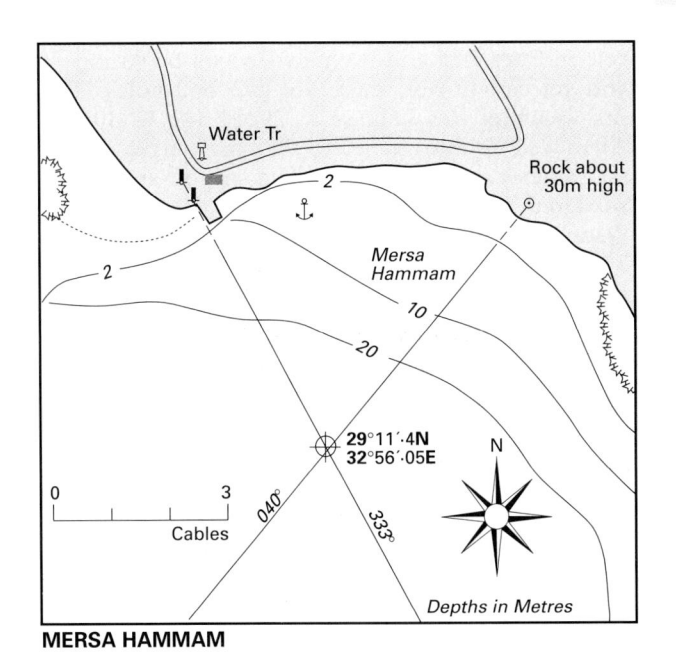

Ras Matarma and Damaran Abu Mieish
Admiralty chart 2090, plan page 234

The headland is low and sandy and the W point is actually detached from the coast by a lagoon. Look out for the road junction at Damaran Abu Mieish as you approach. Coming from the N give Ras Matarma a wide berth. Shoal ground shoves out a good 200m SW. The S'most of the two towers charted inland of the point should bear approximately 080° from the anchorage off the hotel bungalows. This bearing could also be used to run in on. A long wooden pier runs out from the beach.

Anchor in 4–8m, sand, off the coast S of the buildings at approx 29°26'N 32°45'·6E with fair shelter from the NW from sea, though not from wind. In calm weather anchor in the lee of the *ras* though note the deep water close S. An alternative in quiet weather, about 2M SSE, is behind a big bulge of reef at 29°24'·15N 32°48'·35E.

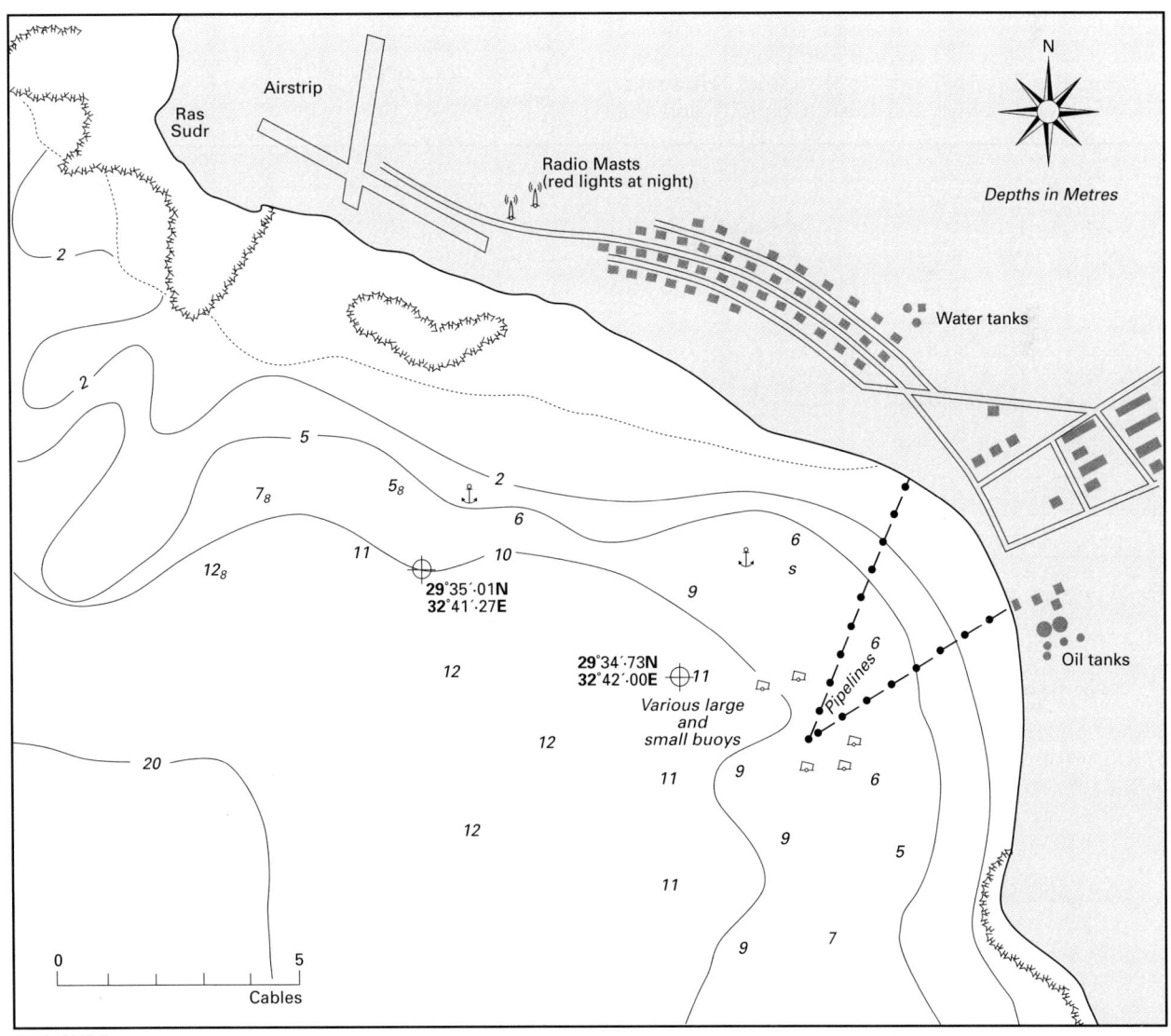

MERSA HAMMAM

RAS SUDR

Ras Sudr
Admiralty chart 2090, plan page 235

Approach
The shoal ground W of the *ras* pushes out at last ½M W and SW, so stay in over 10m. This is an oil loading terminal, keep clear of the moorings at the end of the pipelines. There's another SPM at 29°34'·74N 32°39'·9E in the NW approaches.

Anchorage
6–10m, sand, opposite the radio masts, with good holding, although some swell works in. Alternatively, anchor in about 6m SSW of the water tanks and N of the buoys for better shelter. Some boats have reported mooring to the buoys themselves. It can be rolly.

General
The sea is murky, so watch your echo sounder. The 'spring of Moses', Ayun Musa, is about 17M NNW of here. The water from the wells is reportedly still brackish in spite of Moses' efforts to sweeten it. The soldiery ashore can get very excited if you land, though the nearby tourism university makes for interesting chat. On the other hand, it makes for friendly 'borrowing' of dinghies, so leave someone with yours. Diesel is obtainable and some supplies.

From Ras Sudr it's only 25M to Suez. However, the wind can change and freshen rapidly, and the tide in Suez Bay and the narrows can be strong. If you get caught out, there are two bolt-holes. The first is under Ras Diheisa at 29°45'·1N 32°40'·3E. Tuck well up into the lee of the long tongue of reef and anchor in 3–4m, sand. The second and better sheltered, 1M further N, is under the mile-long Qadd el Tawila, at the E end, near 29°47'·0N 32°38'·0E in 3–5m sand or sand and coral. The latter is just 10M from Port Tewfik.

6. Gulf of Aqaba and approaches

Admiralty charts
Large scale 801
Medium scale 2375
Small scale 159, 12

US charts
Large scale 62191, 62222, 62225
Medium scale 62220
Small scale 62230

French charts
Large scale 6978
Medium scale 6908, 6978
Small scale 7113

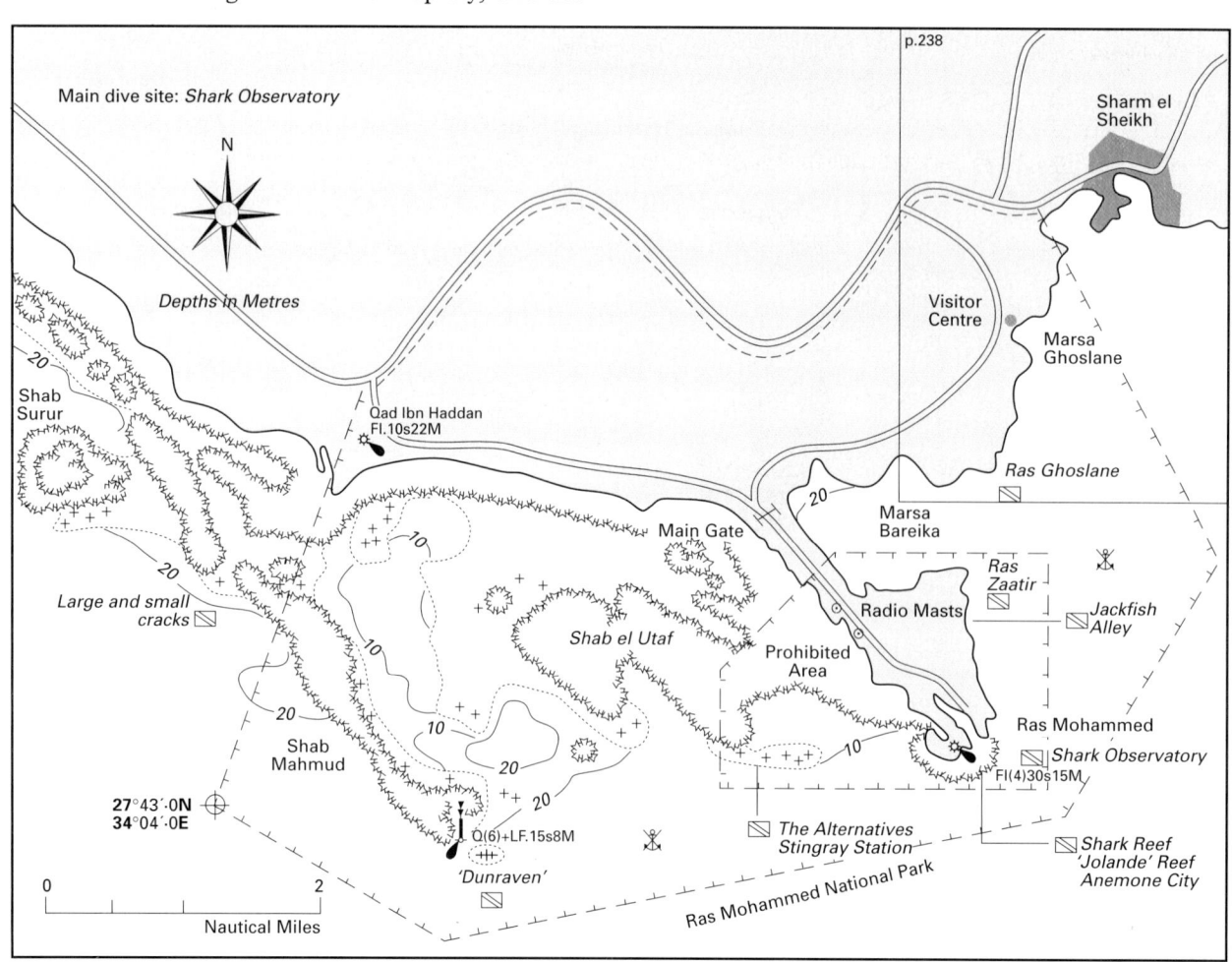

DIVE SITES AT RAS MOHAMMED

Gulf of Aqaba and approaches

Ras Mohammed National Park

27°44'N 34°10'·5E, plan of dive sites page 236

Ras Mohammed Peninsula is a prohibited area and unlawful entry risks arrest. No anchoring in the park is allowed without clearance from the authorities at the customs pier in Sharm el Sheikh. Even if you're very careful and detest bureaucracy anyway, it isn't worth flouting the rules. There is active land patrolling by park rangers. In the absence of official patrols local dive craft protecting their livelihoods are effective enforcers. Assuming that you have checked in at Sharm el Sheikh there are buoys you must use. Frankly, it's probably cheaper and certainly a great deal less hassle to park your boat in El Gouna and, if dive Ras Mohammed you must, buy a berth on a dive safari. Otherwise, enjoy your diving outside the park boundaries, there's plenty of it.

What applies to the area around Ras Mohammed itself applies to the Strait of Tiran and Tiran I. This area (see plan of dive sites) is also part of the Ras Mohammed National Park and the same rules apply. However, as far as we know a stop at the anchorages on Tiran I (see page 239) to wait out a hard N'ly is OK even if you haven't checked in at Sharm el Sheikh. Anchoring or using the dive buoys on the reefs in the strait usually gets you moved on by patrols.

Sharm El Sheikh

Admiralty chart 801

Approach

Call on VHF Ch 16 as you approach. It is obligatory to use an agent and charges are made for anchoring or mooring.

Formalities

This is a port of entry and you must check in with the authorities. The Customs fee system is in operation, though Sharm tends to be a law unto itself – and expensive. For full details of these charges see page 181.

Mooring

It's best to use a mooring. Up to three boats may use most of the buoys at any one time. If that's not possible anchor in 5–10m, sand, clear of any permanent moorings. Tuck well in since depths increase dramatically within a cable of the N shore of the bay.

Facilities

Touristville, though the place is still low-ish key in resort terms. Fuel and water are available but at tourist prices. Along the N shore of Sharm el Moiyia there are bakeries and a market in a large hangar-like shopping area. The banks, GPO, telephone, shops and restaurants are uphill in Sharm el Sheikh town. This is one of the few places on the Egyptian Red Sea coast with a diving-emergency centre equipped with a decompression chamber. Transport is good. Domestic flights from the airport. Buses to Elat, Suez and Cairo. Fast ferry service to Hurghada. A recommended agent in Sharm el Sheikh vouched for by Abu Tig Marina is Sharm El Sheikh Yacht Service *email* wahab–yachtservice@yahoo.com

Merset El At (Naama Bay)

Plan page 238

Anchor in 5m, sand, with good holding. Note that in a S'ly, backwash off the steeply shelving beach can make the anchorage tenuous. There are

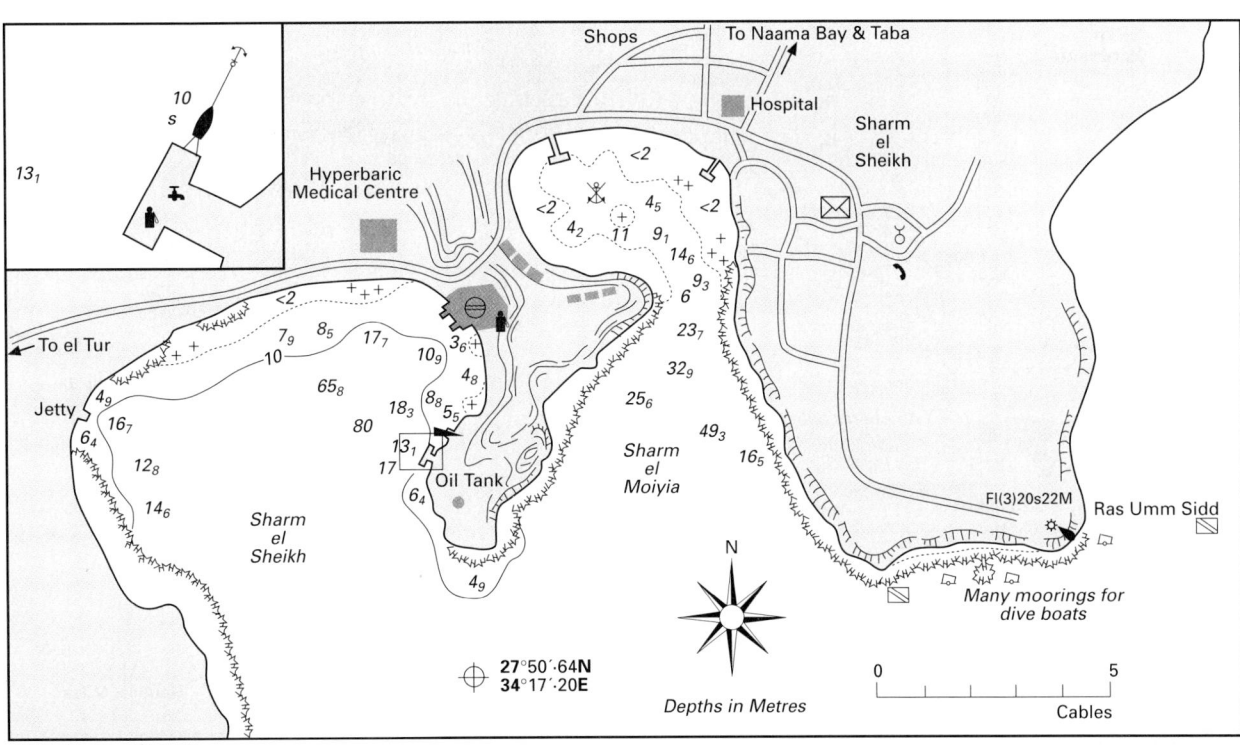

SHARM EL SHEIKH

Red Sea Pilot

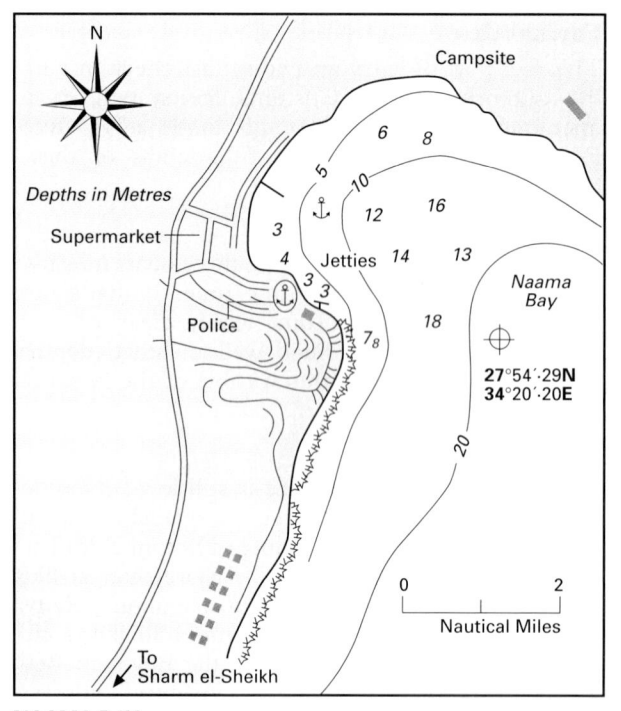

NAAMA BAY

moorings laid near the jetties in the S of the bay. There is a supermarket for shopping. We've had ambiguous reports about the welcome given to visiting yachts. Check first with the authorities at Sharm el Sheikh.

Strait of Tiran and Tiran I

Admiralty chart 801, plans below and page 239

Note National boundaries in this area are not quite settled. Most atlases show the Saudi Arabian/Egyptian border as lying through the Strait of Tiran, making BOTH Tiran I and Jaz Sanafir Saudi Arabian. Hence tread carefully. Tiran I and the Strait of Tiran reefs seem to be accepted as Egyptian and are part of the Ras Mohammed National Park. Jaz Sanafir, which appeared in the first edition of this pilot and has excellent dive sites, is treated as within Saudi Arabian waters and is hence off limits.

Approach

Enterprise and Grafton Passages form the two lanes of the traffic separation scheme, which you should observe whatever you see the local dive boats doing. The reef edges are steep-to and clearly visible.

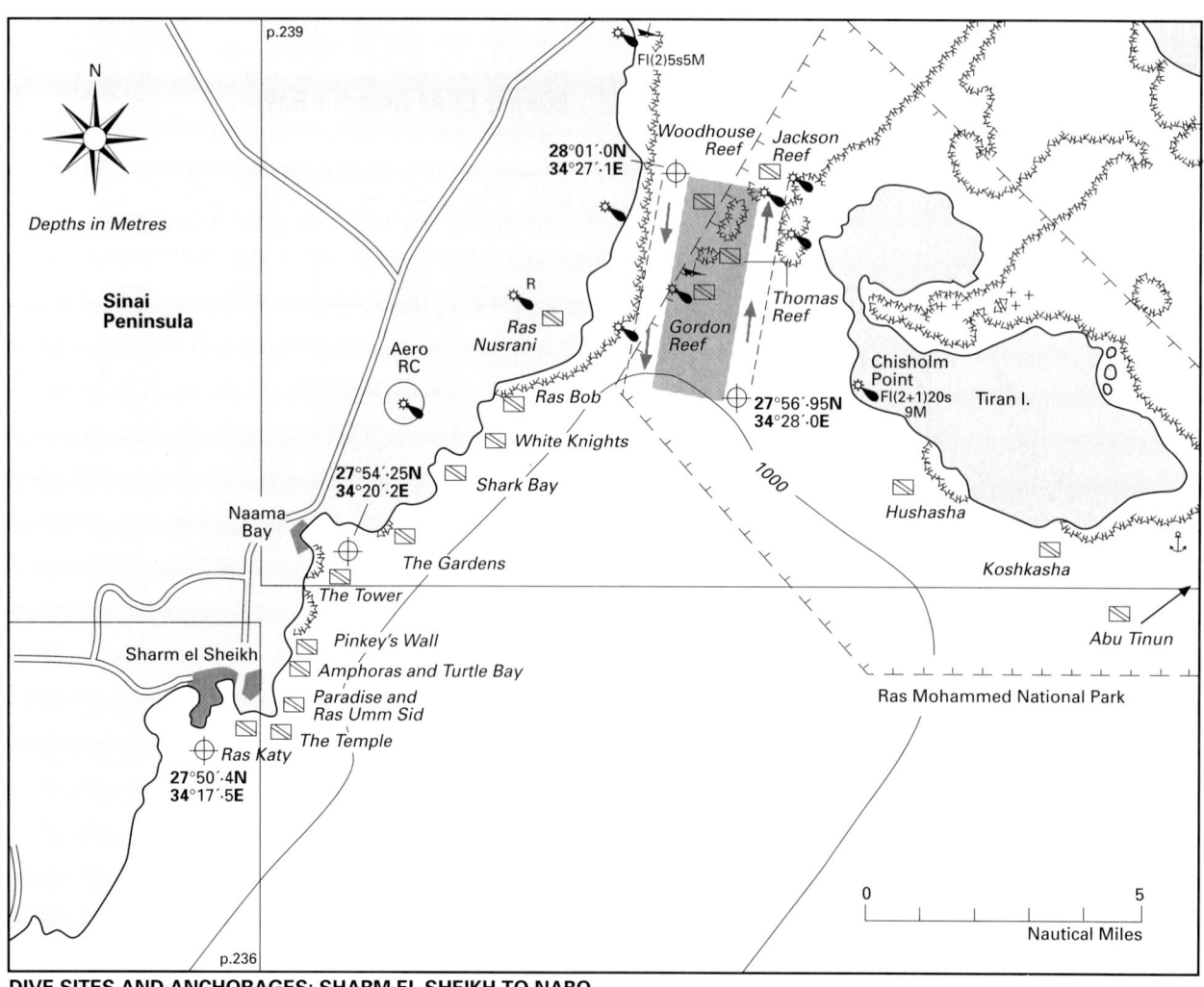

DIVE SITES AND ANCHORAGES: SHARM EL SHEIKH TO NABQ

Gulf of Aqaba and approaches

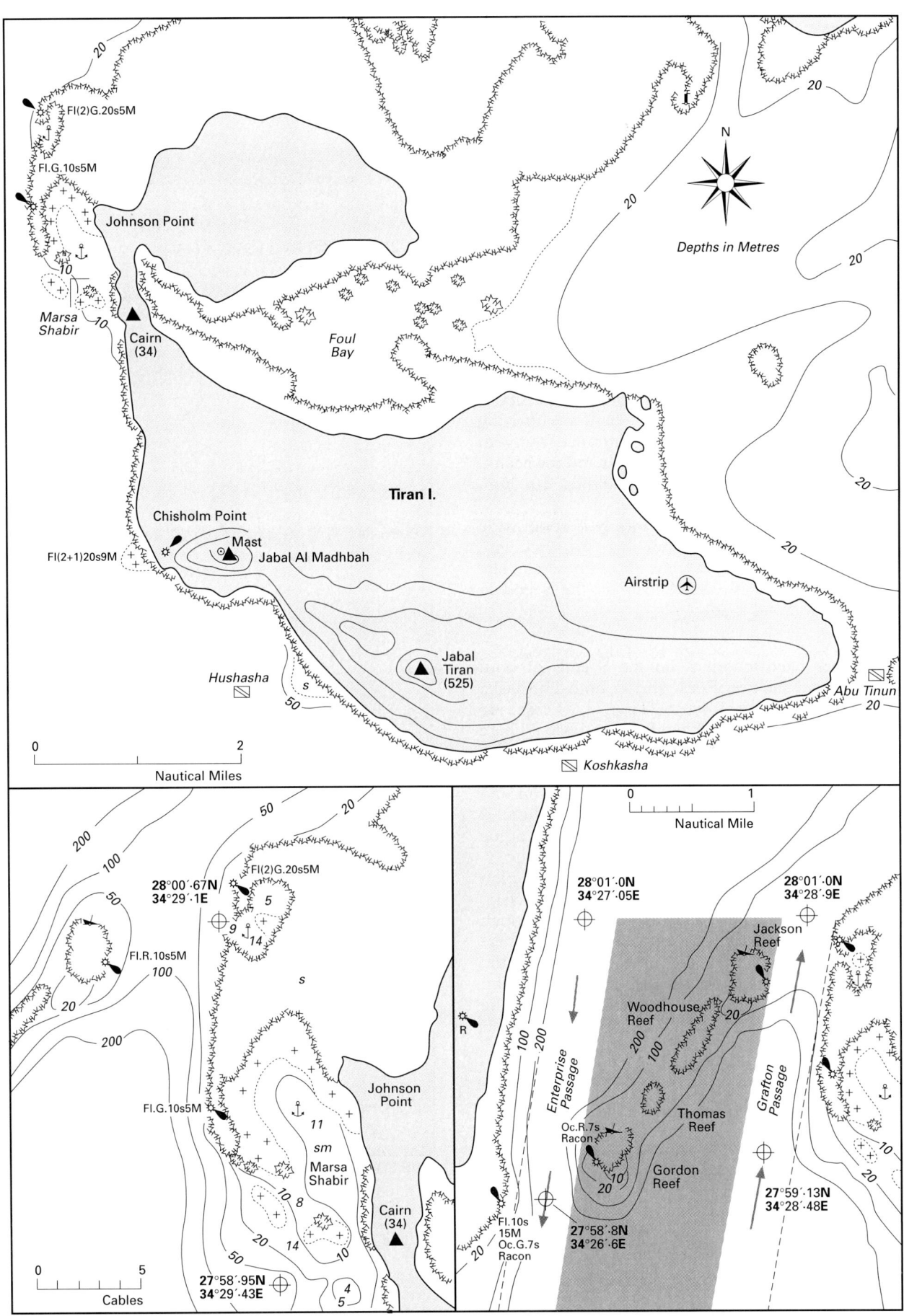

TIRAN I, TIRAN I ANCHORAGES AND STRAIT OF TIRAN

Red Sea Pilot

Dangers
There are strong currents between the reefs in the middle of the strait. In S'lies, the tidal stream has been known to set N at >3 knots. High tide is 1–1½ hours after high water at Shaker I, with a range at springs of 0·6–1·2m. The flood sets N, the ebb S.

Conspicuous
The wrecks on the reef N of Ras Nusrani, on Gordon and Jackson Reefs and on the coastal reef N of the anchorage make good radar targets. The two prominent hills on Tiran I, and a small, cairn-topped rise at the narrowest point of the isthmus are good landmarks.

Anchorages
The two anchorages on the W side of Tiran I, Marsa Shabir and the pool N of it, are good spots for divers or if you're waiting for a wind. Enter Marsa Shabir by holding to the island shore. There are off-lying rocks on the reef patch W of the entrance, but with 200m width of clear water between it and the island, least depth 10m. Anchor in good holding, 10–15m, sand and mud.

The entrance to the pool further N is 100m wide with a least depth of 9m. Inside there's a patch with 2·5m over it about 200m ENE of the entrance. The pool shoals at its N end. Anchor in the S end in 12–15m sand and mud.

General
There are fixed-moorings on the S ends of both Gordon and Jackson Reefs and at both Hushasha and Koshkasha sites on S Tiran I. These are technically reserved for Egyptian dive boats. There is a sandbank on the S side of Gordon Reef, with good holding in 10–20m, sand. However, the Egyptian marine police have moved yachts on and even made arrests for anchoring there. There is active patrolling using RIBs. Be prepared to argue stress of weather if you anchor.

At the SE end of Tiran I there is an Egyptian Army outpost on the coast S of the rough airstrip. Anchoring is apparently allowed just W of it, though do go ashore to make your number. The anchorage is apparently well sheltered in N'lies. Ask at the outpost for permission to land on the beach. Inland is out of bounds because of mines laid during the Arab/Israeli wars.

Nabq

Plan page 241

The beacon and the two wrecks on the reef S of the entrance are easily visible. Keep the mangroves on the shore by the radar station bearing 290° or less. There is an isolated rocky patch with a stranded wreck bearing 035°M from the beacon on the headland SE of the anchorage. Anchor in 4–5m, sand and coral, in the S inlet in the fringing reef, with good shelter from the S. Opinions differ as to shelter here in any sort of N wind, but it looks doubtful.

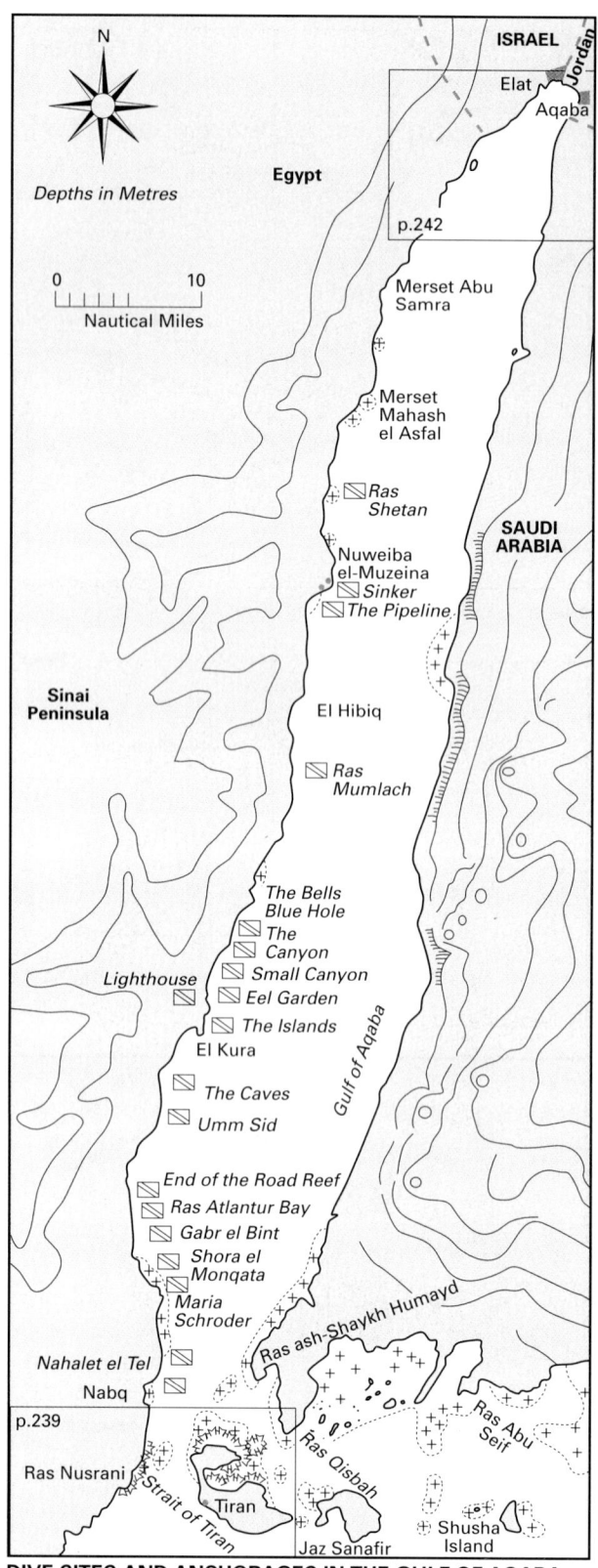

DIVE SITES AND ANCHORAGES IN THE GULF OF AQABA

Gulf of Aqaba and approaches

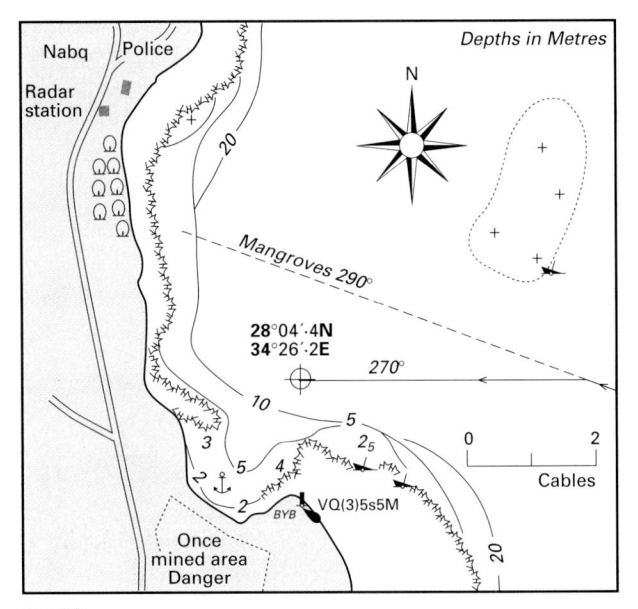

NABQ

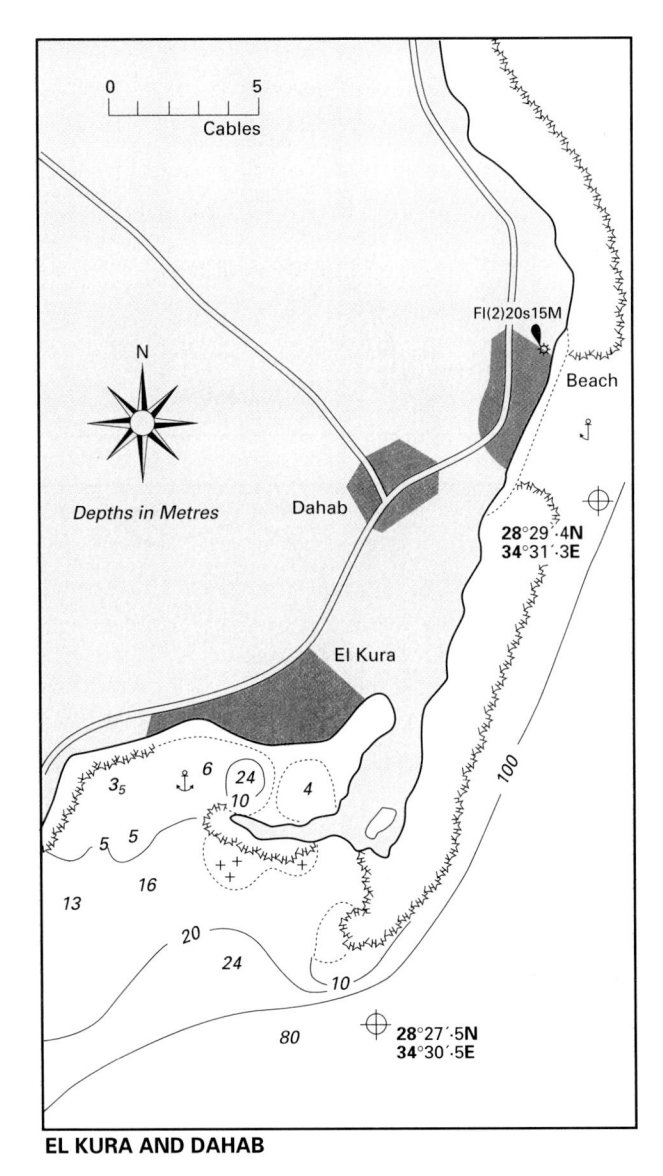

EL KURA AND DAHAB

El Kura (Dahab)
28°28'N 34°30'E

A long tongue of reef extends from the E side of the spit and there are rock and coral outcrops on its S side. Keep well clear of the spit but drop your hook before you reach the deep hole, N of the sand bar SE of the anchorage. Good holding in 6m, sand.

General

El Kura is only one part of a sprawling resort area generally known as Dahab, the dilapidated town, 1km inland from the beach. At the beach by the lighthouse a recent development is a self-contained tourist complex; in quiet weather, you could anchor in the break of the reef there. About 2M N, the 'Bedouin village' of Asla or Assalah is the now much-changed original Dahab backpacker hang-out. Finally, there is El Kura itself, which has the reputation of being the only all-weather anchorage in the Gulf of Aqaba, so don't expect lots of room to manoeuvre. Dahab is a major dive resort but isn't as developed as Sharm or Naama Bay. Most of the popular dives here can be done from shore.

Between El Kura and Nuweiba the E coast of the Gulf is very open. However, there are two possibilities for persistent N winds. The first is in the lee of the reefs S of Ras Abu Galum in 28°37'·1N 34°34'·4 E, or, for real shoal drafters, in the lagoon under the Ras itself at 28°37'·7N 34°34'·85E. The second is in or N of some patchy ground just S of El Hibiq immediately W or N of 28°50'·5N 34°38'·1E.

Nuweiba El Muzeina

There is a ro-ro ferry terminal NE of the anchorage so anchor clear of it in 7–10m, W of the jetty.

This is where the pilgrim ferry for Aqaba berths, and ostensibly it is forbidden to go ashore. In

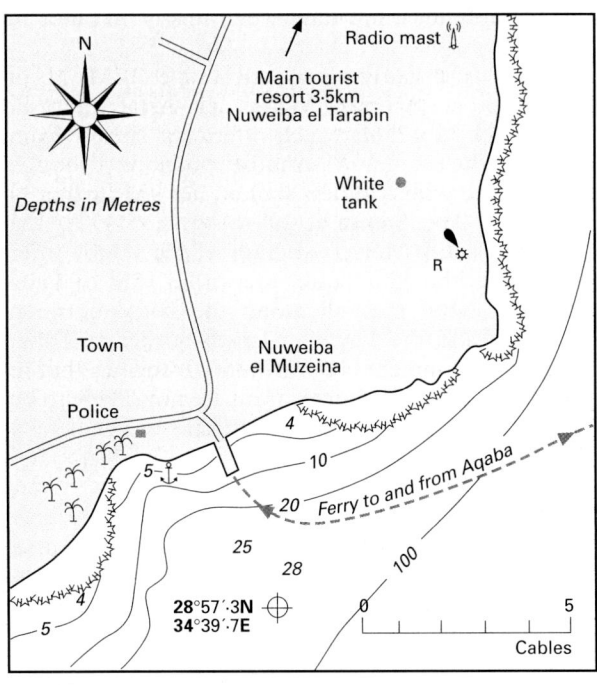

NUWEIBA

241

Red Sea Pilot

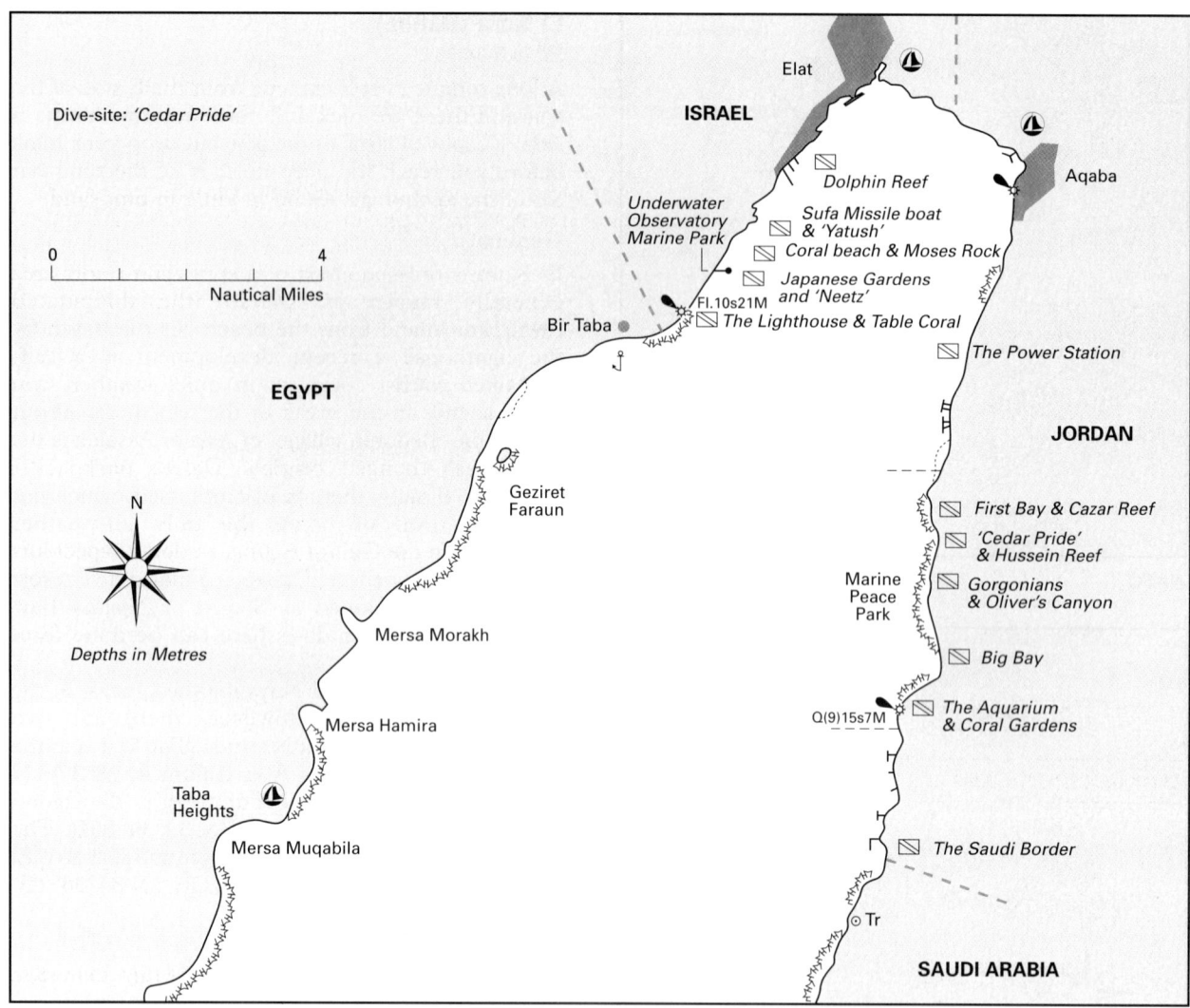

MARINAS, ANCHORAGES AND DIVE SITES IN THE N GULF OF AQABA

January, February, April and October strong S'lies are possible for a few days when this is no place to be.

There is a fairly open anchorage 12M N of Nuweiba at Merset Mahas el Asfal, approx 29°09'·8N 34°42'·6E which offers fair shelter from N'lies in the lee of a *ras*. Another possibility, though smaller and with a bit less shelter, lies 9M further N in Merset Abu Samra at 29°18'·45N 35°45'E. 1M further N still Merset Mahash el'Ala might offer anchorage, though all three are within 15M of Taba Heights. Note that all along the coast between Nuweiba and the Egyptian-Israeli border at Taba, tourist development is burgeoning. It follows that in any of the above anchorages you are now likely to be off a hotel, tourist village or campsite.

Taba Heights (Mersa El Muqabila)

Plan page 243

This large development built by Orascom, the same company that built El Gouna, has a small marina. At present the basin, dredged to 3–4m, has stern-to berths for only 20 boats from 8–30m. Pontoons with 50 berths and a fuelling berth will be installed. The

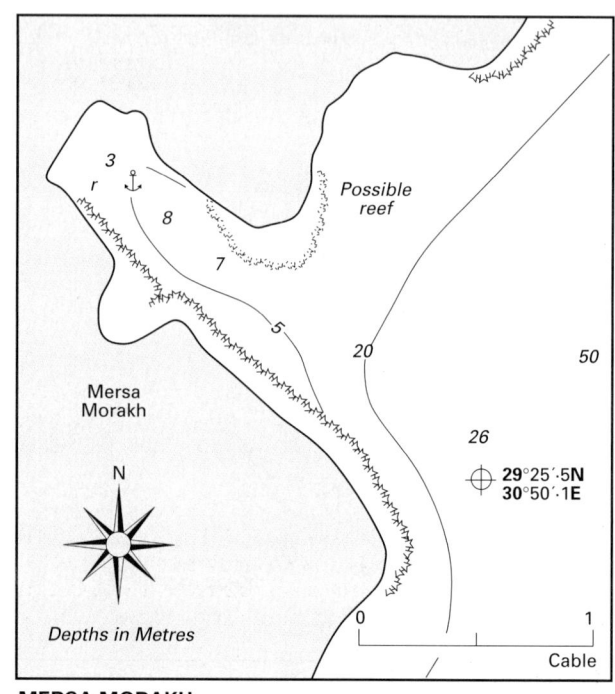

MERSA MORAKH

Gulf of Aqaba and approaches

Marsa Hamira
Plan opposite

Beware the uncharted coral outcrop with >1m over it, bang in the middle of the bay, on the 5m contour. Anchor in 5–6m, sand, clear of the rocks close inshore in the NW corner. Good holding, though it can be gusty. It is a good idea to use two anchors to limit your swing since it's a favourite with tripper boats from Elat. In addition, a tourist village complex now occupies the *marsa's* shoreline. There has been no recent news of their attitude to visiting yachts.

Mersa Morakh
Plan page 242

The channel into the *marsa* is not more than half a cable wide. There is some conflict between the British description of the inlet and the Egyptian survey we have used. The British talk of a very narrow, shallow entrance along the N shore with a least depth of 2·1m. The survey shows a deeper, wider channel. It follows you should take great care through the entrance. Anchor towards the *marsa* head in 4–8m, rock, with poor holding. Use a fisherman anchor if you have one. For the adventurous only.

Geziret Faraun
Plan below

Note It is essential to have permission from the authorities at Sharm el Sheikh or Bir Taba if you want to anchor here. There is an anchorage for reasonably settled weather off the beach at Bir Taba (approximately 29°29'·2W 34°53'·6E), opposite the hotel near the tripper boats, to see the marine police for permission to visit.

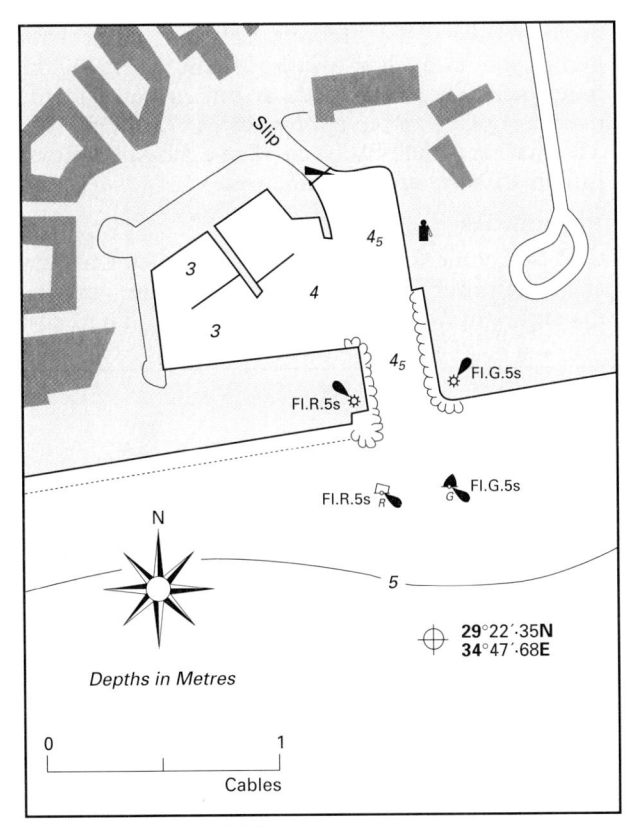

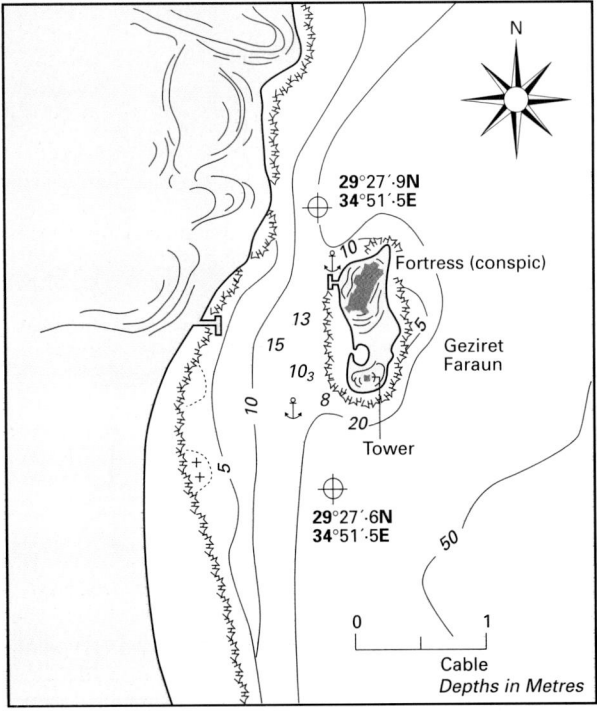

TABA HEIGHTS MARINA

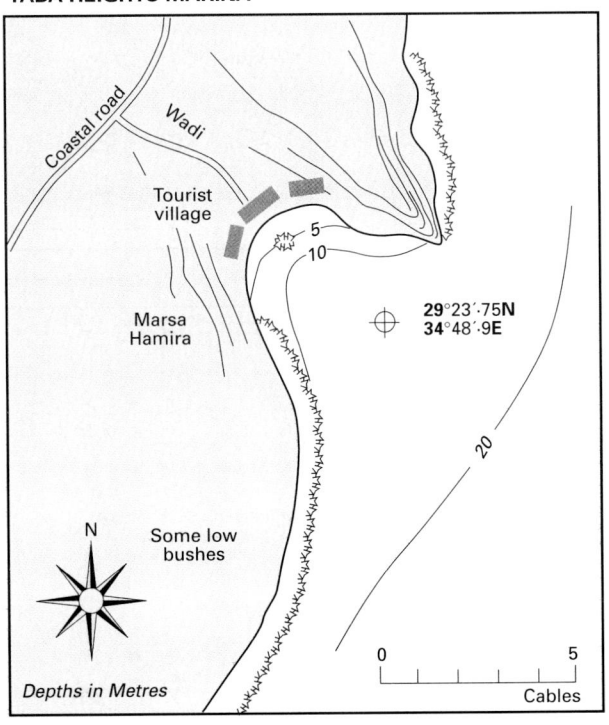

MARSA HAMIRA

resort complex has a Hyatt already on stream and some shops for basics. Several other international hotels are under construction. There is to be a golf course and villas. Existing marina services include water and power. Fuel can be jerry jugged from a service station on the main road about 1km from the marina. For further information *email* marina@orascom.net

GEZIRET FARAUN

Red Sea Pilot

Approach and anchorage
The island is about 7M SW of Elat, separated from the mainland by a channel with a least width of a cable and a least depth of about 8m. The Frankish fort is floodlit at night. Anchor in 7m, sand, good holding about a cable SW of the island, near a mooring buoy. In a S'ly anchor about 50m off the beach N of the jetty and the N of the island in about 10m.

General
The island, otherwise known as Pharaoh's Island, sheltered the Phoenician fleet, which King Solomon sent to explore the Red Sea. The fortress is said to have been built in the 12th century AD by the last Frankish King of Jerusalem but opinions differ and the Egyptians say Saladin (Salah ad-Din Yusuf ibn Ayyub) built it. Subsequently it became a Crusader pirating base before the Christians were ousted from the area. You can see massive ancient water tanks at the S end of the island near the small tower on the S hillock.

Aqaba
Admiralty chart 801, plans below

Approach
You will be called on Ch 16 by the Israeli navy as you approach the coast. Tell them that you are headed for Jordan and then call Aqaba Port Control on Ch 16 when you are about 10M out. They will probably tell you to go direct to the Royal Jordanian Yacht Club and take a mooring to complete formalities. If so contact the yacht club on Ch 67. If not, on closing the harbour, call port control for permission to anchor inshore of the N tip of the main pier. There are lights at the entrance to the marina (Q.G, Fl.R) in approx 29°31'·7N 34°59'·9E. The marina is the NW basin. The SE basin is Royal Jordanian Navy and a secure area.

Formalities
CIQ will come to your boat. It's all quick and easy and definitely no baksheesh! The marina management will help visa formalities. They cost

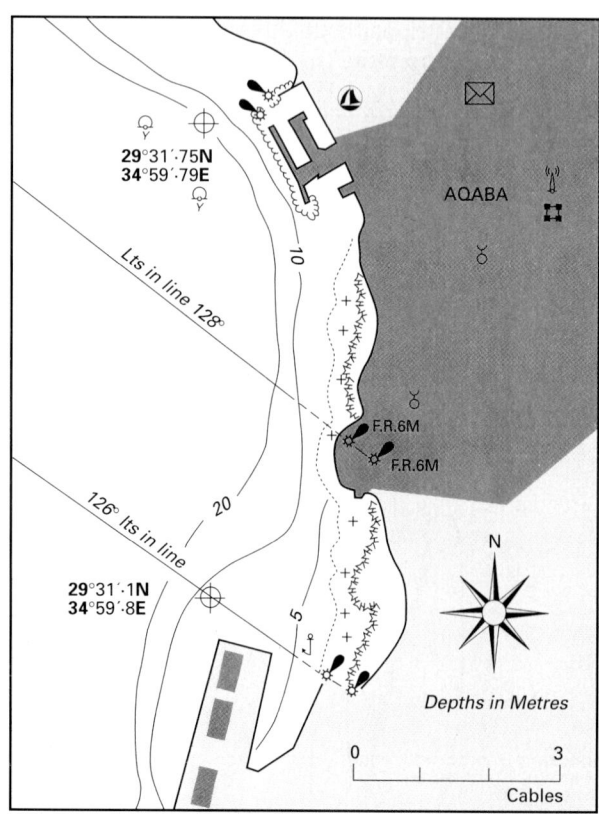

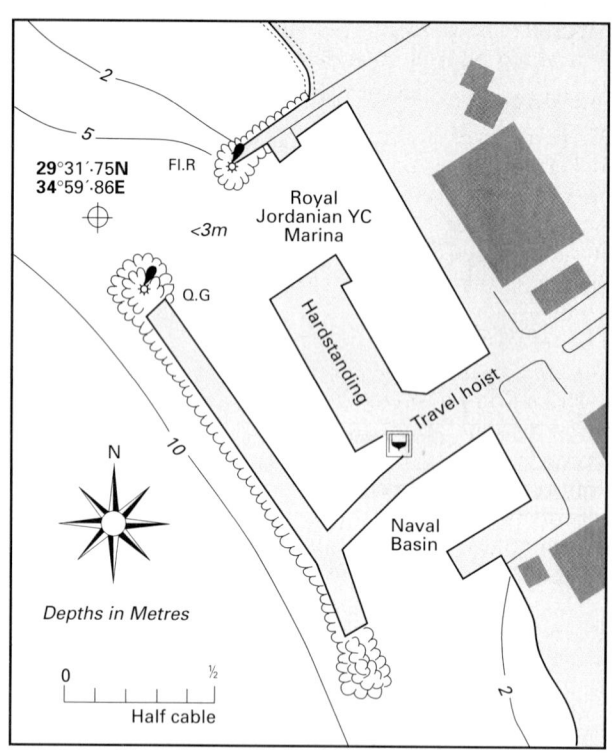

APPROACHES TO AQABA

ROYAL JORDANIAN YC MARINA

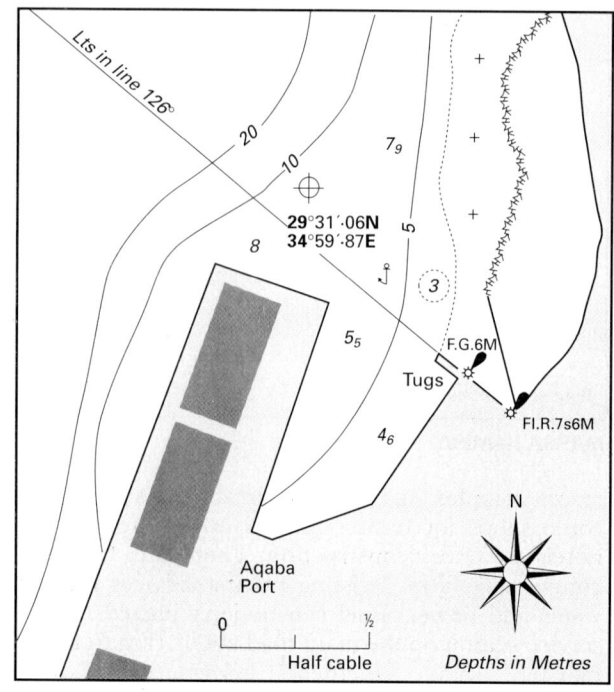

AQABA PORT ANCHORAGE

Gulf of Aqaba and approaches

about US$30, are valid for 2 weeks, extendable to 3 months at any police station at no extra charge. The marina will also organise departure clearance, for which there is a small charge.

Berthing

The marina can take yachts up to 25m LOA and 2·9m draught. Dues are approx US$20 per day. There is also normally a flat fee (in fact an annual charge) of US$55 to cover services and use of facilities, including showers and the swimming pool. Try negotiating if you're only staying a short time. Mediterranean-style or alongside mooring depending on how crowded the marina is.

Facilities

There is good security and protection. There is a fuel dock and each berth has water and electricity. No charge for small amounts of water but if you take a lot you'll be charged for 2 tons at a fixed fee. Haulout is possible with a 25-tonne travel-hoist and there are workshops on site as well as in the town which is a short walk away across the expressway. The tourist information centre is in the Islamic museum in the Old Customs House about 400m S of the marina on the coast. There are several cyber-cafés. One is near the PO and another near the largest mosque. LPG refills are available. Aqaba is a duty-free port with a developing tourist industry. It's a good place to have spares flown in. The club, a friendly place, will hold mail for you, c/o The Royal Jordanian Yacht Club, PO Box 500, Aqaba, Jordan.

Anchorage (if necessary)

4m, sand and mud, off the NE tip of the main cargo piers where shown. Shelter from the N is good, but in S'lies, which are rare, ask for permission to tuck further in if you can.

Diving

There are several sites along the short coastline S of Aqaba. The Marine Peace Park is part of a general environmental initiative in consort with Israel and with luck may involve Egypt and Saudi Arabia too. One result could be that the whole Gulf of Aqaba becomes a conservation area. Meantime anchoring in the park is prohibited and the HEPCA lead around Sinai and Hurghada has been followed, so there are buoys to use. Ask at the Yacht Club, the Royal Diving Centre 15km S of the town (PO Box 21, Aqaba, Jordan, ☎ 962 20 317 035) or one of the dive outfits.

Elat

Admiralty chart 801, plans below

Approach

If you are heading to Elat from Jordan you must sail 5M offshore first, then call the Israeli navy on VHF Ch 10 or 16. Contact Elat Port Control on Ch 14 when you are about 2M out. They will arrange a mooring for you to complete formalities. The international boundary between Israel and Jordan meets the coast about 2M ESE of the marina.

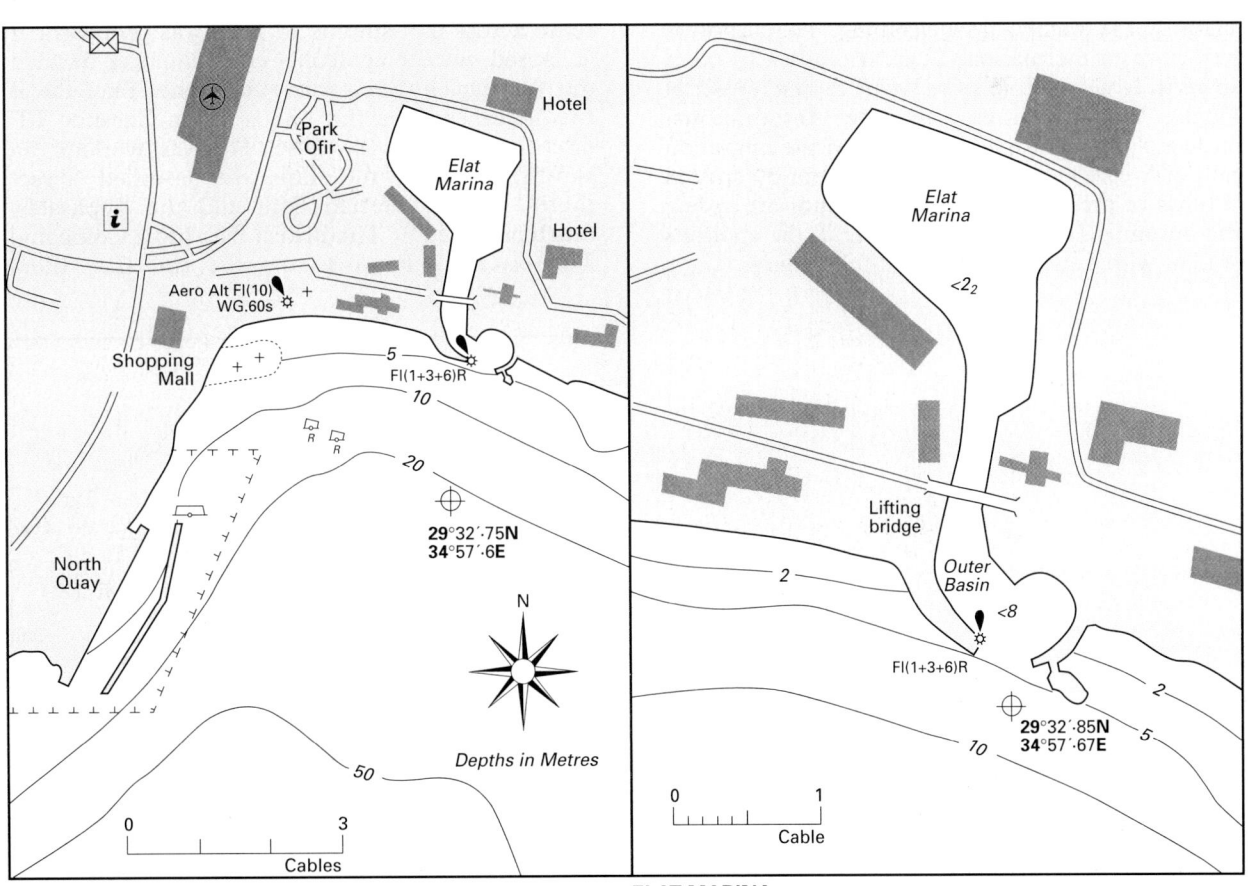

APPROACHES TO ELAT

ELAT MARINA

Red Sea Pilot

Dangers
Entering the marina in strong S'lies is dodgy. Indeed, Elat as a whole is no place to be, unless in the inner marina, if it is blowing hard from the S. There is now a marine farm, approximately 2 cables by 2 cables close E of the marina entrance.

Berthing
Contact the marina on VHF Ch 16 or 11 once you are cleared in to arrange for the bridge, which separates the inner and outer basins, to be opened. Restrictions are max. LOA 25m, max. draught 2·2m. On outside moorings, 40m LOA, 8m draught. Mooring is Mediterranean-style. Charges are approx US$12 per metre per month or US$1 per metre per day.

Facilities
Water, electricity and repairs at the marina but facilities are said to be more primitive than at Aqaba. The management is welcoming and friendly. Good workshops for repairs on the industrial estate. Email at BJ Books in the Tourist Shopping Centre. Other services and facilities in the town, including the PO, fax and telephones. There is a tourist information office near the Caesar Hotel, just outside the marina.

Elat Marina ☎ +972 8 6376761 *Fax* +972 8 6315138.

General
Elat is a duty-free port and, like Aqaba, a good place to have spares flown in. The authorities, who see relatively few yachts, are welcoming. The airport is very close to the marina. There are buses to other towns in Israel, with links to Egypt via the border at Taba. Elat is home to the International Birdwatching Centre because it is on the migration path of over 400 species including about 30 species of birds of prey. Peak migration periods are spring and autumn. There is good walking in the environs of Elat, with outstandingly beautiful scenery.

7. The Suez Canal and approaches

'And here not far from Alexandria,
Whereas the Terren and the red sea meet,
Being distant lesse than ful a hundred leagues,
I meant to cut a channel to them both
That men might quickly sail to India.'
Christopher Marlowe, *Tamburlaine* c. 1588

The history of the Suez Canal

Rameses II was the first to try to link the Red Sea and the Mediterranean in the 12th century BCE. 600 years later, thanks to Darius Hystaspes of Persia, the link got as far as the Bitter Lakes. At that point things hung fire until 274BCE when Ptolemy II got the canal to Arsinoe, somewhere at the S end of the Bitter Lakes. Cleopatra used it to try to reach the Red Sea with the remains of her fleet after her defeat at Actium. When she failed, the great engineering minds of Rome took over but with no immediate success. It wasn't until Trajan, at the beginning of the 1st century CE, that finally a canal reached Suez, which was then called Klusma. All these canals began in the Nile delta and cut across country to modern day Ismailia. Roughly the trajectory of today's Sweetwater Canal. Until the 19th century this early canal was alternately dredged, neglected or deliberately damaged by opposing potentates.

Following Napoleon's invasion of Egypt a new route across the isthmus of Suez was proposed. It collapsed when Napoleon's chief engineer decided the sea level in Suez was 10m higher than the E Mediterranean! In the 1830s Capt Chesney RE discovered this to be false. He was working on another route at the time, the so-called 'direct route', via the Persian Gulf and the Euphrates, much preferred by The British East India Company. Thomas Love Peacock, novelist and East India

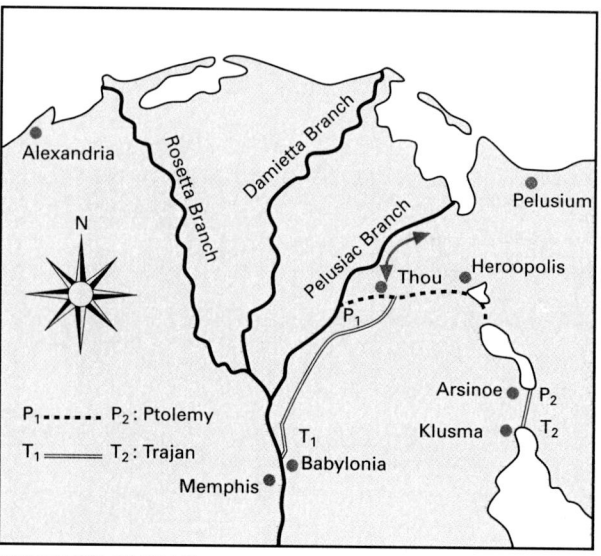

THE FIRST CANALS

246

The Suez Canal and approaches

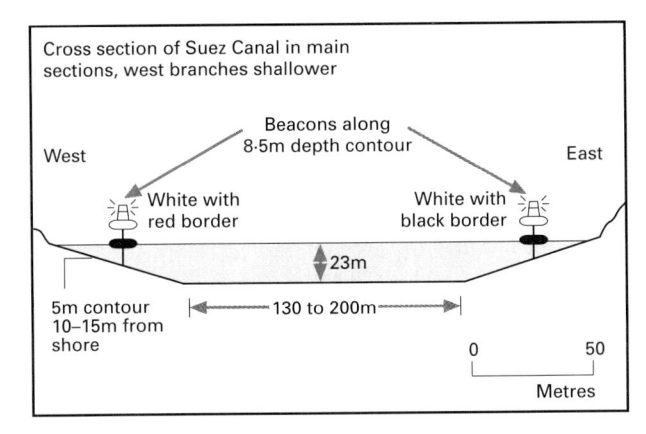

Cross section of Suez Canal in main sections, west branches shallower

Beacons along 8·5m depth contour
West — White with red border
East — White with black border
23m
5m contour 10–15m from shore
130 to 200m

Company bureaucrat, declared it to be the best, although the faster route from the Mediterranean to India, called the 'overland route', was already in full swing, crossing Egypt by river and land caravan from Alexandria to Quseir.

Meanwhile, thanks to the advent of steam ships, the intrepid Lt Wagstaffe of the Indian Navy had pioneered an even faster overland route from Alexandria to Suez. Despite this, the East India Company was still undecided when, in 1854, the French engineer Ferdinand de Lesseps was authorised by the Egyptian ruler, Said Pasha, to begin excavations. Not the least reason for the hesitation was the recently completed British railway which ran from Alexandria via Cairo to Suez! Despite that, the British and French governments gave approval a year later and the Suez Canal Company was formed. Construction work began in 1859. The nearly 90M long canal, passing via the Little and Great Bitter Lakes and Lake Timsah was completed 10 years later.

In 1875, following Said Pasha's near bankruptcy, the British Prime Minister Benjamin Disraeli pushed hard against fierce opposition for Britain to buy the Pasha's shares, giving it a 75% holding. Between then and 1920 the profits on the capital outlay of £4,000,000 were multiplied eightfold, vindicating Disraeli's decision handsomely. The 1888 Constantinople Convention provided for freedom of transit of all vessels through the canal in peace and war and established the tonnage rule upon which your transit fees, some of them anyway, are now based.

An Anglo-French Canal Company ran things until, in July 1956, the Egyptian government nationalised the canal prompting the Suez Crisis and almost a year's closure. Arab-Israeli conflicts between 1967 and 1973 caused a further closure of nearly eight years.

The canal's vicissitudes had implications for the world oil economy. Although the 1956 closure caused an oil shortage in Europe, things recovered so that by 1966, 70% of ships using the canal were oil tankers. But the second, prolonged closure resulted in the emergence of VLCCs and ULCCs using the Cape route. As a result, when the canal reopened in 1975 only 23% of the world's tankers could use it, with obvious consequences for income.

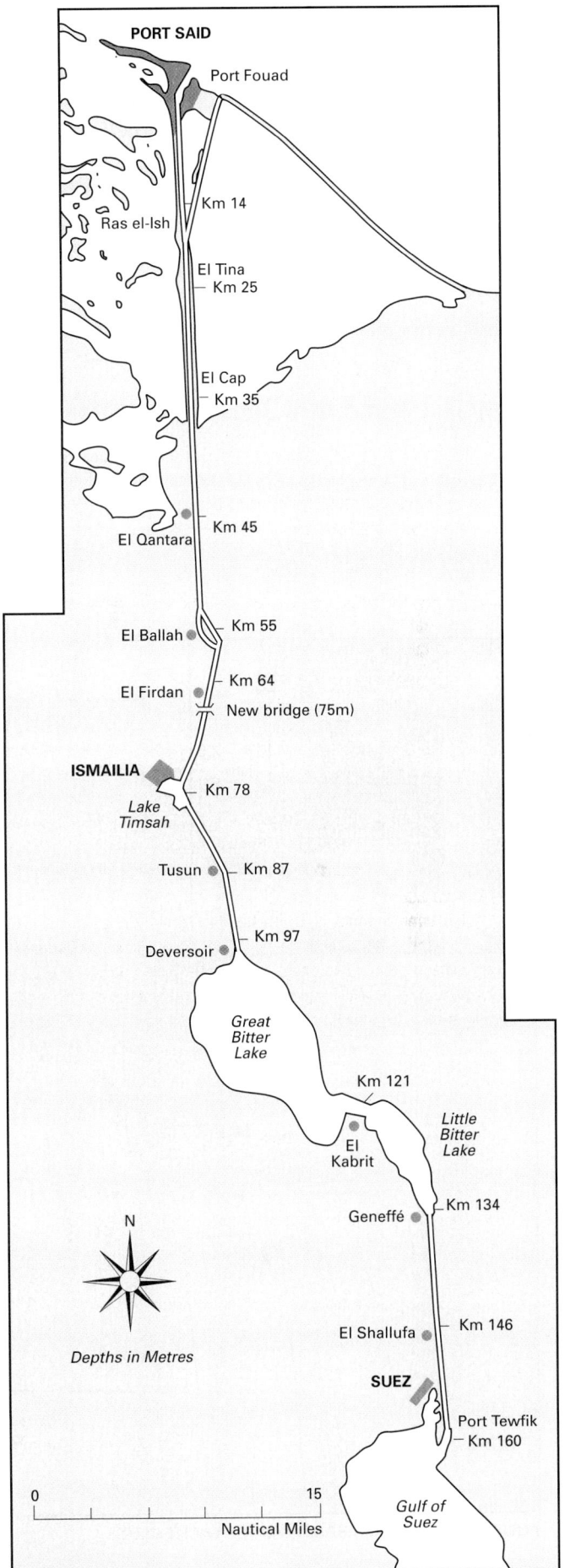

SUEZ CANAL

PORT SAID — Port Fouad
Ras el-Ish — Km 14
El Tina — Km 25
El Cap — Km 35
El Qantara — Km 45
El Ballah — Km 55
El Firdan — Km 64 — New bridge (75m)
ISMAILIA — Km 78
Lake Timsah
Tusun — Km 87
Deversoir — Km 97
Great Bitter Lake
Km 121
El Kabrit
Little Bitter Lake
Geneffé — Km 134
El Shallufa — Km 146
SUEZ
Port Tewfik — Km 160
Gulf of Suez

Depths in Metres

Red Sea Pilot

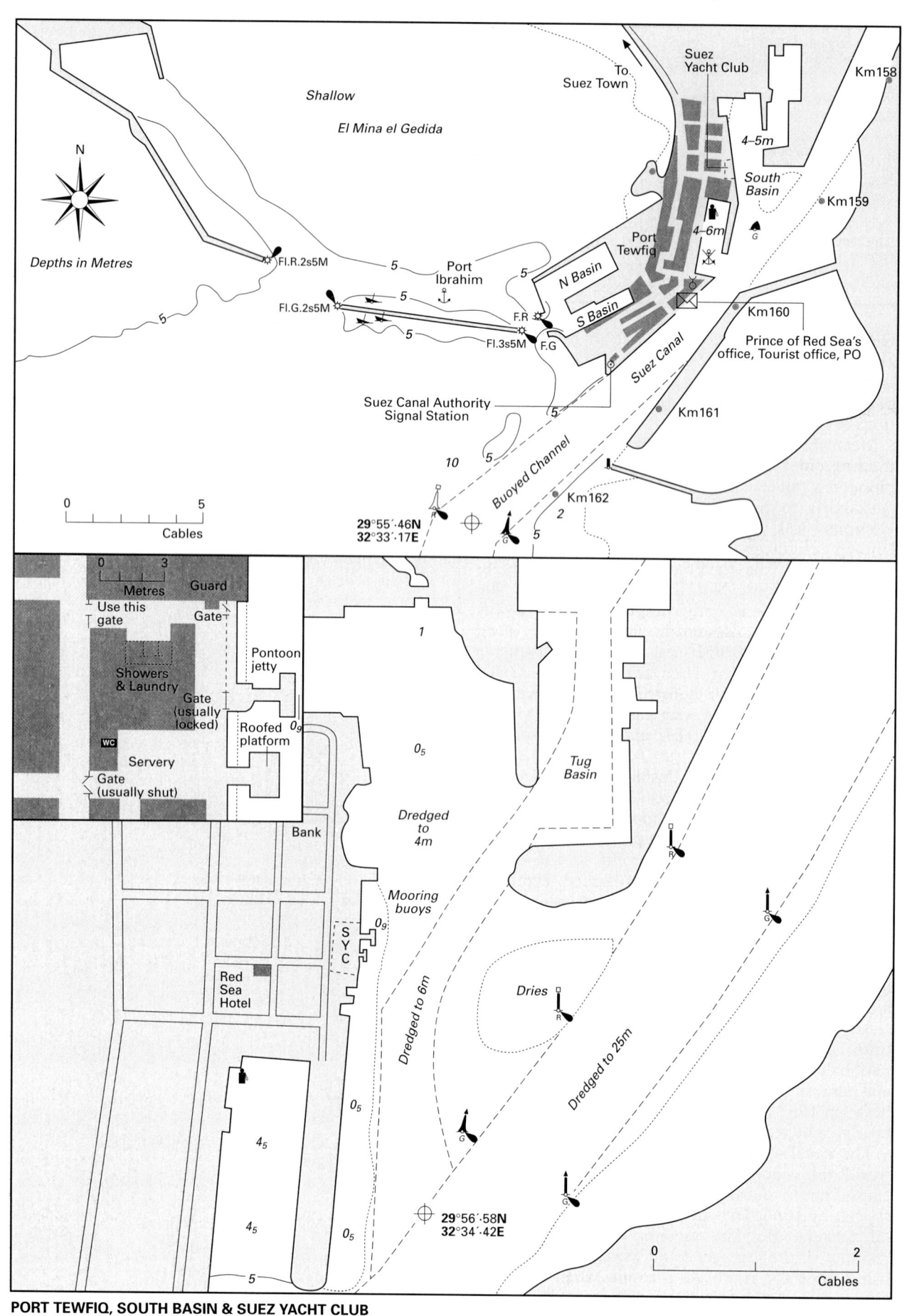

PORT TEWFIQ, SOUTH BASIN & SUEZ YACHT CLUB

The Suez Canal and approaches

In consequence, between 1976 and 1980, the canal was widened and deepened. Since the early '80s oil tankers have again been a large proportion of traffic. A further project to widen the canal to 415m and deepen it to 27m will accommodate oil tankers of 250,000 tons. Income from Canal fees is a vital source of revenue for Egypt. When Lady Brassey made the transit aboard the *Sunbeam* in 1877 there were an average of 4–5 ships passing through it each day. 100 years later there were 60 a day, a number unlikely to be greatly exceeded until the canal can take two-way traffic either by additional width or doubling – a project that is being actively considered.

Admiralty charts
Large scale 2098, 3214, 233
Medium scale 2373

US charts
56082, 56083, 62193, 62194

French charts
7013

Note On the flood tide in Suez Bay there can be strong N-flowing streams. The authors' VMG, for a 5kt boat speed, was over 7kt crossing Suez Bay on the flood. The S-going stream is weaker. Either could make an appreciable difference to ETA.

Port of Suez (Suweis)
Admiralty chart 233

Approach to the port
The approach to Suez is hard to get wrong. The channel is excellently buoyed and the ships parked each side show the way to go. The controlled waters of the canal proper start at No.1 light float, 8 cables before the conspicuous Suez Canal Authority building on the S tip of Port Tewfiq. There are several new marks in the S approaches, be sure to consult a recent chart and *Notices to Mariners*. At night the lights, beacons and buoys can be difficult to spot, especially if there is a lot of shipping. It isn't a good idea to enter the yacht club at night and technically it's not permitted. If you do, do not use a masthead tricolour, use ordinary navigation lights.

Ship movements
N'bound convoys move 0530–1030, and if traffic demands it 0300–0430. S'bound convoys pass Suez between 1400–1700 and 2000–midnight. If you are N'bound and arrive between 0530 and 0600 keep clear of the assembling convoy. The main assembly area is E of the approach channel from approx 29°47'·5N N'wards (see BA2373). To keep out of the way hold to the shorelines. On the W shore cross the shoal (least depth 7m) off Ras Adabiya. To the E hold close around Ras Misalla. If you are S'bound either be clear of the canal approaches before 0600, which means leaving the yacht club before 0500, or wait until about 1030 when the N'bound convoy has passed through and leave before the first S'bound convoy arrives around 1400. There is another window from about 1700–2000 before the second S'bound convoy arrives. Some yachts disregard convoy movements. This is neither courteous nor good seamanship.

Anchorage
If you are in any doubt about your engine or have other mechanical problems, anchor in Port Ibrahim and contact your agent. If you go to the Yacht Club with mechanical trouble, you may create difficulties for your transit. Before entering Port Ibrahim contact the port authorities, especially at night when they get excited about unauthorised small-boat movements.

Suez Yacht Club (Port Tewfiq Yacht Club)
Plan page 248

Note The SCA signal station (see sketch) flies a black flag in S'ly gales when the port may be closed. Berthing at the club (SYC) is obligatory for yachts arranging or awaiting transit of the canal. The club belongs to the Suez Canal Authority (SCA).

Approach
Keep to the starboard side of the canal. Cross the fairway only when you are opposite the second Basin, known as South Basin, which is after the mosque (lit up green at night) and a monument of a battle tank. If you come in at night, remember to use deck-level navigation lights not a tricolour masthead lantern and NEVER both.

Dangers
A green buoy marks the SW end of the shoal N of the entrance. It's lit (Fl.G) but unreliable. If you leave the buoy a boat's length or so to starboard, you'll be fine. Be wary of tugs and service craft entering and leaving the tug basin to the N of South Basin. They don't give way.

Mooring
Moor bow and stern to buoys. Depending on the time of your arrival and on your agent, you may be helped into a berth. Otherwise have your dinghy ready for securing lines. If all buoys are taken you will need bow and stern anchors. There is no swinging room. You'll still pay standard rates.

Facilities
Berthing fees

Boat length (m)	per night	per week	per month
<9·9	US$6	US$30	US$120
10–15	US$8	US$40	US$160
over 15	US$12	US$60	US$240

Fuel Order diesel through the Yacht Club. It costs approx US40c per litre. You must go to the fuel dock S of the club basin. Agents will also arrange fuel delivery by jerry jug at similar prices. If you try to buy at street prices from a gas station you'll end up having to bribe the gate guard because in principle you are smuggling. If you succeed the end price will probably be much the same anyway. If heading N fill up here. It's the most convenient till Ismailia's fuel dock is ready.

Red Sea Pilot

Engine and gearbox oil and small amounts of petrol can be brought in from the town without any trouble. Paraffin is cheap and easily available. Gas bottles can be filled by your agent at a cost. They have to be taken all the way round Suez Bay to the LPG depot.

Water Charges are US$3 per cubic metre. You can go alongside the pontoon to water, but only around HW (<2·8m HWS). There is only 0·9m at chart datum (LAT). If you must water at other states of tide, you can lie alongside a boat on the first line of trots. The hose is long enough to reach. Charges for the washing machine are US$5 per load, when it is in service. Your agent will arrange laundry services for you if you prefer. Showers are free but primitive.

Money Payment for agents, transit fees, fuel and water in US$ is preferred but you may be able to pay your bill in Egyptian pounds at an exchange rate agreed with the Yacht Club. Cash advances at ATM's are available at some banks, e.g. Bank of Egypt. Otherwise you can obtain cash on a credit card over the counter on production of your passport. There is a bank near to the SYC open at weekends but not on Fridays. The PO and a tourist office are marked on the sketch.

Communications Buy phone cards at kiosks for the yellow Menatel phone boxes. For email either use your agent or a cyber-café/computer shop in Port Suez. One, at E£5 per hour, is Future, Suez Gulf Development Co, 4 El Azhar St, Port Suez. This is on the far side of Port Suez on the right, near the Immigration Police building, roughly parallel to the main road out to Cairo but two blocks in. Their hours are 1000–1400 and 1900–2300.

Stores There are several small grocers in Port Tewfiq but Suez town, across the causeway, is much better. Take a minibus for 50 piastres single or a taxi for about E£4–5, into Suez town. Agree on the fare in advance with drivers. It avoids arguments later. There is a good market, grocery stores and reasonable general shopping.

'Duty-free' alcohol is available through agents. Prices are approx. US$13–15 a case for beer, minimum order 10 cases. Worse, on every order there is a customs levy of US$20. Yachts can club together to place a single order. Beer is cheaper in Cyprus and Israel.

General

Euphoria is a common, understandable and infectious reaction to being in Suez if you have come N through the Red Sea. There are, unfortunately, drawbacks. Theft happens at the SYC, so put valuables below. It is safe to leave your yacht at the SYC if you wish to go elsewhere, though it is better if at least one crew member stays with the boat. If not, speak to your agent or the club about a caretaker service.

Agents and formalities

Egypt being Egypt, using an agent avoids an obscurantist, obstructive, bureaucratic nightmare even Kafka couldn't have invented. The agents are in general charming, helpful and, in a way we shall leave you to discover, a good introduction to Egypt. They smooth your passage and your ravelled nerves, if at cost to your cruising kitty. More, should anything go wrong in the canal or even on your way through Egyptian waters, your agent offers vital help. It is fair to pay for this. Fair too to pay more for extra services, such as a quick transit if you arrive late pm and want to leave on a one day fast track run next morning. Choose your agent ahead. Give him several days advance warning of your arrival and a rough ETA.

The agent will come aboard shortly after you arrive at the SYC with several forms for you complete. He takes passports and ship's papers for clearance, returning them in a few hours. Then, as long as you have a visa, you can go ashore. If you have no Egyptian visa but want to go anywhere outside the SYC, agents can get a visa for you. There are no shore passes. Carry your passport, which is checked every time you leave and return to the SYC compound.

Suez Agents

The Prince of the Red Sea Co
40 Gobar El Kaed St, Port Tawfik, Suez
Email princers@gega.net Contact: Heebi or Ashraf
Fax +20 62 330965, ☎ +20 62 222126, 341316, Mobile +20 12-3236530/2291049
The Prince of the Red Sea Agency is used by most boats heading N. There are discounts of 5–6% to members of certain clubs and organisations. Agree on discounts in advance, in writing. Abu Tig Marina may also arrange discounts for berth holders.

Union Ltd
Mr Ibrahim Slama has an office close to the Yacht Club and is well recommended.
Fax +20 62 334301. Office ☎ +20 62 322251, home ☎ +20 62 340982, Mobile +20 123240237. Ibrahim's English is colourful and his gestural mime of unsurpassed eloquence. He is assiduous, generous and ever present and, in addition to working in his own right, sometimes strings for Felix Maritime Agency.

Felix Maritime Agency
See Port Said entry for full details. Representatives in Port Suez are Magdy and Ibrahim Slama. Call on Ch 05 on approach to Suez. Discounts are negotiable.

Other big ship agencies will handle your transit if you ask them (there are lots) but you will pay agents' fees of US$200 and upwards.

The Suez Canal and approaches

DIY

The Suez Canal Authority (SCA) has confirmed to us that yachts can organise their own transit. You must make arrangements with the SCA to issue 'letters' enabling you to obtain or to pay:

- a customs clearance certificate
- transit fees to the SCA at the appropriate bank as directed by the SCA
- an Egyptian insurance policy for the transit (additional to any you already have)
- port and light dues, and harbour-clearance permit
- a pilot
- pilotage dues
- a security-clearance certificate
- a tonnage certificate from the SCA

We are assured this could take 'four days to a week' because 'letters' and invoices for individual applicants stay at the bottom of the heap. You will be actively obstructed by vested interests. Don't forget, if you DIY and anything goes wrong on your transit, you're on your own.

Suez Canal Transit Fees

Yachts are told regularly that fees are about to go up. This is a deliberate attempt to hustle you. Your best weapon is to ask for quotations well in advance, preferably at the same time as several other boats. Give your club affiliations and ask for a discount. Be ruthless. Insist on written itemised receipts. Shameless dishonesty, a sublime ignorance of or indifference to factual truth, a winning charm of manner and a willingness to bluster, bully and delay infinitely are endemic in Egyptian business life. Here are the basic facts. There are two parts to your fee: a) the SCA Tonnage Fee, b) agents' fees and the usual range of customs, immigration and port dues.

SCA Tonnage and measurement

The Suez Canal Authority (SCA) fee based on tonnage means the bigger the boat, the bigger the dues. The SCA tonnage formula also means for boats of the same length, a beamier and deeper draft boat pays more. The SCA's rules require physical measurement by an SCA measurer. Most do a consistent job but there's room for a scam and it is occasionally exploited. A Suez Canal Special Certificate of Tonnage (SCSCT) issued by a classification society (e.g. Lloyds, ABS, Bureau Veritas, etc.), is the safest route, but weigh the cost against your likely savings.

The Suez Canal Gross Tonnage (SCGT) formula, established by the 1888 Constantinople International Tonnage Convention, is the same for yachts as for ships. It measures internal volume as does Registered Tonnage. IT IS NOTHING TO DO WITH DISPLACEMENT. Deductions up to 10% of SCGT are made for crew spaces, engine room, etc. The result is Suez Canal Net Tonnage (SCNT). It ought to be a bit over GRT. In practice it is a lot more because yachts have fin keels which the drafters of the Constantinople Convention knew

	Port Suez (N'bound)	Port Said (S'bound)
Port Authority Fees[1]	30	128
Bank charges[2]	(5)	(10)
Port clearance	none	10
Quarantine[3]	7	2·50
Immigration[5]	10	
Customs[4]	(10)	10
Explosives and drugs check	none	12
Insurance	3	3
Duty stamps	6·50	6·50
Agent's fee	>40	>40
Total in US$	>120	>250

Notes

1. Port Authority fees are 128 units. In Port Suez these are Egyptian pounds (therefore about US$30), in Port Said US$. The Port Said fee was under discussion in autumn 2000 but as far as we can gather has not changed.
2. Since some charges are calculated in Egyptian pounds and some in dollars, bank charges may or may not arise, be prepared.
3. If Port Tewfik/Suez is your first Egyptian port of call, the quarantine fee is payable. You will only avoid it in Port Said if you have previously cleared in at Alexandria or El Arish.
4. An extra customs fee of US$10 per head is payable in Port Tewfik/Suez ONLY if there are crew changes, i.e. if one crew leaves and one joins the customs bill is US$20. The Port Said charge is a flat rate and always applicable.

The breakdown of categories and amounts may not be 'officially' correct. The table is based on information given by the Suez Canal Authority office in Port Said, from the Port Said Port Authority, from Mr Nagib Latif of Felix Agency and Captain Heebi of the Prince of the Red Sea Agency.

nothing about. SCNT is the basis for fees calculated at US$6–US$7 per ton. We don't know how the exact rate is decided.

The relevant SCA rule reads *inter alia*:

'Measure the greatest breadth of the ship to the outside of the outer planking or wales. Then, having first marked on the outside of the ship, on both sides thereof, the height of the upper deck at the ship's sides, girt the ship at the greatest breadth in a direction perpendicular to the keel from the height so marked on the outside of the ship, on the one side, to the height so marked on the other side by passing a chain under the keel.'

(Suez Canal Authority, Rules of Navigation, Pt IV, p172)

That measurement is entered in the formula:

$$\frac{(L \times (0 \cdot 5G + 0 \cdot 5B)^2) \times 0 \cdot 17 \text{ or } 0 \cdot 18}{100 \text{ or } 2 \cdot 83}$$

L = LOA (strictly length between the inside faces of stem post and stern planking)
G = girth measured as above
B = maximum beam
0·17/0·18 = factor depending on whether construction is in 'wood' (0·17) or 'iron' (0·18)
100/2·83 = factor depending on whether measurements are in feet (100) or metres (2·83)

Most SCA measurers measure girth as the rule requires. Only an SCSCT would let you by-pass this, and possibly not even that. If your measurer doesn't set to, insist.

Red Sea Pilot

You can work out your tonnage using the formula and measuring your girth as the rule requires. If you don't want the hassle and the wet rope, here's a short-cut. Enter a figure for Girth calculated as follows:

$G = 2\sqrt{(D+F)^2 + (0.5B)^2}$

where D = draft and F = freeboard at point of maximum beam

To be best prepared:
a. Ask the classification society that measured you for registration, or any classification society, how much a SCSCT would cost for your boat. If this is a reasonable figure get one.
b. Work out rough costs BEFORE you get anywhere near agents. If your boat is GRP, assume the worst and calculate the figures for 'iron'.
c. Have the relevant facts and figures to ensure fair measurement.
d. Have a line long enough to measure your boat's girth under the keel stood by for when the measurer arrives. Indicate that you know the rule and can quote chapter and verse (see above). If he doesn't start measuring insist.
e. If you have an SCSCT and the measurer's tonnage is wildly discrepant, insist on your agent appealing to the SCA. This usually works since the SCA is obliged to accept SCSCTs.

If any problem occurs with your transit fee and you feel it unjustified, complain to your agent and, if you have no satisfaction, write to the SCA via Senior SCA Pilot Captain Farid Roushdy, c/o Ismailia Yacht Club, Ismailia, Egypt, ☎ (0)64 393341 or 394341.

Other fees

Include the cost of visas if you haven't already got them. The minimum is US$15 per person plus duty stamps and any tips. The gross can reach US$25 each. Also allow for your mooring fees in Port Suez, Ismailia and Port Fouad Yacht Clubs. Finally there are dues in US$ as shown in the box above.

Inspection

A day or two before your N'bound transit is scheduled a Canal engineer comes to your boat with your agent. If you are making your own arrangements you must arrange with the SCA for the inspection to be made. He asks basic questions about the engine, safety equipment, etc. The inspection looks for:
1. fire extinguishers (at least one)
2. bilge pumps, number and location (though only one is required)
3. life jackets (sufficient for crew and pilot!) and ascertains:
4. the service speed of the yacht
5. the operation of the engine in ahead and astern

THE ENGINEER'S INSPECTION DOES NOT TAKE PLACE FOR YACHTS MAKING A S'BOUND TRANSIT.

Towing

If you have engine problems an SCA tug is at least US$1,500. If you think you can obtain and fit a sufficiently powerful outboard, this is an alternative. Towing by anything other than an SCA tug is strictly prohibited.

Weather

If strong side winds are forecast your transit may be cancelled. The canal is closed to all traffic in W'lies of 40kts and above. If sandstorms are possible you will be asked if you want to postpone.

Currents in the Suez Canal

The *Admiralty Pilot* gives the following:
November–April N-going from
Bitter Lakes to Port Said   Rate 0·5–2 knots
June–October S-going from
Port Said to Bitter Lakes   Rate 0·5–2 knots

Constancy, rate and direction are dependent on sea levels in the E Mediterranean, which may reverse the normal seasonal direction.

Tides

Spring range
S end of Little Bitter Lake   0·2m
km 146 (El-Shallufa)   0·5m
km 149 (El Kubri)   0·6m
Port Tewfiq   2·1m

Times
HW S end of Little Bitter Lake = HW Suez +50 mins

Tidal Streams

Between Port Tewfiq and Little Bitter Lake:

At Port Tewfiq

1–2hrs before HW Suez N-going stream begins

5hrs after HW Suez N-going stream slackens and S-going stream begins

1–2hrs before HW Suez S-going stream slackens and N-going stream begins

At S end of Little Bitter Lake

10mins–1hr 10mins before HW Suez N-going stream begins

6hrs after HW Suez N-going stream slackens and S-going stream begins

10mins–1hr 10mins before HW Suez S-going stream slackens and N-going stream begins

Strong S winds can affect tidal streams, prolonging N-going streams and delaying or preventing S-going streams.

Rates Fastest near km 149 (El Kubri) during springs, when rate is 1·75kn. With a strong S wind the rate may reach 2·5kn.

Note Tidal streams in the S part of the canal, if allied with a strong NW wind, will markedly affect speed over the ground. A 6kt indicated boat speed could drop as low as 3kt or less VMG. It pays to try to time departure to coincide with a morning flood tide.

Port Said to Bitter Lakes

No perceptible tides or tidal streams between Port Said and Bitter Lakes.

Suez Canal transit – N'bound

Once your yacht has been inspected and measured, you've got and paid for all your paperwork, and documents and receipts have been returned to the SCA, a transit permit is issued and you are told what time your pilot will turn up. The day before your confirmed transit date, you (or your agent) must clear out with immigration and get all the crew's passports stamped.

Day 1

Timing If you arrive late on a Thursday, or on a Friday, the earliest canal transit that will be arranged will be for the following Sunday, since government offices close on Fridays. Transit takes two days, staging at Ismailia. You can get permission to transit in one day if your yacht can sustain 8kts minimum SPEED OVER THE GROUND. For a one day transit you leave at 0200–0300 to keep ahead of the N'bound convoy. A searchlight for transit in the dark isn't necessary unless you are over 200 SCGT. Yachts have completed transit after dark although this is theoretically forbidden.

Departure Collect your pilot from the SYC jetty by dinghy before you drop mooring lines. The pilot will bring his own paperwork with him. Depart Suez from 1000–1100, after the last of the N'bound ship convoy has cleared.

Delays N'bound delays are rare and likely only to be caused by the passage of a laden VLCC which can cause major, hazardous and sudden changes in water level around it (for S'bound delays see page 257).

Speed The normal speed of traffic through the canal is 7½ knots, according to the Admiralty Pilot. Your pilot will want to go as fast as possible and then faster. 5–6 knots is probably the average speed for yachts. The distance to Ismailia is 44M and you can usually expect to arrive in the early afternoon. Don't let the pilot rush you or overtax the engine.

VHF You are required to have VHF and this can be useful for communication with other yachts in your convoy. Let your pilot know that you can talk to the other yachts and ask him to make any communication with signal stations by radio.

Pilots Most pilots will want to helm your boat for at least some of the time they are on board, but you can take charge if you prefer. Generally, the pilots concentrate on their job, are usually reliable and will speak a little English. The SCA has confirmed that you are not required to have a yacht 'pilot' either to enter or leave Port Suez or Port Said. The 'pilots' used by yachts, ranking as SCA boatswain or coxswain, are not authorised for pilotage outside the canal and inner ports. The canal proper starts at km 3·7 at the Port Said end and a similar distance inland at Port Suez. Outside in the approaches only full canal pilots operate and they only work on pleasure vessels the size of superyachts.

It is common for pilots to take you just outside the beacons that mark the 8·5m contour along the sides of the canal. Do watch as you leave the SYC as yachts have gone aground on the shoal patch E of the entrance while pilots have been at the helm. Your pilot may want to take a break during the day, but will not expect to go below unless invited to do so. Refreshment will normally be gratefully received, even during Ramadan! Sometimes the pilots bring food with them. Experiences with pilots vary. You only need two pilots, one for each day of the transit.

Baksheesh in the canal This is a fact of life in Egypt. US$5–10 is more than adequate as a tip. If you want to give anything else like a pack of 20 cigarettes, it's up to you. During the transit, your pilot may tell you to stop at one of the signal stations. This is not compulsory and is usually an attempt to obtain baksheesh for friends ashore. Such baksheesh is in the form of cigarettes. If you are a professed non-smoker who does not carry tobacco, you'll minimise hassle by informing your pilot at the outset. He may ask for his baksheesh some time before he is due to be collected by the pilot launch or dropped off at a jetty. Resist pressure. Give him your tip as he steps off your boat and the argument is shorter. He may ask for cigarettes for the pilot launch. Resist pressure. One carton of 10 packs of 20 cigarettes should more than cover all requirements if you want to be generous.

Sailing Sailing in the canal is forbidden to yachts, unless the pilot thinks it will make you go faster.

Your responsibilities Responsibility for the boat and its handling remains with you at all times. You should have an anchor, mooring lines and fenders ready. Any damage caused to anyone or anything, even if the pilot did it, you pay!

Ismailia

Admiralty chart 233, plan page 254

Approach

There is a compulsory overnight stop for all yachts at Ismailia unless you have made special arrangements (see one-day transit above). Follow the dredged and buoyed channels into the anchorage. Watch out for fishing nets in the bay. Drop off and pick up pilots at the yacht club, 30°35'·1N 32°16'·35E, 50m WSW of the conspicuous signal mast.

Anchorage

The approved yacht anchorage in 4–5m is near 30°34'·9N 32°16'·0E, clear of the approach to the spoil barge berths on the outside of the tug basin, good holding in mud. The mud in places is thin. An anchor must sink in to get a grip. A deck wash is essential. The mud can attack galvanising.

Formalities

If you have a valid visa, and have not cleared out of Egypt, you can go ashore. Be sure to tell your agent in advance to arrange the right day for your next pilot.

Red Sea Pilot

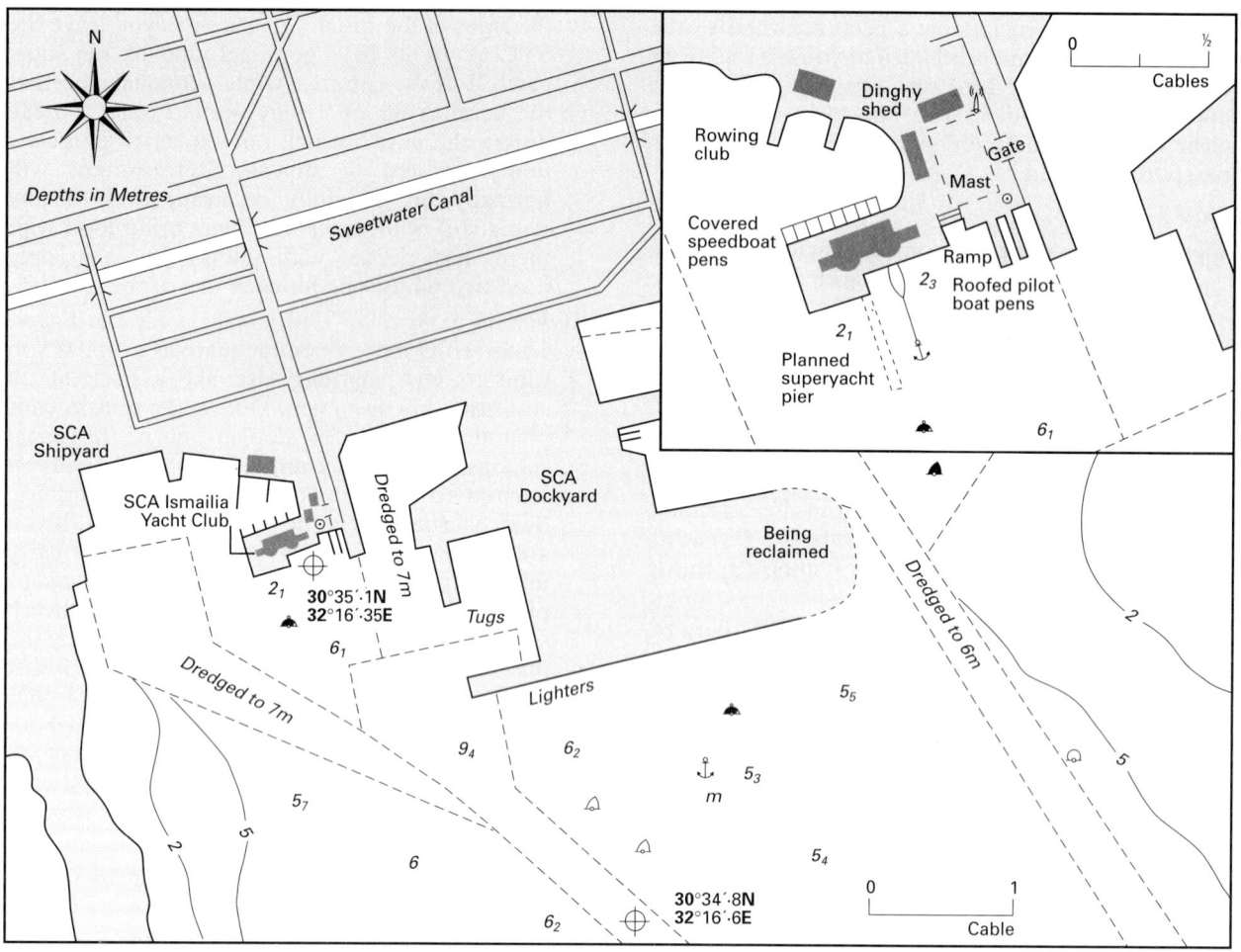

ISMALIA AND SCA ISMAILIA YACHT CLUB (INSET)

Facilities
The SCA Ismailia Sailing/Yacht Club's handsome art deco building is being refurbished and its quay brought up to international marina standards. Call VHF Ch 08 if you want to berth. There will be full service berths for 50 yachts, a coffee shop, two restaurants, showers and toilets. Future plans include additional berthing for 2–4 yachts of 30–50m. All berths have electricity and water supply points. Rates according to LOA, e.g. US$7 a night for an 11m yacht or US$100 a month, power and water included. You can winter over at competitive rates, e.g. for an 11m boat US$200 for 3 months and US$400 for 6 months with power and electricity extra. On board security costs about US$3–4 a day for periods of absence. Haul out for maintenance can be arranged using the SCA shipyard facilities near by. The manager is Senior Pilot Captain Farid Roushdy. The address is SCA Ismailia Yacht Club, Ismailia, Egypt. ☎ +20 (0)64 3933413, 39343413. If you have any complaints, advice or suggestions on canal transit and facilities, including the yacht clubs, write to the SCA via Captain Roushdy at the SCA Ismailia Yacht Club.

Fuel Until the club's planned fuel berth is ready the nearest fuel station is 500m away. From the SCYC keep headed NNW on the road from the gate, over the Sweetwater Canal, over the main road and it's on your right.

Good shopping, markets, email and telecoms in the town. The autumn is the mango season and Ismailia is deservedly Egypt's mango capital. You can also try dates at all stages of maturity; an interesting taste experience. Visas can be extended cheaply at the police station.

General
All the canals have always passed somewhere close to Ismailia. The ancient Heroopolis was nearby though no ruins have been found to establish exactly where. Ismailia was founded by and named after Ismail Pasha, the Egyptian ruler at the time when the Suez Canal was built. The SCA fiefdom on the E side of the town, with all the handsome late-19th-century executive housing, is very attractive. The town centre is a bit scruffy. There are good road communications with Cairo an hour or so away and this is an excellent place to leave your boat to visit the pyramids, etc. If you anchor off, do not leave anything, e.g. buckets and fenders, loose on deck. The fishermen from the lakes on the S side of Lake Timsah are prone to make the rounds of yachts in the small hours. Don't put temptation in their way.

The Suez Canal and approaches

Day 2
The second day of the canal transit, assuming that you are N'bound, takes you to Port Said, and, unless you want to stop, out into the Mediterranean. You will leave at 0800–0830 to stay ahead of the N'bound convoy. Some yachts have had trouble in the final approaches to Port Said. A police boat may come alongside, ostensibly to search for weapons and drugs. The officials, who are unlikely to be in uniform, may ask you to go alongside the customs police wharf (see small-scale map of Port Said). Boats' experiences vary and baksheesh is always the answer.

If you are stopping in Port Fouad the pilot disembarks at the yacht club. If you are not stopping, experiences differ. We dropped our pilot at the pontoon at the SCA building and that's what to plan for. To avoid pilotage hassle, resist dropping your pilot onto a pilot boat. If one approaches insist it is not necessary. If the pilot boat threatens damage, head for the yacht club and squawk loudly on VHF Ch 12 for help from Felix Agency.

Port Said and Port Fouad
Admiralty chart 233, plans page 256

Approach
From the N and E this is easy enough, the only problems are oil rigs, fishing boats and canal convoys. The last are assembling or on the move S'bound 2330–0500 and again 0630–0900. The N'bound convoy exits at any time from 1530–2230. If you are S'bound, time your arrival to be in Port Fouad around dawn. Your arrival waypoint should be in approx 31°25'N 32°15'E, clear NW of the anchorage and assembly area. Skirt the area anticlockwise to reach approx 31°21'·3N 32°21'E where a sub-channel leads to the Port Said entry channel at approx 31°19'·3N 32°21'·9E not far from the El Bahar light tower. It is discouteous and poor seamanship to manoeuvre in the canal or approaches when big ships are assembling at slow speed.

If you are N'bound you usually arrive late afternoon from Ismailia which makes for a trouble free exit unless heavy canal traffic has caused an 0300 early convoy from Port Suez. From Port Fouad Yacht Club time departure either to clear the entry channel after the end of the first S'bound convoy (about 0500) but before the assembly of the second (about 0630), between the S'bound and N'bound convoys (0930–1530), or between the end of the N'bound convoy and the beginning of the first S'bound (2230–2330).

Dangers
Beware of trawlers in the channel within the breakwaters into the inner port. About 50m behind their COURSE MADE GOOD the trawl is marked on the surface by a marker buoy, usually made of a jerry jug. This looks like a bit of awash jetsam until you notice it is moving in synch with the trawler. The marker is not always directly astern of the trawler because of the current. If you are heading to Israel beware the fishing fleet in the approaches to Port Said. It usually congregates around the 20m contour line near 33°E.

Anchorage
See Port Fouad Yacht Club below.

Note: N'bound yachts that have become weather bound have anchored overnight at Port Said, in the basin N of the Port Fouad dockyard, without checking in again.

Formalities
See Agents and Formalities below.

Port Fouad Yacht Club

Approach
From the S look for the very conspicuous twin minarets of the Port Fouad mosque about 150m before the Port Fouad YC basin, immediately inland from the ferry terminal. From the N the shipyard just N of the PFYC is conspicuous for its cranes and the PFYC itself has a tall flagpole. The greenery of the trees around the club also stands out.

Berthing
Keep clear of the S side of the basin which is a busy SCA shuttle route. There are two rusty mooring buoys for the bow lines of large yachts about 50m off the ends of the outer fingers. Avoiding the buoys, lay an anchor towards the mouth of the basin and, given the parlous state of the fingers, aim to put your stern just between the ends of two so you can get ashore. There is safe space for only three or four yachts. The staff, usually on duty till 0800–1800, are friendly and helpful with lines. Holding, in softish mud, is good with enough scope. If all the berths are taken anchor more towards of the shipyard and take a stern line to the wall or anchor off. Either way be sure to check through the Centre when going ashore. There is a constant surge in the basin from the SCA shuttles and traffic in the Canal. The fingers are derelict. It is very dirty with refuse and bunker oil. A minimum duration stay or no stay at all makes best sense until better berthing arrangements are provided.

The office is in the smaller building shown beside the dock.

Berthing fees
Rates at Port Fouad are an outrageous US$10 a day for unsafe moorings and filthy facilities. Chargeable days start at midnight. As a concession, if you stay just one night you will only be charged US$10. But if you stay more than one night, you will be charged US$10 for every midnight you have spent in Port Fouad. That is, arrive at 1900 on Day 1 and leave at 0900 on Day 2 and you pay US$10. But arrive at 1900 on Day 1 and leave at 0900 on Day 3 and you pay US$30! Felix Agency has the concession to run the facilities and collect dues. Security at the PFYC is good but you will be pestered for baksheesh.

Red Sea Pilot

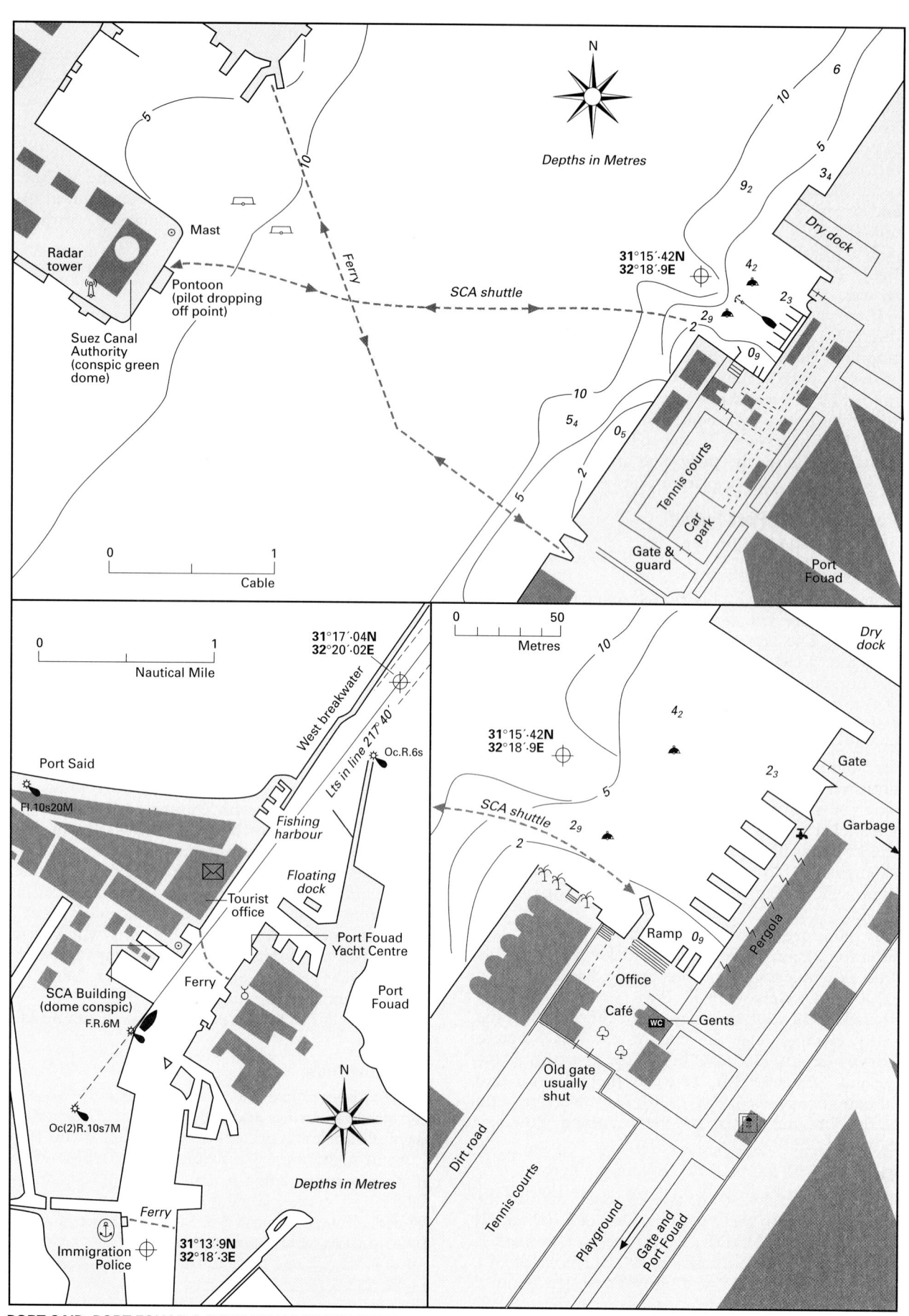

PORT SAID, PORT FOUAD & PORT FOUAD YACHT CENTRE

256

The Suez Canal and approaches

Facilities

Shore power is available from 7 decidedly dodgy looking power points under the eaves of the pergola. At the café only soft drinks and lunch (by arrangement) are available.

Fuel Unless you are a super yacht needing bunker, fuel is jerry jugged from a Port Fouad gas station. Felix Agency will handle this. You will pay street prices plus a tip. If you DIY you will still have to pay a tip to the security guard at the gate and may not succeed in offering enough.

Water Water is available on the dock. There is a noisome shower and WC in the same building as the office (men only unless a burly crew ensures privacy). The main facility (see sketch) is often filthy, in disrepair and neither maintained nor cleaned.

Money There are ATM's at the major banks in Port Said. Sometimes moneychangers offer better rates.

Stores There are good markets ashore, and many duty-free shops with electronic goods, etc. in Port Said. Your agent will provision for you on request.

Communications There are now cardphones everywhere and the phonecards are readily available. The ferry across to Port Said is free. There are good bus, train and service-taxi links with Cairo.

General

Port Said is a scruffy but extremely alluring town with some wonderful, dilapidated old buildings, including the crumbling, exotic fantasy de Lesseps had built for himself. The town was built by Said Pasha when work on the canal began in 1859. The imposing green domed and columned building on the W waterfront is the HQ of the SCA. There are German and American consulates.

Agents and Formalities

Felix Maritime Agency (Mr Nagib Latif), Head Office, Al Gomhoria St, Post Tower Building, 4th Floor Office No 12, PO Box 618 Port Said. Contact Felix Agency on VHF Ch 12 or 7070kHz on approach. ☎ +20 66 333132, 333165, 348772; *Fax* +20 66 333510, 347875, 402443; mobile 012 2119365 *email*: felix@felix-eg.com

Nagib Latif is the only agent dealing regularly with yachts. Even if you have arranged a S'bound transit with the Prince of the Red Sea, Felix Agency will actually handle things in Port Said. Send a fax at least a day or two before arrival with copies of the ship's registration and a crew list with passport details. This is vital if you want to make a swift transit. With Felix Agency payment can be made in Egyptian pounds or US$. Egyptian visas cost US$15 with additional charges depending on duty stamps and service fee. For bona fide members of clubs and associations discounts can be arranged. Try yours and see. You may be asked for evidence of membership and you must apply for a discount in advance. Agree on the fee in writing. Felix Agency has been recommended by most customers. Unfortunately the two principal Suez agents have a duopoly so, unless you are firm and know the system, you can expect to be milked.

Port Police and Immigration do not board yachts. You will meet them with your agent ashore. You'll need lots of crew lists. If you want to go outside the PFYC you will need a visa. A health officer may wish to see a vaccination card on arrival. Exactly what is required is not obvious, and they are generally easily satisfied. All paperwork is completed at the PFYC.

The difference in fees S'bound compared to N'bound is a function of the accounting unit for port dues (see table on page 251). S'bound yachts are treated as commercial vessels and charged in US$. N'bound yachts are treated as tourist craft and charged in £E. Now you know.

Canal-transit arrangements

S'bound from Port Said you leave from 0900–1100 after the last of the second S'bound convoy. S'bound delays depend on N'bound convoy movements. You may be delayed in the El Ballah W channel on Day 1 and in the Great Bitter Lake on Day 2 where you may have to anchor. You will collect your pilot from a launch in the approaches to the Port Fouad Yacht Centre. From Ismailia departure is at 0900 after the first S'bound convoy is past. The pilot will come to the yacht club and you must pick him up in a dinghy if you are anchored off or motor in. See entry on Ismailia above for further information. If you arrive late on a Thursday or on a Friday your agent will not be able to arrange for you to transit the canal until the following Sunday, since government offices close on Fridays. Felix Agency has a good reputation for helping out if you have any problems either in the canal or with the authorities in Egypt. You pay for the service but it can be very useful. See section above on Suez Canal transit for further information.

8. Northern approaches to the Red Sea

For rest, repair and recuperation after sailing up the Red Sea or for final storing and maintenance before a long voyage S' wards the best spots are the closest marinas in Israel and Cyprus. This section includes brief details about them. The 'Countries' section also has background information on Israel because of Elat. We do not include Turkey's excellent marinas, which are two days further N. If that's where you are headed see the latest edition of Rod Heikell's *Turkish waters and Cyprus Pilot*.

The more adventurous may consider further cruising on the Egyptian Mediterranean coast for which we give brief details first.

Red Sea Pilot

Egyptian Mediterranean coast

Admiralty Charts
Medium scale 2573, 2574, 2578, 2681, 3325
Large scale 243, 2578, 3119, 3326, 3400
Small scale 4302, 183, 3400

E and W of Port Said on the Egyptian Mediterranean coast there are several havens which could be used by yachts. Most are shallow, especially at their entrances, which are difficult in fresh to strong onshore winds or swell from bad weather offshore. Too great a reliance should not be placed on aids to navigation.

E of Port Said

El Arish
31°10'N 33°48'·5E

Note The Customs fee (see page 251) is imposed here.

Keep outside the 10m line until N of the town and approach on a roughly SSW line. The coast is very low lying. There is a minaret on a mosque close to the shore NNW of the town. Two beacons on the shore about 1·5M E of the town mark the entry line. Holding is in soft sand. Until the late 1980s this used to be an open anchorage. Since then a small harbour has been built but it suffers badly from ground swell. The East Med Yacht Rally visited here in early 2000 and reports on its comfort as a harbour were not glowing. Only curiosity would suggest a stop. El Arish has a beach resort and can be crowded in summer. There is a Bedouin market on Thursdays.

W of Port Said

Dumyat/Damietta/Ras El-Bar
Damietta Mouth of the Nile 31°31'·6N 31°50'·8E
Dumyat Port 31°30'·5N 31°45'·7E
Admiralty Chart 2578A

Keep outside the 10m line until the harbour's entrance is open. This is a bar harbour and, if there is much of a swell, entry is difficult. It is one of the two main mouths of the River Nile. The Dumyat or Damietta Branch runs for 245km until it joins with the 'Rasheed' or Rosetta Branch 20km N of Cairo. The Damietta Branch carries the lesser flow and is consequently more shoal. Most activity has now shifted to the new commercial port of Damietta (Dumyat) 5M SW, marked by a dredged, buoyed channel 7M long. This would be the place to head if you needed assistance, though it has a nasty reputation for chicanery.

Rosetta
31°28'N 30°21'E
Admiralty chart 2681

It seems likely that the approaches to the river have changed dramatically. Before you consider visiting, ask in Alexandria for the most recent information on accessibility. The eponymous Rosetta Stone was found at nearby Rashid. It had text in three languages and four scripts, hieroglyphic, hieratic or cursive, demotic or Coptic, and Greek.

The Nile
The River Nile, at some 6500km, is one of the longest rivers in the world. 1500km of its length lies within Egypt and, of that, a good 1200km or more

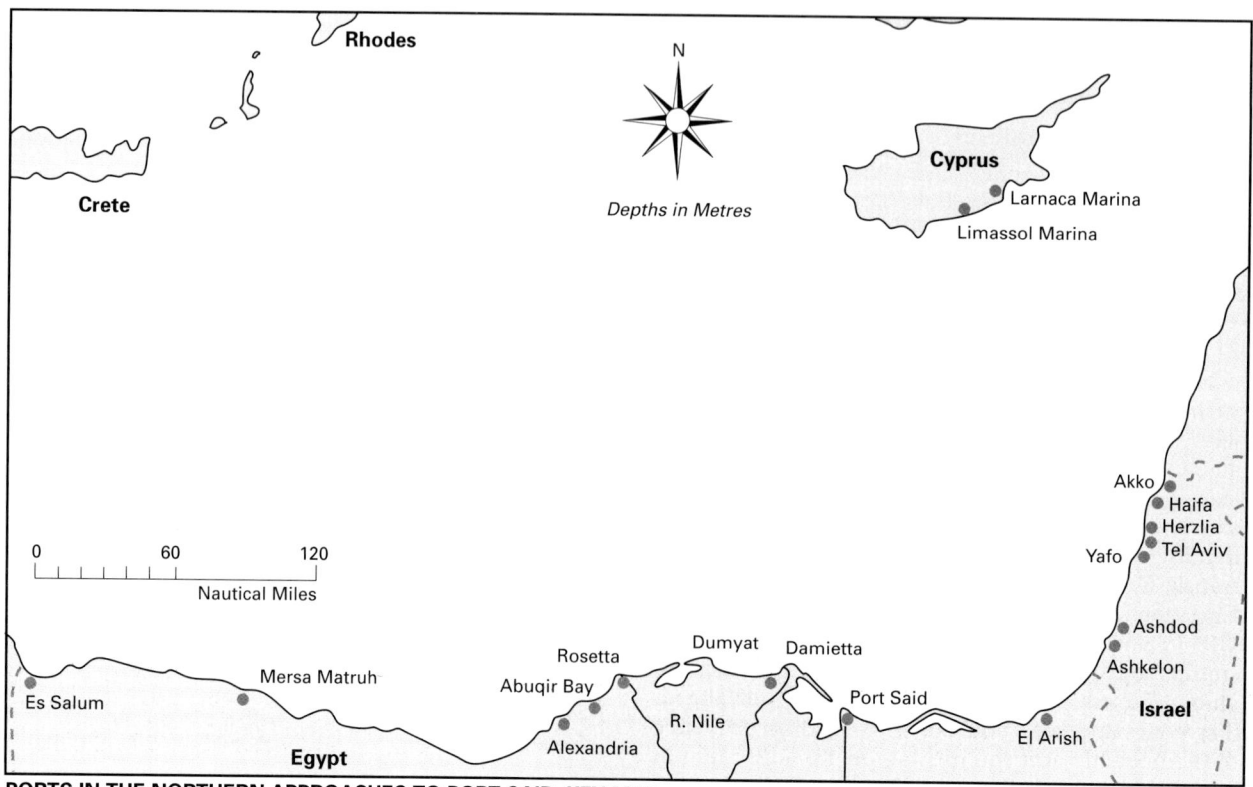

PORTS IN THE NORTHERN APPROACHES TO PORT SAID: KEY MAP

258

Northern approaches to the Red Sea

is navigable by small craft. A boat that enquired about sailing up the Nile in 1999 was asked for a deposit equivalent to the value of the boat to cover the issue of an entry/transit permit and, as we understand it, as a form of insurance to pay for any damage or costs consequent on a breakdown. From all accounts exploring the Nile by yacht at present would at the least be very expensive and entangled in red tape. According to Nagib Latif of Felix Agency it isn't in fact possible but you never know. We do know that one couple did the journey a decade or so ago in a Wayfarer dinghy. So it might be easier to sort out permission for a small, shoal draft boat. All this being the case, we have omitted pilotage details which appeared in early supplements to the first edition.

Abuqir Bay
31°21'N 30°05'E
Admiralty Chart 2681

The approach is open and free of dangers. Like most of the anchorages along this coast, Abuqir Bay should be avoided in onshore weather. A small new harbour has been built in the SW of the bay at El Ma'diya, approx 31°16'·3N 30°09'·2E but pilotage is compulsory. Call authorities on VHF Ch 69 if you need to go in. Otherwise the bay is shallow, depths are 3–6m over sand and mud with a >10m area in the middle. It is now criss-crossed with oil pipelines and there are several production platforms. There is a busy summer resort. The bay is most sheltered and nearest to facilities at the W end, W of Nelson's I. For the historically minded it was here, using typically startling tactics, that the British Admiral Nelson put paid to Napoleon Bonaparte's dreams of eastern dominions. The same battle also gave us the poem *Casabianca* by Mrs Felicia Hemans. It begins: 'The boy stood on the burning deck whence all but he had fled. . .' and commemorates the courage of the 10 year old son of the Comte de Casa Bianca, the flag captain of the French flagship *l'Orient*. Casa Bianca was gravely wounded early in the battle. His son stood by him whilst the *l'Orient* burned and finally blew up with a huge explosion.

Alexandria (El Iskandariyah)
Admiralty Charts 243, 3119

Approach

The approach to the E harbour where the Yacht Club of Egypt (YCE) is to be found is straightforward. Use the W entrance (300m wide) which has a least depth of >3m. The E entrance is shoal. The entrance is difficult in onshore weather. There are two mooring buoys in the middle of the harbour. Around the harbour shore itself there are several conspicuous minarets and a monument. The Abu-el-Abbas Mosque is floodlit at night.

Depths and holding

In the best anchorage on the W side there is 2–5m over sand and mud, but the harbour is very shoal

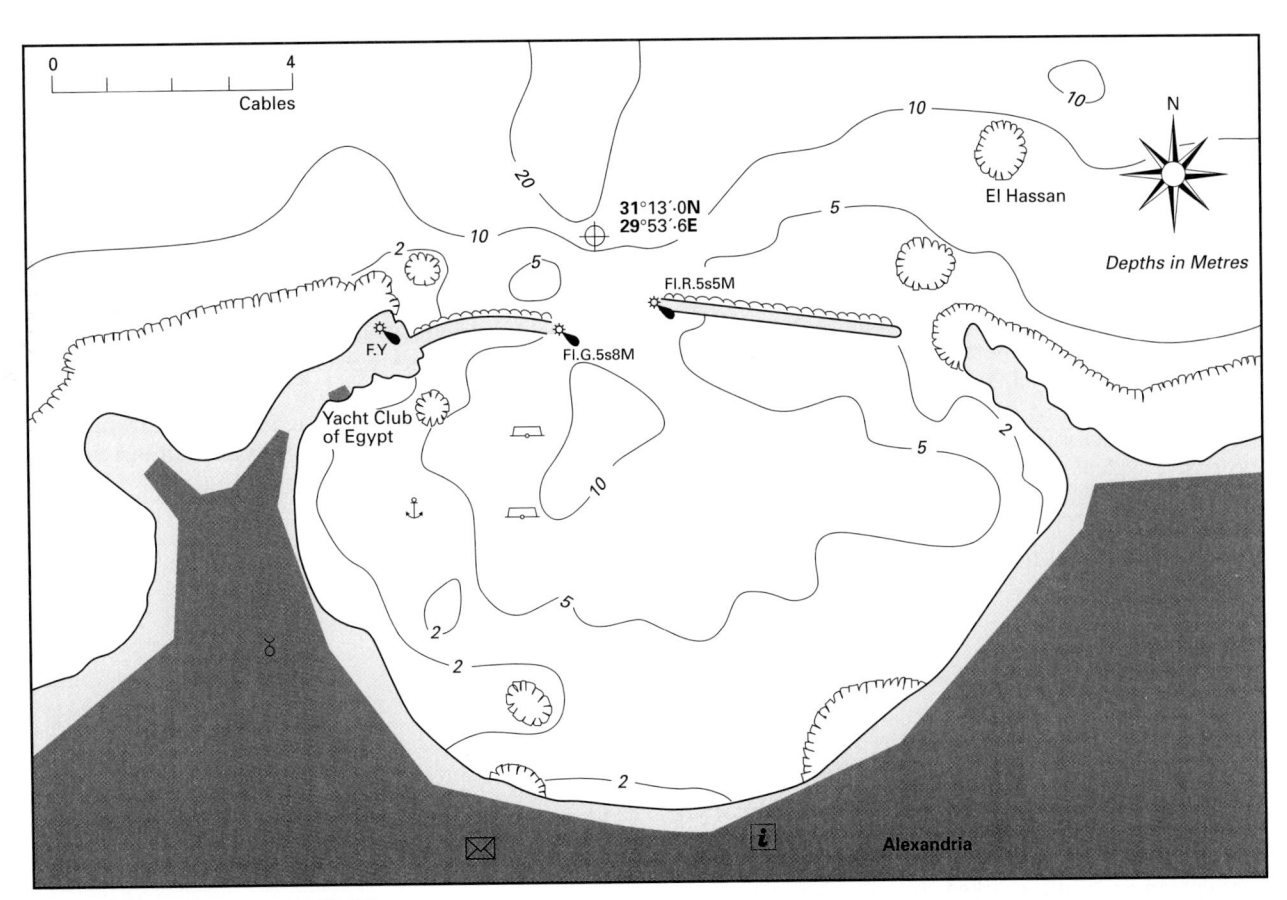

EASTERN HARBOUR, ALEXANDRIA

towards the edges. The E harbour can be markedly unpleasant in N'lies when it would be imprudent to leave your yacht untended.

Formalities

The coastguard is likely to call as soon as you drop anchor. Subsequently you may be visited by the YCE boatman with the visitors' book. If the boatman shows up, enter the names of the crew in the visitors' book and then go with him to the club, taking with you all passports. The YCE Secretary will help you sort out customs, immigration and quarantine. There is a coastguard representative at the club to help. Finally, the law requires that the YCE issues a 'letter of guarantee'. There are fees attached to all these procedures including the new Customs fee.(See page 251). Make sure you establish beforehand what they all are. If you use the YCE's services and facilities you will reasonably be asked to pay a visiting member's subscription. You can, of course, go it alone and sort out all formalities yourself. This would be discourteous and will involve you in finding and paying for an agent and wasting a great deal of time.

Facilities

The YCE has showers and a restaurant. It is open 0900–2100 in winter and 0900–midnight, spring to autumn. You can berth (stern-to) on the YCE jetty for watering ship but, be warned, it is shallow-to. There is a standard charge. Neither water nor fuel are available in unlimited quantities. The YCE slips are for small craft only. Private slips handling larger craft are nearby. Fees are standard market rates, i.e. negotiable. All manner of provisions can be obtained, as can any goods or service you might be likely to require.

General

Alexandria itself is one of the world's great cities, and is very Egyptian. It is Egypt's unofficial summer capital, cleaner and easier to get along in than Cairo. It makes a good base for exploring the Nile delta region or, for those who are so inclined, for visiting the battlefield and cemetery of El Alamein, one of the major turning points of the Second World War. Sadly Alexandria is yet another example of the Egyptian capacity for turning a friendly visit into an expensive ordeal. This ought to be one of the great E Mediterranean ports of call and a must for cruising yachts. Today only think about it if you are deep pocketed and fully battle hardened for coping with Egypt.

Mersa Matruh

31°21'·5N 27°15'E
Admiralty Chart 3400

Approach

Keep well offshore until you have identified the entrance. A waypoint on the 10m line and the leading line is 31°22'·25N 27°13'·86E. The entrance line is on 210° between two beacons, least depth in the entrance >7m, marked by two horizontally-striped beacons on the S shore of the mersa. Once inside, after about 5 cables, the mersa divides. A buoyed and beaconed channel leads on approx 260° to the commercial and military harbour. A more complex buoyed channel threads E to the old harbour off the town. Two buoys mark the first alteration to 169°, its line indicated by two beacons immediately E of a mosque. At the next pair of buoys alter to 129° on a further set of leading marks. A further pair of buoys mark the next alteration to 090° on leading marks with triangle topmarks, one point up, the other point down. A final pair of buoys and leading marks take you on 053° to the old harbour. All or any of these marks may be missing. The entrance is through a narrow pass in the reef surrounded by many shoals and is not feasible in bad weather or poor visibility.

Depths and holding

In the E, old harbour 5–6m, sand and coral or coral. The commercial and military harbour in the W harbour has reasonable depths only at the W end. You may have to go there to complete formalities. Once inside either arm there is good shelter in all weather.

General

Don't head for Mersa Matruh before seeking guidance in Alexandria. Although the Libyan border is 200km W, Mersa Matruh counts as close to it. If Egyptian/Libyan relations are tense, your presence may not be wanted. The mersa is much like the Red Sea equivalents, save the greater development of the town as a resort. Attempts to attract international tourism haven't really worked and most of the visitors are Egyptians.

Es Salum

31°33'N 25°10'E
Admiralty Chart 3400

The approach is clear of dangers. Look for the mosque and lighthouse at the harbour and a pagoda on the point ENE of the town. Anchor in 2–6m sand off the small jetty to the W of the lighthouse, or, if you draw <2m and there's room, go stern-to on the S side. This is a small 'ras' anchorage and hence open to swell if any is making in from N through SE. This is the border town. Don't visit without finding out in Alexandria if it is all right to do so.

Cyprus

Admiralty charts
Large scale 848, 849
Medium scale 850, 851
Small scale 2074, 183

Larnaca Marina
Admiralty chart 846, plan below

Approach
Approach from the SE is straightforward, but see dangers below. Larnaca is hard to miss. The marina is S of the main commercial harbour where the cranes stand out well. Note that the light you are looking for on the marina's E wall is red and may be obscured by the town lights. Call Larnaca Marina on VHF Ch 16 or 8 on approach. Let them know that you want to clear in.

Dangers
Keep at least 3M clear of Cape Kiti, shoal water may extend much further than charted, though in daytime locals use a clear channel close in. Several unlit mooring buoys and wrecks NE of the commercial port need watching if you're coming from E.

Formalities
Port of entry. Mediterranean moor (drop an anchor and go stern-to) at the arrival wharf and wait. Formalities are quick, efficient and courteous but try to arrive well before 1400 on a weekday or customs will charge overtime. The marine police will retain passports and issue shore passes. Guards at the entrance will check the passes until they get to know you.

Facilities
350 berths, maximum draught 3·5m, maximum LOA 35m. The marina is now hopelessly crowded and vacant berths rare. Until there is a vacancy you must wait at the arrival wharf. In even moderate E'lies it is lively, dangerous if they become fresh, and holding is not always good. Use plenty of scope and check your anchor to make sure it is in. If the wind goes E and freshens, move inside immediately and find a gap even if you have to anchor temporarily. Tell the marina staff as soon as you can! Booking ahead is appreciated, though makes no difference to how swiftly you get a berth.

Facilities
All services that you are likely to need are in the town or the marina. Water, fuel, gas, electricity, showers, laundry, telephones, fax and mail. 40-ton travel-hoist (maximum beam 4·9m), repairs, maintenance, chandlery, restaurant, bar. The marina and Larnaca generally are very friendly and the many marina residents are helpful. However, one warning. In the ablutions block there is a counter, called the 'carousel', where people leave unwanted things for others to take. Whatever's on the carousel is assumed to be up for grabs. Turn your back and anything desirable has grown wings and flown.

Contact Larnaca Marina, Larnaca, Cyprus ☎ +357 4 653110/653113; *Fax* +357 4 624110.

General
Larnaca is ancient Kition, home of Zeno the founder of Stoicism. The other Zeno, of the famous paradoxes, came from Elea in Greece. The waterfront of the modern town is a very 'British' tourist resort and something of an acquired taste. Behind it is a bustling Cypriot market town.

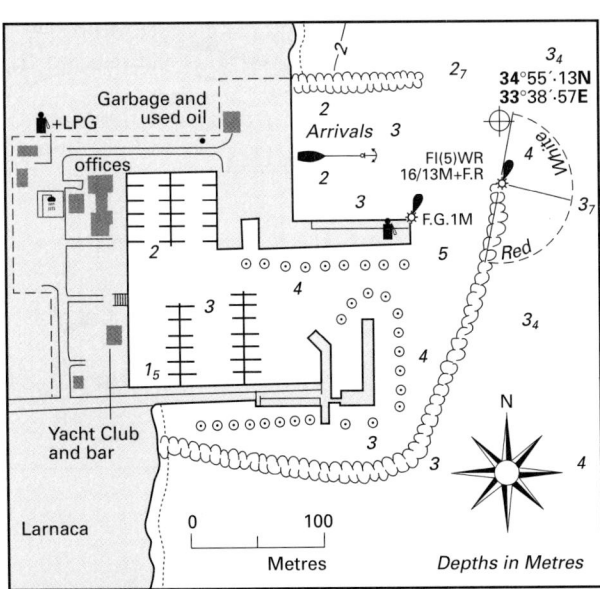

LARNACA MARINA

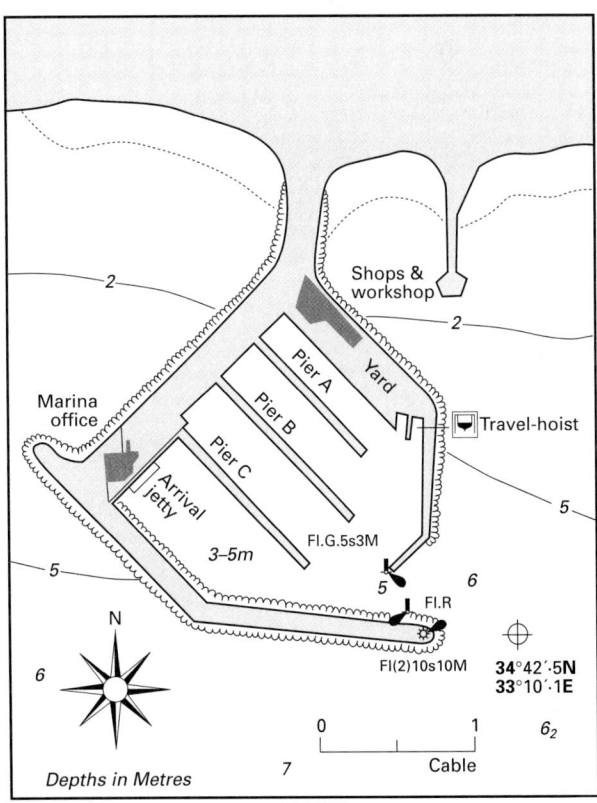

LIMASSOL - ST RAPHAEL MARINA

Limassol – St Raphael Marina
Admiralty chart 846, plan page 261

Approach
The hotel, the breakwater and the masts of yachts in the marina show up well in daytime. At night there are lights on either side of the entrance, though the lights of Limassol town and commercial harbour are brighter. The lights ashore may make the lights on the breakwaters, which are comparatively low-powered, difficult to pick out.

Moorings
227 berths with laid moorings, maximum draught 5m, maximum LOA 30m.

Formalities
Port of entry. You can clear into Cyprus at the marina. Call the marina on Ch 16 on your way in and let them know you want to complete clearance. Go alongside the arrival jetty and check in with the marina office when you arrive.

Facilities
Water, fuel, provisions, showers, laundry, electricity, telephones. 60-ton travel-hoist, repairs, maintenance and chandlery, restaurant, bar etc. 24hr security.

Contact St Raphael Marina, Limassol, Cyprus. ☎ +357 5 321100.

General
This is quieter but more expensive than Larnaca and not as conveniently placed, but haul-out and hard standing may be competitively priced.

Israel (N to S)
Coastal cruising is forbidden in Israeli waters

Admiralty charts
Large scale 1585, 1591
Medium scale 1585
Small scale 2573, 2634

Akko (Acre)
Plan below

Approach
If you are coming from the N (Lebanon), keep 1½–2M off the coast to avoid the Semeireiye Reefs, which lie about 5M N of Akko. Vernon Reef, 1M W of Akko, does not pose any problem. Keep at least ¾M off the point on which the lighthouse stands to avoid the shoal water N of Manara Rock. Call Marina Akko on VHF Ch 16 or 11 on approach. Akko's Ottoman walls make an excellent radar target. The minaret in the town and the tower on Manara Rock are conspicuous.

Marina Facilities
75 berths, maximum LOA 12m, maximum draught 2·5m. Water, electricity, showers, telephone, restaurants, 24-hour security.

Contact Gideon Shmueli, Akko Marina, PO Box 1086, Old Akko, Israel ☎ +772 4 919287/04 917400

General
Akko is one of the world's oldest towns, and has a picturesque Arab quarter adjacent to the port. PO, shops and telecommunication services can be found nearby.

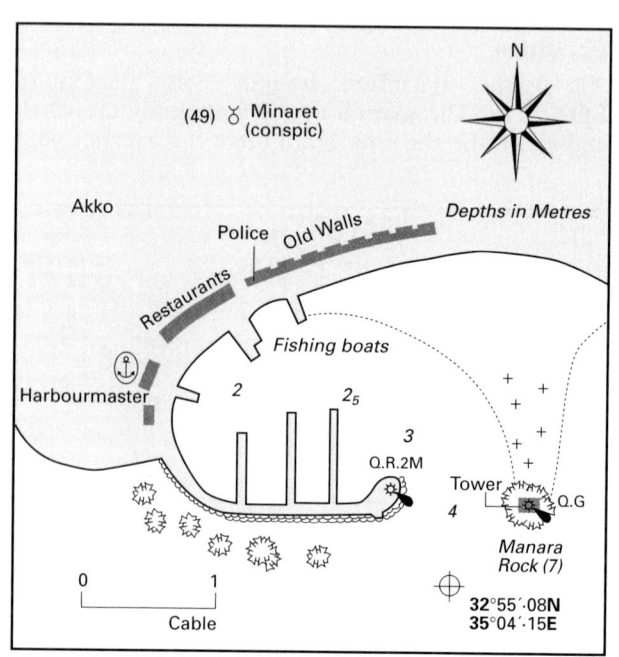

AKKO MARINA

Haifa (Qishon)
Admiralty chart 1585, plan below

Approach
Approaching from the S, keep at least 1M off the coast until Mount Carmel bears approximately SE. Make the turn E only when about 1½M NW of Cape Carmel. Turn E to the N cardinal buoy, then ESE until the N breakwater and conspicuous chimneys in Qishon harbour bear 170°, whence steer 170° into the harbour. From the W, steer to pass 2M N of Cape Carmel, then as above. From the N, keep clear of the headland W of Akko and aim at the E and highest end of Mount Carmel. Contact the harbourmaster on VHF Ch 16, 12, or 14 on approach for moorings in Carmel Yacht Club, Qishon harbour. This is part of Haifa port but ½M ESE of Haifa harbour.

Mooring
The alternative to a mooring at the yacht club is to contact Israel Shipyards Yacht Park, on the pier on the N side of the harbour, at the Shipyards' premises. They have a limited number of moorings available.

Formalities
Port of entry. If you intend to clear in here, call Israeli Navy on VHF Ch 16 as soon as you are within VHF range (about 30–40M out). Ch 11 is the working channel. They'll ask for details of your boat and crew, your last port of call and your ETA and will monitor you as you close the coast. Frontier police, customs and the coastguard will come to you to do clearance when you arrive. You may be subjected to a polite search. Landing permits are usually issued in exchange for passports. Let the officials know if you need your passport for the bank etc, and if you are going S to the Red Sea to avoid getting it stamped.

Facilities
Water, provisions and repairs are available at the yacht club. Hard standing is available for extended periods or for repairs at the Israel Shipyards Yacht Park, of which the authors have heard excellent reports. You can DIY if you prefer. Cranes are used for haul-out. They can lift up to 30 tons with their own lifting straps and up to 50 tons if your boat has her own lifting eyes.

Contact David Avraham, Israel Shipyards Ltd, PO

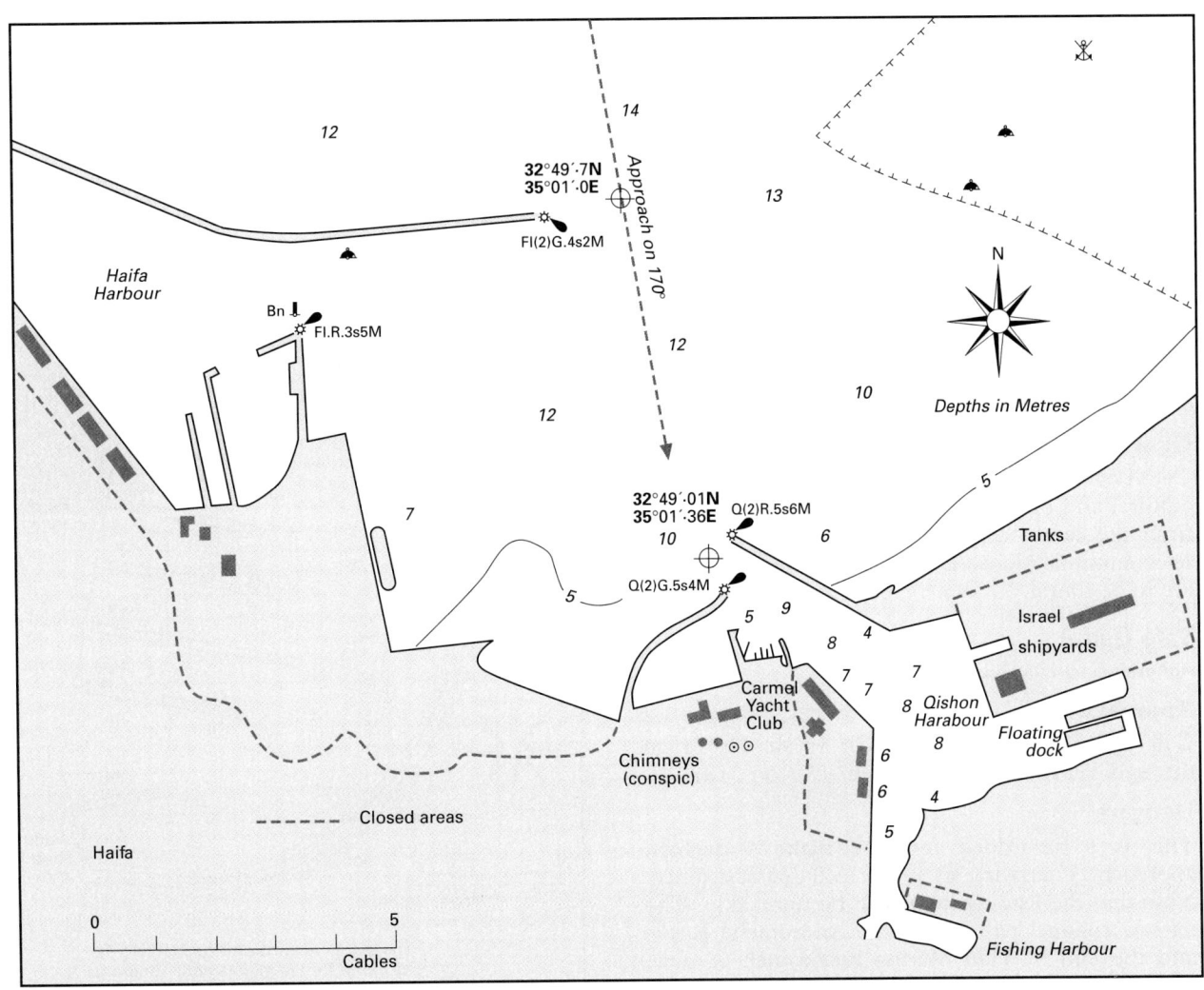

QISHON HARBOUR, HAIFA

Red Sea Pilot

Box 1282, Haifa 31012, Israel ☎ +972 4 460323/460111; *Fax* 4 410572

General

Haifa is Israel's main port; the town's centre is WNW of Carmel Yacht Club. There are regular local buses and good domestic and international transport and telecommunication links.

Herzlia

Admiralty chart 1591, plan opposite

In fresh or strong onshore winds there are breaking seas in the shallow approaches. 800 berths. Facilities include fuel, power, water and a 60-ton travel-hoist with repair services. Max LOA 25m.

Contact Herzlia Marina, PO Box 5881, Herzlia 46100 ☎ +972 9 9565595; *Fax* +972 9 9565593; *email* info@herzliya-marina.co.il

Tel Aviv

Admiralty chart 1591, plan opposite

Approach

In fresh or strong onshore winds there are breaking seas in the shallow approaches. The entrance is hard to spot in daytime and at night the marina's lights are lost against the background of lights ashore. Contact the marina, on VHF Ch 16, 11 during office hours, on approach.

Formalities

Port of entry. See the above entry for Haifa if you are clearing in here. Wait for clearance at North Jetty A after contacting the marina office on arrival.

Facilities

Maximum LOA is 25m. Maximum draught is 2·5m. Water, provisions, repairs, electricity, crane, chandlers, fuel, WC and showers, telephone, restaurant and bar, mail, garbage, security.

Contact Marina Tel Aviv, PO Box 16285, Tel Aviv ☎ +972 2 5271467/2596; *Fax* +972 2 5272466.

General

Tel Aviv is Israel's largest city and its financial, cultural and business centre. The marina is not far from the centre of town, where banks, embassies, telecommunications, PO, shops, information and are to be found.

Yafo (Jaffa)

Admiralty chart 1591, plan page 265

Approach

Call Yafo Marina on VHF Ch 16 or 11, or go through Tel Aviv Radio, call sign *4XT* on 2182.

Dangers

This is a hazardous entry so make a daylight approach. If arriving at night, call for assistance. Note that the leading lights can be unreliable. The narrow channel passes between Andromeda Rocks and the end of Yafo Marina breakwater. A new, safer, dredged channel comes from NNW between

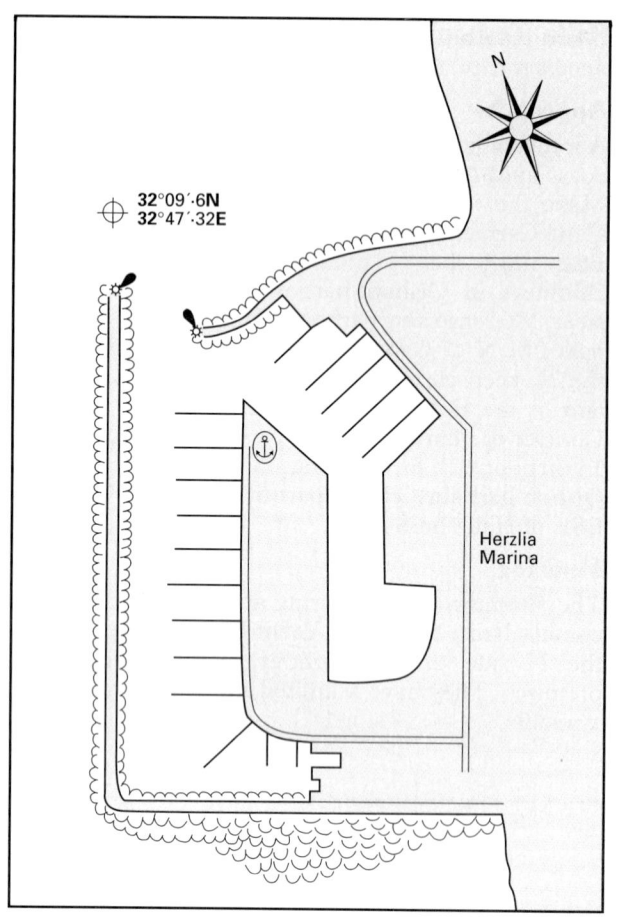

HERZLIA MARINA

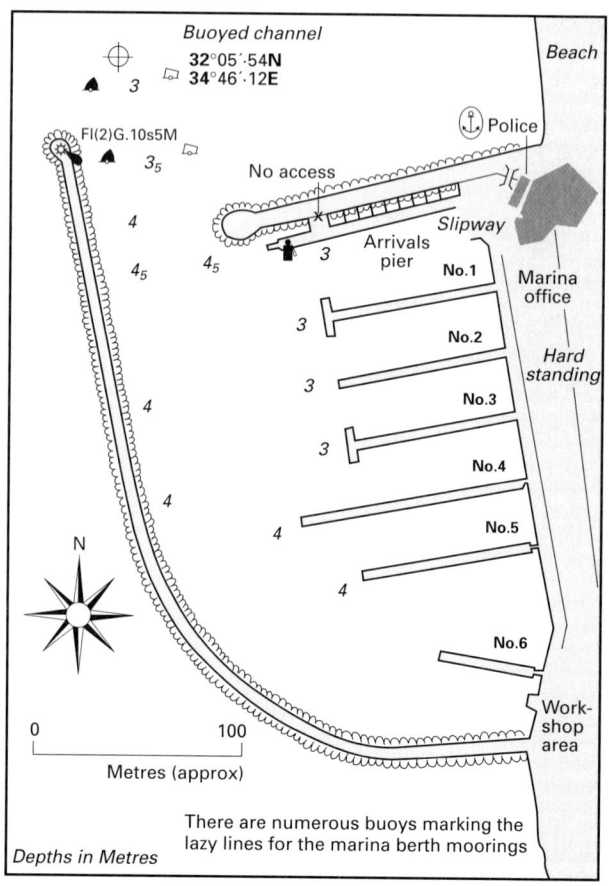

TEL AVIV MARINA

Israel

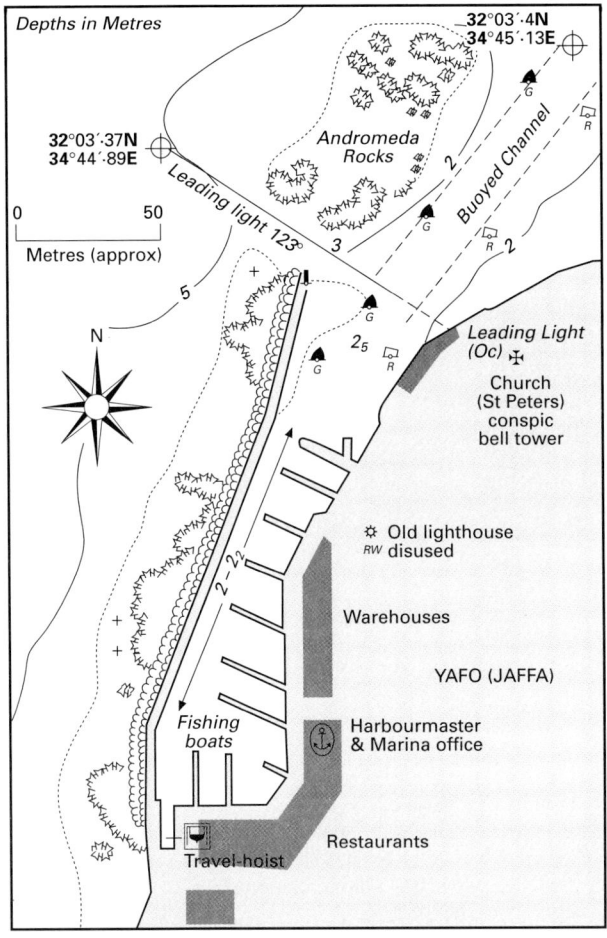

YAFO MARINA

the shore and Andromeda Rocks but be careful even by day. Yafo Marina is also a fishing harbour, so be prepared for fishing-boat traffic.

Facilities
160 berths, maximum draught 2m, maximum LOA 16m. The maximum access depth is 2·2m. Electricity and water are on the pontoon. 35-ton travel-lift, restaurants and customs house.
Contact Yafo Marina, PO Box 6041, Tel Aviv 61010 ☎ +972 3 821102, or 5 227133.

General
The old town around the port is attractive, and has been compared to Greece. This is believed to be the world's oldest working harbour.

Ashdod
Admiralty chart 1585 plan opposite

The commercial harbour is the port of Tel Aviv. There is a small harbour serving a conspicuous power station approx 1M N of Ashdod. Don't confuse the two. The new marina in the S part of the port is reputedly the least expensive in the E Mediterranean. However, Ashdod is NOT a port of entry for yachts. You must go to Ashkelon first to clear.

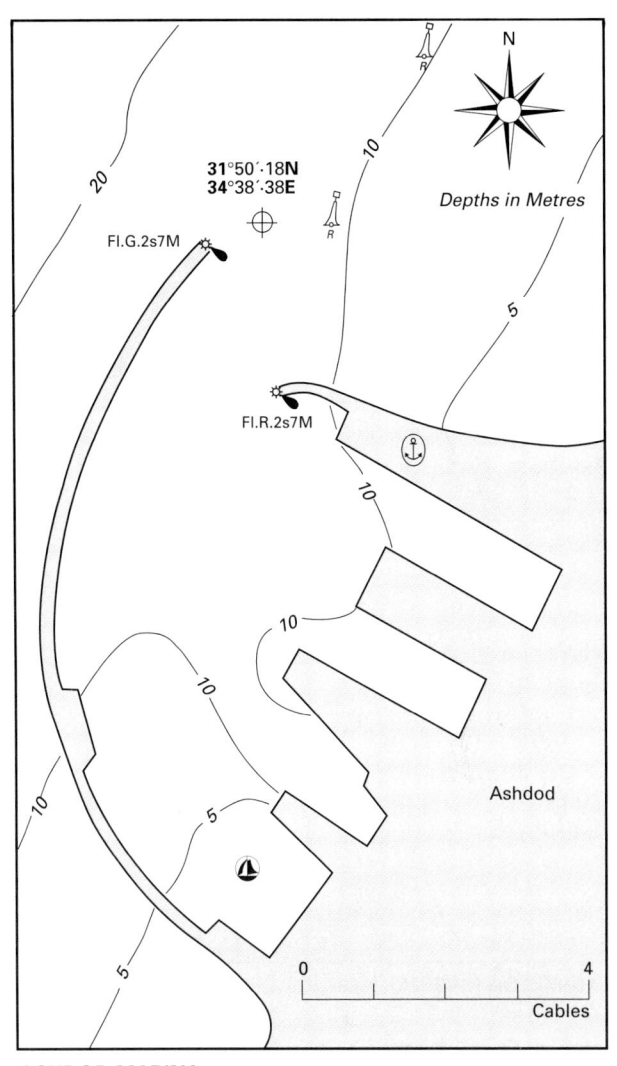

ASHDOD MARINA

Ashkelon
Admiralty Chart 1585, plan page 266

Approach
Call the Israeli Navy on Ch 16 when 30–40M out. They will meet you on your way in. Ashkelon is pretty obvious from the sea, although it is not a large city. There are three small islands off the beach to the S of the town. Hold close to the starboard breakwater on approach and entry to clear a reef to port that is indistinctly marked. A position for the breakwater is 31°41'·15N 34°33'·30E. CIQ procedures are completed at the marina. The marina layout is much the same as the marina at Tel Aviv, the entrance on the NW corner. Your boat may be checked by a military security squad on arrival. Tell the officials not to stamp your passport if you are heading S to the Red Sea.

Facilities
600 berths including visitors berths. Max. LOA 30m. 220V on all berths. Rates, once very competitive, are now similar to those at Larnaca in Cyprus. 100-ton travel-lift (US$10 per m^2 up and down). Repairs, fuel, showers, provisions. Marina

265

Red Sea Pilot

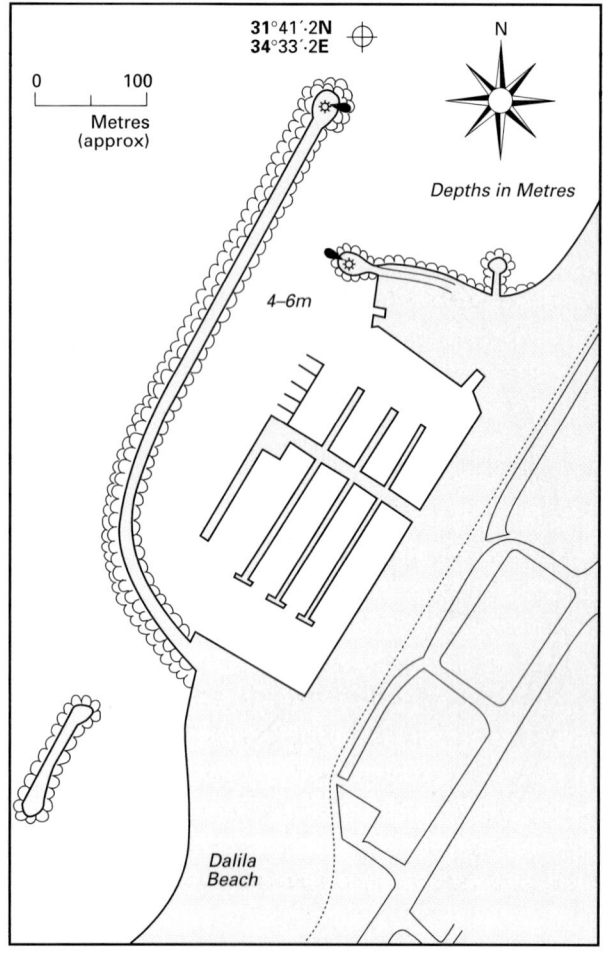

ASHKELON MARINA

staff are very helpful as are Israeli live-aboards but note that boat parts are expensive, as is fuel. Email hook-up is possible at the marina office. The tourist office at the Afridar Centre in the main commercial area opens Sunday, Monday, Wednesday, Thursday 0830–1300, Tuesday 0830–1230, closed Friday and Saturday. PO at 18 Herzl St in the old Arab town of Migdal. There is strict observance of the Sabbath. Nothing is open, even an ATM, from early pm on Friday until late on Saturday.

Contact Ashkelon Marina, PO Box 5335, Ashkelon. ☎ +972 7 733780; *Fax* +972 7 733823, *email* marina@ashkelon.muni.il

General

Ashkelon is 56km S of Tel Aviv. It is a popular beach resort and an expensive one though food prices are similar to Cyprus overall. There are a number of ancient ruins, conspicuously those at the nearby site of a 4,000 year old Philistine city, now part of a national park.

Appendix

I. CHARTS

In this appendix we have listed the charts relevant to this pilot from the British Admiralty (BA), the United States Defense Mapping Agency (USDMA), the French hydrographic office (SHOM). For the exit from the Red Sea into the Mediterranean we have included only the Admiralty charts.

Modern electronic chart packages offer the same coverage of the Red Sea as official paper charts. They also incorporate all the same shortcomings (see Navigation in the Planning section). We have used C-Map for planning and found it a wonderful help. But as we say, if you're navigating with an electronic chart interfaced with GPS, be extremely careful.

Admiralty Charts

Chart	Title	Scale
5	Abd Al Kuri to Suqutrá (Socotra)	350,000
6	Gulf of Aden	750,000
7	Aden harbour and approaches	25,000
12	El 'Aqabah to Dubā and ports on the coast of Saudi Arabia	350,000
	A. Ash Sharmah	75,000
	B. Approaches to Dubā	150,000
	C. Port of Dubā	15,000
	D. Dubā bulk plant terminal	10,000
15	Approaches to Jīzān	200,000
16	Jīzān	30,000
81	Sawākin to Ras Qassār	300,000
	Sawākin	12,500
82	Outer approaches to Port Sudan	150,000
	Sanganeb anchorage	25,000
143	Jazīrat aṭ Ṭā'ir to Bab el Mandeb	400,000
157	Masamirit to Bab el Mandeb	750,000
157	Masamirit to Bab el Mandeb	750,000
158	Berenice to Masamirit	750,000
159	Suez (El Suweis) to Berenice	750,000
164	North and northeast approaches to Mits'iwa	300,000
168	Anchorages on the coast of Ethiopia	
	Dolphin cove	12,500
	Dissei anchorage: Port Smyth	15,000
	Melita bay	25,000
	Entrance to Gubbet Mus Nefit	37,500
	Anfile bay	40,000
	Bera' Isol/ Bahir Selat'/	75,000
171	Southern approaches to Mits'iwa	200,000
183	Ra's at Tin to Iskenderun	1,100,000
233	The Suez Canal. (Qanât el Suweis)	60,000
	A. Port Said (Bûr Sa'îd) to Km52	60,000
	B. Km50 to Km86	60,000
	C. Km75 to Km99	60,000
	D. Great Bitter Lake (El Buheiret El Murra El Kubra) to Suez Bay (Bahr El Qulzum) (Km95 to Km162)	60,000

Chart	Title	Scale
	E. Lake Timsâh (Buheiret El Timsâh)	30,000
	F. El-Kabrît	40,000
253	Golfe de Tadjoura and anchorages	
	A. Golfe de Tadjoura	200,000
	B. Port D'Obock	25,000
	C. Mouillage de Tadjoura	12,500
	D. Mouillage di 'Ambâda	20,000
	E. Entrance to Ghoubbet Kharab	12,000
262	Approaches to Djibouti	50,000
	A. Port of Djibouti	10,000
302	Approaches to El Iskandariya (Alexandria)	37,500
326	Southern approaches to Madīnat Yanbu 'Aṣ Ṣinā 'īyah	75,000
327	Northern approaches to Yanbu'	5,000
	Yanbu' al Bahr	25,000
328	Madīnat Yanbu 'aṣ Ṣinā 'īyah – Minā' al Malik Fahd	25,000
333	Offshore installations in the Gulf of Suez (including Râs Shukheir)	50,000
453	Islands in the southern Red Sea Jabal Zuqar island to Muhabbaka islands	100,000
	Abu 'Alī Channel	40,000
460	Massawa (Mits'iwa) and approaches	
	A. Approaches to Massawa (Mits'iwa)	125,000
	B. Massawa (Mits'iwa) harbour	15,000
542	Madiq Kamarān to Al Hudaydah	100,000
	Port of Hudaydah	15,000
	Approaches to port of Hudaydah	35,000
548	Approaches to Madíq Kamaran and al Luhayyah	100,000
675	Harbours and anchorages on the coast of Sudan	
	Trinkitat harbour	12,500
	Marsa Esh Sheikh Ibrahim	12,500
	Khor Nawarat	30,000
	Marsa Esh Sheikh Ibrahim to Talla Talla Saqir	100,000
801	Plans in the Gulf of Aqaba	
	A. Approaches to Elat and El'Aqaba	75,000
	B. Elat and El 'Aqaba	25,000
	C. El 'Aqaba industrial port	25,000
	D. Approaches to strait of Tiran	75,000
	E. Strait of Tiran	25,000
848	Ports in Eastern Cyprus	10,000
	Larnaca	10,000
	Famagusta	15,000
	Approaches to Larnaca; Gastria Bay	25,000
849	Ports in Western Cyprus	
	Kyrenia, Limassol	10,000
	Vasilikos	15,000
	Moni, Paphos, approaches to Akrotiri	

Red Sea Pilot

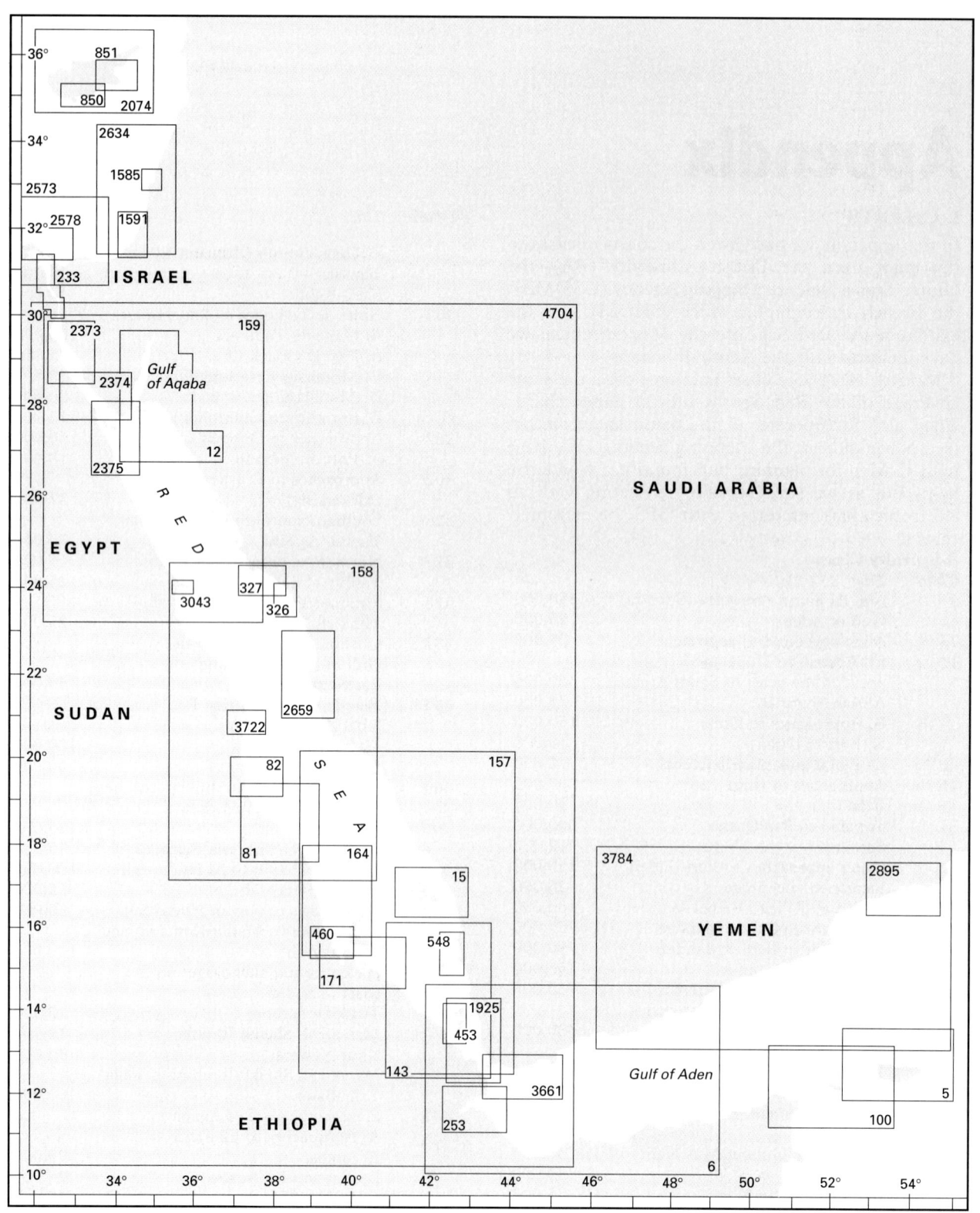

ADMIRALTY CHARTS

Appendix

Chart	Title	Scale
	harbour and Limassol; Karavostasi	25,000
850	Cape Aspro to Cape Pyla	100,000
851	Cape Kiti to Cape Eloea	100,000
1585	Hefa (Haifa) and approaches	70,000
	Hefa (Haifa)	20,000
1591	Ports on the coast of Israel	
	Hadera	25,000
	Tel Aviv-Yafo	25,000
	Ashdod	25,000
	Ashqelon	25,000
	Tel Aviv to Ashqelon	100,000
1925	Jabal Zuqar island to Bab-el-Mandeb	200,000
1926	Aseb bay	75,000
	Aseb	15,000
1955	Ports in the Yemen	
	Kamarān harbour	12,500
	Al Mukhā	25,000
	Southern entrance to Madīq Kamarān	30,000
2074	Cyprus	300,000
2090	Râs Matârma to Ain Sukhna	40,000
	North Ain Sukhna port	20,000
2098	Approaches to Port of Suez (Bûr el Suweis)	40,000
2373	Suez Bay (Bahr el Qulzum) to Râs Sherâtîb	150,000
	A. Râs Budran terminal	50,000
	B. Wâdi Feirân terminal	30,000
2374	Râs Sherâtîb to Ashrâfi Islands	150,000
	Râs Ghârib	15,000
	Zeit terminals	50,000
2375	Ashrâfi Islands to Safâga and Strait of Tiran	150,000
2573	El-Burullus to El 'Arish	300,000
2577	Jeddah (Mīnā' al Jeddah)	15,000
2578	Mhna 'Dumyât (Damietta port) to Bur Sa'hd (Port Said)	100,000
	Mhna 'Dumyât (Damietta port)	35,000
2588	Red Sea – Straits of Bab el Mandeb	75,000
	Mayyūn harbour	25,000
2599	Approaches to Jeddah (Mīnā' al Jeddah)	30,000
2634	Beirut to Gaza	300,000
2658	Outer approaches to Mīnā' al Jeddah (Jeddah)	75,000
2659	Sh'ib Nazar to Qiṭa' Kidan	200,000
2681	Approaches to El Iskandariya (Alexandria) and Khalig Abu Qir	100,000
2895	Outer approaches to Port Salalah (Mīnā Raysut)	100,000
2896	Port Salalah (Mīnā Raysut) and approaches	25,000
	Port Salalah (Mīnā Raysut)	10,000
2950	Plans on the coast of Somaliland	
	Elayu	7,500
	Las Khoreh	12,500
	Bandar Gaan	14,500
	Bosasso (Bandar Cassim)	20,000
	Alula	20,000
	Obbia	30,000
	Dante (Hafun) - north & south anchorages	30,000
	Eil Marina	40,000
	Bandar Beila anchorage	50,000
	Capo Elephante anchorage	50,000

Chart	Title	Scale
	Candala, Oloch and Damo	50,000
	Itala	50,000
	Illigh	60,000
2954	Gulf of Aden – eastern portion, including Socotra island	750,000
3043	Ports on the coast of Egypt	
	Berenice	25,000
	Hurghada	25,000
	Safâga	35,000
	Approaches to Berenice	60,000
	Approaches to Safâga	75,000
	El-Quseir	75,000
3119	Mina el Iskandariya (Port of Alexandria)	12,500
3214	Bûr el Suweis (Port of Suez)	20,000
3325	Approaches to Mersa el Hamra and Sidi Kerir	100,000
3326	Mersa el Hamra	30,000
3400	Ras Al Muraysah to El Iskandariya	500,000
3492	Approaches to Port Sudan	50,000
	Port Sudan	10,000
	Marsa Gwiyai	20,000
3530	Approaches to Berbera	50,000
	Berbera	10,000
3660	Aden Inner harbour	7,500
3661	Straits of Bab-el-Mandeb to Aden harbour	200,000
3662	Little Aden oil harbour	7,500
3722	Approaches to Muhammad Qol	75,000
	Muhammad Qol	10,000
	Marsa Inkeifal	12,500
3784	Ra's al Kalb to Ra's Marbāṭ	750,000
	Nishtūn	10,000
	Al Mukallā	37,500
	Ash Shihr Termina	50,000
3785	Mīnā Raysut to Al Maṣīrah	750,000
	Marbāt bay: Madrakah anchorage	35,000
4302	Mediterranean Sea eastern Part	2,250,000
4703	Gulf of Aden to the Maldives and the Seychelles group	3,500,000
4704	Red Sea	2,250,000
4705	Arabian Sea	3,500,000
5501	Mariners' Routeing Guide. Gulf of Suez Ed.2	

FRENCH (SHOM) CHARTS

Chart	Title	Scale
4792	Golfe d'Aden – Mouillages dans le Golfe de Tadjoura	
	Mouillage du Lac Salé (Gubet Kharab)	12,000
	Mouillage de l'île du Diable (Gubet Kharab)	12,000
	Mouillage de Tadjoura	12,000
	Mouillage de Khor Ambadu	19,000
	Mouillage de l'Etoile (Gubet Kharab)	12,000
	Entrée Gubet Kharab – Mouillage des Boutres	12,000
6265	De l'île Socotra au Ras Asir (Cap Guardafui)	415,000
6266	Abords du Cap Guardafui – De la Pointe Osbolei à Bender-Beila	418,000
	Mouillage de Dante	30,000
6326	Iles Seba	50,000
6388	Abords d'Aden	50,000

Red Sea Pilot

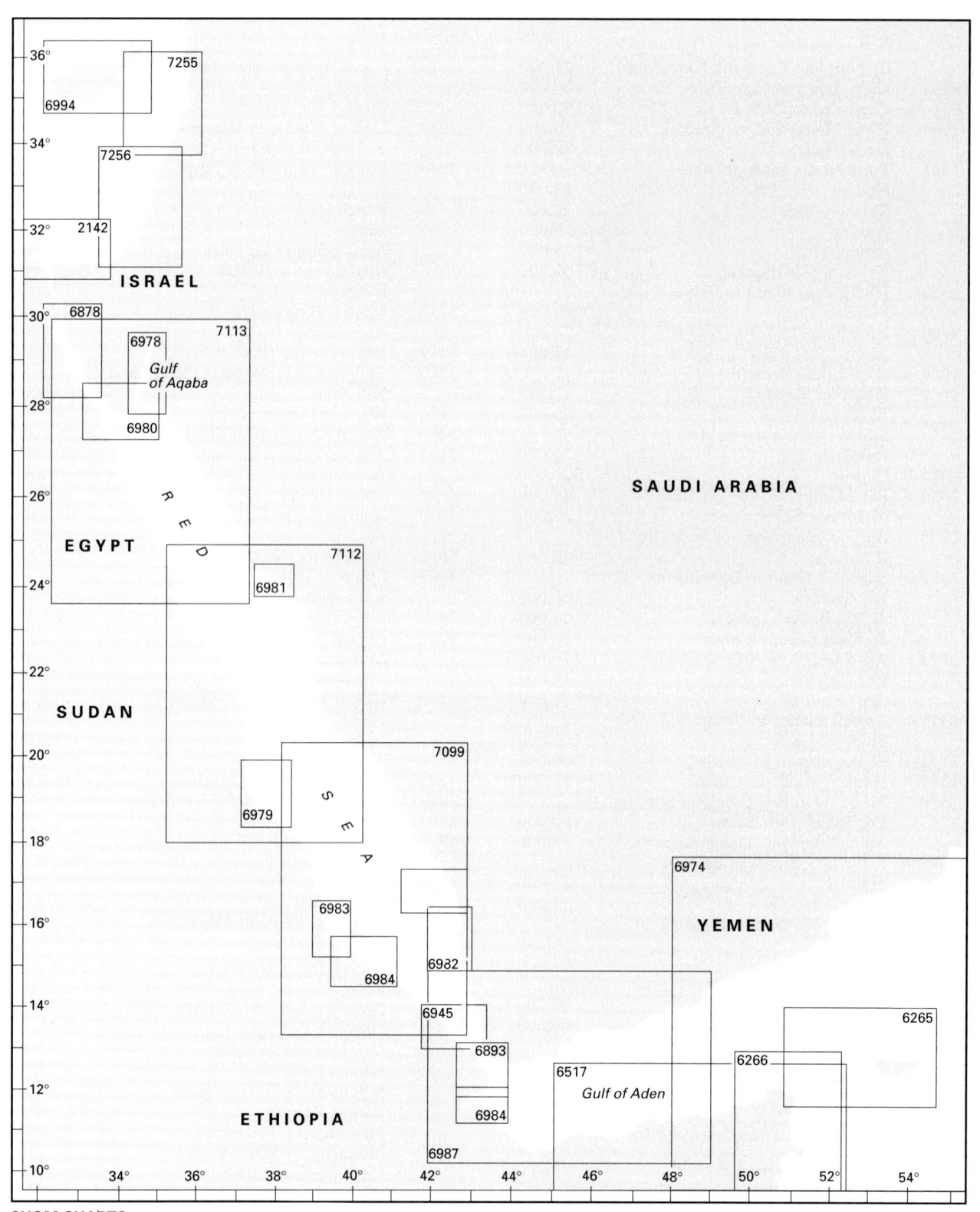

SHOM CHARTS

Appendix

Chart	Title	Scale
	Port d'Aden	10,000
6819	Canal de Suez (Qanât el Suweis) – Isthme de Suez – Carte-index	
	A	50,000
	B	50,000
	C	50,000
	D	50,000
	E	50,000
6878	Golfe de Suez – De Râs Ghârib au port de Suez	175,000
	Mouillage de Wâdi Feirân	20,000
	Ras el-Sudr	25,000
	Baie Abu Zenîma	25,000
6908	Détroits de Jubal et de Tirān	176,000
	Mouillage de Râs Ghârib	15,000
	Tor (At Tūr)	37,500
6947	Abords et partie Est du golfe d'Aden	1,080,000
	Approches de Mīnā' Raysūt et de Salālah	100,000
6965	Approches de Djeddah (Jiddah)	80,000
	Port de Djeddah (Jiddah)	20,000
6978	Golfe d'Aqaba	300,000
	Détroit de Tirān	25,000
	Approches d'Elat et d'Aqaba	75,000
	Elat et Aqaba	25,000
6979	Mer Rouge, côte du Soudan	
	Approches de Būr Sūdān	175,000
	Būr Sūdān et Marsa Gwiyai	27,500
6981	Approches de Yanbu'	100,000
	Madīnat Yanbu' as Sinā'īyyah – Mina Al Malik Fahd	40,000
	Yanbu'al Bahr	25,000
	Accès Sud de Yanbu'	175,000
6982	Des îles Farasān à Hodeida (Al Hudaydah)	175,000
6983	Accès à Mits'iwa (Massawa) (partie Nord)	175,000
6984	Accès à Mits'iwa (Massawa) (partie Sud)	175,000
6987	Partie Ouest du golfe d'Aden – Bab-el-Mandeb	700,000
	Berbera	25,000
7013	Baie de Suez (Bahr al Qulzum)	35,000
	Ports Ibrâhîm et Tawfik (Tawfiq)	15,000
7071	Ports de Mits'iwa (Massawa)	15,000
7099	Mer Rouge – Partie Sud – De l'archipel Sawakin aux îles Hanish	700,000
7111	Approches de Hodeida (Al Hudaydah)	35,000
	Port de Hodeida (Al Hudaydah)	15,000
7112	Mer Rouge – Partie centrale – De Abū el Kizân à l'archipel Sawākin	700,000
	Marsa Halā'ib	35,000
7113	Mer Rouge – Partie Nord – De Râs Muhammad à Abū el Kizân	700,000
	Mouillages de l'île Safâga	75,000
	Hurghada	75,000
	Ash Sharma	100,000
	Mouillage d'Hurghada	25,000
7169	Ports de Chypre	
	Baie d'Akrotiri	100,000
	Limassol	25,000
	Vasilikos	15,000
	Baies de Larnaca et de	

Chart	Title	Scale
	Famagouste	100,000
	Larnaca	25,000
	Famagouste	12,500
7255	De El Lādhiqiyeh à Soûr	250,000
7256	De Soûr à Al Árish	250,000
7306	De Ra's at Tin à Iskenderun et à Būr-Sàīd (Port Saïd)	1,180,000
7333	De Al Árish à Damiette (Dumyāt)	300,000
7334	De Damiette (Dumyāt) á Mersa el Hamra	300,000
7367	Approches de Port-Saïd (Būr Sa'id)	50,000
	Port-Saïd (Būr Sa'id)	12,500
7514	Ports du Liban	
	Aboprds de Tarābulus (Tripoli)	25,000
	Aboprds de Ra's Selaata	20,000
	Aboprds de Saydā (Saïda)	25,000
	Aboprds de Sūr (Tyr)	25,000
7515	Approches de Hefa (Haïfa)	70,000
	Hefa (Haïfa)	20,000
7517	Approches de Jīzān	200,000
	Port de Jīzān	25,000
7519	Bab el Mandeb et Golfe de Tadjoura	200,000
	Mouillage de Tadjoura	10,000
	Entrée du Ghoubbet El Kharâb – Mouillage des Boutres	15,000
	Port d'Obock	15,000
7520	Abords de Djiboutu	50,000
	Port de Djibouti	10,000
7641	Ports de La Côte d'Israël	
	Hadera	25,000
	Tel-Aviv - Yafo	25,000
	Ashdod	25,000
	Ashqelon	25,000
	De Tel-Aviv á Ashqelon	25,000
8003N	Guide pour la preparation de la traversee du golfe de Suez	

US NIMA CHARTS

Chart	Title	Scale
W703	703 Gulf of Aden to the Maldives and the Seychelles Group	3,500,000
W704	704 Red Sea	2,250,000
W705	705 Arabian Sea	3,500,000
56082	Difirswar By-Pass to Bur Said By-Pass (Suez Canal – Northern Part)	30,000
56083	Bur Taufiq to Difirswar By-Pass (Suez Canal – Southern Part)	30,000
62000	Gulf of Aden	1,000,000
62001	Red Sea	1,800,000
	Plan: Khalig El Suweis (Gulf of Suez)	500,000
62024	Al Masirah to Ras Raysut including Suqutra I	1,000,000
62040	Suqutra Island and vicinity	300,000
62046	Approaches to Suqutra	150,000
	Plan: Ghubbah di-Hadiboh (Tamrida Bay) (not shown on index)	50,000
62050	Raas Aantaara to Rass Binna including Abd Al Kuri	300,000
62070	Raas Jilbo to Raas Goragii	300,000
62080	Gulf of Aden – Western Part	300,000
62090	Gulf of Aden – Western Part	300,000
62091	Approaches to Berbera	75,000

271

Red Sea Pilot

Chart	Title	Scale
	Plan: Berbera	15,000
62092	Approaches to Djibouti (Gulf of Aden)	100,000
	Continuation of Djibouti	100,000
62093	Djibouti and approaches	20,000
62095	Republic of Djibouti – Port of Djibout	10,000
62097	Approaches to Bandar at Tawahi (Aden Harbour)	75,000
62098	Bandar At Tawahi (Aden Harbour)	20,000
62100	Jazirat al Hanish as Sashir to Bab al Mandab	150,000
62105	Perim Island and Small Strait	18,000
62110	Ed to Assab including Az Zuqur and Jazair Hanish	150,000
	Plan: Assab Harbour	25,000
62111	Plans in the Red Sea	
	A. Aqiq (Sudan)	27,160
	B. Marsa Shaykh Ibrahim (Sudan)	16,240
	C. Sharm Dumaygh (Saudi Arabia)	7,170
	D. Marsa Halaib Anchorages	29,230
	F. Ed (Eritrea)	21,170
62115	Ras Abu Masarib to Ras Marjah (Saudi Arabia)	150,000
62120	North Massawa Channel	150,000
62121	Approaches to Massawa Harbour	75,000
	Plan: Massawa Harbour	15,000
62130	South Massawa Channel	150,000
62140	Approaches to Al Qadimah (Saudi Arabia)	150,000
62142	Approaches to Bur Sudan	25,000
	Plan: Bur Sudan	10,020
62143	Approaches to Bur Sudan and Sawakin	100,000
62144	Approaches to Sawakin	20,000
62162	Berenice and Approaches	50,000
	Plan: Berenice	20,000
62170	Approaches to Madinat Yanbu As'Sina'iyah (Red Sea – Saudi Arabia)	150,000
62171	Outer approaches to Madinat Yanbu As'Sina'iyah	75,000
	Plan: Yanbu Al Bahr	15,000
62172	Mina Al Malik Fahd (King Fah) at Madinat Yanbu As'Sina'iyah	25,000
	Plan: Northern entrance channel	50,000
62177	Bur Safaga and approaches	75,000
	Plan: Bur Safaga	15,000
62188	Approaches to El-Ghardaqa	75,000
	Plan: El-Ghardaqa	25,000
62191	Madiq Gubal to Ras Gharib	150,000
	Plans: A. Sharm El-Sheikh	25,000
	B. Ras Shukheir	50,000
62193	Bahr el-Qulzum (Suez Bay) Egypt	15,000
62194	Approaches to Bahr el Quizum (Suez Bay)	50,000
	Plan: Ain Sukhna	25,000
62195	Ra's Shukheir to Ra's Sudr	150,000
	Plan: Ras Gharib	25,000
62202	Yemen – South Coast	300,000
	Plan: Ghubbat Al Mukalla	25,000
62220	Gulf of Aqaba	150,000
62222	Strait of Tiran	25,000
62225	Elat and Al Aqabah (Gulf of Aqabah)	25,000
62230	Madiq Gubal to Geziret Zabargad	500,000

Chart	Title	Scale
62241	Mina Jiddah (Saudi Arabia)	15,000
62242	Approaches to Jiddah	75,000
62250	Jaziret Zabarjad to Port Sudan	500,000
62270	Port Sudan to Sajid	500,000
62271	Jaza'ir Farasan and approaches to Jizan	200,000
62276	Port of Jizan	15,000
62285	Approaches to Madiq Kamaran	150,000
	Plan: Madiq Kamaran	35,000
62288	Plans of Al Luhayyah and Al Mukha	
	Plans: A. Al Luhayyah	25,000
	B. Mukha (Mocha)	25,000
62290	Sajid to Siyyan (OMEGA)	500,000
62292	Al Ahmadi and approaches	75,000
	Plan: Al Ahmadi	25,000
62295	Ra's Isa to Ra's Mutaynah including Jazir Az Zubayr and Jazair Hanish	150,000
62306	Ra's Fartak to Ash Shihr	300,000
62310	Ra's Fartak to Ra's Janjali	300,000
62312	Approaches to Ras al Madrakah	100,000
62313	Port Salalah (Mina Raysut) and approaches	75,000
	Plan: Port Salalah (Mina Raysut)	25,000

Bellingham Chart Printers offer a Red Sea folio of high quality, good value reproductions of US charts, *email* sales@tidesend.com, www.tidesend.com; PO Box 1728 Friday Harbour, WA 98250, USA. ☎ +1 360 468 3900; *Fax* +1 360 468 3939

Chart agents for international mail order
Imray Laurie Norie and Wilson Ltd, Wych House, The Broadway St Ives Cambridgeshire PE27 5BT, UK
☎ +44 1480 462114 *Fax* +44 1480 496109,
Email ilnw@imray.com http://www.imray.com
US National Imagery and Mapping Agency (NIMA), FAA, Chart Distribution, AVN-530, National Aeronautical Charting Office, Riverdale, Maryland 20737-1199, USA ☎ +1 301 436 8301; *Fax* +1 301 4366829; http://acc.nos.noaa.gov
Boat Books, Crows Nest, 31 Albany St, Sydney, Australia ☎ +61 2 4391133
Limassol Theseas Savva, 118 Franklin Roosevelt Ave, Limassol, Cyprus ☎ +357 51 55899
Haifa I-M Techno-Marine, Room 731, Haifa Levant Building, Palmer's Gate No. 1, 31000, Israel ☎ +972 4 521906 or 645622
MS Baaboud Trading and Shipping, Al Neel Street, 4AI Saalbah District, Jeddah, Saudi Arabia ☎ +966 2 63677572. (NB In Port Sudan the duty free outlet acts as an agent for MS Baaboud.)
Egypt
Alexandria Hydrographic Office, Shobat Al Misaha al Baharia, Ras el Tin, Alexandria ☎ +20 3 801006 (Domestic charts)
Marinkart, 24 El Nasr Street, Alexandria ☎ +20 3 804387
Marinkart, 29 Ramses Street, Port Said Free Zone ☎ +20 66 220148
Edwardo Marine Services, PO Box 179, 41 El Tour Street, International Shipping, Port Said ☎ +20 66 30031
Suez Marinkart, 38 Gohar Alkaid Street, Port Tewfiq, Suez ☎ +20 62 23183/29045

Appendix

II. BOOKS AND WEBSITES
This list is just a starting point. We have included some of the general reference works that we found useful and a selection of background reading. The website information was good at time of writing.

Reference
Boat owner's mechanical and electrical manual, Nigel Calder (2nd ed McGraw Hill 1995)
Mariner's weather handbook, Steve Dashew (Beowulf 1998)
First aid manual The authorised manual of the St John Ambulance, St Andrew's Ambulance Association and the British Red Cross Society (latest edition)
Advanced First Aid afloat, Peter Eastman and John Levinson (Cornell Maritime 2000)
First aid at sea, D Justins and C Berry (Reed 1993)
Healthy Travel, Africa, Isabelle Young (Lonely Planet 2000)
Ship's medicine chest and medical aid at sea (US Department of Health, latest edition)
Travellers' health, Richard Dawood (3rd ed OUP 1992)
Where there is no doctor, David B Werner (Hesperian Foundation 1998)
Medicines: the comprehensive guide, IKM Morton and JM Hall (Parragon 1997)
Health information online: www.masta.org, www.travmed.com, www.tripprep.com

Other cruising guides
Guide pratique voile et plongée de la Mer Rouge et du Golfe d'Aden Felix Normen (2 vols, Normandie-Yemen, 31 rue Lepic, 75018 Paris, 1992)
Mediterranean Cruising Handbook Rod Heikell (Imray 4th ed 1998)
Turkish Waters and Cyprus Pilot Rod Heikell (6th ed Imray 2001)
Imray Mediterranean Almanac (Bienniel)

Environment
Birds of Eastern Africa, Ber van Perlo, (Princeton 2001)
Birds of Europe with North Africa and the Middle East Lars Jonsson (Christopher Helm 1992)
Coral reef fishes: Indo Pacific and Caribbean, Ewald Lieske and Robert F Myers (Collins 1994)
Corals of the world, Elizabeth Wood (TFH Publications 1989)
Dangerous marine animals, Bruce W Halstead (Cornell Maritime 1995)
Mosquito, Andrew Spielman and Michael d'Antonio (Faber 2001)
Red Sea: Key environments series, AJ Edwards and SM Head (Elsevier Science 1987)
Red Sea coasts of Egypt, Jenny Jobbins (American University in Cairo Press 1990)
Red Sea coral reefs, G Bemert and RF Ormond (Routledge 1981)
Red Sea explorers, Peter Vine and Hagen Schmid (Immel 1995)
Red Sea fishwatcher's field guide (laminated card, Seahawk Press 1982)
Red Sea invertebrates, Peter Vine (Immel 1995)
Red Sea reef fishes: a diver's guide, JE Randall (Immel 1995)
Red Sea safety: a guide to dangerous marine animals, Peter Vine (State Mutual 1995)
Red Sea shells, Doreen Sharabati (Routledge 1985)
Reef fishes of the Red Sea, Mary and Richard Field (Kegan Paul 1998)
Sharks of tropical and temperate seas, RH Johnson and Bernard Salvat (Gulf 1995)

Guides and modern travel literature
If you want information about the Red Sea countries from the perspective of the average traveller the Lonely Planet series is the most useful. At time of writing the one notable lacuna in their list is a recent title for Sudan. For the latest see www.lonelyplanet.com
Other general guidebooks include the Footprint Handbooks, www.footprint-handbooks.co.uk
The following titles make interesting background reading.
Hansen, Eric, *Motoring with Mohammed* (Vintage 1993)
Mackintosh-Smith, Tim, *Travels with a tangerine* (John Murray 2001)
Mackintosh-Smith, Tim, *Yemen: Travels in Dictionary Land* (Picador 1997)
Raban, Jonathan, *Arabia through the looking glass* (Picador 1992)
Theroux, Peter, *Sandstorms: days and nights in Arabia* (Norton 1992)
Young, Gavin, *Slow boats to China* (Penguin 1983)

History, politics and biography
Miles Bredin, *The pale Abyssinian: a life of James Bruce* (Flamingo, 2000)
Dan Connell, *Against all odds* (Asmara, Eritrea, Red Sea Press)
Roy Fullick and Geoffrey Powell, *Suez: the double war* (Leo Cooper 1990)
Richard Hall, *Empires of the monsoon* (Harper Collins 1996)
Bernard Lewis, *The Middle East: a brief history of the last 2000 years* (Touchstone 1997)
Peter Mansfield, *The Arabs* (5th ed Penguin 1992)
I Marston, *Britain's Imperial role in the Red Sea area* (Connecticut 1962)
Roy Pateman, *Even the stones are burning*, (Asmara, Eritrea, Red Sea Press*)
Chris Peters, *Sudan: a nation in the balance* (Oxfam 1996)
Scott Peterson, *Me against my brother: at war in Somalia, Sudan and Rwanda*, (Routledge 2000)
John Pudney, *Suez: De Lesseps' canal* (Dent 1969)
Malise Ruthven, *Islam: a very short introduction* (OUP 2000)
Nawal el Saadwi, *Daughter of Isis* (Zed 1999)
*Red Sea Press www.africanworld.com
email awpsrp@castle.net
Eritrea website http://eritrea.org
Sudan website www.sudan.net

Classical texts
Periplus of the Erythraean Sea, with extracts from Agartharkhides, On the Erythraean Sea, trans and ed by GWB Huntingford (Hakluyt Society 1980)
Periplus Maris Erythraei, trans and ed by Lionel Casson (Princeton UP 1989)
The Book of Francisco Rodrigues (Hakluyt Society 1944)
Aristotle, *Historia Animalium*
Arrian, *Campaigns of Alexander*
Herodotus, *Histories*
Pliny the Elder, *Historia Naturalis II*
Ptolemy (Claudius Ptolemaeus), *The geography*
Strabo, *The geography*

Classic travel literature
Details of the latest known edition are given.
Brassey, Lady Anne, *A voyage in the Sunbeam* (Century 1984)
Burton, Sir Richard, *A personal narrative of a visit to Al Madinah* (Dover 1979)
De Monfreid, Henri, *Adventures of a Red Sea smuggler*

(Hillstone 1974)
De Monfreid, Henri, *Hashish* (Penguin 1985)
Forster, E M, *Pharos and pharillon* (M Haag 1923)
Greenlaw, J P, *The coral buildings of Suakin* (Oriel 1972)
Ingrams, Harold, *Arabia and the isles* (3rd ed Murray 1966)
Lithgow, William, *Rare adventures and painefull peregrinations* (Cape 1928)
Lord Valentia, *Voyages and travels to India, Ceylon and the Red Sea, Abyssinia and Egypt*, (London 1809)
Martineau, Harriet, *Eastern life, past and present* (1848)
Stark, Freya, *Beyond Euphrates* (Century 1989)
The coast of incense (Arrow 1990)
Thesiger, Wilfred, *Danakil diary* (Harper Collins 1996)
Desert, marsh and mountain (2nd ed Motivate 1993)
The life of my choice (Harper Collins 1992)
Waugh, Evelyn, *Remote people* (Harper Collins 1990)

Voyages under sail

There are many more than appear here. The dominant feature of almost all accounts, those below included, is how little space is devoted to the Red Sea transit. To most voyagers the Red Sea is a corridor from one fascinating room to another, and who looks closely at corridors? It is their loss.

The cruise of the Amaryllis, GHP Muhlhauser (Sheridan 1992)
Deep water and shoal, William Albert Robinson (Hart Davis 1949)
High Endeavours, Miles and Beryl Smeeton, by Miles Clark (Grafton 1991)
Hong Kong to Barcelona in the junk Rubia, JM Tey (Harrap 1962)
Macpherson's voyages, ed JS Hughes (Methuen 1944)
Moana returns, Bernard Gorsky (Elek 1959)
Sailing all the seas in 'The Idle Hour', Dwight Long (Hodder and Stoughton 1940)
Sheila in the wind, Adrian Hayter (Hodder and Stoughton 1959)
The world is all islands, Carl Nielsen (Allen and Unwin 1957)

Diving

Dive sites of the Red Sea, Guy Buckles (2nd ed New Holland 1999)
Diving and snorkelling guide to the Red Sea, John Ratterree (Pisces 1995)
Diving and snorkelling, Red Sea, Jean-Bernard Carillet et al (2nd ed (Lonely Planet 2001)
Red Sea diver's guide, Shlomo and Roni Cohen (Underwater World Publications 1994)
Red Sea diving guide, Andrea Ghisotti and Alessandro Carletti (AA Gaddis 1994)
Girl on the ocean floor, Lotte Hass (Harrap 1972)
South from the Red Sea, Haroun Tazieff (Lutterworth)
Under the Red Sea, Hans Hass (Jarrolds 1952)

Admiralty publications
Red Sea and Gulf of Aden Pilot (NP 64)

Admiralty List of Lights
Volume E (NP 78)
Volume F (NP 79)

Admiralty List of Radio Signals
Volume 1 Part 1, Coast Radio Stations in Europe, Africa and Asia (NP 281(1))
Volume 2 Radio Navigational Aids, Electronic Position Fixing Systems and Radio Time Signals (NP 282)
Volume 3 Part 1 Radio Weather Services (NP 283 (1))
Volume 5 Global Maritime Distress and Safety Systems (GMDSS) (NP 285)
Volume 6, Part 3, Vessel Traffic Services, Port Operations and Pilot Services –Mediterranean and Africa (NP 286(3))
Mariner's Handbook (NP 100)
Ocean Passages for the World (NP 136)
Tide Tables, Volume II (NP 202)

III. FOOD

Food takes on more importance than ever when you are miles away from shops and restaurants. There are long stretches of coast where you will be on your own so it really pays to plan well ahead. That said, you can leave provisioning for the Red Sea till you get to Salalah in Oman, where you'll find several western style supermarkets, or Aden, where you can find most things at much lower prices but where shopping is a bit more unpredictable, or fun, depending on your temperament. If you're headed S, stock up in Cyprus. It's cheaper than Israel.

Bread Bread is a staple in Arab countries and is available in all the ports. Oman and Yemen have quite delicious bread. In Yemen the best is called *khubz* and it is served with all restaurant meals in diameters from 20cm upwards. Bakeries sell leavened bread as well, also best when bought hot. After a couple of days you can re-heat it effectively in a pan or oven but none of the bread keeps very long. If you don't bake bread aboard take a stock of crackers, also widely available. Canned dried yeast is available in Yemen and Sudan.

Fruit and vegetables The quality of fruit and vegetables in many Red Sea markets is very good. Oman's produce is excellent, very tempting and reasonably priced but in Aden you will find real bargains. The best market is in Crater. In Massawa, take a yellow minibus to Edaga (which means market in Tigrinya) for the bigger market. Asmara's market is a must if you go to the capital. Suakin's market isn't bad at all but if you can stand the flies (worst in the autumn and early winter), Port Sudan is far superior as are its flies. In Egypt, the best green grocers we found were in Safaga's souk and at Suez. Hurghada's market is also reasonable but El Gouna's prices are relatively sky high. Seasonal fruit such as mangoes are fantastic in Egypt in the autumn, as are dates. In the spring look out for apricots, strawberries and peaches. Olives are also a good buy. Fresh fruit juices are delicious and safe, as far as we know, in Yemen, Eritrea and Egypt but they do tend to be very sweet.

Dairy produce Good feta-type cheese is sold in Egypt in tetra-paks. It keeps very well and stows easily. Yemen's yoghurt and buttermilk are very cheap and worth trying, especially the marvellous, smoky buttermilk. They are usually kept in chillers but the yoghurt will last well without refrigeration. Processed cheese from Europe and Australia can be found in most of the ports. In Eritrea there are good Italian cheeses.

Meat, poultry and eggs Roasted chickens are on sale everywhere but Sudan. The most delicious are in Aden and Salalah. The best eggs we found were in Eritrea. These kept for several weeks with no treatment or refrigeration. Sudanese eggs are poor with very pale yolks but very cheap. Take egg boxes with you. None of the shops or markets have them outside Oman. If you buy eggs from a small stall you may find that they are already cooked.

We rarely buy meat as we have no fridge aboard but if you are a carnivore you won't be disappointed. There is excellent meat and poultry for sale in all the ports. We ate outstandingly good fillet beef in Sudan. It was bought for us by an agent and cost about US$3 per kilo. If we'd gone ourselves we would probably have paid half that but may not have got off first base. The flies in Port Sudan market are indescribable. Stephen was showing symptoms of hepatatis A by the time we reached Eritrea – a mild dose thank goodness. We both ate exactly the same food and only one of us succumbed though neither of us had been vaccinated. High standards of hygiene and vaccination are the best defences though we reckon that being overcautious causes its own problems. If your gut is too sterile you will have far fewer defences than those whose habits are less than immaculate.

Fish and seafood Catching your own is pretty easy. If you are in harbour for a while try the New Market in Salalah. In Massawa you can get excellent fish from the fish co-operative near the anchorage at very reasonable prices.

Supermarkets and non-perishable food Both Aden and Salalah have some very good supermarkets. In Aden go to Khormaksar for the best variety of imported goods, and the highest prices. Alternatively, try the smaller shops in Ma'ala and Crater. They will deliver if you buy enough. Djibouti has all you might want – and wine too – but at prices to make your toes curl. Massawa has several small supermarkets that stock some interesting local food and drink but shop around a bit for the best value. Quality varies considerably and the shops near the port have a noticeable mark-up. Up in Asmara there are very good small supermarkets with delicatessens with Italian as well as local specialties. And Sudan? There are a few grocers in Port Sudan that carry surprisingly exotic items but you'll pay relatively high prices for them. All the ports have products like tomato puree, sardines, beans and evaporated milk, pasta, sugar and sweet biscuits. There is no real trouble finding most basic household items. The more exotic your tastes, the more carefully you should stock up in advance.

Avoid buying rice, flour and pulses from sacks. They look authentic but they're probably full of weevils. Weevils get into sealed packets too somehow. Fortunately they're not poisonous. We found tiny horned beetles in a bag of chickpeas once. They still had plenty left to eat, fortunately, and hadn't migrated elsewhere. See the Medical and health section for more on insects. Remember to take plenty of dried vegetables and seeds for sprouting. Sooner or later you'll run out of fresh and beansprouts make great salads with rehydrated peppers, tomatoes and mushrooms. Dried fruit, especially dates and nuts, are good value in Oman at many of the supermarkets.

Coffee and tea Never refuse an invitation to tea or coffee if you can possibly avoid it. It is often offered ceremoniously and you can just take a tiny amount. It may seem rude if you refuse. Shopkeepers in Sudan and Egypt will offer you a glass of tea if you spend long looking round. Eritrean coffee is a good buy. Instant coffee is expensive everywhere.

Spices and herbs Herbs are not that common in Red Sea markets. The best bet is Eritrea with its Italian links. Spices, gloriously coloured, fresh and plentiful, are everywhere. Treat yourself.

Alcohol The only Red Sea countries where we would buy alcohol are Djibouti and Eritrea. Stock up before you leave Malaysia, Sri Lanka or the Mediterranean. French wine is available in Djibouti and Eritrea has plenty to choose from. Eritrean beer is cheap only if you return the bottles but it's good stuff. So is the gin and other spirits. No need to return those bottles. The wine is made from grapes imported from Italy and we didn't rate it. In the Yemen and Sudan you will find alcohol-free beer in the supermarkets. The Klausthaler in Aden is surprisingly good if you're desperate. Real beer is available but expensive, in a few bars in Aden. Beer in Egypt is expensive but good and El Gouna has a brewery and winery.

Restaurants and takeaways The gastronomic highlights of our times in the Red Sea countries were in Yemen, Eritrea and Egypt. Yemen is a great place to eat out if you're not too fussy. Standards of hygiene vary but we have heard of very few cruisers who fell ill after eating out in Aden or Sana'a. Spit roast chicken is delicious and cheap. Charcoal baked fish is more expensive but worth it. There is a whole range of spicy vegetable and bean or lentil dishes. The best plan is to go inside and have a look round before you choose. Nobody seems to mind. Falafel, the deep-fried chickpea rissoles are cooked and sold at road-side stalls in many of the ports. Very cheap and delicious and cooked in oil hot enough to kill any greeblies. One notable feature of Yemeni restaurants is their table cloths. Unless you go to an expensive or hotel restaurant they will be of standard size and shape and are very informative, if a bit out of date. Why? Because they're old, English language newspapers. Usually from Singapore or Malaysia. Somebody aboard a shipping line that plies the Red Sea must be making an extra bob or two delivering them!

Eritrea has good value restaurants and with more variety. Ethnic cuisine features *injera*. This is a large, slightly bitter but otherwise rather tasteless, thin spongy pancake-type bread that comes in a greyish brown colour. The best is made from a local cereal called *tef*. Others are made from millet or sorghum. Not all that appetising but it does a very good job of

Red Sea Pilot

Sample prices 2000-2001
All prices in US$ and cents

	Oman	Yemen	Eritrea	Sudan	Egypt
Bread equiv 6 rolls	26c	20c	15c	30c	15c
6 fresh eggs	75c	75c	60c	60c	50c
36 shot film processing[1]	$7	$6	$10	$8	$7
Email (1hr)	$2·50	$1·50[2]	$1[3]	$2	$2·8
1 minute overseas phone call	$1·50	$1·25	$2·50	$2	$1·50
1 litre – diesel	35c	25c	25c	30c	45c
1 kilo – oranges	$1·50	60c	$1	$1	75c
1 kilo – potatoes	90c	70c	60c	$1	60c
10 pieces laundry		$2·50	$2	$2·50	$4

1. unreliable except in Oman
2. unreliable connections
3. slow and unreliable

soaking up the favourite fiery stews like *zegni*. Other spicy meat dishes are *tibsi* and *zilzil*. Vegetables include *silsi* and *ades*. Restaurants in Massawa, Asmara and Assab also offer Italian food. The best are up in the capital. Try the cafés for home-made pastries, both sweet and savoury.

Egyptian restaurants have many familiar Middle eastern dishes. Try the mezzas for starters then kofta and kebab. Sweet pastries and cakes are also delicious.

Cooking fuel Filling gas bottles can be problematic in the Red Sea. Make sure you have a good selection of pigtails and adapters. Kerosene is relatively easy to come by but if you use meths to light your stove Eritrea will be the last stop on the way north where you will find it easily. It's worth carrying some kind of stand-by cooker. A camping gaz stove or a Coleman's multifuel cooker are easily stowed. A good alternative, rather more stable to use while underway, is a gas cartridge fuelled caravanning stove. For all of these take spare fuel as it will be hard to find later. Failing all else, economise on cooking fuels if you're running low by using a pressure cooker and rarely using the oven. Fill a pump pot with boiling water for hot drinks rather than boiling a kettle every time. Carry standby meals of pre-processed food and take plenty of instant soups, noodles, dehydrated vegetables. Pour boiling water over rice and legumes in a vacuum flask and leave for a few hours to pre-cook.

IV. GLOSSARIES

Arabic

A little Arabic is very useful and an enjoyable asset although English is widely spoken in the Red Sea countries. Several dialects of Arabic are spoken in the region, but Modern Standard Arabic (MSA) is understood by all educated people in Arab countries. As far as possible we use transliterated MSA below but some words and phrases are understood only in certain areas. Common alternatives are included. There is no agreed system of transliteration, hence the variants of many names found on charts. We have chosen an informal notation intended to make pronunciation easier. The choice of vocabulary and phrases is geared towards your likely needs.

Introductions etc.
Arabic *Ahrabee*
English (language) *Ingleezee*
Gently does it/slowly slowly *Shwoy shwoy*
God willing *Inshallah*
Good morning *Sabah al khayr*
Good morning (reply) *Sabah an nur*
Goodbye *Ma salaama*
Hello *Salaam al haykum*
Hello (reply) *Al haykum salaam*
No problem *Mish mushkallah*
OK, good *Mesh* (Egypt) *Tamam* (Yemen)
Please *Min fadlak* (to man)
 Min fadlik (to woman)
Passports/papers *Waraka/wataiq/basboor/jawaaz*
Rope/painter *Habl*
Thank you *Shukran*
Thank you very much *Shukran jazeelan*
Welcome *Marhaban*
What's your name? *Maa ismak*
My name is *Ismee*...
Where are you from *Min ain ente* (fem)/ *enta* (masc)
I'm from *Ana min*... Britain *Britaniya*, Australia *Usturaliya*, USA *Amrika*, Europe *Irop*, France *Francia*, Germany, *Almani*
Yes *Aywa*
You're welcome/don't mention it *Afwan*

Problems
Forbidden *Mamnooh*
Get lost *Maafi fooloos/yalaa*
Get lost (ruder) *Imshi!*
I don't understand *Laa afam/mush fahim*
I haven't got any *Ma fi*
It doesn't matter/ never mind *Ma lesh*
Maybe *Rubbama*
No *La*
No money *Ma fooloos*
Problem *Mushkallah*
Sorry *Aasif*
Stop! *Yuaf!/waif!*
Your mother will know! (an amicable Yemeni warning against shameful actions) *Oomakdaria!*

Ashore
Where is the... *Ain al*...
Bank *El-bank*
Bathroom *Hammam*
Customs *Jamaarik/jumruk*
Diesel *Deezil*
Doctor *Tebeeb/doktoor*
Hospital *Mustashfa*
Engine *Motur/muharrik*
Market *Souk/bazaar*
Near *Kareeb*
Office *Maktab*
Petrol *Banzeen*
Visa *Fisa/tasheera*
Water *Maa/mayya*
Post office *Maktab al bareed*
Telephone *Tilifoon*
Left *Yasaar*
Right *Yameen*
Straight on *Alatool*
Please may I take a photo? *Mumkin sura?*
Tomorrow *Bukra, ghadan*

Appendix

Shopping
How much/how many? *Bekam?*
Just a little bit *Bas qaleel*
Half *Nusf*
Expensive *Ghalee/kteer*
Food *Taham/ochel*
bananas *mauz*
bread *kubz/aysh*
butter *zibda*
cabbage *malfoof*
carrot *jazar*
cheese *jibna*
chicken *dajaj*
coffee *qahwa*
courgette *koosa*
cucumber *khiyaar*
eggs *bayd*
fish *samak*
fruit *fawaakih*
matches *kibreet*
meat *lahma*
milk *haleeb/laban*
oil *zayt*
onion *basal*
orange *bortogan/burtuqaal*
pepper *filfil*
potatoes *batatas*
rice *roz*
salt *mal*
sugar *suker*
tea *shaay*
tea with milk *shaay ma haleeb*

Navigational terms found on charts, (*kareeta*) and other useful words
ain *freshwater spring*
alama (Yemen) *buoy, beacon*
amiq (Egypt) *deep*
ari *shoal, shallow*
awama (Egypt) *buoy, beacon*
bahr *sea*
bahri (Yemen) *south*
bir *well*
darba (Sudan) *squall*
ganub (Egypt) *south*
gebel (Egypt) *mountain*
gezira (Egypt) *island*
gharb (Egypt, Yemen) *west*
ghariq *deep*

Numbers	Arabic numerals
0 *sifir*	
1 *wahid*	1 ١ 9 ٩ 17 ١٧
2 *ithneen*	2 ٢ 10 ١٠ 18 ١٨
3 *thalatha*	
4 *arbaa*	3 ٣ 11 ١١ 19 ١٩
5 *khamsa*	
6 *sitta*	4 ٤ 12 ١٢ 20 ٢٠
7 *saba*	
8 *thimania*	5 ٥ 13 ١٣ 30 ٣٠
9 *tisa*	
10 *ashra*	6 ٦ 14 ١٤ 50 ٥٠
20 *ishreen*	
50 *khamseen*	7 ٧ 15 ١٥ 100 ١٠٠
100 *imia*	
1000 *alf*	8 ٨ 16 ١٦ 150 ١٥٠

ghauba (Yemen) *sandstorm*
ghazir (Sudan) *deep*
ghubbet *bay*
habba, habub *squall*
haboob/ hubub (Djibouti/Egypt) *sandstorm*
ishara (Sudan) *beacon*
jebel (Sudan) *hill, mountain*
jezirat (Sudan, Yemen) *island*
jinub, janoob (Sudan) *south*
kebir *great, big*
khalfa (Sudan) *sandstorm*
khor *creek, inlet*
majra *channel*
markib/marakib *boat/boats*
marsa, mersa *anchorage, harbour*
mina *harbour, port*
miqra (Egypt) *channel*
nahr *river*
qad *reef*
qibli (Yemen) *north*
ras *cape, headland*
saahil *coast*
saghir/seghir *small*
satah (Egypt) *shallow*
shaati *beach*
shab *reef*
shamal (Egypt) *north*
shargi (Sudan) *east*
sharm *creek, inlet*
sharq (Egypt, Yemen) *east*
shib (Yemen) *reef*
shimal (Sudan) *north*
wadi *watercourse*
yakht *sailing boat*

Tigrinya

Tigrinya is the native language of many Eritreans living around Massawa and Assab but again, most of the people you are likely to need to talk to will speak English. Here are just a few useful words and phrases for use ashore. If you can produce them they will make the Eritreans smile and you will feel even more welcome. Anybody will help you with pronunciation if you ask.

Introductions etc.
Don't mention it/you're welcome *genzubka* (to man) *genzubki* (to woman)
Excuse me *yikray ta*
Good morning *keme hadarka* (to man) *keme hadarki* (to woman)
Goodbye *dehan kun*
Hello *kemaiala/selam*
I understand *yirede anee*
I don't understand *ayeterede anen*
Please *bejaki* (to woman) *bejaka* (to man)
Thank you *yekanyerlay*
Today *lomanti*
Tomorrow *tsebah*
Wonderful *girum*
Yes *oowe*; No *aikunun*

Ashore
Where is ...*abey alo*...
diesel *nafta*
gas (LPG) *srindar*
hospital *beit hikmena*
market *shook/edaga*
paraffin *lamba*
petrol *benzin*
post office *beit pusta*
toilet *shiraq*

Shopping
How much...? *Kenday*...?
beer *bira*
bread *banee*
cheese *asema/formajo*
coffee *boon*
meat *ciga*
milk *tseba*
orange *aranci*
potato *dineesh*
rice *ruz*
tea *shay*
sugar *shukar*
tomato *komeedero*
water *mai*

Numbers
1 *hade*
2 *kelite*
3 *seleste*
4 *arbate*
5 *hamushte*
6 *shedushte*
7 *shewate*
8 *shemonte*
9 *teshaate*
10 *aserte*
11 *aserta hade*
20 *isra*
100 *mi itee*

Index

Abail Is, 91
Abd Al Kuri, 80
Abhur, Sharm, 125
Abingdon Reef, 155, 160, 161
Abou Maya, 84
Abu, Khor, 121
Abu Ali Is, 102
Abu Asal, Khor, 172-3
Abu Dabbab, Marsa, 200, 201
Abu Dara, Ras, 184
Abu Duda, 123
Abu Durba, 229
Abu Fanadir, Khor, 170
Abu Fendera, Shab, 184
Abu Galawa, 191
Abu Galum, Ras, 241
Abu el Ghosun, 195
Abu Imama, Marsa, 167, 168
Abu Kazan, 207
Abu Al Kizan, 148
Abu El Kizan, 198
Abu Madafi, 125
Abu Madd, Marsa, 186-7
Abu Makhadiq, Marsa, 209
Abu Minqar, 209
Abu Naam, Marsa, 183
Abu Nigara, Shab, 218
Abu Nuhas, Shab, 221
Abu Rithami, 209, 211
Abu Rudeis, Ras, 231
Abu as Saba, Khor, 121
Abu Samra, Merset, 242
Abu Shagara, Ras, 172
Abu Shagrab, Ras, 162, 163, 165
Abu Shar Bay, 212, 216
Abu Shauk, 123
Abu Soma, Ras, 207-9
Abu Suweira, Ras, 234
Abu Tig Marina, 213-15
Abu Zenima, Ras, 234
Abuqir Bay, 259
Abydos Marina, 216
Acre (Akko), 262
Ad Dosan I, 116
Adabiya, Ras, 249
Aden, 75-7
  routes, 18, 20, 21, 22
Aden, Gulf of, 38, 40, 41
Adjuz, 93
Adulis, 93, 98, 128
Agig (Aqiq), 130
Aidheb, 181
AIDS, 65
Akbar Uqayli, 119
Akhawein, El (The Brothers), 204
Akko (Acre), 262
Al = definite article, *see proper name*
Alam, Marsa, 199, 201
Alexandria (El Iskandariyah), 259-60
Ali, Shab, 224-5
Alleda, Shab, 228-9
Alueda, Sharm, 180-81
amateur (HAM) radio, 30, 34
Ambada, 84
Amid, Marsa, 141

Amiq, Khor, 123
Anbar, Shab, 128, 141, 143
anchorages, 46
anchors, 49-50
Andeba Ye Midir Zerf Chaf, 93
Andebar Deset, 96
Anfile Bay, 92
Angarosh, 155, 160, 161
Anse = Bay, *see proper name*
Antar, Sharm, 126
Aqaba, 244-5
  routes, 23
Aqaba, Gulf of, 26, 40, 41, 53, 56
Aqiq (Agig), 130
Ar Ruays, Ghubbat, 125
Arabian Sea, *currents*, 36-7, 40
Arabic language, 276-7
Arah, Ras Al, 79
Arakiyai, Marsa, 152-3
Arish, El, 258
arms, 61
Aruk Giftan, El, 211
Arus, Marsa, 148-9
Ash Shayk Mirbat, 126
Ashdod, 265
Ashkelon, 265-6
Ashrafi, Shab, 223
Asis, Ras, 130-31
Asla (Assalah), 241
Asmara, 98
Assab, 88-90
Assalah (Asla), 241
astronavigation, 47-8
At, Merset El (Naama Bay), 237-8
At Tair, Jazirat, 109-10
At Tarfa, Ras, 121
Ata, Marsa, 141
Aweitir, Marsa, 152
Aydob (Heidub), Marsa, 137
Ayun Musa, 236
Aziz, Jazirat, 79

Bab = Strait, *see proper name*
Baghdadi, Ras, 195
Bahari, Sharm El, 203
Bahdur I, 130
Baie (de, d', du) = Bay (of) *see proper name*
baksheesh, 63, 253
Balihaf, 75
Banas (Baniyas), Ras, 179, 187, 191
Bandar = Bay, Harbour, Port, *see proper name*
Baniyas (Banas), Ras, 179, 187, 191
banks' opening hours see country information in Part I
Bar, Ras El-, 258
bargaining, 62-3
barometric pressure, 25-7, 30
Barra Musa Kebir, 128
Barra Musa Seqir, 128
barracuda, 65
Bashir, Marsa, 141
Batuga, Jebel, 187

Bayda, Shib al, 125
Bayer, Jazirat, 160-61
beacons, 46-7
begging, 63
Berailsole Bahir Selate, 91
Berbera, 83
Berenice, Port, 187
Bihar (Sharm Abhur), 125
bioluminescence, 34
Bir Ali, 74
Bir Ghadir, 197
Bir Quei, 205
Bir Taba, 243
bird life, 57, 170, 172, 246
Birk, Khor Al, 121
Bitter Lakes, 247, 252, 257
Blind Reef, 221
Bluff Point, 222
Bodkin Reef, 186
Boosaaso, 81
Boutres, Ilot Des, 84
Brothers, The (El Akhawein), 204
Brothers, The (SW of Socotra), 80
Budran, Ras, 205
Burns Reef, 128

Caluula, 81
Canqor, 81-3
Capo = Cape, *see proper name*
Carless Reef, 211, 218
Caseyr, Ras, 81
cell phones, 52
Centre Peak I, 109
Chagos, *routes*, 18, 20
charts, 41, 267-72
Chiltern Shoal, 111
ciguatera, 64
Claude, Shab, 191
climate 24-30, *see also country information in Part I*
cloud formations, 30
coast radio stations, 54
Cochin, *routes*, 18, 20, 21
communications see country information in Part I
cone shells, 64
convergence zones, 27
cooking fuel, 276
coral cuts, 64
coral reefs, 42, 43-5, 55, 56-7
Corinth Canal, 20
Corner Reef (Jinniya), 128
Crazirat Irj, 130
crime, 58, 59
crown-of-thorns, 64
crustaceans, 56
Curdumiat I, 91-2
currency see country information in Part I
currents, 36-40
cyclones, 28
Cyprus, 261-2
  coast radio, 54
  marinas, 55, 257, 261-2
  routes, 20, 21
Daba Libah (Ras Eiro), 84
Dabadib, Jezair, 155
Daedalus Reef, 198

Dahab (El Kura), 241
Dahlak Bank, 56, 97, 100-101
Dahrat Abid, 128
Dakliyat, Khor, 96
Damaran Abu Mieish, 235
Damath, Shab, 139-41
Damietta, 258
danger zones, 60
Dangerous Reef, 185-6
Dar Ah Teras, 127, 141
Darad, El, 83
Daror, Marsa, 147
Darra, Marsa, 152
Debel Ali, 93
Dehalak Deset, 96
Delwein, Khor, 168, 172
depressions, 27
Derebsasa Deset, 92
Dergammam, 93
Dergammam Kebir, 93
Des Boutes, Ilot, 84
desert, 57-8
Dha al-Fauf I, 116
Dhanab al Qirsh, 127, 128, 141-3
Dhu Lalam, 101
Dhu-I-Fidol, 101
Dhubab, Ras, 111
Dhul'lawa (Dullow), Khor, 167, 172
Dibsel, 128
Difnein, 101
Diheisa, Ras, 236
Dilemmi, 93
Dirra, Ras, 197
diseases, 65
Dissei Anchorage, 93-6
distress, 52
diving and dive sites, 56-7
Djibouti, 83-7
  Club Nautic (YC), 83, 84
  coast radio, 54
  diving, 56
  general information, 5-6
  piracy prevention, 61
  Port, 83-4
  port radio, 53
  routes, 18, 20, 22
Dohrab I, 119
Dolphin Cove, 93
Dolphin House, 218
Dolphin Reef, 191
Duba, Port of, 126
Dudo, Mersa, 91
Dugah Its, 130
Dullow (Dhul'lawa), Khor, 167, 172
Dumaygh, Sharm, 126
Dumeira I, 87
Dumyat, 258
Dungunab Bay, 159, 161-2
Dunraven (wreck), 218
Durwara, Jezirat, 135

Eagle I, 131
echo sounders, 47
Ed Domesh Shesh, 128
Eeles Cove, 174
Egypt, 20, 175, 178, 179, 181-260
  coast radio, 54
  general information, 9-11
  marinas, 55, 213-16, 242-3, 254, 255-7

  port radio, 53
Eight Degrees Channel, 18
Eiro, Ras (Daba Libah), 84
Eitwid Reefs, 127
El-Bar, Ras, 258
El-Erg, Shab, 218
El-Qad Yahya, Merset, 225
El = definite article, *see proper name*
Elat, 245-6
  routes, 23
Elba Reef, 167, 176-8
electrics, 50
electronic charts, 41
electronic logs, 47
Elefante, Capo, 81
Elphinstone Reef, 200
embassies see country information in Part I
Endeavour Harbour, 219-21
engine, 50
Entaasnu, 101
environment, 55-8
EPIRBs, 52
Er Koweit, 139
Erg, Shab El-, 218
Erg Marsa Alam, 191, 199
Erg Riyah, 227
Eritrea, 88-101
  general information, 6-7
  piracy prevention, 61
  port radio, 53
Ernesto Reef, 200-201
Eroug Diab, 191
Es Salum, 260
escorts, 5, 60, 61, 81
Esh Sheikh, Marsa, 135-6
Esh Sheikh Ibrahim, Marsa, 136-7
Estam Aghe, 92
Etoile, Baie de l', 85

Fairway Patch, 131
Fairway Reefs, 209
Fajrah, Marsa Al, 113
False Ras Gharib, 228
Fara Is, 123
Farasan, Shib, 119
Farasan Is, 57, 116-19
Faraun, Geziret, 243-4
Fare, Ile du (Le Fare du President), 85
Fare du Ghoubbet, 84-5
Farrajin I, 130
Fatuma Deset, 88, 91
Fawn Cove, 130
Fejera, Marsa, 113
Felix, Shab, 163
Felug, Ras, 81
Fiddler's It, 93
Fijab, Marsa, 148, 151
Filadhu I, 20
first aid, 63-4
fish, 55-6
  dangerous, 63-4
fishermen, 62
flags, national see country information in Part I
food and drink, 2, 274-6
  health precautions, 63
Fouad, Port, 255-7
Foul Bay, 167, 183-7
Freedom Anchorage, 93

Index

Fujaim, Marsa, 113
Fury Shoal, 191
Gafatir, Marsa, 169-70
Galle, *routes*, 18, 21
Gazeirat, Gazirat = Island, *see proper name*
Gebel = Hill, *see proper name*
Geziret, Geziret = Island, *see proper name*
Ghadeira, Shab, 197
Ghalib, Port, 199, 201
Gharib, Ras, 227-8
Ghoubbet, Ghub, Ghubb, Ghubbat, Ghubbet(ar) = Bay, *see proper name*
Ghulaifiqa, Khor, 113
Gifatin (Giftun) Is, 209
Gihan, Ras, 228-9
Girid, Marsa, 179, 180-81
glossaries, 276-7
GMDSS, 52-3
Goa, *routes*, 18, 20
Gordon Reef, 240
Gota White Rock, 187
Gouna, El, 212, 213-16
routes, 24
GPS, 47, 66
Greece, *routes*, 20, 21
Green Reef, 127, 141-3
GSM, 52
Guardafui, Capo, 81
Gubal I, 221
Gubal, Strait of, 216-25
Guinni Koma, 85
Gulf (of) *see proper name*
guns, 61
Gurna Reef, 152
Gwilaib, Marsa, 174
Gwiyai, Marsa, 143, 145

Habban, Sharm, 126
Habili Hamada, 191
Hai Dugai Its, 130
Haifa (Qishon), 263-4
Halaib, Marsa, 173, 179-80
Halaka, Marsa, 167-8, 172
Hali, Ras, 123
Hamada (wreck), 194
Hamira, Mersa, 243
Hammam, Mersa, 234
Hamra Dom, Gebel, 184
Hamrawein, 204
Hamsiat, Marsa, 170-72
Handa Deset, 92
Hanish Is, 102-9
Hant Deset, 92
Hara, Shab al, 162, 163
Harat I, 101
Harbanaikwan, Khor, 175
harbours, 46
Harmil, 101
Harorayeet (Two Its), 127, 141
Harr, Al, 126
Hart Group, 5, 60, 81
Hasa, Marsa, 185
Hasa, Shab El, 230
Hasani, Al, 126
Hasr I, 121
Hasy, Sharm, 126
Hatiba, Ras, 125
Hawatib Kebir, 101
Hawk (wreck), 189
health and medical, 63-5
Heidub (Aydob), Marsa, 137
Herzlia, 264
Hibiq, El, 241
Himeira, Marsa, 186
Hindi Gidir, 128, 141, 143
Hofrat el Malh, Marsa, 184
holidays see country information in Part I

Honkorab, Ras, 195
Horseshoe Reef, 183, 187
Howakil Bay, 93
Hudaydah, Al, 113-14
Humara, Ras Al, 123
humidity, 29-30, 63
Hurghada, 181, 183, 209-12
hurricanes, 28
Hushasha, 240
Husn al Ghurab, Bandar, 74
Hyndman Reefs, 207
Ibn Abbas, 130
Ibrahim, Marsa, 123
Ibrahim, Port, 249
Igli, Marsa, 199-200
Ile(s), Ilot(s) = Island(s), *see proper name*
Imran, Ras, 79
Inkeifal, Marsa, 158-9
Inmarsat, 51
insects, 63, 64-5
Internet (information websites), 273-4
Internet (weather forecasts), 31-3
Intraya Deset, 96
Iri, Gazirat, 128
Iskandariyah, El (Alexandria), 259-60
Islam, 1-2
Ismailia, 253-4, 257
Israel, 245-6, 262-6
coast radio, 54
general information, 11-13
marinas, 55, 245-6, 257, 262-6
port radio, 53, 54
routes, 20, 21
Isratu, 101
Istahi, Ras, 130

Ja'afirah, Khor Al, 121
Jabal Zubayr I, 109
Jabal Zuqar I, 106-9
Jabal, Jebel = Mountain, Hill, *see proper name*
Jackson Reef, 240
Jaffa (Yafo), 264-5
Jalajil, 123
Janabah Bay, 116
Jazair, Jazirat = Island, *see proper name*
Jebal, Jabel = Mountain, Hill, *see proper name*
Jeddah, 123-5
routes, 22
jellyfish, 64
Jezirat(az) = Island, *see proper name*
Jibna, Shab, 128
Jinniya (Corner Reef), 128
Jizan, 121
Jordan, 244-5
coast radio, 54
general information, 13-14
marinas, 55
port radio, 53
Royal Diving Club, 245
Royal Jordanian YC, 244, 245
Juzur(at, az) = Island, *see proper name*

Kadda Dabali, 86
Kafr el Gouna, 216
Kalafiyya, Gazeirat, 130
Kamaran, Jazirat, 114
Karam Masamirit, 128
Karb Its, 127, 130, 141
Karin, 83
Kasar, Ras, 129
Katib, Ras al, 113-14
Kemer Marina (Turkey), 21

Kenisa, Ras, 225
Khaisat, 70
Kharab, Ghoubbet (Pass), 84
Khawkhah, Al, 113
Khawr
  Sharm Al, 126
  Khoor, Khor = Inlet, Cove, *see proper name*
  khors/sharms/marsas (explanation), 46
Khotib, 116
King Fadh Port, 125
Kingston (wreck), 225
Klën, 186
Kordumuit Deset, 91-2
Koshkasha, 240
Kura, El (Dahab), 241
Kuwai, Marsa, 139-41

Laasqoroy, 81
Lac Sale, Baie du, 85
languages 276-7, *see also country information in Part I*
Larnaca, 261
routes, 20, 21
Lavanzo (wreck), 178
Leoni anchorage, 163
Leukos Hormos, 204
Libya, 20
lights, 46
Limassol, 262
routes, 20, 21
lionfish, 64
Little Inkeifal, 158
Littlefoot, Marsa, 139
Long I, 128, 135
Loran C, 47
Low I (Hanish Is), 104-5
Low I (Zubayr Group), 109
lubricants, 48
Luli, Sharm, 195

Madfa, Sharm el, 185
Madinah, Al, 125
Madinat Yanbu as Sinaiyah, 125
Madiq Gubal, 216-25
Madote, 93
Magarsam, Jaz, 158
Mahabis Is, 194
Mahas el Asfal, Merset, 242
Mahasin, Ghubbet Al, 123
Mahasin, Ras, 123
Mahe, *routes*, 20
Mahmud, Shab, 218
Majdahah, Ras, 74
Malab, Ras, 234
Malahi, El, 191
Malcatto, Ras, 93
Malcomma, Ras, 96
Maldives, *routes*, 18-20
Male, *routes*, 20
Malik Fahd, Mina al, 125
Malta, *routes*, 20
Mandeb, Bab el, 79
routes, 22, 23, 24
tidal streams, 41
mangroves, 57
Mansour, Shab, 191
Manta Point, 218
Maqatin, 75
Maqdam, Marsa, 131
marinas, 55
marine band SSB (MF/HF), 51, 52
marine life, 55-7
  dangerous, 63-4
Marine Peace Park, 245
Marmaris, *routes*, 21
Marob, Khor El, 172, 175
Marsa = Anchorage,

Harbour, *see proper name*
marsas/khors/sharms (explanation), 46
Masamirit It, 143
Masdud, Marsa, 172
Massawa, 96-101
  diving, 56
Matarma, Ras, 235
Matruh, Mersa, 260
Maydh, 81
Mayteb, Jazirat, 159-60
Mecca, 125, 204
medical and health, 63-5
Mediterranean Yacht Rally, 21
Melita Bay, 93
Melita Pt, 133
Merlo Reef, 155, 160, 163
Mersa, Merset = Anchorage, Harbour, *see proper name*
Mesharifa I, 162
Messina, Straits of, *routes*, 20
Mezraiya, Gebel, 223, 225
Middle Reef (Safaga), 209
Mina = Harbour, Port, *see proper name*
Minqar Channel, 209
Mintaka anchorage, 135
Mirear I Reef, 185
Misalla, Ras, 249
mobile phones, 52
Mohammed, Ras (National Park), 218, 237
Moiyia, Sharm el, 237
Mojeidi, 93
Mombasa, *routes*, 20, 21
money see country information in Part I
monsoons, 25
moon, 30
Morakh, Mersa, 243
moray eels, 65
mosques, 26
mosquitoes, 63, 64-5
Moucha, Is, 84
MRCC and MRSC, 52
Mubarak, Marsa, 199, 201
Muhammad Qol, 161, 162
Mujamila, Ras, 113
Mukalla, Al, 70-74
Mukawwa, Gezirat (Gez), 183, 187
Mukawwar I, 158
Mukha, Al, 112-13
Muos Hormos, 216
Muqabila, Mersa El (Taba Heights), 242-3
Mus Nefit, Ghubbet, 96

Naama Bay (Merset El At), 237-8
Nabq, 240
Nabqiya It, 126
Nahud, Khot, 121-3
Naman, Sharm, 126
Narawat, Khor, 126, 127, 128, 129-30
routes, 22-3, 24
Nasiracurra, 93
navigation, 41-8
navigational aids, 46-7
NAVTEX, 33
Near I, 106
Nekari, Marsa, 198
Neptune (wreck), 189
Nile, River, 258-9
Nimra Talata, 143
Nine Degrees Channel, 18
Nishtun, 70
Nokra Deset, 96
Nuweiba El Muzeina, 241-2

Obock, 85

Harbour, *see proper name*
marsas/khors/sharms (explanation), 46
Masamirit It, 143
Masdud, Marsa, 172
Massawa, 96-101

Oman
  and piracy prevention, 60
  coast radio, 54
  general information, 2-3
  pilotage, 66-70
  port radio, 53
Orca Village Dive Resort, 206
Oreste Point, 116
Oseif, Marsa, 172-3
Owen Reef, 128

Palk Strait, 16, 18
Panorama Reef, 209
Pearly Gates, 119
Pelican I, 123
Peshwa, 128
Pharoah's I, 244
Philotera, 207
Pinaule, 128
piracy, 52, 59-62
Piraeus, *routes*, 21
Point, Pointe = Point, *see proper name*
pollution, 58
population see country information in Part I
Port Berenice, 187
Port Fouad, 255-7
  YC, 255-7
Port Ghalib, 199, 201
Port Ibrahim, 249
port radio, 53-4
Port Said, 255
  routes, 20, 21
Port Salvadora, 153
Port Sharma, 126
Port Smyth, 93
Port Sudan, 143-5
  diving, 56
  routes, 22, 23, 24
Port Tewfiq, 252
Port Tewfiq YC (Suez YC), 249-50
ports of entry see country information in Part I
Poseidon's Garden, 218
position fixing, 41-3
pressure distribution, 25-7, 30
problems, 58-65
Ptolomeis Theron, 128
public holidays see country information in Part I
Puntland, 60

Qab Miyum, 128
Qad, Gez El, 184
Qad, Marsa El, 183
Qad Yahya, Merset El-, 225
Qadaf, Marsa, 163
Qadd el Tawila, 236
Qadimah, Mina Al, 125
Qandala, 81
Qirsh, Dhanat Al (Green Reef), 127, 141-3
Qishon (Haifa), 263-4
Qishran, Marsa, 123
Qita El Banna, 155, 158, 161
Qoow (Bandar Ziada), 81
Quaria Kalitat, 187
Qubbat Isa, Ras, 179
Quei Reefs, 204-5
Queis, Shab, 211
Qulan, Ras, 194
Qumeira, Shab, 165-6, 167
Qunfida, Al, 123
Quoin Hill, 167
Qusayir, Ras, 70
Quseir, El, 181, 203-4
Quseir, Shab, 128

Raas = Cape, Point, *see proper name*
Rabegh, Sherm, 125

279

radar, 47, 61
radio skeds, 30
radio weather forecasts, 31, 32, 33
radio-navigation aids, 47
radios and services, 51-54
Raka, Marsa, 123
Ramadan, 26
Rambler Shoal, 131
Ras anchorages, 46
Ras = Cape, Point, *see proper name*
Rashid, 258
Ratl, Ras ar, 74
Rayis, Shab, 225
rays, 56
Raysali, Khor, 85
RDF, 48
Red Sea Hills, 139
reef anchorages, 46
Reef I, 131
reef sailing, 43-5
reefs, 42, 43-6, 55, 56-7
refrigeration, 50
Renato Reef, 197
rescue services, 52
Rhodes, *routes*, 21
Rhounda Dabali, 86
Ribda, Marsa, 173-5
rig, 49
risks, 58-65
Rocky I, 183, 187, 188
rodes, 50
Romiya, 101
Rosetta, 258
routes, 15-24
Ruahmi, Ras, 230
Ruays, Ghubbat Ar, 125
Rubetino Channel, 88
Ruemi, Shab, 148
Rumi, Shab, 143, 148
running gear, 49
Ruwabil Is, 183

Saba I, 109
Sabaya, Jabal, 123
Sabir, Marsa, 125
Sables Blancs, Les, 85
Safaga, Mina, 181, 183, 205-7
    routes, 23, 24
sails, 49
St John's I, 187-91
St John's Reef, 183
St Raphael Marina (Limassol), 262
Sakhra el Beida, El, 187
Sal Hashish It, 209
Salak, Marsa, 153-5
Salak, Shab, 155
Salalah, Mina, 66-70
    routes, 18, 20, 21, 22
salinity, 34, 48
Salubah I, 119
Salum, Es, 260
Salvadora, Port, 153
Samadai, Ras, 198
Samadai Reef, 199
Samhah, Jazirat, 80
Sana'a, 77
Sanafir, Jaz, 238
Sanahor Deset, 91
Sandy Cape, 172
Sanganeb Reef, 143, 145-7
SAR, 52
Sarah H (wreck), 218, 225
Sarso (Sarad Sarso) I, 119
Sataya, 191
Saudi Arabia, 116, h119-26, 238
    coast radio, 54
    general information, 7-8
    port radio, 53
Saylac, 83
Scout Anchorage, 186
sea snakes, 64

sea, tide and currents, 34-41
sea urchins, 64
seasonal winds, 27-8
seasons (routes), 15-18
Sebil, Ras el-, 226
Segid I, 116, 119
Seil Norah, 101
Sekala, 212
Seven Brothers, 86
Seychelles, *routes*, 18, 20
Shaab, Marsa, 184-5
Shab = Reef, *see proper name*, 295
Shabir, Marsa, 240
Shadwan Channel, 219
Shag Rock, 218, 224-5
Shaker (Shadwan) I, 218-19
Shambaya I, 160
Shammah, Jazirat al, 79
Shark I, 106
sharks, 56, 63-4
Sharm, Shab, 197
Sharm = Cove, *see proper name*
Sharma, Port, 126
Sharma, Ras, 70
sharms/marsas/khors (explanation), 46
Shatira It, 130
Shear, Shab, 207
Sheikh, Marsa Esh, 135-6
Sheikh el Abu, 101
Sheikh Ibrahim, Marsa Esh, 136-7
Sheikh Riyah Harbour, 227
Sheikh Sad, Marsa, 135-6
Sheikh, Sharm El, 183, 237
shellfish, 56
shells, 64
Sheratib, Ras, 230
Sherm = Cove, *see proper name*
Shib = Reef, *see proper name*
Shinab, Khor, 166-7
Shinab, Shab, 167
shipping traffic, 45-6
Shoora, Khoor, 83
shops (opening hours) see country information in Part I
Showarit, Gez, 194
Shubuk, Shab, 127, 128, 131
Shubuk Channel, 127, 133-6, 172
Shukheir, Ras, 226
Shumma I, 93
Sinai Peninsula, 226
    diving, 56
Sindi Sarso I, 119
Sinkat, 139
Sirrain I, 123
Sisters, The, 80
Siyal Is, 183, 184
Siyul, Gezirat, 194-5
Siyul Kebira, 219, 221
Siyul Saghira, 221
Siyyan, Ras, 86
Smyth, Port, 93
Somalia, 5, 52, 58, 60, 81-3
South Qeisum, 221-2
spares, 50
Sri Lanka, *routes*, 18, 21
stingrays, 64
stings, 64
stonefish, 64
Strait (of) *see proper name*
Suadi, Shab, 151
Suakin, 137-9
Suakin el Qadim (Suakim the Old), 181

Suakin Group, *routes through*, 126-8
Sudan, 126-81
    coast radio, 54
    general information, 8-9
    port radio, 53
Sudan, Port, 143-5
    diving, 56
    routes, 22, 23, 24
Sudr, Ras, 236
Suez Canal, 52, 246-57
    agents, 250, 257
    port radio, 53
Suez, Gulf of, 40, 41, 53
Suez (Suweis), Port of, 249
    routes, 22, 23, 24
Suez YC (Port Tewfiq YC), 249-50
Sumaima, 123
Sumar I, 133
Sumar Inlet, 135
sun precautions, 63
Suqutra (Socotra), 79
Suweis *see* Suez
Suyul Hanish, 102
Taba, 246
Taba Heights (Mersa El Muqabila), 242-3
Tadjoura, 85
Tadjoura, Gulf of, 84-6
Taila Is, 155
Tair, Jazirat At, 109-10
Talla Talla Kebir, 127, 128, 141
Talla Talla Saqir, 127, 128, 141
Tamarshiya, 128
Tamr Henna, 216
Tankefal, Marsa, 158
Tarafi, Marsa, 199-200
Tarfa, Ras At, 121
Tawil, Shab, 128
Tawila Channel, 213
Tawila I, 219-21
Tel Aviv, 264
    routes, 20, 21
Telat, Juzur, 155
telephones 52, *see also country information in Part I*
temperatures, 24, 29-30
    sea, 34-6
tenders, 50
Terma, Ras, 91
terrorism, 58-9
Tewfiq, Port, 252
Tewfiq, Port, YC (Suez YC), 249-50
Thelemet, Marsa, 232-3
Theron, 128
Thistlegorm (wreck), 218, 225
tides and tidal streams, 40-41
Tientsin (wreck), 191
Tigrinya language, 277
time see country information in Part I
timing (routes), 15-18
Timsah, Lake, 247, 254
Tiran I, 237, 238-40
Tiran, Strait of, 237, 238-40
    routes, 23
Tongue Island, 105
Tor Harbour (El Tur), 181, 227
Toronbi, Ras, 202
Toubib, Anse d'al, 85
Towartit, Shab, 128, 141
traffic separation scheme, 21, 102
triggerfish, 65
Trinkitat Harbour, 131

tropical revolving storms, 28
Tubya, Gez, 209
Tundaba, Marsa, 198
Tur, El (Tor Harbour), 181, 227
Turfa, Ras, 121
turtles, 56
Tutana Reef, 179
Two Its (Harorayeet), 127, 128, 141
typhoons, 28

Uligan I, 18-20
Umbeila, Marsa, 175, 179
Umbria (wreck), 145
Umm Agawish el-Kebir, 209
Umm el Burush, 161
Umm el Kiman, 223
Umm Qamar, 211
Umm es Sahrig, 93
Umm Usk, Shab, 221
Uqban I, 116
US Navy, 61
Uwainidhiya It, 126

vaccinations, 63, 65
VHF, 52
visas see country information in Part I
visibility, 28

Wadi Feiran, 231
Wadi Gimal, Gezirat, 195-7
Wadi Kalitat, 187
Wadi Lahami, Marsa, 191-4
Wahlan, Khor, 121
war, 58, 59
Wasi, Marsa, 170
Wasm, Khor al, 121
water (health precautions), 63
weather, 24-31
weather (forecasting services), 31-4
weather (rules of thumb), 30-31
weatherfax, 31
websites (information), 273-4
websites (weather forecasts), 31-3
Wejh, Sharm, 126
White Rock, 183, 187
Widan, 121
winds, 24-8, 30, 31
    local, 27-8
    prevailing, 27-8
Wingate Reef, 143
Wishka I, 116
Wizr, Marsa, 203
Wreck Pt anchorage, 155
Wreck recovery anchorage, 163
Wughadi It, 126
Wusta, 101

Xabo, 81
Xiis, 81

yacht and equipment, 48-52
Yafo (Jaffa), 264-5
Yahar, Sharm (Al Harr), 126
Yanbu, Sharm, 125
Yanbu Al Bahr, 125
Yemen, 70-80, 102-10, 111-19
    coast radio, 54
    general information, 3-5
    piracy prevention and protection, 60
    port radio, 53
Zaam, Marsa, 126
Zabara, Mersa, 223-4
Zabargad, Geziret, 183, 187-91

Zafarana, Ras, 233
Zebara, Marsa, 199
Zeit Channel, 222-3
Zeitiya, Marsa, 222
Ziada, Bandar (Qoow), 81
Zubaydah, Marsa, 126
Zubayr, Jazair Az (Subayr Group), 109-10
Zula Bahir Selate, 93
Zula, Gulf of, 93, 128

إسم الباخرة :

رقم التسجيل : العلم :

الجنسية : الظابط :

الجنسية : جماعة الباخرة :

آخر ميناء نزول أو زيارته :
تاريخ وصول النزول :
تاريخ الرجوع تقريبا :
الميناء المقبل نزول فيه :

سبب رسوة الباخرة : أسطب على لاغير صحيح
الجو سيء جداً
مشاكل تقنية
فرض للإستراحة
الباخرة عاطلة
نتمنى على إستقبالكم المرحب ونتمنى أن تقضي أيام هنا ونشكركم على معرفتكم ومهامتكم.

الإمضاء :

المكان : التاريخ

التاريخ والساعة للزيارة الرسمية :

إمضاء الظابط المسؤول :